AF877471

Historical & Cultural Astronomy

Series Editors

WAYNE ORCHISTON, Adjunct Professor, Astrophysics Group, University of Southern Queensland, Toowoomba, QLD, Australia

MARC ROTHENBERG, Smithsonian Institution (retired), Rockville, MD, USA

CLIFFORD CUNNINGHAM, University of Southern Queensland, Toowoomba, QLD, Australia

Editorial Board Members

JAMES EVANS, University of Puget Sound, Tacoma, WA, USA

MILLER GOSS, National Radio Astronomy Observatory, Charlottesville, USA

DUANE HAMACHER, Monash University, Melbourne, Australia

JAMES LEQUEUX, Observatoire de Paris, Paris, France

SIMON MITTON, St. Edmund's College Cambridge University, Cambridge, UK

CLIVE RUGGLES, University of Leicester, Leicester, UK

VIRGINIA TRIMBLE, University of California Irvine, Irvine, CA, USA

GUDRUN WOLFSCHMIDT, Institute for History of Science and Technology, University of Hamburg, Hamburg, Germany

TRUDY BELL, Sky & Telescope, Lakewood, OH, USA

DAVID DEVORKIN, National Air and Space Museum, Smithsonian Institution, Washington, USA

The Historical & Cultural Astronomy series includes high-level monographs and edited volumes covering a broad range of subjects in the history of astronomy, including interdisciplinary contributions from historians, sociologists, horologists, archaeologists, and other humanities fields. The authors are distinguished specialists in their fields of expertise. Each title is carefully supervised and aims to provide an in-depth understanding by offering detailed research. Rather than focusing on the scientific findings alone, these volumes explain the context of astronomical and space science progress from the pre-modern world to the future. The interdisciplinary Historical & Cultural Astronomy series offers a home for books addressing astronomical progress from a humanities perspective, encompassing the influence of religion, politics, social movements, and more on the growth of astronomical knowledge over the centuries.

The Historical & Cultural Astronomy Series Editors are: Wayne Orchiston, Marc Rothenberg, and Cliff Cunningham.

Thierry Montmerle • Danielle Fauque
Editors

Astronomers as Diplomats

When the IAU Builds Bridges Between Nations

Springer

Editors
Thierry Montmerle
Institut d'Astrophysique de Paris
PARIS, France

Danielle Fauque
Groupe d'Histoire et de Diffusion des Sciences d'Orsay (GHDSO-EST),
Université Paris-Saclay
ORSAY, France

ISSN 2509-310X ISSN 2509-3118 (electronic)
Historical & Cultural Astronomy
ISBN 978-3-030-98624-7 ISBN 978-3-030-98625-4 (eBook)
https://doi.org/10.1007/978-3-030-98625-4

© The Editor(s) (if applicable) and The Author(s), under exclusive license to Springer Nature Switzerland AG 2022
This work is subject to copyright. All rights are solely and exclusively licensed by the Publisher, whether the whole or part of the material is concerned, specifically the rights of translation, reprinting, reuse of illustrations, recitation, broadcasting, reproduction on microfilms or in any other physical way, and transmission or information storage and retrieval, electronic adaptation, computer software, or by similar or dissimilar methodology now known or hereafter developed.
The use of general descriptive names, registered names, trademarks, service marks, etc. in this publication does not imply, even in the absence of a specific statement, that such names are exempt from the relevant protective laws and regulations and therefore free for general use.
The publisher, the authors and the editors are safe to assume that the advice and information in this book are believed to be true and accurate at the date of publication. Neither the publisher nor the authors or the editors give a warranty, expressed or implied, with respect to the material contained herein or for any errors or omissions that may have been made. The publisher remains neutral with regard to jurisdictional claims in published maps and institutional affiliations.

Cover Text: French delegation on the boarding stairway of a flight from Paris to Moscow via Copenhagen to attend the IAU General Assembly in 1958. André Danjon, IAU President, is among the passengers. (Credit: photographer unknown, © Bibliothèque de l'Observatoire de Paris)

This Springer imprint is published by the registered company Springer Nature Switzerland AG
The registered company address is: Gewerbestrasse 11, 6330 Cham, Switzerland

Foreword

Several initiatives were organized to commemorate the centenary of the of the International Astronomical Union (IAU), including a symposium and its proceedings,[1] a traveling exhibition,[2] various competitions and projects open to worldwide participation,[3] a commemorative ceremony in the same place where the IAU was founded in 2019,[4] etc. These initiatives mainly focused on the scientific and outreach aspects of the IAU's activities, but certainly one, and a very important one, was missing: the IAU as a mediator around the world. So when Thierry Montmerle approached me with the proposal to organize a conference on "Astronomers as Diplomats" as part of the centenary celebrations, I immediately supported the idea.

The Conference, which took place at the Institut d'Astrophysique de Paris, the host of the IAU Office, on Oct. 4, 2019, included many interesting invited talks and involved a large number of participants. As a result of Thierry's enthusiasm, this book initially envisioned as the conference proceedings, later evolved into a more ambitious work, compiling not only the invited talks but also other complementary contributions relevant to understanding the evolution of the IAU over the century, and, in particular, of its role in diplomacy.

On behalf of the IAU, I would like to express to Thierry and Danielle, editors of this work, my gratitude for all the effort involved. I agree with them, science diplomacy by scientific organizations, the IAU in particular, is one of the most relevant roles of the Unions for the humanity in general. It goes beyond sharing knowledge and opportunities, encouraging collaboration and stimulating discovery. It constitutes an informed statement of the deep understanding of the humanity's shared one home, regardless of borders, traditions, and beliefs. Although at the founding of the IAU, exclusion was an accepted principle and goal, as described in

[1] Under One Sky: The IAU Centenary Symposium, Eds. Sterken, C., Hearnshaw, J., Valls-Gabaud, D., Cambridge University Press (2019).

[2] The IAU100 Above and Beyond traveling exhibition

[3] see https://www.iau-100.org/ for the IAU100 Final Report

[4] Ceremony at the Academy Palace in Brussels, on April 11, 2019

one of the chapters of this book, fortunately inclusion has become a must in recent times. All types of inclusion. And this is one of the reasons that motivates me and makes me feel proud to be part of the IAU.

IAU General Secretary (2018–2021) Teresa Lago

Preface

Astronomers and a Century of Building a Community Worldwide in Spite of Wars and Revolutions

The Historical Context

If the sound of rifles, cannon, and airplanes stopped on Nov. 11, 1918 at 11 am in Europe, following an armistice agreement ending combat in what was going to be named "the Great War," it was not until the Treaty of Versailles, signed on June 28, 1919, that this tragic war formally ended, bearing the seeds of future conflicts because of the harsh conditions imposed on Germany and its absence at Versailles.

Just three weeks later, in the *Palais des Académies* of Brussels, refurbished after having been transformed into a makeshift hospital for German troops during the war, an "Inter-Allied Conference" gathered over 200 scientists from the Allied powers. Apart from the Belgian delegation, which made up about half of the participants (106 out of 224), the majority came from the leading victorious countries: France (48), the USA (27), Great Britain (19), and Italy (15), followed by Canada (2), Japan (1), New Zealand (1), Poland (1), Portugal (1), Rumania (1), and Serbia (2) (Schuster, 1920). In spite of their relatively small number, the participants represented most of the important fields of science, some of them having directly contributed to war developments, like chemistry, geodesy, and radio communications. Being intrinsically international—and peaceful—, astronomy was also well represented (at least 31 astronomers attended the meeting), and the community was ready to set up a formal organization.

Of course, the Brussels Conference, the Third Interallied Conference organized by the scientific academies, after London and Paris in 1918, had been prepared well in advance, in fact even months before the armistice, partly as a consequence of the American entry into the war in April 1917. The goal was to officially approve the creation of an *International Research Council* (IRC) and of its Statutes, and examine proposals to create international scientific unions, during the Conference if possible, or later, at successive IRC General Assemblies. These unions would be supervised by the IRC, and their statutes aligned on those of the IRC (Fauque & Fox, 2020).

Beginning on July 18, the Brussels Conference ended on July 28, 1919. On the positive side, the IRC was put on track, its provisional Executive Committee confirmed. Three International Unions were created on the spot: IAU (astronomy), IUGG (Geodesy and Geophysics), and IUPAC (Pure and Applied Chemistry); two others a little later, after ratification by their member countries: IURS ("radio science," its Union being better known under its initial French acronym, URSI) and IUBS (Biological Sciences). The most advanced one was undoubtedly the International Astronomical Union (IAU), with statutes that had largely inspired those of the IRC (rather than the other way around), and which was structured around 32 thematic "standing committees" (soon to be re-named "commissions"). On the downside were the questions that would plague the birth and early development of the IRC, and as a consequence that of the Unions, i.e., the conditions for membership of neutral countries, and more importantly, of the defeated "Central Powers" (Germany and its allies). In 1919, it was inconceivable to the Allies to be at the same table than their German colleagues, notably those who had signed the famous "Manifesto of the Ninety-Three."[1] Their exclusion for several years was voted.

It was only in 1926, at a special meeting, that the IRC accepted to invite Germany, Austria, Hungary, and Bulgaria. But the days of the IRC were counted. Since 1927, a number of German scientists had been invited to attend several General Assemblies of the Unions, influencing their reforms. In 1931, the IRC was terminated, and replaced by the *International Council of Scientific Unions* (ICSU), in which every Union was free to invite any country it wanted. This freedom for the Unions gave a little space to the ICSU in their activities. To mitigate these constraints, however, there were diplomatic and scientific efforts by some individual scientists in their respective specialty, and for our purpose, some astronomers.

After WWII, ICSU and the IAU (as one of its members) entertained relations with the United Nations and some of its agencies since their very creation in late 1945. One of them, the United Nations Education, Science and Culture Organization (UNESCO), played a crucial role in the development of the scientific Unions after WWII. In fact, UNESCO retained several aims of the *International Institute of Intellectual Cooperation*, founded in 1924 and just dissolved, but with larger spectra of interests. This organization was born on Nov. 16, 1945, and the following year (November 1946) concluded a convention with ICSU in 9 articles, notably on the freedom of scientific information, facilities of traveling, financial subsidies for the different activities of the scientific Unions, etc. At the same time, ICSU

[1] Also known as a proclamation "To the Civilized World, by Professors of Germany" (in German: *Aufruf an die Kulturwelt*). It was issued on Oct. 4, 1914, as a patriotic protest to the outside world against accusations of atrocities being committed by the German armies during the invasion of (neutral) Belgium. The signatories included several Nobel Prizes, past or future.

acknowledged UNESCO as the principal body of the UN in the field of international scientific relations (UNESCO, 2006).

This convention had a major impact on the re-start of the Unions after WWII. The ICSU and the Unions immediately undertook reforms. If the first subsidies of UNESCO (via ICSU) initially represented a large fraction of their budget, they diminished rapidly, mainly because of the creation of more Unions, and of the growing general inflation. Adding to these practical difficulties, in the context of the Cold War, political antagonisms became more and more acute. Under the authority of the UN, UNESCO was thus requested to endorse the UN national members, then to ask ICSU to do the same. After The People's Republic of China (PRC) was admitted to the UN in 1971 and Taiwan (the Republic of China) was expelled, the case of China was by the far the worst problem to resolve with two representatives, *Academia Sinica Nanking* (based in Taipeh, Taiwan, since 1949) (Taiwan) and *Academia Sinica Peking* (PRC). This led to a major crisis between UNESCO and ICSU, when the majority of the Unions, applying a strict reading of the statutes, did not follow the UN injunctions to exclude Taiwan when the PRC was admitted to the UN in 1971.[2]

The activities of ICSU members were also affected by political tensions. The freedom of circulation for scientists was often obstructed because of political reasons, notably concerning the visas. The ICSU issued a series of recommendations to help finding solutions, on a case by case basis (e.g., Greenaway, 1996). Through these principles, the Unions could act with the governing bodies. Some delicate situations could find a positive solution over the long term, thanks to the concerned actors' diplomacy. In fine, beginning in the sixties, a larger opening to the developing countries was favored by direct actions of UNESCO, especially in education and teaching, through ad hoc commissions of several Unions.

Celebrating the IAU Centenary

In any case, the IAU, as well as the other "first-generation" scientific Unions, was founded in 1919, so in 2019, there were many initiatives taken worldwide which offered the opportunity to celebrate the centenary of their existence and reflect on their evolution and present role in modern society. For the IAU, a key factor behind this evolution is, simply put, the demographics of astronomers: at the time of its first General Assembly in Rome, Italy, in 1922, i.e., three years after the Brussels Conference, the IAU had 207 individual members (83 of them attending), representing 19 countries; in 2018, at the time of the 30th General Assembly in Vienna, Austria, the IAU had nearly 14,000 individual members (over 3000 attending), representing 110 countries (85 being "national members"). A 70-fold increase in the number of astronomers over a century! (Meanwhile the world

[2] See below, and for details in the case of the IAU, Montmerle & Zhou (2022).

population grew by a factor of "only" 4, from about 2 billion in 1920 to nearly 8 billion today...)

Yet, in spite of this tremendous evolution, while the history of astronomy is the subject of innumerable books, very little has been written about the history of the IAU *per se*. For a long time, the only reference work has been the book by Adrian Blaauw, former IAU President (1976–1979), entitled "*History of the IAU: The Birth and First Half-Century of the International Astronomical Union,*" written at the request of the IAU Executive Committee and published by Kluwer in 1994 (in spite of the fact that the 50th anniversary took place in 1969!), and rich with international political context (Blaauw, 1994). The IAU Centenary in 2019 offered a new opportunity to fill this gap, with two books, both published that same year. One was the Proceedings of "*Under One Sky: the IAU Centenary Symposium*" (Sterken et al., 2019), held during the Vienna General Assembly in 2018, and gathering many historical aspects of the IAU, including its administration and contributions by many countries. The second one is entitled "*The International Astronomical Union: Uniting the Community for 100 years*" (Andersen et al., 2019) and describes the progress of astronomy" seen through the eyes of the IAU," with less emphasis on the political background and more on science, and including numerous interviews of IAU members having held offices in the IAU executive.

While their contents and approaches are different, these three books clearly underline a common characteristic: "the Birth of the IAU," "Under One Sky," "Uniting the community," all point to a key tool of international relations: *diplomacy*. Yes, in some circumstances, especially in periods of international tensions and crises, astronomers (and scientists in general!) can be "diplomats," even if not professional ones. In his book (p. 252), Blaauw quotes his compatriot Jan Oort (decade-long General Secretary and President of the IAU in difficult times), writing on his own experience at the time of the IAU's 50th Anniversary, as published in the "Golden" *IAU Information Bulletin* of June 1969 (on the occasion of its 50th anniversary): "*... [I have] become involved in building up international understanding and fertile contact between countries where sometimes relations outside of astronomy were extremely poor. Politicians have rightly envied us for the natural ties between peoples we had been able to establish and to conserve*."

Of course, diplomacy has its limits: we have mentioned above the problem of membership of non-Allied countries during the early years of the IRC (and thus of the IAU), and in this book, we will consider the difficulties of the "Divided countries" during the Cold War, as a result of WWII, and how they were overcome, at least partly. We also describe more recent episodes of collaboration (or conflict) with the UN and UNESCO.

As former IAU General Secretary (2012–2015), one of us (TM) followed closely the organization of the IAU Centenary. Much like the "International Year of Astronomy" of 2009 (the 90th anniversary of the IAU...), the 2019 Centenary celebrations were concentrating on the role of the IAU in attracting public attention on astronomy worldwide, symbolized by the IAU logo "*100 Years: Under One Sky,*" a theme which was adopted also, as mentioned before, by professional astronomers for the "IAU Centenary Symposium" held in Vienna.

IAU

100 Years: Under One Sky

This Symposium proceedings contain 63 contributions, of typical length 6–8 pages (according to the editorial instructions), reflecting well the great diversity and evolution of the activities of the IAU over a hundred years. Of these, and grouped in two parts (Part IV: "the IAU and its history," and Part V: "The IAU and its relation to diverse geographical regions"), about 20 describe a variety of diplomatic roles played at different epochs by the astronomical community, but as a result of the editorial choice, none of these contributions could explore these roles in detail.

This was the origin of the idea to go beyond the Vienna Centenary Symposium and organize a Colloquium focusing on the theme *"Astronomers as Diplomats,"* adding the subtitle *"When the IAU builds bridges between nations,"* as one of the so-called IAU@100 events held in 2019.[3]

A very appropriate opportunity arose when the French *Académie des Sciences* decided to organize a joint celebration for the "pioneer Unions" that were created in 1919: IAU, IUGG, IUPAC, IUBS, and URSI. This was particularly appropriate, since the Académie was, along with the Royal Society, one of the key organizers of the two Inter-Allied Conferences preceding the Brussels meeting. In addition, the first President of the IRC, as well as the ones of the first Unions, were all French Academicians: Emile Picard (IRC), Benjamin Baillaud (IAU), Charles Lallemand (IUGG), Charles Moureu (IUPAC), Gustave Ferrié (URSI), and Yves Delage (IUBS).

The Académie formed a scientific organizing committee, and issued a call for proposals to hold events linked with the Centenary. Both of us (TM, as former IAU General Secretary, and DF, as a science historian) were members of this Committee, and joined forces to propose together an international Colloquium on the theme *"Astronomers as Diplomats."* The venue would be the Observatoire de Paris (of which Benjamin Baillaud was the Director at the time) and the Institut d'Astrophysique de Paris (TM's institution and home to the IAU Secretariat since November 1988). The proposal was accepted, and endorsed by the IAU;[4] this was to be a one-day colloquium with invited speakers only, aimed at the interested public in astronomy and beyond (science historians, space law experts, UNESCO

[3] https://www.iau-100.org/

[4] https://www.iau.org/news/announcements/detail/ann19056/

representatives, etc.). Coincidentally, the chosen date was Oct. 4, 2019, thus offering the opportunity to celebrate as well, and to the day, the 60th anniversary of the launch of the Earth's first artificial satellite, *Sputnik-1*—the first step into the space age, space being an area where astronomers still have to be diplomats!

The Colloquium gathered 10 speakers (from various countries in Europe, and from Russia and China), and attracted about 120 participants. Yet, such a one-day Colloquium was clearly insufficient to explore many "diplomatic" issues in detail, so from the start the idea was to produce a book based on the Colloquium, keeping its spirit but written not as a usual proceedings volume, but rather as a set of "*case studies*," completed by inviting additional contributions later to fill gaps or develop some issues. This was done in a "dynamic" way, some authors having been invited late in the editorial process to enhance the previous contributions as we, the Editors, considered useful or necessary. In the process, over the past two years since the Paris colloquium, nine more contributions were added, with authors from the Netherlands, Korea, China (mainland and Taiwan).[5] In addition, this approach proved so successful that, with the encouragements of Springer, it generated two "satellite" books, precisely about China (Montmerle & Zhou 2022; Montmerle et al., 2022).

The present book is the end result of this process. A large part of its 18 chapters is based on previously unpublished and untapped archival documentation (IAU, RAS, UN, etc., even Chinese Internet blogs), yielding a fresh approach to events that were supposed up to now to be well known and understood. In many cases, the emphasis has been on individual astronomers that played a central role in diplomacy, some of them being unknown or poorly known in the West.

Outline of the Book

The outline of the book is based on the four-part architecture of the Paris Colloquium.

Part I: The Creation of the IAU and the Aftermath of WWI

The chapters show that the IAU had its seeds of international collaboration long before the outbreak of WWI (Chap. 1: the *Carte du Ciel*; Chap. 2: the *Bureau International de l'Heure*). Previously unknown details about the creation of the

[5] The current membership of China in the IAU has two components, called "China-Nanjing" and "China-Taipei" (https://www.iau.org/administration/membership/national/). See Chap. 5 of this book for a summary, Blaauw's (1994) Chapter 9 for more details, and Montmerle & Zhou (2022) for an extensive study.

IAU (Chap. 3) have emerged, in particular from the archives of the Royal Society in London, the *Bureau des Longitudes* and the *Académie des Sciences* in Paris. As we have seen, the Central Powers (Germany, Austria, Hungary, and Turkey) were initially banned from joining the Unions for several years: A.S. Eddington, a vocal pacifist, was opposed to these decisions, and tried to maintain his pre-war network of scientists from opposite sides (Chap. 4). We had also hoped to present a special chapter on the international activities of the first IAU president, Benjamin Baillaud, that could explain the particular role he played in the constitution and first years of the IAU, but the pandemic circumstances did not permit the access to his personal archives, kept at the *Archives nationales* (France), in Pierrefitte-sur-Seine, near Paris.[6] One can, however, have a flavor of this role by consulting the proceedings of the "IAU Centenary Symposium" (Bougeret, 2019).

Part II: After WWII: Divided Countries

During the decade following WWII, the world became politically separated into two opposite blocs, "East" and "West" or "North" and "South", depending on the border rough orientations... Concerning the IAU, this happened to three countries, with very different situations today: China, Korea, and Germany.

The Republic of China, born on Jan. 1, 1912, became geographically separated in 1949 between mainland China and the island of Taiwan, a situation which, as we have seen above, provoked a dramatic crisis within all the scientific Unions, when China withdrew after the admission of Taiwan as a separate member,—and which, in the case of the IAU, began in 1959 and ended in 1982. To illustrate the specificity of the Chinese situation at that time (echoing today's political tensions), and in order to go deeper than could the presentation in Paris (Chap. 5), we have chosen to invite more contributions for the book, perhaps at the cost of some imbalance: revisiting the conditions in which China withdrew from the IAU (Chap. 6), investigating the life of the main Chinese interlocutor of the IAU in those times, Zhang Yuzhe (Chap. 7), illustrating how some contacts could be maintained between Chinese and Western astronomers (Chap. 8), and focusing also on "astronomy across the Strait," i.e., in Taiwan, a story that began under Japanese occupation and continued under the nationalist regime (Chap. 9). In doing so their authors have used an impressive amount of new, previously unpublished or inaccessible material that is presented here for the first time.

After having been a colony of Imperial Japan for 35 years (1910–1945), Korea was split after a murderous war (1950–1953) and is still geographically divided today but between two mutually recognized nations, both independent IAU

[6] Since 2016, the Benjamin Baillaud Collection is kept in the *Archives Nationales* under the shelf mark 759 AP. An inventory is available online. The contact for information is <dap.an@culture.gouv.fr>.

members: their common history is told in Chap. 10. As for Germany, which was split in 1949, first with free-crossing borders but becoming separated by walls and fences in 1961, it reunited in 1990. This led to odd IAU affiliations over time, as explained in Chap. 11.

Part III: During and After the Cold War

While the case of the "divided countries" represented a proxy political confrontation between East and West after WWII, a more direct confrontation was looming between the two "superpowers," the USA and the USSR. With the stockpile of nuclear weapons and missiles growing every day, the "Cold War" was threatening to degenerate into WWIII.

Yet scientists, and astronomers in particular, tried to keep peace afloat. Radioastronomy included a technique called "*Very-Long Baseline Interferometry*" (VLBI), which is based on multiple antennas spread so far apart that they could be combined across borders. It offered particularly rich opportunities for scientific collaboration, notably between Western Europe and the USSR, as told in Chap. 12. But scientific and technological competition, too, was a marker of the Cold War era, in particular during what was called "the Moon race." After the revelation of the "new Moon" (its far side in 1959 and later its small-scale surface features, visible only from space), the IAU, through the seemingly innocent field of "lunar nomenclature," became involved in serious conflicts, both "diplomatic" (with the USSR), and even internal, as told in Chap. 13. In the post-*Apollo* era, however, the international climate tended to cool down, collaborative efforts resumed, and innovative methods were found by the Soviets to explore the Moon's surface (Chap. 14). In parallel, with the advent of evermore powerful computers, astronomical data from telescopes worldwide could be assembled together in huge databases available to everyone, and as Chap. 15 explains, the "*International Virtual Observatory*" even helped astronomers from former "socialist countries" to overcome the fall of the USSR in 1991.

Part IV: The IAU, the United Nations, and UNESCO

Over time since WWII, the relations between the IAU and the UN have had a number of ups and downs, and a whole book could be written about it. Here we have focused on three "back-of-the-stage" episodes. Chapter 16 reveals an untold conflict (albeit not with UNESCO, but with the UN Department of Economic and Social Affairs) about the so-called extraterrestrial names, that happened in the post-*Apollo* era and took a decade to resolve. With the Cold War out of the way since 1991, in more recent years, the IAU and the UN (via UNESCO) have very fruitfully collaborated, but even that was not easy, as the background story of the *International Year of Astronomy 2009* shows (Chap. 17).

The Editors chose to close the book on the very positive note: the ongoing IAU-UNESCO "Astronomy and World Heritage Initiative" (Chap. 18), which vividly demonstrates that astronomy is universal, uniting mankind "*Under One Sky*," not only in space (all over the globe), but also in time (over centuries).

Paris, France Thierry Montmerle
Orsay, France Danielle Fauque

References

Andersen, J., Baneke, D., & Madsen C. (2019), *The International Astronomical Union, uniting the community for 100 years* (376 p.). Springer Nature Switzerland.

Blaauw, A. (1994). *History of the IAU, the birth and first half-century of the International Astronomical Union* (296 p.). Kluwer Academic.

Bougeret J.-L. (2019). [IAUS 349], pp. 139–146.

Fauque, D., & Fox, R. (2021). Binding the wounds of war. Internationalism, national interests, and the order of world science, 1919–1931. In W. U. Eckart & R. Fox (Eds.), *Blockades of the mind – Science, academies, and the aftermath of the Great War*. In *Acta Historica Leopoldina*, *78*, Halle, pp. 41–68.

Greenaway, F. (1996). *A history of the International Council of Scientific Unions* (280 p.). Cambridge University Press.

[IAUS349]: Sterken, C., Hearnshaw, J., & Valls-Gabaud D. (Eds.). (2019). *Under one sky: the IAU centenary symposium* (IAU Symposium 349, 542 p.). Cambridge University Press.

Montmerle, T., & Zhou, Y. (2022). *China and the IAU: Divorce, separation, and reconciliation (1959–1982)*. Springer Nature Switzerland, in press.

Montmerle T., Zhou Y., & Gomas, Y. (2022). *The two lives of Cheng Maolan: From the French "Silk Road" towards astronomy to the meanders of Mao's China*. Springer Nature Switzerland, in press.

Schuster, A. (Ed.). (1920). *Constitutive Assembly held at Brussels, July 18th to July 28th, 1919. Reports of Proceedings*. Harrison & Sons.

UNESCO. (2006). "UNESCO and ICSU: Sixty years of cooperation" in *Sixty years of science at UNESCO 1945–2005* (pp. 496–506). Online at: https://unesdoc.unesco.org/ark:/48223/pf0000149231?29=null&queryId=3b13f6be-39ca-4fc1-ad08-fa0cf3eb8469

Acknowledgements

The Editors wish to thank the *Académie des Sciences*, and in particular Claude Boucher, Chair of the Organizing Committee of the *Centenaire des Unions Scientifiques*, for financial support to invite speakers. The "Astronomers as Diplomats" Conference took place in the Auditorium of the Institut d'Astrophysique de Paris, and a welcome cocktail was kindly offered by the nearby Observatoire de Paris: we warmly thank their Director and President (François Bournaud and Claude Catala, respectively, at the time) for their hospitality, and for the support of the Observatoire's *Direction de la Communication* (Sabrina Thiéry and Frédérique Auffret).

We also thank the librarians of the Paris Observatory Library, and the curators of the following archives: Académie des Sciences in Paris, Royal Society and Royal Astronomical Society in London, Accademia Nazionale dei Lincei in Rome, as well as the librarians of the United Nations' Dag Hammarskjöld Library in New York.

Contents

Part I
The Creation of the IAU and the Aftermath of WWI

Chapter 1
Precursors to IAU: Paris Observatory and the *Carte du Ciel* Project

Ileana Chinnici

Abstract The *Carte du Ciel* was the most extended project of international cooperation in the field of astronomy, from the end of the nineteenth to the middle of the twentieth century. Launched by Paris Observatory in 1887, it was aimed at photographing the whole sky vault in order to produce a Chart and a Catalogue of all visible stars. The required effort was underestimated for several reasons and the project could not be achieved in the form originally intended. However, its scientific and historical value is remarkably important, as it provided reference stars catalogues for well-known space astrometric missions in the twentieth century and paved the way to the establishment of the wide-world cooperation promoted by the International Astronomical Union.

1.1 Introduction

> The enterprise of the Astrophotographic Chart and Catalogue [has] paved the way to the international scientific cooperation and introduced the practice of astrophotography, revealing its obstacles, its methods, its limits.[1]

These words, pronounced in 1970 at the IAU General Assembly by French astronomer Paul Couderc (1899–1981), perfectly explain the role played by the *Carte du Ciel* project[2] in the history of astronomy. Launched by Paris Observatory in 1887, it was the first international scientific enterprise aimed at photographing the sky of both hemispheres and conceived as a worldwide effort, by the concurrent

[1] *L'entreprise de la carte et du catalogue photographiques du ciel [a] jeté les bases de la coopération internationale dans la science, introduit dans la pratique l'emploi de la photographie, montré ses difficultés, ses méthodes, ses limites.* (Couderc, 1971, p. 178)

[2] On the development of the project, see Chinnici, 1999, 2008b and Lamy, 2008.

I. Chinnici (✉)
INAF-Osservatorio Astronomico di Palermo, Palermo, Italy
e-mail: ileana.chinnici@inaf.it

© The Author(s), under exclusive license to Springer Nature Switzerland AG 2022

T. Montmerle, D. Fauque (eds.), *Astronomers as Diplomats*, Historical & Cultural Astronomy, https://doi.org/10.1007/978-3-030-98625-4_1

cooperation of various observatories in many countries, under the leadership of France.

During the nineteenth century, scientific cooperation in astronomy was a recurrent idea; astronomers became more and more aware that it was necessary to distribute their work to obtain important results requiring extensive campaigns of observations.

An early example was the Lilienthal Society, also known as *Himmelspolizei*, established in Germany in 1800 to survey the ecliptical charts, in order to find the missing planet between Mars and Jupiter.[3] In 1871, the Italian Spectroscopical Society proposed a program of solar monitoring,[4] open to the international cooperation, that later inspired the establishment of the International Union for Cooperation in Solar Research.[5] In the last quarter of the century, the *Astronomische Gesellschaft* (established in 1863), by impulse of Friedrich Argelander (1799–1875), collaborated to the revision of the star catalogue known as *Bonner Durchmusterung*.

Of course, the development of astrophysics, which happened in the second half of the nineteenth century, thanks to the application of both techniques of spectroscopy and photography, increased the amount of collected data, and the need of a cooperative approach became more and more evident. Moreover, the resounding results of the astrophysical research were detracting the attention of the scientific community from traditional astronomy, based on celestial mechanics and astrometry.[6] In this context, the *Carte du Ciel* project intended to synthesize modernity and tradition, as it was an interesting attempt to extensively apply the modern technique of photography in the field of classical astronomy.

1.2 Preliminary Actions

Early astronomical photography was practiced in many countries around the half of the nineteenth century, by professional and non-professional (amateur) astronomers, who took daguerreotypes and photographs of celestial bodies and phenomena (Moon, Sun, eclipses, comets). The first proponent of a photographic map of the sky, in 1857, was a well-known English pioneer of astronomical photography, Warren de la Rue (1815–1889).[7] In the decades 1860–1880, early astrophotography con-

[3] It was Ceres, discovered in Palermo in 1801; the existence of the planet was expected, according to Titius-Bode empirical law (see Hoskin, 1993; Chinnici, 2011).

[4] See Chinnici, 2008a.

[5] See, for instance, Andersen et al., 2019, p. 6.

[6] Many classical astronomers were prejudiced against astrophysical studies: see Meadows, 1984, p. 61.

[7] See Chinnici, 1999, pp. 3–4.

Fig. 1.1 The Great Comet of 1882, photographed by David Gill at the Cape Observatory. The photograph shows many faint stars in the background, and stimulated the use of photography for astrometric studies. From Berry, 1898, facing p. 393

tributed to the development of solar physics[8] and stellar spectroscopy[9] and started to be quite regularly practiced in several important observatories of both hemispheres. In 1882, Edward Pickering (1846–1919), director of Harvard College Observatory, obtained some successful photographs of the northern sky hemisphere, with stars down to magnitude 6. The director of the Cape of Good Hope Observatory, sir David Gill (1843–1914), took excellent photographs of the Great Comet of 1882,[10] with many faint stars visible in the background (Fig. 1.1).

We may look at their studies like at seeds of the future *Carte du Ciel* enterprise, as they started to develop the necessary expertise in the field of astrophotography in their countries.

In France, Ernest B. Mouchez (1821–1892) (Fig. 1.2), director of Paris Observatory, decided to invest in this new technique. In 1879, he established an "atelier photographique" at the observatory[11] and asked famous opticians brothers Paul

[8] It is worth mentioning here the photographs of the 1860 total solar eclipse, obtained by De la Rue and Angelo Secchi (1818–1878) in Spain, which definitely confirmed the solar nature of the prominences (see Chinnici, 2019, pp. 183–186).

[9] In the 1870s, Henry Draper (1837–1882) in New York and William Huggins (see below) in London obtained early stellar spectrograms (see Hughes, 2013).

[10] The photography of the Great Comet of 1882 is considered a key event, that altered the character of astronomy and the nature of the observatories, by introducing the practice of photography in the astronomical research (see Dewhirst & Hoskin, 1999, pp. 252–253).

[11] It is worth mentioning that Paris Observatory did have a tradition in this field, as the first daguerreotype of the Sun was obtained there (see Meadows, 1970, p. 7).

Fig. 1.2 Admiral Ernest B. Mouchez, director of Paris Observatory, eminent organizer of the *Carte du Ciel* project. (Portrait by Nadar, © Bibliothèque de l'Observatoire de Paris)

(1848–1905) and Prosper Henry (1849–1903) to build high-quality instruments for astronomical photography; in 1884, they obtained excellent results with plates reproducing stars down to magnitude 12.

In the southern hemisphere, Sir David Gill was informed by another renowned English pioneer of astrophotography, William Huggins (1824–1910),[12] about the results obtained in Paris and wrote to Mouchez, asking for technical details of the lenses made by the Henrys.[13] A month later, he asked for a sample of the photographs; he wanted to evaluate their quality, in view of his purpose:

[12] Huggins and Gill were collaborating for obtaining photographs of the solar corona out of the eclipses; in 1885 Gill installed a Grubb coronograph at Cape Observatory for this aim (see Becker, 2011, pp. 208–210).

[13] See Gill to Mouchez, Dec. 23, 1884, in Chinnici, 1999, p. 4.

Fig. 1.3 The Henry-Gautier astrograph at Paris Observatory, type instrument for the *Carte du Ciel* photographic work. From Mouchez, 1886, p. 11

> I propose to make a complete and uniform series of Photographic Maps of the Southern Heavens, making also a Catalogue of approximate places and magnitudes from these maps. I feel sure that I can count upon your aid in this matter.[14]

Mouchez willingly sent him some sky photographs by the Henrys and, after praising their photographic work, expressed his conviction that "we shall obtain the complete solution for sky charts thanks to photography."[15]

For this aim, Mouchez commissioned the construction of an instrument of higher performance. The Henry brothers designed and built an astrograph consisting of two joined telescopes, one for visual observations (0,24 m aperture, 3,60 m focal length) and the other for photographic work (0,33 m aperture, 3,43 m focal length). It required a special mounting ("anglaise" type), permitting to completely rotate the instrument around the polar axis; the mounting was fabricated by a French maker, Pierre Gautier (1842–1909). The so-called "équatorial photographique Henry-Gautier" started to be operating in April 1885 at Paris Observatory (Fig. 1.3).

In the meantime, on February 23, Gill sent an important letter to Mouchez, in which he wrote:

---

[14] Gill to Mouchez, Jan. 18, 1885, in Chinnici, 1999, p. 4. Of course, Gill was mainly interested in southern hemisphere sky, whose best catalogues contained no more than about 130,000 stars - an exiguous number if compared with northern hemisphere star catalogues.

[15] ... *nous allons obtenir la solution complète des cartes célestes par la photographie* (Mouchez to Gill, Jan. 22, 1885 in Chinnici, 1999, p. 85).

> ... now that you have shown at Paris what can be done, we owe a duty to posterity to form a photographic astronomical library of the Heavens by the methods you have worked out. [...] The great point however is that the work should be undertaken on a carefully prepared plan and be continued on a uniform system with the same instrument, persistently and continuously [...] It is only in a great National Establishment that we can hope for such a long sustained systematic effort.[16]

Mouchez interpreted this letter by Gill as an encouragement. Enthusiastic about the first results obtained with the new instrument,[17] on May 11, 1885, the Director of Paris Observatory presented them to the Académie des Sciences, announcing:

> Today we may consider the long-lasting problem concerning the application of photography to obtain a sky chart of stars down to magnitudes 14 and 15 as being completely solved. [...] The first great problem that could be solved in a few years is the exact construction of a Carte du Ciel, namely the counting, classification and position of all stars visible with large instruments ...[18]

Mouchez also listed other studies which would have benefitted from celestial photography, such as those on asteroids, double and multiple stars, photometry, etc. and launched the idea of a sky chart as a project proposed by Gill:

> The capable and very active director of the Cape of Good Hope Observatory, Mr. Gill, was so impressed by the great importance of the advancements obtained by Mr. Henrys [...] that he immediately sent me a project to establish a cooperation of several observatories, in order to undertake together and as soon as possible the Carte du Ciel, which will be easily executable in six or eight years.[19]

In June, Gill wrote to Mouchez to be "very glad to enter into a scheme for a 'carte du ciel' in cooperation with you."[20]

Gill's support was a determining factor for the development of the project. Behind the scenes, he played a crucial role in guiding decisions and suggesting actions with skillful diplomatic ability, and prevented the failure of the project as a whole, succeeding in the completion of what was his own main aim—namely, the Astrographic Catalogue.

[16] See Gill to Mouchez, Feb. 23, 1885, in Chinnici, 1999, pp. 83–85.

[17] See Mouchez to Gill, May 1, 1885, in Chinnici, 1999, p. 85.

[18] *On peut considérer aujourd'hui le problème longtemps cherché, d'appliquer la Photographie à la construction de la* Carte du ciel *jusqu'aux étoiles de 14e et 15e grandeur, comme certainement résolu. [...] Le premier grand problème qui va pouvoir être résolu dans quelques années est la construction exacte de la* Carte du Ciel, *c'est-à-dire le dénombrement, le classement et la position de toutes les étoiles visibles avec les grands instruments ...* (Mouchez, 1885a, p. 1178–79)

[19] *L'habile et très actif directeur de l'Observatoire du Cap de Bonne-Espérance, M. Gill, a été si frappé de la grande importance des progrès réalisés par MM. Henry [...] qu'il m'a adressé immédiatement un projet pour établir une entente entre divers observatoires, afin d'entreprendre ensemble le plus tôt possible la* Carte du ciel, *qu'il serait facile d'exécuter ainsi en six ou huit années.* (Mouchez, 1885a, p. 1180) It is unclear whether Mouchez refers to the letter of February 23 or to another letter, probably lost. The February letter, actually, did not contain an explicit project of cooperation, but rather some suggestions about selected priorities in the photographic work, probably in order to pragmatically reduce the eventual colossal work.

[20] Gill to Mouchez, June 3, 1885, in Chinnici, 1999, p. 86.

1.3 In Search of Other Partners

Between June and July 1885, Mouchez started to contact other potential collaborators, looking for scientific support to his project. He sent an exploratory letter (Fig. 1.4a–b) to the main astronomical societies[21] as well as to his colleagues Edward C. Pickering (1846–1919), William Huggins, Luiz Cruls (1848–1908) and Otto W. Struve (1819–1905), respectively directors of the prestigious Observatories of Harvard College, Tulse Hill (London), Rio de Janeiro and Pulkovo, in order to ask their advice about the realization of a photographic sky chart. He joined a copy of the test plates that the Henrys had obtained, to show their excellent results. This important letter confirms that the project was conceived, since the beginning, as a widely international enterprise:

> ... I send you a photographic plate of the Milky Way obtained with our new instrument [Henry-Gautier Astrograph] [...] In one-hour exposure, we obtain all the stars down to magnitude 15 on plates and 14 on paper [...] Since the plate covers a sky area of 7 square degrees, we would need 6 000 plates to cover the entire sky vault [...]
>
> Today the photographic sky chart is easy to achieve if 5 or 6 observatories well positioned in both hemispheres agree to carry out this extended and important work. It could be completed in six or eight years and we shall leave to future astronomers the exact representation of our sky at the end of the 19th century, with 20 or 25 million stars. [...]
>
> Please, consider this matter and let me know your advice.[22]

The recipients of the letter were appropriately chosen,[23] from a scientific and a geographic point of view. In Europe, Huggins was an expert of celestial photography and a friend of Gill and could support the British participation. Struve was a scientific authority, one of the most influential figures in the field of astrometry and could secure an important academic support. In America, Pickering was an expert and reputed astronomer, also interested in celestial photography. In Rio de Janeiro, Cruls had already offered his collaboration to Mouchez, after reading the

[21] See, for instance, Mouchez, 1885b; Gill suggested to Mouchez, step by step, how to proceed in sending official invitations (see Gill to Mouchez, July 26, 1886, in Chinnici, 1999, p. 91).

[22] *... je vous envoie une épreuve photographique obtenue dans la Voie lactée avec notre nouvel appareil [l'équatorial photographique Henry-Gautier] [...]. En une heure de pose, nous obtenons toutes les étoiles jusqu'à la 15e grandeur sur le cliché et jusqu'à la 14e pour la reproduction sur papier. [...] L'épreuve représentant une superficie de 7 degrés carrés, il en faudrait 6000 semblables pour toute la voûte céleste [...]. La carte photographique du ciel est donc devenue aujourd'hui facilement exécutable si 5 ou 6 observatoires, bien choisis dans les deux hémisphères peuvent s'entendre pour entreprendre ce vaste et si important travail. Elle pourrait être exécutée en 6 ou 8 ans et nous lèguerions aux astronomes de l'avenir un état exact de notre ciel à la fin du 19eme siècle où figureraient 20 ou 25 millions d'étoiles. [...] Je serais heureux si vous vouliez bien étudier un moment cette question et me faire connaître votre opinion.* (Mouchez to Pickering and Huggins, Paris, June 26, 1885; to Cruls and Struve, Paris, July 15, 1885, in Chinnici, 1999, pp. 4–5)

[23] Even if no documentary evidence has been found until now, it is highly possible that Gill contributed to compile the list of recipients; the letters were, in fact, sent with two different dates—this circumstance suggests that the original list was extended.

(a)

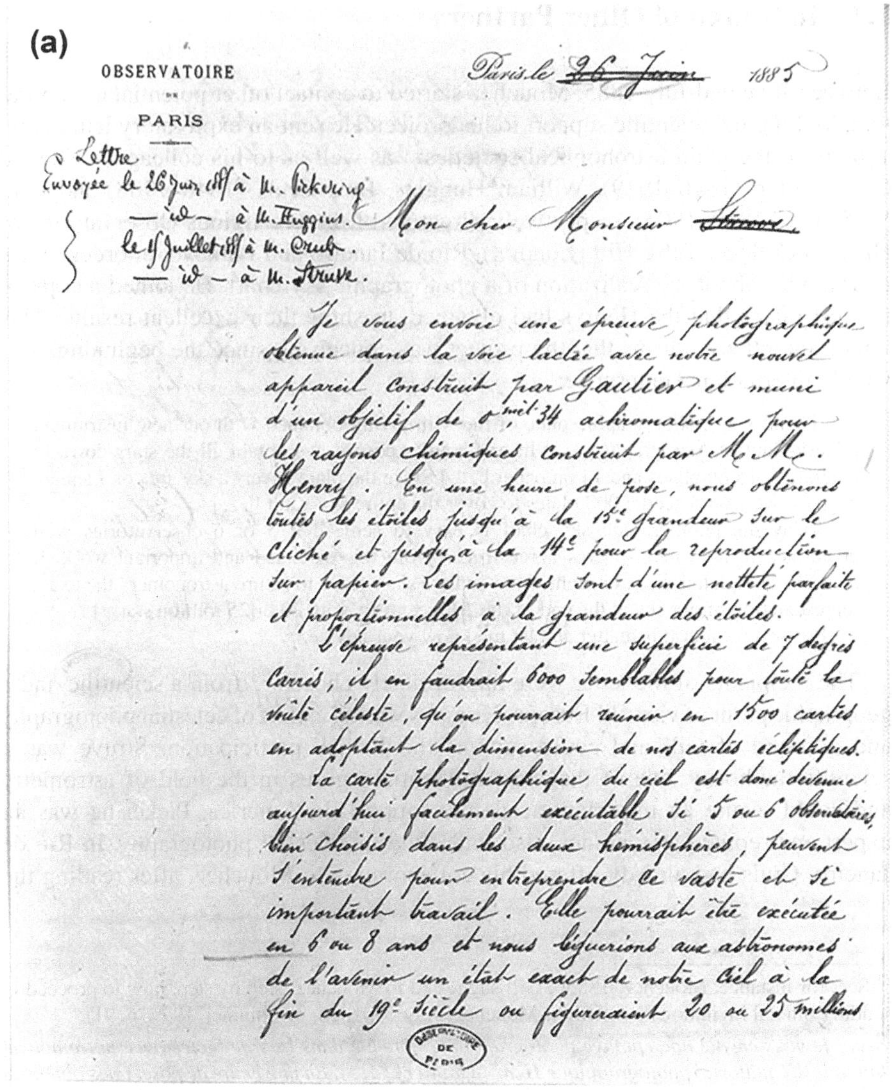

OBSERVATOIRE DE PARIS

Paris, le ~~26 Juin~~ 1885

Lettre envoyée le 26 Juin 1885 à M. Pickering
— id — à M. Huggins.
le 15 Juillet 1885 à M. Cruls
— id — à M. Struve.

Mon cher Monsieur ~~Struve~~

Je vous envoie une épreuve photographique obtenue dans la voie lactée avec notre nouvel appareil construit par Gautier et muni d'un objectif de 0 mil. 34 achromatique pour les rayons chimiques construit par MM. Henry. En une heure de pose, nous obtenons toutes les étoiles jusqu'à la 15e grandeur sur le cliché et jusqu'à la 14e pour la reproduction sur papier. Les images sont d'une netteté parfaite et proportionnelles à la grandeur des étoiles.

L'épreuve représentant une superficie de 7 degrés carrés, il en faudrait 6000 semblables pour toute la voûte céleste, qu'on pourrait réunir en 1500 cartes en adoptant la dimension de nos cartes écliptiques.

La carte photographique du ciel est donc devenue aujourd'hui facilement exécutable si 5 ou 6 observatoires bien choisis dans les deux hémisphères, peuvent s'entendre pour entreprendre ce vaste et si important travail. Elle pourrait être exécutée en 6 ou 8 ans et nous léguerions aux astronomes de l'avenir un état exact de notre ciel à la fin du 19e siècle où figureraient 20 ou 25 millions

Observatoire de Paris

Fig. 1.4 (**a**) First page of the draft of the explorative letter by Mouchez about the proposal of making a photographic chart of the sky. (Mouchez to E. Pickering, Huggins, Cruls and Struve; © Bibliothèque de l'Observatoire de Paris)

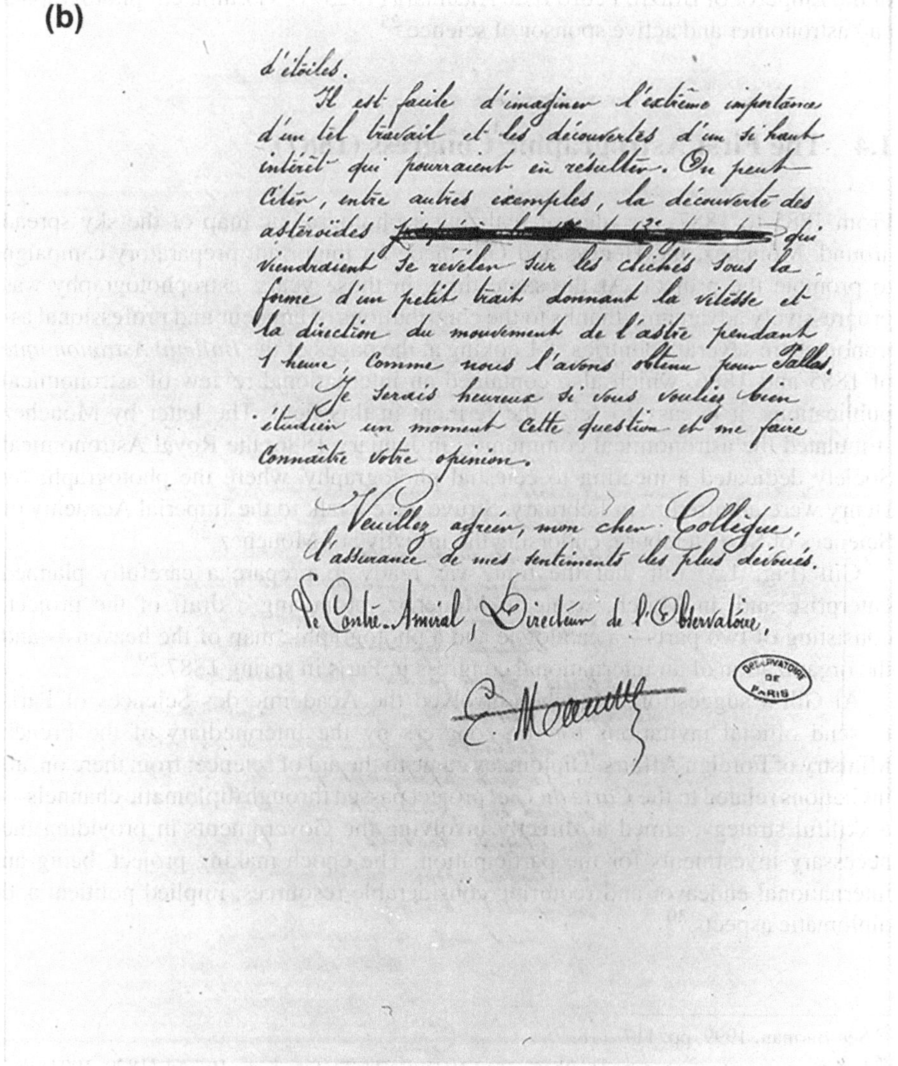
(b)

d'étoiles.

Il est facile d'imaginer l'extrême importance d'un tel travail et les découvertes d'un si haut intérêt qui pourraient en résulter. On peut citer, entre autres exemples, la découverte des astéroïdes [illegible] qui viendraient se révéler sur les clichés sous la forme d'un petit trait donnant la vitesse et la direction du mouvement de l'astre pendant 1 heure, comme nous l'avons obtenu pour Pallas.

Je serais heureux si vous vouliez bien étudier un moment cette question et me faire connaître votre opinion.

Veuillez agréer, mon cher Collègue, l'assurance de mes sentiments les plus dévoués.

Le Contre-Amiral, Directeur de l'Observatoire,

E. Mouchez

Observatoire de Paris

Fig. 1.4 (**b**) Second page of the same draft. (Mouchez to E. Pickering, Huggins, Cruls Cruls and Struve; © Bibliothèque de l'Observatoire de Paris)

latter's communication to the Académie des Sciences[24]— his offer was certainly considered well grounded, as Mouchez knew that Cruls could count on the support

[24] See Cruls to Mouchez, June 12, 1885, in Chinnici, 1999, p. 318.

of the Emperor of Brazil, Pedro II de Alcantara (1825–1891), amateur photographer and astronomer and active sponsor of science.[25]

1.4 The First Astrographic Congress (1887)

From 1885 to 1887, the idea of realizing a photographic map of the sky spread around. Mouchez, the Henrys and Gill made an important preparatory campaign to promote the project. At the same time, in those years, astrophotography was progressively advancing, thanks to the contributions of amateur and professional astronomers in several countries.[26] Looking at the pages of the *Bulletin Astronomique* of 1885 and 1886, which also contained an international review of astronomical publications, it is easy to seize the ferment in this field. The letter by Mouchez stimulated the astronomical community; in January 1886, the Royal Astronomical Society dedicated a meeting to celestial photography, where the photographs by Henry were admired;[27] in February, Struve gave a talk to the Imperial Academy of Sciences of St. Petersburg, endorsing the initiative of Mouchez.[28]

Gill (Fig. 1.5) felt that the time was ready to prepare a carefully planned enterprise and, in March, wrote to Mouchez, proposing a draft of the project, consisting of two parts—a catalogue and a photographic map of the heavens—and the organization of an international congress in Paris in spring 1887.[29]

At Gill's suggestion, Mouchez involved the Académie des Sciences of Paris to send official invitations for the congress by the intermediary of the French Ministry of Foreign Affairs. Diplomacy came to the aid of science: from there on, all invitations related to the *Carte du Ciel* project passed through diplomatic channels—a skillful strategy, aimed at directly involving the Governments in providing the necessary investments for the participation. The epoch-making project, being an international endeavor and requiring considerable resources, implied political and diplomatic aspects.[30]

[25] See Barman, 1999, pp. 117–118.

[26] It is worth mentioning, among others, the amateur astronomers Isaac Robert (1829–1904) and Andrew A. Common (see below) in England and the professional astronomer Wilhelm Oswald Lohse (1845–1915) at Potsdam Astrophysical Observatory, in Germany.

[27] See "Revue des publications astronomiques," *Bulletin Astronomique,* t. III (1886), pp. 151–153.

[28] See Struve, 1886.

[29] Gill to Mouchez, Mar. 1, 1886, in Chinnici, 1999, p.5. The letter by Gill was translated and published in order to be circulated (see Gill, 1886) and it provided a base for the discussion at the first congress.

[30] It is to be remarked that all overseas countries participating in the project had political connections with France; on the participation of these observatories in the *Carte du Ciel*, see the controversial book by Pyenson (1993): for Santiago Observatory, pp. 248–256; for La Plata and Cordoba Observatories, pp. 262–267; for Tacubaya Observatory, pp. 280–282; for Rio de Janeiro Observatory, pp. 288–289. A special mention is to be deserved to the Vatican Observatory,

Fig. 1.5 David Gill, Royal Astronomer at the Cape of Good Hope, "éminence grise" of the *Carte du Ciel* project. From *The Observatory*, vol. XXXVII, 1914

The original list (of about 30–35 invited participants) was prepared by Mouchez and Struve, who spent a week in Paris in September 1886.[31] It also included the names of the most important instrument-makers, as suggested by Struve, and some names provided by Gill. Struve also suggested to send a provisional program of the questions to be discussed. The First Astrographic Congress—to be held at Paris Observatory (Fig. 1.6)—was conceived and organized as a scientific, politic, and social event (including gala dinner, dancing, decorations, etc.) under the enthusiastic direction of Mouchez.[32]

Fifty-five participants (twenty of them were French) from a dozen countries attended the meeting (Fig. 1.7).[33] It was an important diplomatic success for France, which played the role of leader country, with Paris Observatory as leading

which was established on purpose to participate in the *Carte du Ciel* project; this provided a scientific pretext (supported by France) for a political move to reaffirm the Vatican sovereignty and independence from the Italian Government (see Chinnici, 1999, pp. 460–464).

[31] See Mouchez to Gill, Sept. 20, 1886, in Chinnici pp. 92–93.

[32] For further details about the Congress, see Motais de Narbonne, 1988. Other information on the *Carte du Ciel* project can be found in Débarbat et al., 1988 (section "Historical Research").

[33] See Chinnici, 2008a, 2008b, pp. 23–24.

Fig. 1.6 A photograph of Paris Observatory at the beginning of the twentieth century. From *The Observatory*, vol. XXXVI, 1913

institution. Mouchez's opening speech may appear quite overblown, but it really expressed the awareness of doing an unprecedented endeavor:

> ... it is a great honor to host the first assembly opening this new era for the astronomical science. It will be a glorious and unforgettable date in the history of astronomy, as will be unforgettable the magnificent work that we want to transmit to the future generations, a work that can be defined as the most exact and complete inventory of the detectable universe at the end of the 19th century.[34]

After one week at work (16–25 April), the congress members established an International Permanent Commission including 11 members,[35] plus the directors

34 *... C'est un grand honneur (...) de recevoir la première assemblée où va s'inaugurer cette ère nouvelle pour la Science astronomique. Ce sera une date glorieuse et inoubliable dans son histoire, comme sera inoubliable dans l'histoire de l'Astronomie l'œuvre grandiose que nous voulons léguer aux générations futures, œuvre qu'on pourra définir comme l'inventaire exact et aussi complet que possible de l'univers perceptible à la fin du XIX*^{e} *siècle* (Congrès astrophotographique, 1887, p. 6)

35 The first members of the International Permanent Commission, elected in 1887, were: David Gill from Cape of Good Hope Observatory, William H. Christie (1845–1922) from Greenwich Observatory, Carl H. Vogel (1841–1907) from Potsdam Observatory, Edmund Weiss (1837–1917) from Vienna Observatory, Otto W. Struve (1819–1905) from Pulkovo Observatory, Maurice Loewy (1833–1907) and Prosper Henry (1849–1903) both from Paris Observatory, Pietro Tacchini (1838–1905) from Collegio Romano Observatory in Rome, Nils Dunér (1839–1914) from Lund Observatory, Edward C. Pickering from Harvard College Observatory, and Jules C. Janssen (1824–1907) from Meudon Observatory.

Fig. 1.7 The first International Astrographic Congress, held in Paris in 1887; group photograph of the participants. (Photo by Nadar, © Bibliothèque de l'Observatoire de Paris)

of the participant observatories, and an Executive Committee of 9 members.[36] The 11 members of the Commission came from 7 European countries, plus South Africa (then a British colony) and the USA. The participation of observatories of different continents was crucial, since the project was intended to be global. France reinforced its leading position with five members in the Commission and three in the Committee, including the president of both, who was, of course, Mouchez.[37]

It was also decided that all observatories being able to obtain financial support from their governments to acquire the necessary instruments would have joined the enterprise.

[36] The first Executive Committee, elected in 1887, was composed by Mouchez (President), Christie, Dunér, Janssen, Struve, Tacchini (members), Gill, Loewy, Vogel (secretaries).

[37] This leadership was further reinforced by the participation of 4 French observatories over 18 participant observatories from other countries.

1.5 A Double Project: Chart and Catalogue

The second meeting of the Commission took place in 1889, again in Paris. Shape and sizes of the glass plates were fixed[38] as well as exposure times.[39] However, the execution of the *Carte du Ciel* project appeared to be more complex than expected, due to the additional work for the Catalogue, a much longer and expensive effort, which included the measurement of star coordinates in the plates, their reduction at the year 1900, and their publication. The original proposal by Mouchez was the execution of the Chart, but the Catalogue was an addition strongly required by Gill, who considered it more feasible and useful than the Chart itself. Actually, the second meeting was mainly devoted to the catalogue preparatory works. Gill wrote to Mouchez:

> I think you will find that the point capital will become not la carte photographique du ciel—in the sense of so many photographic plates of the sky—that will become effete long before the work is finished—[. . .]. It is the Catalogue, the organization for its execution, computation and publication which must cause the Astrophotographic Congress of Paris to be an Epoch in the History of Astronomy. That must be kept in view if the Committee desires to fulfil its functions in the most complete manner.[40]

Gill's proposal certainly made sense, because the best catalogues available at that time contained just a few hundred thousand stars: in comparison, the Astrographic Catalogue, which was expected to contain about four million stars, would have been a real breakthrough.[41]

This extension of the work implicated as well to extend the participation in the project to other observatories, which did not carry out the photographic work, but were involved in the preparation of a reference star catalogue (for instance, this was the case of Pulkovo Observatory, which in 1909 provided a list of stars in northern sky[42]), in the plates measurement operations and in the reduction calculations (this was the case of Edinburgh Observatory, whose astronomers measured and reduced

---

[38] The photographic plates were squares 16 x 16 cm; it was decided that a grid of small squares 5 x 5 mm, covering the central 13 x 13 cm area, would be impressioned in the plate, in order to facilitate the measurements of the star positions later.

[39] Each plate should have been exposed two or three times, so that each star could be identified by two or three points and not confused with defects or dust grains on the plates. The Commission also discussed how to preserve the exposed plates to be transmitted to the future generations of astronomers; the final decision was to use gravure on copper plates, but this operation was later carried out by only a few Observatories.

[40] Gill to Mouchez, Aug. 17, 1889, in Chinnici, 1999, p. 105.

[41] Bonner Durchmusterung, the most complete star catalogue of the northern sky, contained about 320,000 stars down to magnitude 9. For the southern sky, the most extended catalogues were the Uranometria Argentina (1879) by Gould, with about 50,000 stars down to magnitude 7 (declinations −23° to 90°) and the Sudliche Durchmusterung (1886) by Schonfeld, with about 120,000 stars (declinations −1° to −23°).

[42] The list was compiled by Oskar Backlund (1846–1916) and resulted from a previous catalogue, made in 1900, with an appendix of stars having been observed at the Odessa branch (see Dadaev, 1978, p. 51). For the southern sky, the list was compiled by Samuel S. Hough (1870–1923), who

the plates taken at Perth in 1900). Directly or indirectly, whether through the photographic work or the catalogue preparation, almost all national observatories of the participant countries (and even some private observatories, like the D'Abbadia "Castle-Observatory," near Hendaye in France) were involved in this worldwide project.

1.6 Work Plan and Guidelines

At the end of the third meeting, held in Paris in 1891, the project was sufficiently defined and almost ready to be carried out. It was hence officially announced the participation of 18 observatories and assigned the division of the sky zones for the photographic work (Appendix A); the beginning of the operations, for most of the participant observatories, was expected in summer 1891.

Astrographs fitting the *Carte du Ciel* purposes were acquired for the participant observatories by the respective governments. If we try to draft a geopolitical map of the project (Appendix B), it clearly appears a mixing of nations with a consolidated astronomical tradition, stable political situation and flourishing economy, and nations lacking of these important preliminary conditions. Consequently, this map indirectly shows which observatories were candidate to fail, due to the considerable financial effort that their participation implied in comparison with the scarcity of resources, training, expertise, facilities.

At the 1891 meeting, after four years of tests, the work plan was also fixed.[43] It was decided to take two series of photographic plates, one for the Chart, reproducing stars down to magnitude 14 (40 minutes exposure), and an additional series for the Catalogue, with stars down to magnitude 11 (5 minutes exposure): the resulting Astrographic Catalogue should have been the most accurate and complete star catalogue ever made.

The plates covered a sky field of $2°10'$,[44] (Fig. 1.8a, b) but only the central part was free of optical distortions and the usable field was consequently reduced to about 1°. Every participant observatory had to take over a thousand photographs for the Chart series and as many for the Catalogue. By supposing a hundred

succeeded Gill as director of the Cape Observatory. The Backlund-Hough list was a reference in astrometry for many years.

[43] The photographic work was carried out by astronomers (who learnt to use photographic means) or by assistants, possibly having some experience in photographic operations. Among the astronomers who greatly helped in the resolution of many technical questions, are to be mentioned Charles Trépied (see below), who studied the problem of establishing the photographical magnitude, and Hermann Carl Vogel, quite expert in photographic work, who provided a method for the impression of the grid in the plates (see note 38).

[44] In order to assemble the plates and make a double check of the star positions, it was originally decided that, the two Chart series were centered respectively on even and odd numbered declinations, so that the corner of one plate was coincident with the center of the other plate.

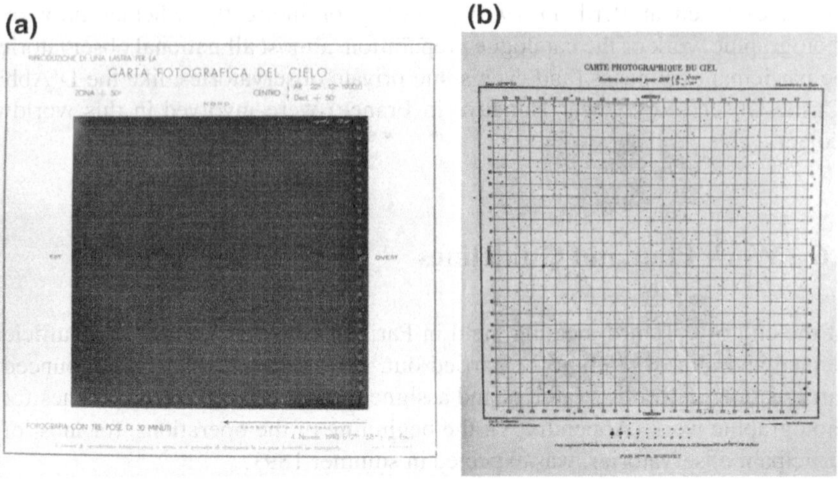

Fig. 1.8 (**a**) Reproduction on paper of a *Carte du Ciel* plate taken at Catania Observatory. From: Chinnici, 1999, pl. 16; (**b**) Typical *Carte du Ciel* chart; the printed negative was doubled in size. Courtesy of INAF-Palermo Astronomical Observatory LIbrary

Fig. 1.9 The Henry-Gautier photographic equatorial installed at the Vatican Observatory, one of the most efficient participants in the *Carte du Ciel* project. From *Pubblicazioni della Specola Vaticana,* fasc. II, 1891

average useful nights per year, the whole photographic work was expected to be achieved in about six years. The Henry-Gautier photographic equatorial was chosen as instrument-type (Fig. 1.9): this was a significant success for the French optical industry.

The fourth meeting, in 1896, still in Paris, was also mainly devoted to the Catalogue; the Chart, little by little, drifted away, in the background, losing its original appeal.

1.7 Nationalism and Financial Interests

The countries having the highest number of observatories involved in the *Carte du Ciel* project were France and Great Britain.

In France, the *Carte du Ciel* project was an important opportunity to reinforce the decentralization process that led to the establishment of the Department Observatories (*observatoires de départements*[45]) in the last quarter of the nineteenth century.[46] Paris Observatory had always played a leading role, and old minor observatories (like Marseilles) were considered no more than simple astronomical stations. The establishment of the observatories of Bordeaux, Besançon, Lyon and Toulouse was an attempt to promote the departments and provide them some degree of independence from Paris. On the colonial side, the Algiers Observatory, strictly controlled by Paris, obtained considerable increases in the budget and a measure of independence from the local educational authorities, thanks to the participation in the Carte du Ciel project.[47] Consequently, the involvement of three of these observatories (Bordeaux, Toulouse, and Algiers) in the *Carte du Ciel* project was an opportunity to somehow legitimate their existence, provide them with first-class instruments and obtain additional resources.

While Paris Observatory was the pulsating heart of the whole project, the other three French Observatories would have supported it, in some sense ensuring the success of the project by the execution of a large part of the work. Of course, this strategy reinforced the leadership of France in the project—a leadership that remained unquestioned until the end of the project. All the Commission meetings were held at Paris Observatory and, at Struve's proposal,[48] the President of the Commission was, by default, the Director of Paris Observatory.

Of course, France had not only scientific but also financial interests in the project, especially for instrument-makers. Henry-Gautier astrographs were ordered for all participant French Observatories, as well as for Vatican, La Plata, Córdoba, Santiago, San Fernando, Rio de Janeiro and Uccle Observatories.[49] Photographic

[45] The French word "département" is taken here in the geographical sense ("Département de la Seine"), somewhat comparable to "counties" in the UK.

[46] See Chinnici, 2011.

[47] On the relevant participation of Algiers Observatory in the Carte du Ciel project, see Pyenson, 1993, pp. 96–99.

[48] See Beuf to Tisserand, Jan. 21, 1893, in Chinnici, 1999, p. 256.

[49] The Henrys also built the lenses for Helsinki astrograph, whose mount was constructed by German firm Repsold (see Donner to Mouchez, Sept. 12, 1889, in Chinnici, 1999, p. 226). Repsold

plates made by the Lumière brothers, which were very sensitive, were widely used, even outside of France.

The effort of the French observatories for completing the work assigned for both Chart and Catalogue was considerable (see Appendix C). Paris Observatory also paid a very high cost in terms of human resources. In about fifteen years, from 1892 to 1907, three Directors died—Mouchez (aged 71), Félix Tisserand (1845–1896) (aged 51) and Maurice Loewy (1833–1907) (aged 74), who guided the development of the project for about a decade, giving a determining impulse to its advancement.[50] This list can be extended, by adding in 1909 Auguste Fraissinet (1846–1909) (aged 63), very active secretary of the Observatory and invaluable assistant[51] in facilitating the execution of the project. It cannot be excluded that the prolonged stress resulted in a fatal outcome to their health. Of course, these close changes also affected the steadiness and continuity of the *Carte du Ciel* work.

1.8 Communication Strategies and Protectionism

Across the Channel, to secure the contribution from the British observatories, Gill was aware that he had to overcome two main difficulties. Firstly, it was necessary to convince the Government to finance their participation in the project. He succeeded in this purpose, by adopting a winning outreach strategy. Gill captured the public opinion by a series of lectures and publications, illustrating the successful results of astrophotography and its potentialities. He wrote to Mouchez:

> My lecture [on star charting by photography] was considered successful. The photographs [by the Henrys] were projected by electric light in a field 20 feet in diameter, and showed splendidly. I repeat the lecture at Oxford tomorrow. I think that with a little pressure from the outside public and the public papers the matter will now go forward.[52]

Gill obtained funds for the participation of the Royal Observatories of Greenwich and The Cape. Moreover, Oxford University Observatory as well as Australian colonial observatories[53] (Fig. 1.10) took part in the project: together with France, having four participant Observatories, the UK was the nation that mostly invested in the *Carte du Ciel* plan (see Appendix C).

also built the mounting of the Potsdam photographic refractor, whose optics were commissioned to the German optical firm Steinheil (see Tsvetkova et al., 2009, p. 879).

[50] See Chinnici, 2007.

[51] See Baillaud to Thome, Oct. 17, 1908, in Chinnici, 1999, p. 201.

[52] Gill to Mouchez, June 5, 1887, in Chinnici, 1999, p. 97.

[53] The participation in the *Carte du Ciel* had a remarkable impact on the Australian observatories, and somehow secured their survival and independence (see Haynes et al., 1996, pp. 94–95: on Perth Observatory, pp. 91–92; on Melbourne Observatory, pp. 80–81; on Sydney Observatory, pp. 61–66).

Fig. 1.10 Sydney Observatory, among the Australian observatories the one that mostly contributed to the *Carte du Ciel* project. From Haynes et al., 1996

Secondly, it was necessary to convince the British Government to buy (French) refractors. Now, the purchase of those instruments was a rather complicated matter. First of all, Gill admitted that "there is unquestionably a prejudice in favor of the reflectors in England"[54]—but he was sure that the unsurpassed quality of the photographs obtained at Paris Observatory would have been decidedly convincing about the necessity of using refractors. Nevertheless, it was also necessary to take into account the competition between the optical industries of the two countries.

The purchase of French telescopes was in fact blocked by the intervention of Howard Grubb (1844–1931), Dubliner instrument-maker who intended to build the instruments for the British participating observatories. Grubb wrote to Gill:

> I shall be glad to hear from you as soon as possible about the details of the photographic telescopes. Are there not one or two of them to be made for Australia as well as for Greenwich and the Cape? Of course, the greater number I have to make, the less price I can make them for.[55]

Fearing to miss a big business, Grubb polemically questioned the acquisition of French astrographs, and threatened to rise a question in the Parliament if the instruments were ordered in France:

> Now I don't think there can be any reasonable doubt that these instruments can be made in our own Country at last as well, probably decidedly better, than in France and my estimate is lower than that of the Frenchmen. If therefore the Government send their orders to France they will have to be prepared with a good reason to do so as I shall certainly take steps to have the question raised in the House. I cannot however think it likely that the Government would go off hand and order instruments [...] without waiting for the advice of the two

[54] Gill to Mouchez, Mar. 23, 1887, in Chinnici, 1999, p. 95.

[55] H. Grubb to Gill, May 4, 1887, in Glass, 1997, p. 133.

> societies most interested; however, this matter has put me upon my guard and I think I had better be on the spot whenever any discussion comes on.[56]

Grubb obtained to build the astrographic telescopes for all participant British Observatories.[57] He quickly designed the mount in 1888,[58] but had to make several attempts for building 13-inch. Objective lenses, which could be appropriated for the photographic work.[59] Only in 1890, he could provide his astrographs with proper photographic objective lenses[60], which were delivered to British and colonial observatories participating in the *Carte du Ciel* project.[61]

1.9 Criticism, Disagreement, Competition

Gill's insistence on the importance of the Catalogue made clear to the astronomical community that the project was diverging from its original goal, which was the Chart. This provoked reactions of disapproval, especially from British astronomers Andrew A. Common (1841–1903) and Herbert H. Turner (1861–1930), editors of the journal *The Observatory*, that violently attacked Gill's position:

> We consider that the scope of the work has been illegally magnified, and a catalogue illegally substituted for a chart; and the meaning of the word illegal is that the decisions of the Committee on these points are, in our opinion, contrary to those of the real legislative body – the Congress of 1887 [...] « Ce qui me préoccupe surtout c'est de voir la tendance qui se manifeste dans la Conférence à vouloir toujours augmenter un travail déjà si considérable en lui-même » (Procès-verb., p. 51). We would respectfully associate ourselves with M. Cornu [Marie Alfred (1841–1902), French physicist] in his anxiety.[62]

In order to contrast these criticisms, in 1891 it was decided to give priority to the Chart plates; nevertheless, it was the work of the Catalogue that advanced, being strongly pushed by Gill, who considered that resolution just a compromise. In his opinion, maintaining the ambiguity of the Chart/Catalogue project was a strategy of French astronomers to obtain financial support from their government, being a chart

[56] H. Grubb to Gill, June 1, 1887, ibid., p. 134.

[57] Regarding the astrograph for Sydney Observatory, Grubb only built the optics, while the mechanic parts were made by local firms (see Haynes et al., 1996, p. 62).

[58] See Grubb, 1888, in Glass, 1997, p. 242.

[59] See Pritchard to Mouchez, July 18, 1889, in Chinnici, 1999, p. 276; also see Gill to Mouchez, Aug. 17, 1889, in Chinnici, 1999, p. 105. On Grubb's difficulties with the making of astrograph lenses, see Glass, 1997, pp. 146–149.

[60] The final version of Grubb astrograph's design was published in *Engineering* in 1890; see Glass, 1997, p. 150.

[61] A Grubb astrograph was also ordered for the (non-colonial) Tacubaya Observatory in Mexico; it required some optical corrections, which were made by Alvan Clark & Sons, renowned optical firm in Cambridge, Massachusetts (see Bartolucci, 2000, pp. 65–68).

[62] *The Observatory,* 1891, pp. 184–5.

Fig. 1.11 Edward C. Pickering, director of the Harvard College Observatory, the "great absent" in the *Carte du Ciel* project. From *Memorie della Società degli Spettroscopisti Italiani,* vol. VIII, 1919

much more appealing and inspiring than a catalogue, in front of the public opinion. He wrote to Pickering:

> … our true function is the Catalogue – indeed there was a motion very nearly carried out that we should leave the Chart aside and push on the Catalogue plates – And this would have been carried I believe but for the fact that the French astronomers want money from the Government for "personnel" to carry on the work, and that whilst the Chart appeals to the general public – and millions of stars appeal to the popular imagination, the solid work of a catalogue does not do so. Thus a compromise was reached which permitted the Frenchmen to say that the Chart was to be commenced at once, although the Catalogue plates were first to be pushed on.[63]

Additional doubts arose in the international astronomical community because of a somehow competing project. If we look at the list of participant Observatories (see Appendix A), it is striking the absence of North American Observatories. This can be explained by the fact that a similar project was ongoing at Harvard College Observatory, under the direction of Pickering (Fig. 1.11).[64]

[63] Gill to Pickering, Apr. 8, 1891, in Jones & Boyd, 1971, p. 210.

[64] As early as 1883, Pickering had planned to make a chart of the sky by photography, but he lacked funds. Two years later, however, he was able to buy an 8-inch photographic doublet, corrected and mounted by Alvan Clark & Sons, photographing stars as faint as the magnitude 14 (see Jones & Boyd, 1971, pp. 206–207). A few years later, he acquired a similar instrument for the station at Arequipa in Peru (see Holden, 1893).

In 1887, though invited, Pickering could not attend the Paris Congress but wrote a letter to the participants, proposing to use refractors of short focal length, with double object-glass (consisting of four lenses) for photographing the whole sky; the employment of such large-field cameras would have considerably reduced the effort and duration for the completion of the Chart.

This proposal was, however, rejected by the Congress participants, at Pickering's regret; Gill explained him why:

> Your letter to the Congress received much more consideration than you appear to think. [. . .] The question of the adoption of combinations giving very large fields, composed of 4 lenses was fully talked out [. . .]
>
> The scale of magnitude and accuracy to be attained were really the points which went against the 4-glass combination. Nothing less than 14th magnitude stars would satisfy a large number of members of the Congress and it was felt that if we go beyond 15 minutes of exposure we ran some risk of failure. To combine these conditions would have involved the use of lenses of 12 or 13 inch. aperture and 4 glass lenses of this aperture would have cost too much – indeed I do not know if we could get them made at all. [. . .] I hardly think you can use so large a field to advantage because of the changes in the differential refraction during exposure – the correction of a 4-glass objective is also a species of compromise and the law of distortion is probably troublesome to deal with. In fact the Congress was determined to do the work on a scale of accuracy and magnitude which appeared to exclude the use of double combinations.[65]

Actually, it is not to be excluded that Gill himself dissuaded the members of the Congress from accepting Pickering's proposal, being his main aim the Catalogue of stars down to magnitude 11—a result that could not be obtained with large-field combinations.

Pickering hence decided to do by himself, somehow challenging the Commission, and started to work at his competing project. In November 1888, he strongly appealed for "A Large Photographic Telescope" to be funded with 50,000 dollars, and in June 1889, he obtained a gift from Miss Catherine Wolfe Bruce (1816–1900), a wealthy philanthropist. Due to the extensive publicity given to Miss Bruce's contribution, the news reached the European astronomical community; the reaction of the British astronomers in the pages of *The Observatory* was vitriolic:

> . . . if Prof. Pickering is right in thinking that he can chart the whole sky in two or three years with a single telescope, then the combination of seventeen observatories to do the same work is all wrong . . .
>
> He has thrown out a distinct challenge, and matched himself against the rest of the world in the same undertaking. [. . .]
>
> . . . he has tacitly condemned the work of the International Conference before it is commenced, and claimed success for himself; whereas it is comparatively certain that the former work will be carried out, based as it is on definite work with a known instrument; while his own plan depends on that assumption, which has so often proved delusive, that a telescope may be multiplied by three.[66]

[65] Gill to Pickering, June 22, 1887, in Jones & Boyd, 1971, p. 209.

[66] Turner & Common, 1889, pp. 310–11.

Pickering, however, did not believe in the successful execution of the complex *Carte du Ciel* project and preferred to follow his own plans:

> . . . seventeen observatories have undertaken to photograph the entire sky with telescopes of the usual form thirteen inches diameter. Several years will be required for this work. They propose to photograph stars to the 14th magnitude. Our 8-inch telescope photographs fifteenth magnitude stars. The Bruce telescope can hardly fail to take as faint stars or to surpass the Paris photographs. If it succeeds as I anticipate it should take five to ten times as many stars [. . .] I do not feel that I am to blame if with one telescope we do more and better work than with seventeen of the usual form.[67]

Gill declared himself "disgusted"[68] by the attack in *The Observatory* and was quite convinced that Pickering's photographic map was more effective than the *Carte du Ciel* map; he wrote to him from Paris in 1891: "*. . . there is a pretty strong feeling that your chart will be a much better and more practical one that ours*"[69]—a further reason to insist on being the catalogue the true aim of the project.

The Bruce Photographic Telescope (24-inch aperture, 11-foot focal length), made by Alvan Clark & Sons, was ready in 1893.[70] Pickering planned to use it at Harvard to photograph the northern sky, then to move it to Arequipa, the southern branch of Harvard Observatory, to complete the photographic survey of the sky. Plates would be stored at Harvard in a fireproof building for examination and measurements.[71]

Pickering produced a Photographic Map of the sky of both hemispheres in 1903[72]; though small-scale, this is the first complete photographic sky map ever made. Pickering was so diplomatically cautious, however, as to avoid explicitly entering into competition with the *Carte du Ciel* project. He contented himself with less ambitious results: his 55 plates covered sky regions of about 30 x 30 degrees and reproduces stars down to magnitude 12. Moreover, Pickering clearly stated that the utility of these charts laid more on the measurement of photometric variations than of accurate positions:

> These photographs have proved unexpectedly useful here for determining the past as well as the present changes in light of variable stars, new stars and similar objects [. . .]. Of course,

[67] Pickering to Bruce, in Jones & Boyd, 1971, p. 275.

[68] Gill to Pickering, Nov. 4, 1889, in Jones & Boyd, p. 276.

[69] Gill to Pickering, Apr. 8, 1891, in Jones & Boyd, 1971, p. 210.

[70] See Holden, 1893. The Bruce telescope was operating until the half of the twentieth century and gave important services to astronomy, especially in the photograph of southern sky (see Jones & Boyd, 1971, pp. 285–286). The digitization of the collections of all photographic plates kept at Harvard (including the ones by Pickering) is currently ongoing (see DASCH project: https://projects.iq.harvard.edu/dasch).

[71] See Jones & Boyd, 1971, p. 272.

[72] The successful results obtained with the Bruce telescope induced Turner to retreat his position; once he saw that the southern observatories were delaying the photographic work, in 1897 he proposed to leave the photograph of the southern sky to Pickering—a recognition for the Harvard astronomer (see Jones & Boyd, 1971, p. 276).

> the small scale diminishes their value for measuring positions, although the minuteness of the images in part compensates for this difficulty.[73]

In the end, however, Pickering's doubts about the Carte du Ciel execution will prove to be well-founded. As we shall see below, the *Carte du Ciel* project, for the part concerning the Chart, will remain unachieved—only the astrographic catalogue was completed, though in a different version compared to the original plan, and with an incommensurably longer time (the last volume was published in 1964).

1.10 Desertions and Replacements

In about a decade, many observatories of Latin America abandoned the project and were replaced by others, that abandoned, on their turn, in the following decade. As it has been previously mentioned, this can be explained by the fact that some countries underestimated the effort, especially in terms of financial resources. Through their participation, they wished to seize the opportunity to acquire visibility and prestige in the international scientific (and political) context. On their turn, thanks to this opportunity, the participant Observatories hoped to obtain financial resources from their governments. They failed for many reasons: instability of their governments, financial crisis, difficulties in provisions, insufficient resources, inadequate directorships, etc.

This situation was pragmatically foreseen by Gill, who kept ready to relieve the work of all Southern Hemisphere observatories.[74] He wrote to Mouchez in 1891:

> I am not in the least disquieted by the civil war in Chile[75] nor by the other troubles which you describe in Buenos Ayres. We shall get from these places precisely what I expected, viz. nothing at all.
>
> But why disquiet yourself? [...] It will not be a very serious matter if we have to undertake all the Southern Hemisphere ...[76]

It is here evident the different attitude of the two "fathers" of the *Carte du Ciel* project: Gill's approach was more realistic and pragmatic, whereas Mouchez had a real diplomatic approach; he wanted to maintain an ideal wide international participation, headed by France, and invested much of his time in this effort, though it costed delays and inefficiencies for the whole endeavor.

---

[73] Pickering, 1903, p. 70.

[74] In 1909, after the death of Thome (see Appendix A), the completion of the work of Córdoba Observatory was assigned to the Cape Observatory.

[75] In 1891, a conflict between President José Manuel Balmaceda (1840–1891) and the Parliament members provoked a Civil War: it was won by the Parliament members who established a parliamentary republic. The country, however, experienced political instability until World War II.

[76] Gill to Mouchez, Aug. 30, 1891, in Chinnici, 1999, p. 107.

A similar search for visibility and prestige also drove the participation of some British colonial observatories; in a letter by astronomer William Ernest Cooke (1863–1947) to Colonial Secretary about the participation of Australian observatories in the *Carte du Ciel* project, this is clearly stated:

> The importance of this matter can scarcely be over-estimated.
> It is a direct invitation from the entire scientific world to participate in the greatest scientific work that has ever been undertaken in the world's history [...] ... we have the opportunity of acquiring scientific lustre.[77]

In the fifth Commission meeting, in 1900, no progress was reported from Santiago, La Plata[78] and Rio de Janeiro[79] Observatories; the last two ones were respectively replaced by Córdoba[80] and Perth[81] Observatories. In 1904, Uccle Observatory went in help of Potsdam Observatory, which could no more take long exposure Chart plates, because of lacking resources.[82]

In 1909, the Santiago Observatory seemed to be almost ready to start the work,[83] but in that year, during the sixth Commission meeting, the Santiago zone was

[77] Cooke to Colonial Secretary, 1896 in Haynes et al., 1996, p. 91.

[78] The situation at La Plata was alarming, as the photographic lens was broken and the Observatory was in abandoned conditions (see Thome to Loewy, Nov. 8, 1899, in Chinnici, 1999, pp. 171–172), probably even before the death of Director Beuf (see Appendix A), in 1898, who suffered from mental breakdown (see Beuf to Mouchez, Oct. 4 and 29, 1893, in Chinnici, 1999, pp. 252 and 256).

[79] In the 1890s Cruls had been engaged in the cartography of the central Brazil and in the project to move the Observatory from Rio de Janeiro to Santa Cruz, which was never achieved. In 1901 he was commissioned to determine the boundaries between Brazil and Bolivia; this was probably the reason that prevented him to carry out the *Carte du Ciel* work. The astrographic telescope was never mounted (see Paolantonio & García, 2019, p. 495).

[80] On the work carried out at Córdoba and, in general, on the participation of Latin American Observatories in the *Carte du Ciel* project, see Paolantonio & García, 2019.

[81] The participation of Perth Observatory was proposed in 1897 (see Cooke to Loewy, Sept. 14, 1897, in Chinnici, 1999, p. 300) and supported by the two Royal Astronomers at Greenwich and Cape, Christie and Gill (see Loewy to Cooke, Aug. 9, 1900, in Chinnici, 1999, pp. 302–303). The Commission assigned to Perth the sky zone previously assigned to Rio de Janeiro (see Loewy to Cooke, Aug. 9, 1900, in Chinnici, 1999, pp. 302–303).

[82] See Lecointe to Loewy, Nov. 21, 1904, and Fraissinet to Lecointe, Nov. 30, 1904, both in Chinnici, 1999, pp. 393–394. The entire declination zone, previously assigned to Potsdam, was re-photographed and the plates were measured by Oxford, Uccle and Hyderabad Observatories. For a global evaluation about the participation of Potsdam Observatory in the *Carte du Ciel* project, see Bigg, 2008.

[83] See Ristenpart to Baillaud, s.d. [1909], in Chinnici, 1999, p. 348. In 1900 the Santiago zone was assigned to a new Observatory to be established in Montevideo (see Loewy to Cooke, Aug. 9, 1900, in Chinnici, 1999, p. 302; letters to Loewy by Enrique Legrand [1861–1939], professor of cosmography at the University of Montevideo, are listed in Chinnici, 1999, p. 32). However, in a note dated 1909, Fraissinet wrote that nothing had been done by the Republic of Uruguay and the zone was hence ready to be reassigned to Santiago Observatory (Fraissinet to Loewy, s. d. [1909], *ibid.*). See also Paolantonio & García, 2019, p. 496.

Fig. 1.12 Nizamiah Observatory at Hyderabad (India); in 1909 it took on the photographic work of the sky zone previously assigned to Santiago Observatory. From *Nizamiah Observatory Platinum Jubilee Souvenir 1908–1983*, 1983

assigned to Hyderabad Observatory (Fig. 1.12), which was realistically considered most reliable.[84]

1.11 The Establishment of a Central Bureau for Measurements

The photographic work was just a part of the entire project. A major effort, regarding the Catalogue, was the measurement of the star coordinates on the plates, their conversion into equatorial coordinates and their reduction to the year 1900. This required to recruit additional personnel and instruments—a special apparatus, the macromicrometer[85] (Fig. 1.13), was devised to be used for these measurement operations—but many Observatories could not afford further costs. Moreover, it

[84] The participation of Hyderabad Observatory was solicited by Oxford Observatory (see Prashad to Turner, Aug. 15, 1908, in Chinnici, 1999, p. 240).

[85] There were several types of macromicrometers, designed by Gautier, Turner, and Repsold, which were respectively supplied for French, British, and German participant observatories (see *Bulletin* ... 1909, p. 513; on the evolution of the instruments for plates measurements, see Fauque, 1988). The most used macromicrometer, by Gautier, is described in the annual report of Paris Observatory for 1892 (see Tisserand, 1893, pp. 15–16).

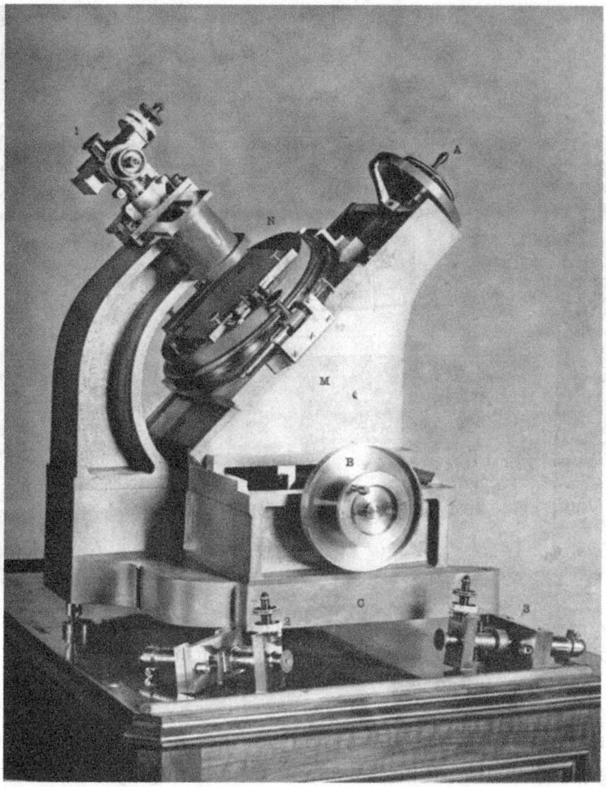

Fig. 1.13 Gautier macromicrometer, designed in 1892 for measuring the coordinates of the stars on the plates. From *Pubblicazioni della Specola Vaticana,* vol. IV, 1894

was important that this accurate and delicate work was well standardized and coordinated.

For this reason, Gill suggested the establishment of a Central Bureau for the measuring operations and proposed to entrust this work to women:

> The measurement of the plates is another matter—and I feel convinced it can be best and most economically done in a Central Bureau. Of course, if we can get the money there is no reason why we should not make the measures here, but women's work for this purpose can be procured in Europe at so much less cost than here, and could be so effectually organized and so economically supervised in a single establishment, that I am convinced we shall all be ultimately compelled to adopt that method of working.[86]

This suggestion by Gill was taken into account by many observatories, who recruited female staff to measure the star coordinates on the plates.[87] Being trained

[86] Gill to Mouchez, Aug. 30, 1891, in Chinnici, 1999, p. 107.

[87] On the employment of women in French astronomical observatories, see Lamy, 2006.

Fig. 1.14 Nuns measuring the Carte du Ciel plates with macromicrometers at the Vatican Observatory. Courtesy of Vatican Observatory Historical Archives

in accurate and precise work (embroidery and needlework were part of the usual female background at that time), women assured highly reliable results at low cost. Even the Vatican Observatory employed nuns (Fig. 1.14) for this work;[88] female staff was also employed at Harvard College Observatory, for measuring and classifying stellar spectra for another major project, the Henry Draper Catalogue.[89]

A tribute is here to be paid to Dorothea Klumpke (1861–1942), the first director of the Bureau des Mesures, which was established in 1891 (Fig. 1.15), and to all women of Paris Observatory and other observatories, who contributed to the *Carte du Ciel* Astrographic Catalogue as well as to the development of the Harvard classification scheme of stellar spectra, thanks to their patient, repetitive, accurate work.

As it has been previously mentioned, many observatories which were not involved in the photographic work, helped with plates measurement and calculations for the conversion from rectangular (Cartesian) into equatorial coordinates as well as with their reduction to the year 1900, while others provided the reference star catalogue. All these circumstances explain why the *Carte du Ciel* can really be defined a "global" scientific enterprise.

[88] The four Sisters of the Holy Child Mary congregation who measured the star positions on the plates were Emilia Ponzoni, Regina Colombo, Concetta Finardi and Luigia Panceri (see Glatz, 2016).

[89] On women in US astronomical observatories, see, for instance, Mack, 1993 and Sobel, 2016.

Fig. 1.15 Dorothea Klumpke (sitting), director of the female staff of the Bureau Central des Mesures at Paris Observatory, ca. 1894, with three of her collaborators (© Bibliothèque de l'Observatoire de Paris). See also Randazzo, 1998, p. 26

1.12 The Fragmentation of the Project

At the beginning of the twentieth century, new interests were emerging in astronomy, notably in the fields of galactic astronomy and cosmology. In 1906 Kapteyn's selected areas,[90] in 1909, Eros parallax measurements drove the attention of the international astronomical community. The initial enthusiasm about the *Carte du Ciel* was fading away, the work revealed to be too expensive and time-consuming; it was necessary to reshape the project to re-motivate

[90] The Dutch astronomer Jacobus Cornelius Kapteyn (1851–1922) selected areas of sky (approximately $1° \times 1°$) to study of the structure of the Milky Way and the galactic pole regions. In each area the magnitudes, colours, spectral types, and proper motions of the stars were measured to as faint an apparent magnitude as possible.

the participant observatories. This was the challenge that expected Benjamin Baillaud (1848–1934), new director of Paris Observatory after the death of Loewy in 1908. The appointment of Baillaud, former director of Toulouse Observatory, was far from being haphazard. Having efficiently directed one of the French participant observatories, he had the necessary expertise, relationships, and scientific authority to lead the project. At Gill's proposal,[91] Baillaud was chosen as President of the Astrographic Commission, likewise his predecessors. As usual, Gill suggested invitations, date and program of the next meeting, to be held in 1909, hoping to advance at least in the achievement of the Catalogue.[92]

The new topics which interested the astronomical community were somehow discussed and included in the *Carte du Ciel* plans, with the purpose of updating and restyling the project, which appeared by then obsolete. Reinserting the project in the astronomical research mainstream of that time, of course, implied to further delay the achievement of the original project—actually, to drift away from its initial program. The project fragmented into subprojects, thus losing its consistency and unity. The numerous resolutions taken by the Commission at the 1909 meeting—including the establishment of three sub-commissions on reference star catalogue, Eros parallax and special (galactic) studies—clearly showed this fragmentation. Definitely, these resolutions can be regarded as a compromise, aimed at giving a new impulse to the project, by enlarging its horizon.

However, after a considerable financial investment, in terms of instruments, staff and volumes publication (about 30 volumes), the *Carte du Ciel* project was interrupted by the catastrophic event of World War I. At that moment, only the Oxford Observatory had almost entirely concluded its work, including the publication of its section of the Astrographic Catalogue. Almost all participant Observatories reduced or interrupted their activities; mostly of the staff was called up to arms, and finances were cut to sustain military operations. Only the Vatican Observatory, not being involved in the conflict, could regularly work during the War.[93] In those years, due to the heavy economic consequences for many countries, being the Observatories lacking of financial and human resources, the project completion was seriously compromised.

---

[91] See Circular by Gill, s. d. [1908] in Chinnici, 1999, pp., 122–123; see also Gill to Baillaud, Mar. 4, 1908, in Chinnici, 1999, p. 125.

[92] See letters by Gill to Baillaud in Chinnici, 1999, pp. 125–129.

[93] During the Great War, the Vatican Observatory advanced very much in the photographic work, taking advantage from the daily blackout in Rome to capture excellent quality plates. On the participation of the Vatican Observatory in the *Carte du Ciel* project, see Chinnici, 1999, pp. 460–464.

Fig. 1.16 Benjamin Baillaud, director of Paris Observatory, first President of the IAU in 1919. From *L'Astronomie* 1934, p. 537

1.13 Actions of the IAU

After the war, in 1919, the International Astronomical Union (IAU) was established, with the purpose to promote the international cooperation in the field of the astronomical research. In the reports of the Interallied Conference held in London in 1918, which preceded the establishment of IAU, it was explicitly stated that the new association "will doubtless result from the union" of the *Carte du Ciel* Commission, the Solar Research Union and other bodies.[94]

The *Carte du Ciel* had effectively created an international network that facilitated the IAU establishment. Its organization methods, with almost regular periodical meetings, resolutions, vote by nation—as in a sort of international 'parliament'[95] of astronomers—was a model that could be easily adopted by the IAU, which hence "benefited from the established reputation of the *Carte du Ciel*."[96]

It was therefore quite obvious that the first President of IAU would be Benjamin Baillaud (Fig. 1.16)—a choice that expressed continuity with the *Carte du Ciel* Commission. He had acquired expertise to guide and coordinate a wide community of astronomers—and could also ensure that the project itself be completed.

[94] See Blaauw 1994, p. 49.

[95] See Lamy 2011, p. 197.

[96] See Andersen et al., 2019, p. 14.

The completion of the work was entrusted to Commission 23 "*Carte du Ciel*" (with Herbert H. Turner as the first president), which replaced the International Permanent Commission, and decided to give priority to the achievement of the Catalogue. The support by IAU was determining: the publication of the last twenty volumes of the Catalogue was entirely financed by IAU. Thanks to this financial support, in 1964, the last volume of the catalogue was published—at last, the Catalogue had been completed.

The Chart met with a different fate. After many decades elapsed, the technical development and the new astronomical frontiers had rendered the completion of the Chart obsolete—and it was abandoned. In the IAU General Assembly, held in 1970 in Brighton, it was definitely decided to leave the Chart unachieved (see Appendix D).

Before its discontinuation, IAU Commission 23 presented an important summary report about the project,[97] and identified in some technical aspects (exposure duration, reproduction on metal of the plates, etc.) the main reasons for the delay accumulated during the execution of the Chart. The Commission recognized, at the same time, that its conclusion was by then out of purpose:

> Undoubtedly, the situation of the Chart would have been better if, in the decade 1920-30, we were not becoming aware that the structure of the Universe is revealed thanks to stars much fainter than those recorded in the Chart: in the plates of the Chart we find very few galaxies. Moreover, the invention of Schmidt telescopes, giving better images in a field of 36 square degrees, than those of the Chart in a field of 4 degrees, has allowed to rapidly publish atlases like the Sky Atlas, where stars down to at least magnitude 20 are countable. [. . .] These atlases fulfill the aims of the founders much better than the original Chart.[98]

The report about the Catalogue was more favorable:

> Fortunately, the Astrographic Catalogue, which at the beginning was considered ancillary, has had a better destiny, though late; today it promises many applications and its collection of ancient plates is a unique documentary source.[99]

Consequently, as a result of the IAU General Assembly in Brighton, Commission 23 merged with Commission 24 ("Parallaxes and proper motions") into a new Com-

[97] See Couderc, 1971.

[98] *La situation de la Carte aurait sans doute été meilleure si, dès la décennie 1920–30, il n'était devenu évident que la structure de l'Univers se découvre grâce à des astres beaucoup plus faibles que ceux que la Carte enregistre: sur les clichés de la carte on trouve bien peu de galaxies. Enfin, l'invention des télescopes de Schmidt, qui donnent dans des champs de 36 degrés carrés des images meilleures que celles des champs de 4° de la Carte, a permis de publier rapidement des atlas comme le Sky Atlas, où figurent les astres jusqu'à la magnitude 20 au moins (. . .). Ces atlas remplissent les désirs des fondateurs de la* Carte du Ciel *beaucoup mieux que la Carte initiale.* (Couderc, 1971, p. 173)

[99] *Heureusement, le Catalogue photographique, qui, au début, paraissait une opération accessoire, a connu un destin meilleur, quoique tardif; il donne aujourd'hui la promesse d'applications nombreuses et la collection de ses clichés anciens est une source de documents encore unique.* (Ibid.)

mission 24 ("Photographic Astrometry"). About sixty years later than expected, the *Carte du Ciel* enterprise finally come to an end.

1.14 Later Issues

The *Carte du Ciel* project appears today as an ambitious but premature endeavor, undermined by a number of weak points (see Sect. 1.15).

However, later issues have not frustrated this impressive effort. In 2000, Commission 24 merged with Commission 8 ("Positional Astronomy") into a new Commission 8 ("Astrometry") having a Working Group on the *Carte du Ciel*, that in 2006 was merged with the IAU Preservation and Digitization Photographic Plates (PDPP) Task Group.[100]

In fact, at the turn of the current century, the *Carte du Ciel* became again of interest for astronomers, as the historical plates acquired a high scientific value for astrometric studies. In the 1990s, the Astrographic Catalogue was digitized at US Naval Observatory—paradoxically in that country that did not participate in the *Carte du Ciel*—and made available as AC 2000;[101] it found important applications in the construction of ACT (Astrographic Catalog/Tycho) reference catalogue[102] and Tycho-2,[103] both resulting from ESA's astrometric satellite *Hipparcos*. The scientific interest of the *Carte du Ciel* plates is still standing, due to the current mission of ESA's astrometric satellite *Gaia,* launched in 2013.

Today, however, only a few observatories still preserve their *Carte du Ciel* plates in good conditions. In many others, most of the photographic plates were reported as damaged or destroyed since the 1950s. A few groups have tried to apply modern methods to digitize the *Carte du Ciel* plates and extract useful information from them.[104] This means that preserving the *Carte du Ciel* plates is still important and useful for science, and probably it is time for a survey of the existing plates[105] and a joint project for their conservation, hopefully to be promoted by IAU.

---

[100] See Griffin, 2009. Due to the recent IAU reorganization, PDPP is currently a Working Group of IAU Commission B2 (Data and Documentation).

[101] See Urban et al., 1998a.

[102] See Urban et al., 1998b.

[103] Tycho-2 contains proper motions derived from a comparison with the Astrographic Catalogue and over a hundred ground-based astrometric catalogues. See Hog et al., 2000.

[104] It is worth mentioning that at Paris Observatory, in the 1980s, *Carte du Ciel* plates were scanned to study proper motions (with MAMA machine - "Machine Automatique à Mesurer pour l'Astronomie" - now discontinued; see Guibert, 1988); digitizations of *Carte du Ciel* plates have been also made at Bordeaux, Vatican, Uccle, Potsdam, San Fernando, Córdoba, Helsinki and encouraged by interesting recent studies (see Lehtinen et al., 2018).

[105] See Randazzo & Isaksson, 2020.

1.15 Concluding Remarks

The *Carte du Ciel* project seemed to be an affordable challenge for the astronomers of the late nineteenth century, and it arose a worldwide interest and enthusiasm in the astronomical community. Nevertheless, it was marked by an increasing complexity, due to several—expected and unexpected—factors, that led to its partial failure.

We try here to identify some weak points of the project (partly predictable from the beginning), and some unexpected diverging factors:

- Ambiguity and diversity of purposes (the *Carte du Ciel* project was somehow torn in two by a sort of internal tension/competition between Chart and Catalogue).
- Non-homogeneous equipment (the participant Observatories were not equipped with one and the same astrograph, because of the interests of local makers).
- Management failures and open technical questions (many logistic and technical problems were under-evaluated, and the preparatory tests were probably insufficient; moreover, the instruments in use became soon obsolete, as a result of the development of new technologies—i.e., Schmidt cameras).
- Different political and economic situations of the participant Countries (in many cases, the adhesion to the project was given without sufficient disposable resources, due to the political instability of the governments).
- World conflicts (they provoked the interruption of the work and the loose of many human and financial resources).
- New scientific frontiers (galactic and extra-galactic astronomy and astrophysics became the most advanced research topics at the beginning of the twentieth century—astrometric catalogues and maps were focused on selected areas, rather than on the entire sky vault).

All these elements—and probably others too—were responsible for the unacceptably long time of execution and the incompletion of the original project, which seems to have been conceived too early to be fully successful.

Moreover, the huge cost that the *Carte du Ciel* execution required (see Appendix C) impacted on the development of the astrophysical research in the participating Observatories (and countries) by distracting resources. This increased the gap with the USA, where astronomy turned into astrophysics earlier that elsewhere, facilitated by the lack of a tradition in classical astronomy, cultivated for centuries in the European Observatories. In some sense, the *Carte du Ciel* was a dream of the last European classical astronomers, formed in the tradition of cataloguing and mapping the sky; through this immense shared effort, they have left a sort of scientific monument to the future generations and, mostly important, a model of organization for the international astronomical community. Being the first example of international scientific cooperation on a worldwide scale, the *Carte du Ciel* did in fact prefigure the IAU, which will be animated by the same spirit of collaboration in the scientific research.

We may conclude by coming back to Couderc's words:

The *Carte du Ciel* enterprise has established the foundations of the international scientific cooperation, has introduced the application of photography, showing its difficulties, its methods, its limits. Our gratitude is to be deserved to those who have generously cooperated to it, and the results that the Catalogue is expected to provide in a near future, will certainly justify their perseverance.[106]

May the intuition and tenacity of the men who collaborated in this daring scientific project be always kept in memory.

Acknowledgments I wish to thank Thierry Montmerle (IAP, Paris) for his invitation and encouragement to write this contribution. I am grateful to Nicole Capitaine (Observatoire de Paris), for welcoming my training stay at Paris Observatory in the 1990s, and mostly indebted to Suzanne Débarbat (Observatoire de Paris), who was my supervisor during that stay: nothing about the *Carte du Ciel* would have been written by me without her friendly support.

Let me thank as well Olga B. Dluzhnevskaya (Institute of Astronomy RAS, Moscow) for her kind help in finding information about the contribution of Russian Observatories.

Appendix A

List of participant Observatories and distribution of the photographic work (1891)

Sky zones	Observatories	(*)	Directors at the moment of the assignation
+90° +65°	Greenwich	(°)	William Henry Mahoney CHRISTIE (1845–1922)
+64° +55°	Vatican	(//)	Fr. Francesco DENZA (1834–1894)
+54° +47°	Catania	(1)	Annibale RICCÒ (1844–1919)
+46° +40°	Helsingfors (= Helsinki)	(2)	Anders Severin DONNER (1854–1938)
+39° +32°	Potsdam	(3)	Hermann Carl VOGEL (1841–1907)
	(later Uccle)	(//)	Georges LECOINTE (1869–1929)
+31° +25°	Oxford	(°)	Rev. Charles PRITCHARD (1808–1893)
+24° +18°	Paris	(//)	Ernest Barthélemy MOUCHEZ (1821–1892)
+17° +11°	Bordeaux	(//)	Georges RAYET (1839–1906)
+10° +05°	Toulouse	(//)	Benjamin BAILLAUD (1848–1934)
+04° −02°	Algiers	(//)	Charles TRÉPIED (1845–1907)
−03° −09°	San Fernando	(//)	Cecilio PUJAZÓN Y GARCÍA (1833–1891)
−10° −16°	Tacubaya (= Mexico City)	(4)	Ángel ANGUIANO (1840–1921)
−17° −23°	Santiago	(//)	Jean Albert OBRECHT HUBER (1858–1924)[a]

(continued)

[106] *L'entreprise de la Carte et du Catalogue [. . .] n'en a pas moins jeté les bases de la coopération Internationale dans la science, introduit dans la pratique l'emploi de la photographie, montré ses difficultés, ses méthodes, ses limites. A ceux qui y ont coopéré avec désintéressement, il convient de penser avec gratitude et les résultats que fournira, dans un proche avenir, le Catalogue, justifieront assurément leur persévérance.* (Couderc, 1971, p. 178)

Sky zones	Observatories	(*)	Directors at the moment of the assignation
	(later Hyderabad)	(5)	Arthur Brunel CHATWOOD (1866–1915)
−24° −32°	La Plata	(//)	Francisco BEUF (1834–1899)
	(later Córdoba)	(//)	John Macon THOME (1843–1908)
−33° −40°	Rio de Janeiro	(//)	Louis Ferdinand CRULS (1848–1908)
	(later Perth)	(°)	William Ernest COOKE (1863–1947)
−41° −51°	Cape of Good Hope	(°)	David GILL (1843–1914)
−52° −64°	Sydney	(6)	Henry Chamberlain RUSSELL (1836–1907)
−65° −90°	Melbourne (later Sydney)	(°)	Robert Louis John ELLERY (1827–1908)

(*) *Type of Astrographs:*
(//) Henry-Gautier
(°) Grubb
(1) Steinheil (optics), Salmoiraghi (mechanics)
(2) Henry (optics), Repsold (mechanics)
(3) Steinheil (optics), Repsold (mechanics)
(4) Grubb (lens corrected by Alvan Clark & Sons)
(5) Cooke (guide telescope by Grubb)
(6) Grubb (optics), Mort Docks Engineering Co. & Atlas Engineering Co. (mechanics)
[a]The interest in the participation of Santiago Observatory was manifested in 1887 by the preceding director, José Ignacio Vergara Urzúa (1837–1889), who died two years later. In 1908, the Santiago Observatory again offered its collaboration, under the directorship of Friedrich Wilhelm Ristenpart (1868–1913).

Appendix B

Geopolitical map of the **Carte du Ciel** *project: nations, regencies, observatories involved in the photographic work (Fig. 1.17)*

- UK (Queen Victoria)
  - Royal Observatory, Greenwich
  - Oxford University Observatory
  - Royal Observatory, Cape of Good Hope
  - Melbourne Observatory, Australia
  - Sydney Observatory, Australia
  - Perth Observatory, Australia
- Kingdom of Italy (King Umberto I)
  - Osservatorio Astrofisico di Catania
- Vatican State (Pope Leo XIII)
  - Specola Vaticana

- Russian Empire (Czar Alexander III)
  - Kejserliga Alexanders-Universetets Astronomiska Observatorium, Helsingfors (today Helsinki—in Finland, at that time part of the Russian Empire)
- German Empire (Emperor Frederick III, in June 1888, succeeded by Wilhelm II)
  - Astrophysikalische Observatorium, Potsdam
- French Third Republic (President Jules Grévy, in December 1887, replaced by Sadi Carnot)
  - Observatoire de Paris
  - Observatoire de Bordeaux
  - Observatoire de Toulouse
  - Observatoire d'Alger
- Kingdom of Spain (King Alfonso XIII of Spain)
  - Instituto y Observatorio de Marina de San Fernando (near Cadiz)
- Republic of Mexico (President Porfirio Díaz)
  - Observatorio Astronómico Nacional Mexicano, Tacubaya (near Mexico City)
- Republic of Chile (President José Manuel Balmaceda)
  - Observatorio Astronómico Nacional, Santiago del Chile
- Republic of Argentina (President Miguel Juárez Celman, in 1898 Julio Argentino Roca)
  - Observatorio Astronómico de La Plata
  - Observatorio Astronómico, Córdoba
- Empire of Brazil (Emperor Pedro II)
  - Imperial Observatorio, Rio de Janeiro
- Kingdom of Belgium (King Leopold II)
  - Observatoire Royal de Belgique, Uccle
- Hyderabad State, India (under British Paramountcy) (Nizam Asaf Jah VI)
  - Nizamiah Observatory, Hyderabad

Fig. 1.17 Non-European Observatories participating in the *Carte du Ciel* project: in red, those included in the original 1891 list; blue dots, those which adhered later. Background map: https://fr.wikipedia.org/wiki/Fichier:A_large_blank_world_map_with_oceans_marked_in_blue_planisferio_en_blanco.svg

Appendix C

Summary of the main Carte du Ciel *costs*

Final Report:

- 22 participant Observatories (photographic work)
- Time employed: 20–40 years per person
- 150 volumes published
- 4,600,000 star positions measured
- 8,600,000 coordinates *x, y* measured

COSTS (on average) per Observatory (FF = French Francs)

- Instruments: 40,000 FF (in 1899)
- Photographs and measurements: 250,000 FF (12 plates = 7 FF, 1893)
- Publications: 500,000 FF
- Staff average cost = 15,000 FF (in 1898)
- Publications, materials, and measurements annual costs = 21,000 FF (in 1899)

Total Cost

- Photographs and measurements: 6,000,000 FF
- Publications (Chart and Catalogue): 12,000,000 FF

For comparison: *Average annual salary of an astronomer: 300 FF (in 1893); cost of a measured plate: 200 FF*

French Observatories

- Total number of plates: 5000
- Total amount of Chart costs: 1,000,000 FF
- Total cost for printing 12 volumes x 4 Observatories: 500,000 FF

British Observatories

- Price of Grubb Astrograph: £. 1500 (37,500 FF)
- Catalogue: 2,000,000 stars, estimated duration: 25 years

  Estimated cost: 250,000 FF (£. 10,000) per year (£. 1 = 25 FF)

Appendix D

List of printed Chart plates: (from: http://www.astropa.inaf.it/carte-du-ciel/by Donatella Randazzo)

Sky zone	Assigned to	Carried out by	Sky charts
+90° to +65°	Greenwich		All charts printed
+64° to +55°	Vatican		Heliogravures for 1/3 of +55°, +56°, +57°; the remaining printed on photographic paper
+54° to +47°	Catania		NONE
+46° to +40°	Helsingfors		NONE
+39° to +32°	Potsdam	Uccle	Heliogravures for the entire zone printed and distributed by Uccle
+31° to +25°	Oxford		NONE
+24° to +18°	Paris		Heliogravures for the even zones +18°, +20°, + 22°, +24°
+17° to +11°	Bordeaux		Heliogravures for the even zones +12°, +14°, +16°
+11° to +05°	Toulouse		Heliogravures for the odd zones +5°, +7°, +9°
+04° to −02°	Algiers		Heliogravures for the odd zones −1°,+1 °,+3°
−03° to −09°	San Fernando		Heliogravures for the odd zones −5°, −7 °, −9°
−10° to −16°	Tacubaya		Heliogravures for the zones −11°, −13 °, −15°, −16 °
−17° to −23°	Santiago	Hyderabad	NONE
−24° to −32°	La Plata	Córdoba	Heliogravures for the zone −25° only
−33° to −40°	Rio de Janeiro	Perth	NONE
−41° to −51°	Cape of Good Hope		NONE
−52° to −64°	Sydney		NONE
−65° to −90°	Melbourne		NONE

Bibliography

Andersen, J., Baneke, D., & Madsen, C. (2019). *The International Astronomical Union: Uniting the community for 100 years*. Springer.

Barman, R. J. (1999). *Citizen Emperor: Pedro II and the Making of Brazil, 1825–1891*. Stanford University Press.

Bartolucci, J. (2000). *La modernización de la ciencia en México. El caso de los astrónomos*. Plaza y Valdés Editores.

Becker, B. J. (2011). *Unravelling starlight. William and Margaret Huggins and the Rise of the New Astronomy*. Cambridge University Press.

Berry, A. (1898). *A short history of astronomy*. J. Murray.

Bigg, C. (2008). La *Carte du Ciel* vue de Potsdam. In Lamy (Ed.), *Bulletin de la Société belge d'astronomie*, vol. 14, pp. 129–154, 1909.

Chinnici, I. (1995). Il contributo dell'Italia all'impresa della Carte du Ciel. *Giornale di Astronomia, 3*, 11–22.

Chinnici, I. (1999). La Carte du Ciel. *Correspondance inédite conservée dans les Archives de l'Observatoire de Paris*, Observatoire de Paris & Osservatorio Astronomico di Palermo G. S. Vaiana, Palermo.

Chinnici, I. (2007). "Maurice Lœwy et la *Carte du Ciel*", *L'Astronomie,* 121, juillet-août, pp. 360–363.

Chinnici, I. (2008a). The Società degli Spettroscopisti Italiani: Birth and evolution. *Annals of Science, 65*, 393–438.

Chinnici, I. (2008b). La *Carte du Ciel*: genèse, déroulement et issues. In Lamy (Ed.), pp. 19–43.

Chinnici, I. (2011). "*La Carte du Ciel* et les observatories français", *La (re)fondation des observatoires astronomiques sous la IIIe République* (ed. J. de La Noë et C. Soubiran), Pessac, Presses Universitaires de Bordeaux, pp. 379–394.

Chinnici, I. (2019). *Decoding the stars: A biography of Angelo Secchi, Jesuit and scientist*. Brill.

Congrès astrophotographique international tenu à l'Observatoire de Paris pour le levé de la Carte du ciel, Paris, 1887.

Couderc, P. (1971). "Historique de la Commission de la *Carte du Ciel*", *Transactions IAU*, vol. XIV, pp. 172–178.

Dadaev, A. N. (1978). *Pulkovo Observatory: An Essay on its History and Scientific Activity* (translated from Russian by K. Krisciunas), NASA, .

Débarbat, S., Eddy, J. A., Eichhorn, H. K., & Upgren, A. R. (Eds.), *Mapping the Sky. Past Heritage and Future Directions*, Proceedings of IAU Symposium No. 133, Kluwer Academic Publisher, 1988.

Dewhirst, D., & Hoskin, M. (1999). The message of starlight: the rise of astrophysics. In M. Hoskin (Ed.), *The Cambridge Concise History of Astronomy*. Cambridge University Press.

Fauque, D. (1988). "Protohistoire de la machine à mesurer les clichés astronomiques" in Débarbat et al., pp. 53–57.

Glass, I. S. (1997). *Victorian Telescope Makers. The lives and letters of Thomas and Howard Grubb*. Institute of Physics Publishing, Bristol and Philadelphia.

Glatz, C., "Mapping with the stars: Nuns instrumental in Vatican celestial survey", *Catholic News Service*, 28 April 2016 (in https://www.ncronline.org/blogs/vatican/eco-catholic/mapping-stars-nuns-instrumental-vatican-celestial-survey).

Gill, D. (1886). "Photographie Astronomique", *Bulletin Astronomique,* tome III, pp. 161–164.

Griffin, E. (2009). "Division XII/Commission 5/Task Force Preservation and Digitization of Photographic Plates – Triennial Report 2006–2009", Transactions IAU, Volume XXVIIA *Reports on Astronomy 2006–2009* (Karel A. van der Hucht, ed.).

Grubb, H. (1888). Telescopes for stellar photography. *Engineering, 45*, 402.

Guibert, J. (1988). "The MAMA machine: proper motions from digitization of centenary plates" in Débarbat et al., pp. 425–426.

Haynes, R., Malin, D., & McGee, R. (1996). *Explorers of the Southern Sky. A History of Australian Astronomy*. Cambridge University Press.

Hog, E., Fabricius, C., Makarov, V. V., Bastian, U., Schwekendiek, P., Wicenec, A., Urban, S., Corbin, T., & Wycoff, G. (2000). Construction and Verification of the Tycho-2 Catalogue. *Astronomy and Astrophysics, 357*, 367.

Holden, E. S. (1893). The Bruce Photographic Telescope. *Publications of the Astronomical Society of the Pacific, 5*, 220–222.

Hoskin, M. (1993). Bode's Law and the Discovery of Ceres. In J. F. Linsky & S. Serio (Eds.), *Physics of Solar and Stellar Coronae* (pp. 35–46). Kluwer Academic Publishers.

Hughes, S. (2013). *Catchers of the Light. The forgotten lives of the men and women who first photographed the Heavens*, 2 vols., Art-de-Ciel Publishing.

Jones, B. Z., & Boyd, L. G. (1971). *The Harvard College Observatory*. Harvard University Press.

Lamy, J. (2006). La *Carte du Ciel* et la création du 'Bureau des Dames' à l'Observatoire de Toulouse. *Nuncius, 21–1*, 101–120.

Lamy, J. (Ed.), *La Carte du Ciel*. EDP Sciences, 2008.

Lamy, J. (2009). Adjusting Astronomical Practices: The '*Carte du Ciel*', the democratic rules and the circulation of opinions at the end 19th century. In D. Valls-Gabaud & A. Boksenberg (Eds.) *The role of astronomy in society and culture*, Proceedings IAU Symposium No. 260, IAU 2011, pp. 195–201.

Lehtinen, K., Prusti, T., de Bruijne, J., Lammers, U., Manara, C. F., Ness, J.-U., Siddiqui, H., Markkanen, T., Poutanen, M., & Muinonen, K. (2018). Digitization and astrometric calibration of Carte du Ciel photographic plates with *Gaia* DR1. *Astronomy & Astrophysics, 616*, A185.

Mack, P. E. (1993). Straying from their orbits: Women in astronomy in America. In G. Kass-Simon & P. Farnes (Eds.), *Women of science. Righting the Record* (pp. 72–116). Indiana University Press.

Meadows, A. J. (1970). *Early solar physics*. Pergamon Press.

Meadows A. J. (1984). The new astronomy. In O. Gingerich (Ed.), *The general history of astronomy*, vol IV, part A. Cambridge University Press, pp. 59–72.

Motais de Narbonne, A. M. (1988). "'Petite histoire' du Congrès astrophotographique de 1887", in Debarbat et al., pp. 129–134.

Mouchez, E. B. (1885a). Carte photographique du ciel à l'aide des nouveaux objectifs de MM. P. et Pr. Henry. *Comptes-Rendus de l'Académie des Sciences, 100*, 1177–1181.

Mouchez, E. B. (1885b). Proposal for Photographing the Heavens. *Monthly Notices of the Royal Astronomical Society, 48*, 1–2.

Mouchez, E. B. (1886). *Rapport annuel sur l'état de l'Observatoire de Paris pour l'année 1885*. Gauthier-VIllars.

Randazzo, D., & Isaksson, E. (2020). "Carte du Ciel original plates: call for a joint survey by librarians" (poster submitted to LISA IX Congress).

Paolantonio, S., & García, B. (2019). The *Carte du Ciel* and the Latin American Observatories. In C. Sterken, J. Hearnshaw & D. Valls-Gabaud, (Eds.), *Under One Sky: The IAU Centenary Symposium Proceedings*, IAU Symposium No. 349, pp. 494–501.

Pickering, E. C. (1903). A photographic map of the entire sky. *The Astrophysical Journal, 18*, 70–74.

Pyenson, L. (1993). *Civilizing Mission. Exact Sciences and French Overseas Expansion, 1830–1940*. The John Hopkins University Press.

Randazzo, D. (1998). Le donne in Astronomia nell'800: il fenomeno Klumpke. *Giornale di Astronomia, 24*, 22–33.

Sobel, D. (2016). *The Glass Universe*. Viking Books.

Struve, O. W. (1886). "La photographie au service de l'Astronomie" (in German). *Bulletin de l'Académie Impériale des Sciences de St.-Pétersbourg, 30*, 484–500.

Tisserand, F. (1893). *Rapport annuel sur l'état de l'Observatoire de Paris*. Imprimerie Nationale.

Tsvetkova, K., Tsvetkov, M., Böhm, P., Steinmetz, M., & Dick, W. R. (2009). The Potsdam plates of the *Carte du Ciel* project. I. Present inventory and plate catalogue. *Astronomische Nachrichten, 330*, 878–884.

Turner, H. H., & Common, A. A. (1889). Another Photographic Chart of the Heavens. *The Observatory, 12*, 308–311.
Urban, S. E., Corbin, T. E., Wycoff, G. L., Martin, J. C., Jackson, E. S., Zacharias, M. I., & Hall, D. M. (1998a). The AC 2000: The astrographic catalogue on the system defined by the *Hipparcos* catalogue. *The Astronomical Journal, 115*, 1212–1223.
Urban, S. E., Corbin, T. E., & Wycoff, G. L. (1998b). The ACT reference catalog. *The Astronomical Journal, 115*, 2161–2166.

Ileana Chinnici is astronomer at INAF-Palermo Astronomical Observatory. Her main research field is history of nineteenth-century astronomy and astrophysics. She has published the correspondence related to the *Carte du Ciel* project that is preserved in Paris Observatory archives.

She is currently member of the Organizing Committee of IAU Commission 3 ("History of Astronomy") and from 2006 to 2012 chair of the Working Group "Archives" of the Commission.

She is author of books and articles in specialized journals; her recent biography of Angelo Secchi, Jesuit and pioneer of astrophysics, has been awarded the Osterbrock Book Prize 2021.

She is also scientific advisor of Palermo Observatory heritage, including the museum and the historical library and archive, and adjoint scholar at Vatican Observatory.

Chapter 2
The Foundation of the *Bureau International de l'Heure*

Nicole Capitaine, Pierre Baüer, and Suzanne Débarbat

Abstract The history of the foundation of the *Bureau International de l'Heure* is a fascinating story that relied on huge efforts and works at the end of the nineteenth century and the beginning of the twentieth century. This was made possible thanks to scientific and technical innovations of renowned scientists as well as to an effective international cooperation. It is closely linked to the history of the creation of the International Astronomical Union (IAU), of the International Union of Geodesy and Geophysics (IUGG) and of the Union Radio-Scientifique Internationale (URSI), all of these organizations being officially created in 1919 thanks to several common scientific personalities.

Keywords Determination of time: history · worldwide unification · Universal time · International Conference on Time · International Astronomical Union (IAU) · Bureau international de l'Heure (BIH) · Foundation

2.1 The Determination of Time at the Beginning of the Twentieth Century

2.1.1 From Apparent Solar Time to Universal Time

The time used until the beginning of the nineteenth century was the *apparent solar time*, defined as the *hour angle of the Sun*. This time, based on the apparent daily motion of the Sun and displayed by sundials, has two major defects. First, it is

N. Capitaine (✉) · S. Débarbat
SYRTE, Observatoire de Paris, Université PSL, CNRS, Sorbonne Université, Paris, France

Bureau des longitudes, Paris, France
e-mail: nicole.capitaine@obspm.fr; Suzanne.Debarbat@obspm.fr

P. Baüer
Bureau des longitudes, Paris, France
e-mail: pierre.bauer3@orange.fr

© The Author(s), under exclusive license to Springer Nature Switzerland AG 2022

T. Montmerle, D. Fauque (eds.), *Astronomers as Diplomats*, Historical & Cultural Astronomy, https://doi.org/10.1007/978-3-030-98625-4_2

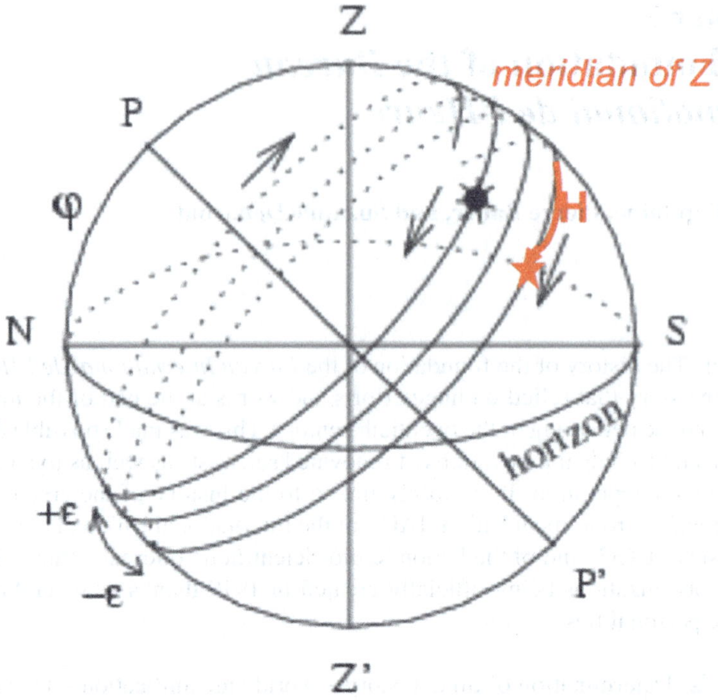

Fig. 2.1 The apparent motion of the Sun (in black) and the ("fictitious") mean Sun (in red) on the local sphere at Z, H is the hour angle of the mean Sun from the local meridian, and ε is the obliquity of the ecliptic

not uniform due to the eccentricity of the Earth's orbit around the Sun and of the obliquity of the ecliptic; therefore, it cannot be directly used for setting the clocks. Second, it depends on the longitude of the location.

A regularized time, known as *mean solar time*, was used since the middle of the nineteenth century; it was derived from the apparent solar time by adding the *equation of time* to correct its inequalities, and it can be related to the motion of a "fictitious mean Sun" (see Fig. 2.1). This time is, like the apparent time, associated with the local meridian.

During the second part of the nineteenth century, railway operations required the use of a single definition of the hour at the national level or of a few ones for countries extended in longitude. Many countries adopted the mean time (increased by 12 hours) of the meridian of their capital and, if necessary, of other major cities.

The adoption of the Greenwich meridian as the prime meridian at the international level for longitude and time, in 1884, made possible the worldwide unification of time through the definition of the *universal time* (see Sect. 2.2.2).

2.1.2 Astronomical Determinations of Time and Time-Keepers

Until 1930, the Earth's rotation was considered as being uniform, and until 1960, the unit of time, the second, was defined as the fraction 1/86,400 of the mean solar day. Thus, until the middle of the twentieth century, the time was based on astronomical determinations of the Earth's rotation by the method of "meridian transit." As the measurements of meridian transits of the Sun cannot be precise—due to the high luminosity and the large diameter of the Sun—they were replaced by measurements of the meridian transits of stars using meridian circles (for more details on these instruments, see Le Guet-Tully 2012). The timing of the passage of stars across the local meridian provides *sidereal time* (i.e., time relative to the stars), which can be easily converted to mean solar time (i.e., time relative to the Sun). Clocks used for astronomical observations indicated sidereal time or both mean time and sidereal time (see Fig. 2.2).

In addition to the astronomical determination of time, it was necessary to use mechanical clocks, also called "time-keepers," for keeping the time between the observations (see Fig. 2.3). By the mid-twentieth century, the best clocks were able to keep time to around a few seconds a year (for more details, see Débarbat 2006). The observatories where positional astronomy was combined with a time service were in charge of distributing the accurate time (see Fig. 2.4).

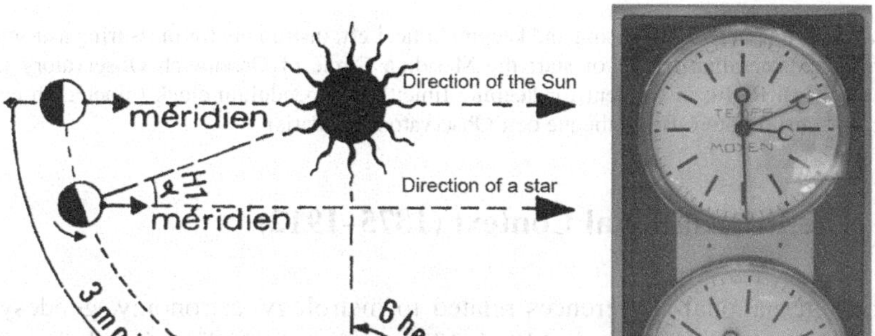

Fig. 2.2 Solar time versus sidereal time. Left: difference over a period of three months of the apparent directions from the Earth's meridian of the Sun and of a star. Right: clock used for astronomical observations (mean time: top clock and sidereal time: bottom clock) [Photo ©Bibliothèque de l'Observatoire de Paris]

Fig. 2.3 Instruments for measuring and keeping time. Left: instrument for measuring astronomical time from the meridian transit of star: the Meridian Circle of Greenwich Observatory (André & Rayet 1874). Right: instrument for keeping time: Leroy pendulum clock (nineteenth century; height: 152 cm) [Photo ©Bibliothèque de l'Observatoire de Paris]

2.2 The International Context (1875–1912)

Several international conferences related to metrology, astronomy, geodesy, and time took place during the period 1875–1912, the list of which is given below. These resulted in international agreements and organizations in various scientific domains, which impacted and inspired to a large extent the international organization on time.

- In 1875, the *Metre Convention* was signed in Paris with the aim of coordinating international metrology and the development of the metric system (see Sect. 2.2.1).
- In 1884, the "International Meridian Conference" met at Washington to adopt a common zero longitude and standard of time (see Sect. 2.2.2).
- In 1886, the "International Geodetic Association," gathering 20 governmental delegates, was created as a worldwide extension of a firstly created central European organization related to a few triangulation networks. The IGA defined a research program and adopted an organizational structure and financial mode on

Fig. 2.4 Paris watchmakers taking the time at the Paris Observatory clock. Drawing by Claverie published in *Le Journal Illustré*, 1877 [©Bibliothèque de l'Observatoire de Paris]

the model of the "Metre Convention," with a Central Bureau (in Potsdam), a Permanent Commission, and General Conferences at three-yearly intervals. These renowned conferences were giving rise to important international discussions. In 1919, this activity was transferred to the section of geodesy of the newly created "International Union of Geodesy and Geophysics" (IUGG) (see Levallois 1980 and Torge 2005).

- In 1887, the "International Astrographic Congress" was held in Paris to establish the *Carte du Ciel* project with the aim of mapping the entire sky through photography (see Chinnici 2022). The project was launched by Ernest Mouchez (1821–1892), director of the Paris Observatory, and David Gill (1843–1914), director of The Cape Observatory. This undertaking required the joint work of 18 observatories in the world, as well as an impressive international organization coordinated by a Permanent International Committee.
- In 1896, after the request of Arthur Downing (1850–1917), the Superintendent of the (British) Nautical Almanac, to Maurice Lœwy (1833–1907), the director of the "Connaissance des temps," the "Conférence internationale des étoiles fondamentales" met in Paris under the responsibility of the Bureau des longitudes (see [Bureau des longitudes] 1897), with the participation of the directors of the astronomical almanacs of several countries of the world (for more details, see Débarbat 2009). The aim was to adopt a uniform system of fundamental stars and astronomical constants for astronomy, navigation, and geodesy.

- In 1899, the *International Latitude Service* (ILS) started its regular operations under the authority of the IGA to monitor the variations of latitude at the world level in order to determine *polar motion* (see Sect. 2.2.3).
- In 1911, upon invitation of Benjamin Baillaud (1848–1934) (see Fig. 2.8), then director of the Paris Observatory (see Bougeret 2019), the "Congrès international des éphémérides astronomiques" was held in Paris (23–26 October). Agreements were reached at an international scale for dividing the computation of ephemerides among the principal national almanac offices and for the use of the Greenwich Meridian; this was in effect on January 1, 1916 in the six principal almanacs (see [Bureau des longitudes] 1913 or [Bureau des longitudes] 1914).
- During the period 1903–1912, three "International Radiotelegraph Conferences" assembled in Berlin (1903, 1906) and then London (1912), for adopting an "International Radiotelegraphic Convention" aimed at regulating the radiotelegraphy worldwide (see Sect. 2.2.4).

The following sections give more details on the conferences of the above list which were the most essential for the international organization on time.

2.2.1 The Metre Convention (Paris 1875): A Model for the International Organization on Time

The "Metre Convention" (also called "the treaty of Metre") was signed in Paris on May 20, 1875, by representatives of 17 nations to establish a permanent organizational structure to enable member governments to act by common agreement in all matters relating to units of measurement and to promote their use throughout the world. For that purpose, the Convention included 14 articles (see below) which led to the creation of an International Bureau (the BIPM) placed under the authority of a General Conference (the CGPM) and the supervision of an International Committee (the CIPM). This Convention stipulated the way in which the activities of the BIPM are to be financed by the contributions of the contracting States and managed. The organization of the Bureau, as well as the composition and attributions of the CIPM and of the CGPM, is determined by the Regulations annexed to the Convention.

THE METRE CONVENTION (1875): Articles 1–4 & 9
(see [Bureau International des Poids et Mesures] 1921)

- FIRST ARTICLE
  The High Contracting Parties undertake to create and maintain, at their common expense, a scientific and permanent *International Bureau of Weights and Measures* with its headquarters in Paris.
- ART. 2
  The French government will take the necessary steps to facilitate the acquisition or, if the case arises, the construction of a building specifically dedicated to this purpose, in accordance with the conditions set out in the Regulations annexed to the present Convention.
- ART. 3

The International Bureau shall operate under the exclusive direction and supervision of an *International Committee for Weights and Measures*, itself placed under the authority of a *General Conference on Weights and Measures*, consisting of the delegates of all the contracting governments.

- ART. 4

  The Presidency of the General Conference on Weights and Measures is assigned to the then current President of the Paris Academy of Sciences.

  (...)
- ART. 9

  All the costs of the establishment and installation of the International Bureau of Weights and Measures, as well as the annual expenses for its maintenance and those of the Committee, shall be covered by the contributions of the contracting States established according to a scale based on their current population.

2.2.2 The International Meridian Conference (Washington, 1884) and the Definition of Universal Time

The need for an international agreement on longitude and time became more and more crucial in the 1870s due to the huge reduction of the delays for connecting distant regions of the world through transoceanic submarine telegraph cables (installed since the 1850s) and the improvement of the accuracy of the determinations of marine positions with stable chronometers. Following a worldwide consultation by the President of the United States Congress in 1882, the suggestion of the IGA Congress of Rome, in 1883, to adopt the same fundamental meridian of origin for all countries, and the adoption of preliminary time zone systems in a few countries, the President of the US issued an invitation to meet in Washington the following year.

On October 1, 1884, the conference, known as "The International Meridian Conference," brought together delegates from 27 states in the diplomatic room in Washington under the presidency of Admiral Rodgers, Superintendent of the US Naval Observatory. The purpose was to establish a common prime meridian for longitude and a standard of time-reckoning throughout the whole world (see Fig. 2.5).

After a full discussion, seven resolutions, known as "The Final Act," were passed (see [Royal Astronomical Society] 1885).

FINAL ACT (1884)

I. That it is the opinion of this Congress that it is desirable to adopt a single prime meridian for all nations, in place of the multiplicity of initial meridians which now exist.
II. That the Conference proposes to the Governments here represented the adoption of the meridian passing through the center of the transit instrument at the Observatory of Greenwich as the initial meridian for longitude.
III. That from this meridian longitude shall be counted in two directions up to 180 degrees, east longitude being plus and west longitude minus.

IV. That the Conference proposes the adoption of a universal day for all purposes for which it may be found convenient and which shall not interfere with the use of local or other standard time where desirable.

V. That this universal day is to be a mean solar day, is to begin for all the world at the moment of mean midnight of the initial meridian, coinciding with the beginning of the civil day and date of that meridian, and is to be counted from zero up to 24 hours.

VI. That the Conference expresses the hope that as soon as may be practicable the astronomical and nautical days will be arranged everywhere to begin at mean midnight.

VII. That the Conference expresses the hope that the technical studies designed to regulate and extend the application of the decimal system to the division of angular space and of time shall be resumed, so as to permit the extension of this application to all cases in which it presents real advantages.

The Final Act thus defined the "universal time" as the time determined at the Greenwich meridian, beginning at midnight (rather than at midday, i.e., deduced from Greenwich mean time, GMT, by adding 12 hours) and counted from 0 to 24 hours.

Later on (1892), the time zone system, based on the concept of international time, was established to facilitate international relations; the Earth was divided into 24 time zones of 1 hour, the Greenwich meridian being the zero Meridian.

INTERNATIONAL CONFERENCE

HELD AT WASHINGTON

FOR THE PURPOSE OF FIXING

A PRIME MERIDIAN

AND

A UNIVERSAL DAY.

OCTOBER, 1884.

PROTOCOLS OF THE PROCEEDINGS.

WASHINGTON, D. C.
GIBSON BROS., PRINTERS AND BOOKBINDERS.
1884.

Fig. 2.5 International Meridian Conference—Protocols of the Proceedings. 1884

2.2.3 *The International Latitude Service (ILS) (1899) to Monitor Polar Motion and Improve the Determination of Astronomical Time*

From the late nineteenth century, the astronomical determination of latitude and universal time were obtained from observations of stars assumed to have known coordinates in a reference catalog. But the determinations of universal time, which is based on Earth's rotation, suffered from the fact that it did not take into account the motion of the Earth's axis of rotation relative to the crust, called "polar motion" (see Fig. 2.6), which was discovered in the 1890s (Chandler 1891) and was found to be responsible for slight variations in latitude and longitude of stations on the Earth and consequently on the astronomical determinations of time.

Polar motion, which is not predictable, should be studied and monitored from a set of stations around the world. In 1899, after having been created for that purpose by IGA and several years of international discussions between geodesists and astronomers, the "International Latitude Service" (ILS) started its operation. It was composed of six observatories using the same instruments located on a

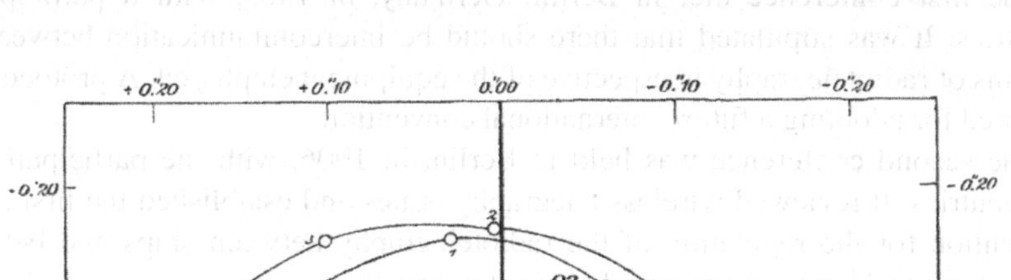

Fig. 2.6 Plot of the motion of the Earth's rotation pole (polhody): values determined by ILS from 1900 (noted 00.0) to 1905 (noted 05.0) and given in arcseconds at 0.1 year interval on the plane tangent to the mean North pole, the vertical line in bold being along the Greenwich meridian; the dots mark every 0.1 year from the beginning of each year (Elbert 1908)

common parallel across the globe in order to allow uniform measurement methods and to minimize systematic errors. A Central Bureau in charge with the observations program and data reductions was installed at the Geodetic Institute of Potsdam (Höpfner 2000). ILS has been funded by IGA until the creation of the IUGG and the International Astronomical Union (IAU).

Polar motion monitoring programs were organized and the corresponding corrections were applied to the determination of universal time.

2.2.4 The International Radiotelegraph Convention (1903–1912): Essential Regulations for the Unification of Time

The growing progress of telegraphy imposed in some areas international regulations in the use of this ultrafast mode of communication. Successive "International Radiotelegraph Conferences" were organized in that purpose.

The first conference met in Berlin, Germany, in 1903, with 8 participating countries. It was stipulated that there should be intercommunication between all systems of radiotelegraphy irrespective of the equipment employed. A protocol was prepared for adopting a future international convention.

The second conference was held in Berlin, in 1906, with the participation of 27 countries. It reviewed wireless telegraphy issues and established the first major convention for the regulation of the radiotelegraphy between ships and between ships at sea and between ships and shore stations.

During the third conference in London (4 June–5 July, 1912), with the participation of 37 States, "The International Radiotelegraph Convention" and the annexed Service Regulations were revised. The principle of freedom of intercommunication was absolutely recognized. This conference also undertook to strengthen the measures relating to the improvement of safety at sea by means of radiotelegraphy. The Convention, with the final Protocol and Service Regulations (see [Bureau International de l'Union Télégraphique] 1912), was concluded between the participants and signed by the Plenipotentiaries of the adhering States on July 5, 1912, to be implemented on July 1, 1913.

This Convention was essential for the unification of time based on wireless telegraphy.

2.3 The International Organization of Time (1910–1913)

The international organization of time at the beginning of the twentieth century, which led to the creation of the "Bureau International de l'Heure" (BIH), was the result of various factors: progress in several technical and scientific aspects related to time, the development of radio time signals, and the existence of international agreements and organizations as described in the previous section.

2.3.1 The First Step in the Foundation of the BIH

At the end of the nineteenth century, the determination of universal time was carried out independently by the various observatories responsible for measuring time and therefore suffered systematic effects. The signals of emission of time were originally optical, electrical, or sound signals.

An important innovation occurred in the late 1840s, when Great Britain used (electrical) telegraph lines to establish standardized time throughout the country.

In 1899, time signals by wireless telegraphy were successful, first in Hamburg and then at the Paris Observatory. It was the start of a new era for the transmission and coordination of universal time around the world.

2.3.1.1 The Beginning of the Worldwide Unification of Time

The Art of Wireless Telegraphy
At the end of the nineteenth century, wireless telegraphy consists in sending and receiving dots and dashes using the "Morse" coding system. The transmitting system is producing sparks whose electromagnetic properties correspond to "wave packets" embracing a wide range of frequencies in the megahertz domain. The receiving system is based on the Branly coil or coherer, a device that is turned into a conducting one when submitted to electromagnetic waves. As far as receiving antennas are concerned, they consist in simple wires or, based on the early induction experiments, a simple loop with a gap. By analogy with other types of waves like those of the sound, it is believed (and this until the 1930s) that electromagnetic waves propagate through an undefined "ether." Gustave Ferrié (1868–1932) (see Sect. 2.3.1.2 and Fig. 2.7), as a well-trained engineer, will develop robust operational systems and invent in 1903 a new electrolytic receiver.

The Bureau des longitudes' Project on the Transmission of Wireless Time Signals
In December 1903, G. Ferrié obtained the installation of a military radiotelegraphy station at the Eiffel Tower including an underground laboratory located below the Champ de Mars and an antenna at the top of the tower. The first wireless link tests were carried out with a range of over 400 km. At the same time, Guillaume Bigourdan (1851–1932), astronomer at the Paris Observatory (see Fig. 2.8), was experimenting with the transmission of time signals using this new technique, including tests over a distance of 2 km. Most of the Observatory's activity then focuses on the diffusion of time as part of a collaboration between the director, B. Baillaud, and G. Ferrié to emit time signals from the Eiffel Tower (see Fig. 2.7).

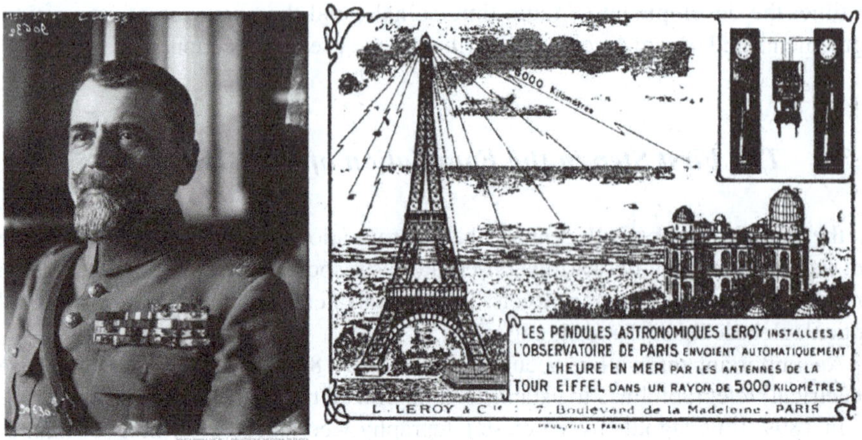

Fig. 2.7 Left: Gustave Ferrié. Credit: gallica.bnf.fr, Bibliothèque nationale de France. Right: long-distance transmission of Paris Observatory time signals from the Eiffel Tower by wireless telegraphy. Credit: Bibliothèque de l'Observatoire de Paris *Bulletin de l'association amicale du personnel scientifique des observatoires français, (1910–1914)* [©Bibliothèque de l'Observatoire de Paris]

Fig. 2.8 Left: Benjamin Baillaud. Credit: Helio.Schützenberger, Paris. Right: Guillaume Bigourdan [©Bibliothèque de l'Observatoire de Paris]

In May 1908, following decisive experiments in the transmission of wireless time signals, carried out in France as well as in other countries, especially in Germany and the United States, the Bureau des longitudes launched a project to disseminate the time signals provided by the Paris Observatory from the Eiffel Tower and requested that a time signal service be temporarily installed on the Eiffel Tower in order to determine the longitudes (Bigourdan 1923; Lambert 1935). Under G. Ferrié's direction, a transmitter was set up at the tower submit, and its effective range increased from an initial 400 km to 5000 km by 1910. On May 23, 1910, the Paris Observatory and the Eiffel Tower radiotelegraph station began transmitting daily time signals. Sending the time of the initial meridian to all reception stations within reach of the Eiffel Tower in conjunction with the Paris Observatory ensured ships the precise determination of their positions at sea and revolutionized the determination of differences in longitudes. As shown in Fig. 2.7, the Leroy astronomical clocks, installed at the Paris Observatory, automatically sent the time at sea via the antennas of the Eiffel Tower within a radius of 5000 km.

Two sorts of signals were transmitted: some daily, at a fixed hour, with 1/2 second precision, for naval requirements, and others, occasional, for the determination of differences in longitude, aiming at a 1/100 second precision.

The Need for Unification of Time
Accurate values of time were sent from Paris Observatory to the Tower daily at fixed hours. Other countries subsequently instituted similar broadcasts. The exchange of long-distance radio time signals then showed that these signals could disagree for up to 1 or 2 s, while the uncertainties in local measurements of universal time were of the order of a few hundredths of a second. The main causes of disagreement were due to various effects: star position errors, instrumental delays, perturbations in wave propagation in the atmosphere, perturbations of the master clocks due to temperature and pressure variations, and especially longitude errors.

Better synchronization was required in order to achieve the universal time as recommended by the International Meridian Conference. Meanwhile, programs for the determination of longitudes by wireless telegraphy were undertaken. The need for a single "timescale" was recognized, as well as the need to create a body responsible for unifying international time (see Guinot 2000).

The following year, March 9, 1911, France adopted the universal time and the international system of 24 time zones centered on Greenwich meridian.

2.3.1.2 Gustave Ferrié and Its Prominent Role for the Unification of Time and Longitudes

Gustave Ferrié and the Wireless Telegraphy
G. Ferrié graduated in 1889 from the prestigious Ecole Polytechnique. Showing a great scientific interest and a high engineering capability, the young army officer is very well acquainted with the work of Ampere, Faraday, Maxwell, Hertz, Tesla, Marconi… He is rapidly convinced of the importance of the wireless telegraphy and undertakes the development of various applications. He manages to convince the military authorities and, in particular the Ministry of War, of the strategic impact of such a technique, insisting on the vulnerability of Earth based and submarine telegraphic cables and, on the need to communicate with vessels (see Baüer 2020). He is the coauthor in 1899 of the first French book on TSF (Télégraphie sans fil,

or wireless telegraphy). Director of the French military telegraph services during World War I, he was promoted to general in 1919.

The Eiffel Tower Rescue
Having convinced in 1900 the Ministry of War of the importance of wireless telegraphy, G. Ferrié is given the duty to set up the French military wireless telegraphy service. At that time, the Eiffel Tower, constructed in 1889 for the world Paris Exhibition in order to celebrate the 100th anniversary of the French revolution, was supposed to be dismantled in 1909. With the strong support of his friend Gustave Eiffel, he launches the idea of installing the major military wireless telegraphy system on the Eiffel Tower. The major objective is to transmit and receive time signals worldwide in conjunction with the Observatory of Paris. The Eiffel Tower is rescued!

Gustave Ferrié and the Bureau des longitudes
From 1908 onward, G. Ferrié is closely associated with the work of the "Bureau des longitudes," initially as an expert (military radiotelegraphy station chief) before being elected as corresponding member of the Bureau in February 1911. Between 1908 and 1919, his name appears more than 300 times in the minutes of the weekly Bureau sessions! Indeed, wireless telegraphy is at the time considered essential by the Bureau in order to achieve its fundamental objectives in terms of time and longitudes. The Bureau develops a large activity about time signals, longitudes determination, meteorological bulletins, seismological records, wave propagation velocity determination, ..., all of these implying the Eiffel Tower equipment and the Paris Observatory.

An international dimension is given by the Bureau to this activity with the organization of an International Conference on Time in 1912, the conduct of experiments worldwide (America, Africa, ships, ...), and the planning of a worldwide longitudes network (Paris, Washington, Shanghai, ...). In August, Ferrié becomes President of the international radio commission on longitudes (1912) and later (1919) the first President of the "Union radioscientifique internationale" (URSI). For obvious reasons, G. Ferrié rarely attends the sessions of the Bureau during World War I; he is then fully involved in modernizing the French military communication system.

2.3.1.3 Preparation of the International Conference on Time by the Bureau des Longitudes

In 1911, a report by G. Ferrié on the discrepancies observed by navigators between received time signals concluded that there was a need to create a central office, allowing the unification of international time. Taking up this suggestion, the Bureau des longitudes asked the government to convene in Paris an international conference to study the means of achieving practical time unification. B. Baillaud offered to host this conference at the Paris Observatory of which he was the director.

The following paragraphs give extracts from the minutes of a few meetings of the Bureau des longitudes (Procès-verbaux du Bureau des longitudes, see http://purl.

oclc.org/net/bdl/) during which this issue is discussed. Henri Poincaré (1854–1912), the famous mathematician,[1] President of the Bureau during the period 1909–1910, was asked to contact personally the Minister of Foreign Affairs[2] to insist on the urgency of a solution. The official support from the Minister was mentioned in the minutes of the May 15, 1912 meeting.

- Procès-verbal de la séance du 31 Janvier 1912/Minutes of the meeting of January 31, 1912 (M. Bigourdan Président)

Original version—Mr. le Ct Ferrié a reçu une lettre de M. Benoist,[3] de laquelle il résulte qu'il est très difficile au Bureau International des Poids & Mesures d'assurer la charge du Service Central de l'Heure. Il annonce que les signaux de la Tour Eiffel ont été reçus par le paquebot "la Touraine" à 5200 km de distance; ces signaux sont même entendus quelquefois à des stations telles que Pernambouc, à une distance de 7000 km par mer, et dans l'Oural, où la transmission se fait par-dessus les continents, ce qui la rend plus difficile. Enfin, il donne lecture de l'avant-projet relatif au Service Central International de l'Heure, destiné à être envoyé au Ministre, pour être présenté au prochain Congrès de Londres.[4] Après une courte discussion, ce projet est adopté avec quelques modifications de détail.

English translation—Mr. le Ct Ferrié received a letter from Mr. Benoist[3], from which it results that it is very difficult for the International Bureau of Weights and Measures to ensure the charge of the Central Time Service. He announces that the signals from the Eiffel Tower have been received by the liner "La Touraine" 5200 km away; these signals are even heard sometimes at stations such as Pernambuco, at a distance of 7000 km by sea, and in the Urals, where transmission is over continents, which makes it more difficult. Finally, he reads the preliminary draft relating to the Central International Time Service, intended to be sent to the Minister, to be presented at the next Congress in London[4]. After a short discussion, this draft was adopted with some minor modifications.

- Procès-verbal de la séance du 27 Mars 1912/Minutes of the meeting of March 27, 1912 (M. Bigourdan Président)

Original version—M. Poincaré donne lecture de la lettre et de la note préparées par la Commission nommée à cet effet, pour être envoyée à M. le Ministre, en vue de la réunion d'une Conférence spéciale, faisant suite au prochain Congrès de Londres, pour discuter la question d'un Service International de l'Heure. Après adoption, il est décidé en outre que M. Poincaré voudra bien faire une démarche personnelle auprès de M. le Ministre des Affaires Etrangères pour insister sur l'urgence d'une solution.

English translation—Mr. Poincaré reads the letter and the note prepared by the Commission appointed for this purpose, to be sent to the Minister, in view of the meeting of a special conference, following the next Congress of London, in order to discuss the issue of an International Time Service. After adoption, it is decided that Mr. Poincaré would like to make a personal approach to the Minister of Foreign Affairs in order to insist on the urgency of a solution.

- Procès-verbal de la séance du 8 Mai 1912/Minutes of the meeting of May 8, 1912 (M. Bigourdan Président)

[1] Henri Poincaré's death was to be announced in the minutes of the meeting of July 17, 1912.

[2] At the time, Raymond Poincaré, Henri Poincaré's cousin, was also President of the Council.

[3] Jean-René Benoît, then Director of the BIPM.

[4] The 1912 *International Radiotelegraph Conference*, see Sect. 2.2.4.

Original version—Le Président profite de la présence de M. le Directeur de l'exploitation télégraphique pour le mettre au courant des démarches faites par le Bureau pour obtenir la convocation, à Paris, d'une conférence internationale à l'effet de faire adopter la station de la Tour Eiffel comme poste de transmission des signaux horaires. Il est donné lecture à M. Frouin de la note adressée au Ministre de l'Instruction publique et de la lettre d'envoi. M. Frouin dit que le sous-secrétaire d'Etat des Postes et télégraphes a répondu au Ministre des Affaires Etrangères qu'il serait préférable de réunir une conférence spéciale à Paris que de porter la question devant la conférence qui doit se réunir à Londres le 4 juin. Il serait à désirer que les invitations fussent reçues par les puissances invitées avant l'ouverture de la conférence de Londres.

English translation—The President took advantage of the presence of the Director of Telegraph Operations to acquaint him with the steps taken by the Bureau to obtain the convening, in Paris, of an international conference for the purpose of adopting the Eiffel Tower as a transmission station of time signals. The note addressed to the Minister of Public Instruction and the covering letter are read to Mr. Frouin. He said that the Under-Secretary of State for Posts and Telegraphs replied to the Minister of Foreign Affairs that it would be better to convene a special conference in Paris than to take the matter to the Conference which is to meet in London on June 4. It would be desirable that the invitations be received by the invited powers before the opening of the London Conference.

- *Procès-verbal de la séance du 15 Mai 1912/Minutes of the meeting of May 15, 1912 (M. Bigourdan Président)*

Original version - M. le Ministre de l'Instruction Publique écrit qu'il est d'accord avec son collègue des Affaires Etrangères et M. le Sous-Secrétaire d'Etat aux Postes et Télégraphes pour réunir prochainement une Conférence Internationale chargée de discuter l'installation d'un service général de l'heure: il demande au Bureau de préciser le programme et d'indiquer un aperçu des frais nécessaires. M. Lallemand et Ferrié sont immédiatement chargés de rédiger la réponse, qui est adoptée en fin de séance à l'unanimité.

English translation—The Minister of Public Education writes that he agrees with his colleague of Foreign Affairs and the Under-Secretary of State for Posts and Telegraphs to convene soon an International Conference to discuss the installation of a general time service: he asks the Bureau to specify the program and to indicate an outline of the necessary expenses. Mr. Lallemand and Ferrié are immediately responsible for drafting the response, which is unanimously adopted at the end of the meeting.

2.3.2 The First International Conference on Time: Paris, October 15–23, 1912

In June 1912, the French government informed the foreign governments of its intention to convene in Paris, on October 15, an international conference in order to study the improvements of all types likely to be brought to the sending of radiotelegraph signals by the Eiffel Tower. The bases for an effective cooperation between the various observatories would also be fixed, where appropriate, and a permanent link would be established between the current French organization and similar organizations already created or planned abroad.

The various governments were invited, if necessary, to be represented at this meeting. This invitation was accepted by a number of States; sixteen of them were officially represented.

2.3.2.1 The Work of the Conference

This "Conférence Internationale de l'Heure" (International Conference on Time) was held in October 1912 at the Paris Observatory; it was placed under the auspices of the Bureau des longitudes (see ([Bureau des longitudes] 1912).

The Conference was chaired by G. Bigourdan, then President of the Bureau des longitudes, with G. Ferrié as General Secretary; it brought together more than one hundred participants (delegates and invited) from sixteen different countries in Europe and America. The language of the meeting and the minutes were in French. Each of the sixteen countries was represented by one or more delegates:

Austria (2 delegates), Belgium (11), Brazil (2), France (18), Germany (8), Great Britain (4), Greece (1), Italy (6), Monaco (1), the Netherlands (3), Portugal (1), Russia (3), Spain (1), Sweden (1), Switzerland (1), and the United States (2).

On October 15, during the first plenary session of the Conference, the provisional program was handed over to the various delegates, and the work was subdivided into four commissions.

The Vice Presidents of the Conference were MM. O. Backlund (1846–1916), Director of the Pulkovo Observatory (Russia), F.-W. Dyson (1868–1939), Astronomer Royal in Greenwich (UK), W. Foerster (1832–1921), former Director of the Berlin Observatory (Germany), and A. Righi (1850–1920), Professor at the University of Bologna (Italy), each of them being President of one commission.

The commissions had the charge to work on the following issues:

First commission—President: W. Foerster
Astronomical determination of the time or the correction of a timepiece. Time keeping.

Second commission—President: A. Righi
Radiotelegraphic transmission of time. Radiotelegraph apparatus to be used for the transmission and reception of time signals.

Third commission—President: F.-W. Dyson
Degree of precision that the clock signals must reach for various applications.

Fourth commission—President: O. Backlund
Collaborations of various astronomical centers to ensure better knowledge of the time. Study of the general organization to be provided both for the transmission and for reception of time signals, so as to satisfy all needs.

A project for the organization of an International Time Service was presented by Charles Lallemand (1857–1938) on behalf of the Bureau des longitudes (see Fig. 2.9), which was the basis of the discussions.

The work of the commissions lasted 8 days and was closed on October 23 by a second plenary session of the Conference.

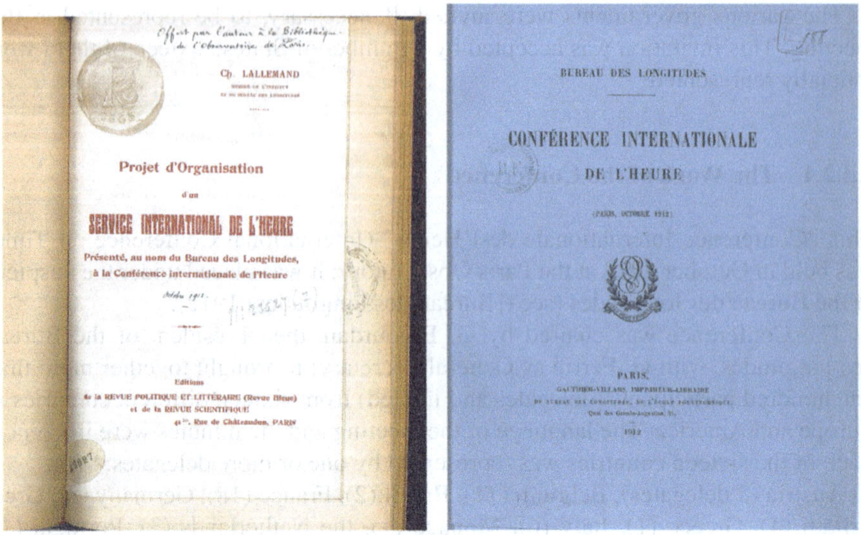

Fig. 2.9 Left: project for the organization of the "Service International de l'Heure". Right: "Conférence Internationale de l'Heure," Paris, octobre 1912 ([Bureau des longitudes] 1912) [©Bibliothèque de l'Observatoire de Paris]

During the time of the Conference, a number of instruments and installations directly related to the work of the Conference were exhibited at the Paris Observatory.

In addition to the scientific meetings, receptions, celebrations, and excursions had been organized in honor of the members of the Conference, including a reception by the President of the French Republic and an evening at the Paris Opera.

2.3.2.2 The Recommendations of the Conference

During the plenary session of October 23, the recommendations of the fourth commissions were presented to the Conference.

A number of recommendations were proposed by Commissions I to III concerning technical aspects of the radiotelegraphic transmission of time and the apparatus to be used, the desirable accuracy of these signals for astronomy, geodesy, navigation, railways, public services, etc., the repartition of stations and the regulation of the emission of time signals from the stations, etc.

For example, it was recommended to aim at covering ultimately the whole globe with a network of stations so arranged in position that at every point of the globe it should be possible for at least two time signals to be received in the course of every 24 hours, one by night and one by day. The signals should be sent out according to a detailed scheme at fixed hours of Greenwich time.

The recommendations of Commission IV were as follows:

A. *Création d'une Commission Internationale de l'Heure—Creation of an International Time Commission*

1. It is useful to seek to achieve the unification of the time, by the sending of radiotelegraph signals, whether ordinary signals or scientific signals.
2. The Universal hour will be that of Greenwich.
3. It will be useful to create a *Commission Internationale de l'Heure* (International Time Commission), in which each of the adhering States will be represented by delegates.
4. It will be useful to create, under the authority of the *Commission Internationale de l'Heure*, an executive body: the *Bureau International de l'Heure*, whose seat will be in Paris.
5. For ordinary signals, the results of the time determinations will be transmitted to this Bureau by the national centers, which will centralize themselves the determinations made by the observatories of their countries and will deduce from it the most exact time.
6. For scientific signals, the Bureau's mission will be to centralize the time determinations made in the associated observatories and to deduce from it the most exact time.
7. The *Bureau International de l'Heure* will communicate the results of the comparisons which would not be promptly published, to the Central Office of the *International Geodetic Association* in Potsdam, which will be asked to undertake an in-depth discussion. These results will also be communicated to other official international associations which request them.
8. Until the circumstances allow this program to be carried out, a Provisional Commission, appointed by the Conference, could organize, on a trial basis, the cooperation in question and study any improvements to be made in this project before officially submitting it to governments for approval.

The proposal of the Conference to establish the headquarters of the Bureau International de l'Heure (BIH) in Paris resulted from the initiative of W. Foerster, head of the German delegation. As for the beginning of the centralization of time determinations by the BIH, it was expected that time signals would be sent out regularly, commencing January 1, 1913.

B. *Communication to the International Association of Academies*

The Conference requests the Academy of Sciences of Paris to kindly submit to the International Association of Academies, with its support, the project for the creation of a *Commission Internationale de l'Heure*, in accordance with the wish expressed by this International Association gathered in London in 1904.

2.3.2.3 Preparation of Draft Statutes for the Time Commission

At the plenary session of October 23, 1912, the proposal for the constitution of a Provisional Time Commission, responsible for implementing the decisions of the Conference, was adopted (see Fig. 2.9). This Commission, made up of one delegate from each State represented, was nominated. Its mission was to prepare draft statutes in accordance with the recommendations of the Conference for the establishment of an International Time Commission, to be submitted to the various governments to definitively create this Commission. An Executive Committee was formed, made up of the President, O. Backlund (delegate of Russia), the director of the Bureau international de l'Heure, B. Baillaud, the Vice President, G. Lecointe (delegate of Belgium), the General Secretary, E. Kohlschütter (delegate of Germany), and the secretary, A. Berget (delegate of Monaco). Ch. Lallemand, rapporteur for the Bureau des longitudes project, was added to the Commission in an advisory capacity and also G. Ferrié, as the director of the Eiffel Tower radiotelegraph station. The Commission began its work the same day and held meetings from October 23 to 25, 1912. B. Baillaud, F.-W. Dyson, Ch. Lallemand, and G. Lecointe were asked to kindly prepare a draft statute for the following day, which was then unanimously adopted by the members present or represented, with certain reservations on the part of the delegate of Germany. The draft statutes, composed of 24 articles, were signed by the sixteen members of the Commission (see Fig. 2.10) declaring that they were unanimous in proposing that this draft be submitted to the various States for their consideration. The Executive Committee was instructed to ask the Ministry of Foreign Affairs of the French Republic, through the intermediary of B. Baillaud, to transmit this draft to the various governments, asking them to make known any observations to which it might give rise on their part. At the same time, these governments should be informed that a meeting of State delegates will be convened in Paris, within about six months to adopt the final text of the statutes, taking into account any observations that may have been made.

The Proceedings of the Conference were published as early as 1912, including the minutes of the plenary and technical sessions, all the notes distributed during the meeting as well as the draft statutes.

Then, the French government, after having studied the project, transmitted to the foreign Governments a draft comprising a "Diplomatic Convention" (distinct from the statutes) and the statutes, this draft being accompanied by the observations from a few States (Germany, Belgium, United States, and Italy) concerned by the project and the Notes to which this gave rise.

Les Membres de la Commission provisoire nommés, à Paris, par la Conférence internationale de l'Heure, dans sa séance plénière du 23 octobre 1912, acceptant cette mission sous réserve de l'approbation de leurs Gouvernements respectifs, sont unanimes à proposer que le projet de Statuts ci-dessus soit soumis à l'examen des divers États.

Ont signé :

MM. Ernst Kohlschütter (Allemagne).
Émile Descovich (Autriche).
Georges Lecointe (Belgique).
Francisco Bhering (Brésil).
Edouardo Mier (Espagne).
H.-H. Hough (États-Unis).
B. Baillaud (France).
F.-W. Dyson (Grande-Bretagne).
D. Éginitis (Grèce).
Augusto Righi (Italie).
Alphonse Berget (Monaco).
Van de Sande Bakhuyzen E. F. (Pays-Bas).
José-Lambertini Pinto (Portugal).
O. Backlund (Russie).
C.-V.-L. Charlier (Suède).
Raoul Gautier (Suisse).

Fig. 2.10 First International Conference on Time (Paris, 1912): final wish and signatures of the members of the Provisional Time Commission ([Bureau des longitudes] 1913)

2.3.3 The Second International Conference on Time: "Conférence Diplomatique" (Paris, 20–26 October 1913)

Finally, one year after the First Conference, on October 20–26, 1913, a Second International Conference on Time assembled in Paris at the invitation of the French government.

2.3.3.1 The Work of the Conference

This Conference was held at Paris Observatory and was organized as a "Diplomatic Conference," in which thirty-two States (sixteen States more than in 1912) were officially represented: Republic of Argentina, Austria, Belgium, Belgian Congo, Brazil, British India, Chile, Cuba, Denmark, Ecuador, Egypt, France, Germany,

Great Britain, Greece, Guatemala, Italy, Kiaochau, Liberia, Mexico, Monaco, Nicaragua, the Netherlands, Portugal, Russia, Serbia, Spain, Sweden, Switzerland, Turkey, the United States of America, and Uruguay.

The sixteen additional States represented in the 1913 Conference were Republic of Argentina, Belgian Congo, British India, Chile, Cuba, Denmark, Ecuador, Egypt, Guatemala, Kiaochau, Liberia, Mexico, Nicaragua, Serbia, Turkey, and Uruguay.

Each State was represented by a single delegate, with the exception of the following States which had more than one delegate: Austria (2 delegates), Belgium (2), Chile (2), France (7), Great Britain (2), British India (2), Germany (4), Italy (5), Netherlands (2), Russia (3), Serbia (2), Switzerland (2), and Uruguay (2).

The language of the meeting as well as of the minutes was French. The Conference was chaired by Gaston Darboux (1842–1917), Secrétaire perpétuel of the Académie des Sciences and Member of the Bureau des longitudes. The Vice Presidents of the Conference were O. Backlund (Russia), W. Foerster (Germany), A. Righi (Italy), as for the first conference, and G. Lecointe (Belgique), and the secretaries were C. Bardeloni (Italy), R. Gautier (Switzerland), G. Ferrié (France), and E. Kohlschütter (Germany).

The major change of the draft submitted by the French government to the 1913 Conference compared to the draft statutes prepared by the 1912 Conference was the division of the 1912 draft statutes into a diplomatic part, the "International Convention on Time," and the regulation part, the statutes.

The terms of the draft Convention and the draft statutes were discussed during five plenary sessions. The observations received and the corresponding Notes, as well as the observations presented by the delegates during the Conference, were taken into account, including an amendment corresponding to a request transmitted by the International Geodetic Association. A few changes and additions were made, such as the replacement of the term "Commission" (of Time) used in 1912 by "Association" (of Time), following the proposal of G. Darboux, President of the Conference (cf. [Bureau des longitudes] 1933). The modifications in the content and presentation were agreed. The Bureau international de l'Heure was maintained in the premises of the Observatory.

This Conference resulted in the drafting of a revised "Diplomatic Convention" for the creation of the "Association Internationale de l'Heure" (International Time Association) and the establishment of statutes. These documents were finalized and accepted at the 5th session (October 24) and the minutes of signatures were signed at the sixth session (October 25) by those of the delegates who had received the necessary powers for it (see Fig. 2.11). The Conference unanimously suggested that, immediately after ratification of the Convention, the already appointed members of the Provisional Committee should be invested with the functions of President, Vice President, General Secretary of the Committee of the Association internationale de l'Heure, and director of the Bureau international de l'Heure, i.e., O. Backlund, G. Lecointe, E. Kohlschütter, and B. Baillaud, respectively (see Sect. 2.3.2.3).

In addition to the plenary sessions, a number of delegates met on several occasions in committees to study the modifications and technical improvements made necessary in the transmission of time signals by the international mission that

SIXIÈME SÉANCE.

PROCÈS-VERBAL DE SIGNATURE

(25 octobre 1913).

Le samedi vingt-cinq octobre mil neuf cent treize, la Conférence internationale de l'Heure s'est réunie en séance plénière à 6 heures de l'après-midi à l'Observatoire de Paris.

M. le Président présente à la Conférence le texte authentique du projet de Convention et de Statuts y annexés, où sont consignés les résultats des travaux de l'Assemblée. Il invite les délégués qui sont munis des pouvoirs nécessaires à signer cette Convention dont l'instrument diplomatique a été préparé en un seul exemplaire conformément à l'usage. Cet exemplaire restera déposé dans les Archives du Gouvernement de la République et une copie certifiée conforme en sera remise par la voie diplomatique à chacune des Puissances signataires.

Par application des stipulations de l'article 9 de ladite Convention les signatures des plénipotentiaires pourront être apposées jusqu'au 1er février 1914.

Fig. 2.11 The Second International Conference on Time (Paris, 1913): 6th session, minutes of signatures ([Bureau des longitudes] 1933)

this institution had to assume. A visit of the Time Services at Paris Observatory and the Eiffel Tower was organized for the delegates; a banquet at the Observatory, under the presidency of the Minister of Public Works, closed the Conference.

The minutes of each plenary session were distributed each day to the delegates during the time of the Conference, but in contrast, it was decided that the minutes of the Conference would be published only when the Convention is ratified because they included diplomatic documents. Due to World War I, the Convention was never ratified and the minutes of the Conference, including all the documents distributed before or during the Conference, only appeared in 1933 (see [Bureau des longitudes] 1933).

The Director of the Paris Observatory, B. Baillaud, was invited to organize immediately the Bureau, although the international convention had not been yet ratified by the member States.

2.3.3.2 Procès-verbal de Signature: Signature Minutes

The Convention was signed by eighteen States: Austria, Belgium, Chile, Denmark, Ecuador, France, Germany, Great Britain, Guatemala, Italy, Liberia, Monaco, the Netherlands, Russia, Serbia, Spain, Switzerland, and the United States. The Convention was then submitted for ratification by the governments concerned, the limit for other signatures being February 1, 1914. The Convention was to come into force on December 31, 1920.

English translation (see the original version in Fig. 2.11)

Sixth Session (October 25, 1913)

On Saturday, October 25, 1913, the International Conference on Time met in plenary session at 6 o'clock in the afternoon at the Paris Observatory. The President presents to the Conference the authentic text of the draft Convention and of the statutes annexed thereto, in which the results of the work of the Assembly are recorded. He invites the delegates who have the necessary powers to sign this Convention, the diplomatic instrument of which has been prepared in a single copy in accordance with custom. This copy will remain deposited in the Archives of the Government of the [French] Republic, and a certified true copy will be delivered through the diplomatic channel to each of the signatory powers. By application of the stipulations of article 9 of the said Convention, the signatures of the plenipotentiaries may be affixed until February 1, 1914.

2.3.3.3 L'Association Internationale de l'Heure (The International Time Association)

The Convention included a preamble and nine articles, extracts of which are given below, while the statutes, annexed to the Convention, included twenty-four articles which specified the constitutional regulations of the Association.

The International Diplomatic Convention on Time

The Sovereigns, Heads of State, and Governments of the Powers hereafter designated, having considered it useful to organize an International Association charged with the task of achieving the unification of time, by the sending of radiotelegraphic or other signals, have resolved to conclude an agreement to this effect and have agreed as follows:

First article— The contracting governments undertake to establish and maintain an **International Time Association**, made up of the delegates of the participating governments.

Article 2— The composition of this Association and of the bodies that derive from it, namely, the General Assembly, the Permanent Council, the Committee, and the **Bureau International de l'Heure**, whose seat is in Paris, their functioning, and their powers, are determined by the statutes appended to the present Convention.

Article 3— The installation costs, as well as the annual operating and maintenance expenses of the International Association and its organs, are covered by the contributions of the contracting States established under the conditions currently provided for in the statutes.

(..)

Article 8— The present Convention is concluded for a period ending on December 31, 1920. The Permanent Council of the International Association shall consider the conditions under which this Convention may be extended.

(..)

Article 9— The present Convention, which will bear the date of October 25, 1913, may be signed in Paris, until the following February 1, by the Plenipotentiaries of the Powers represented at the Conference meeting in Paris in 1913 for the organization of an international Association responsible for to achieve the unification of time, by sending telegraphic signals or others.

The organizational structure and the funding mode adopted in this Convention on Time were similar to those of the Metre Convention for metrology or of the International Association of Geodesy for geodesy.

The Statutes

The first nine articles of the statutes specify:

- That the objective of the International Time Association is "to unify the time by sending radiotelegraph and other signals, whether they are high-precision scientific signals or ordinary signals, meeting the needs of navigation, meteorology, seismology, railways, posts and telegraphs, public administrations, watchmakers, individuals, etc."
- That the Association is made up of the following bodies:
  - The General Assembly (GA), composed of delegates from the States adhering to the Convention
  - The Permanent Council, composed of the delegates having the power to vote
  - The Committee, which consists of a President, a Vice President, and a General Secretary, to be chosen among the GA members
  - The Bureau international de l'Heure (International Time Bureau), which consists of the director, scientific collaborators, scientific assistants, and assistants. Its headquarters are in Paris.
  - The President, Vice President, General Secretary, and the Director of the BIH should be elected by the GA.
- That the purpose of the BIH is:
  - For ordinary signals: to centralize the results of determinations of universal time, expressed in Greenwich Mean Time, transmitted by national time centers, and to communicate the results to emitting stations and national centers
  - For scientific signals: to centralize the determinations made by the associated observatories and determine the most exact time. It should then publish the results of its comparisons.

The following articles specify the rules for the functioning and powers of the bodies of the association, the conditions established for the contributions of the contracting States (i.e., annual dues prorated to the population of the States) and the rules for the use of the sums paid by the States and the income from other sources.

2.4 The "Non-Official" BIH at Paris Observatory (1912–1919)

The 1913 Convention on Time was signed ad referendum, concluded for a period ending December 31, 1920 (Art. 8) and provided for a Bureau international de l'Heure in Paris (Art. 2). This draft Convention was to enter into force as soon as the ratifications had been exchanged, but due to World War I (1914–1918), it was never ratified.

Under these conditions, this Bureau did not have a legal existence, but it nonetheless functioned without interruption from 1912 to 1919 under the direction of B. Baillaud (see Baillaud & Chandon 1922; Bigourdan 1923; [Bureau des longitudes-Observatoire de Paris] 1992).

2.4.1 The BIH During the Period 1912–1919

After the first conference on Time (1912), the BIH was provisionally set up in Paris. B. Baillaud had to assume its direction without the official title. He took the effective management and operation of the BIH from 1912 up to 1919 at Paris Observatory.

The responsibility of the Bureau was to centralize different measurements of universal time in order to unify time by synthesizing the information sent by the various national and international observatories. The Bureau also played an important role in the research on time keeping.

The Bureau worked in cooperation with observatories worldwide (about twenty) and broadcasted radiotelegraphy time signals for the benefit of astronomers and navigators. It was also responsible for the publication of data resulting from international longitude measuring operations.

Following the International Time Conference, B. Baillaud concentrated the instrumental effort of the Observatory on the exact determination of the time at the large Meridian Instrument and the meridian circle of the garden (see Le Guet-Tully 2012). Thus, the large Meridian Instrument of Paris Observatory, which had been used until then for the observation of reference stars for the elaboration of the sky map and observations of the Moon and big planets, was devoted to determining the time. This work began on November 1, 1912.

Observations were made every night, including Sundays and holidays when the sky permitted. The astronomers also calculated the collimation of the optical axis, levelling at the start and end of the series, the chronograph offset, the reading at nadir, etc. The reductions of the observations were made soon after and transmitted for the control of the main sidereal pendulum.

Other activities were devoted on the one hand to observations of pendulums and their comparisons and on the other hand to the emission and reception of time signals. One of the primary applications of this transmission of time signals was the determination of longitudes. The signal transmissions and the pendulum

Fig. 2.12 Left: a fundamental constant-pressure clock manufactured by Leroy. Right: the Pendulum Room of the BIH at Observatoire de Paris (1912) [©Bibliothèque de l'Observatoire de Paris]

comparisons were made morning and evening, every day without exception. The broadcasts were carried out with the help of the Eiffel Tower's military engineers, under the direction of Commandant Ferrié.

The premises occupied at the Paris Observatory by the BIH were as follows (see Baillaud 1913, Bigourdan 1923, Bencker 1945):

1. The vaults and galleries of the catacombs, belonging to the cellars of the Observatory, where the timekeeper clocks available to the BIH (see an example on Fig. 2.12) were installed. The time of these clocks was transmitted remotely by electric wires, which connected these timepieces to the next room.
2. The Pendulum Room (see Fig. 2.12) where the preceding electric wires ended and where there were various other pendulums and auxiliary devices, the office of the astronomer who directed the work of sending and observation of signals, who controlled their passage, etc.; the office of the telegraph operator who assisted the astronomer with telegraph communications, etc.
3. The Recording Room, in the East Tower of the Observatory, where the recording devices were installed (see Fig. 2.13) and which also served as a calculation office, etc.
4. The meridian pavilion which houses a meridian refracting telescope.

The beginnings of the BIH were difficult as World War I broke out before the BIH was organized. But B. Baillaud, who initialized the transmission of a time signal from the Eiffel Tower, maintained the Observatory and time signal throughout World War I.

Fig. 2.13 Left: automatic time signal transmitter manufactured by Brillé-Leroy (1913). Right: antenna receiving time signals at BIH [©Bibliothèque de l'Observatoire de Paris]

Until August 2, 1914, the time service was entrusted partly to astronomers and partly to auxiliary employees. From this date and throughout the war, it was exclusively provided by the reduced staff of the Observatory. B. Baillaud personally assured the discussion of the clock measurements and the determination of the extrapolated corrections.

The measurement of the difference in longitude between Paris and Pulkovo observatories was interrupted, but the counting of the longitude signal bands was continued and the determination of the difference in longitude between Paris and Washington could be completed.

The BIH routinely published the difference between the time signals broadcasted by radio stations and the astronomically determined time, which relied on astronomical observations of star transits to set clocks.

B. Baillaud implemented a backup site in Lyon in 1918, connected with a local radio broadcasting station, in case the Paris site would have to be discontinued.

As in many other fields, with men having left for the front, women came to the frontline. And it was a woman, Edmée Chandon, who was responsible for the observations for the determination of time, assisted by her female colleagues from the Bureau of Measurements (see Barthalot 2009).

2.4.2 *BIH Instruments for Time Measurements and Time Diffusion*

The equipment required for the Paris Observatory to accomplish the mission of the BIH was the following:

- Instruments for the astronomical determination of time: transit instrument, zenithal tube, etc.
- Electric chronographs between the astronomical instruments and the clock to register the instant of contact
- One fundamental clock for keeping time between the astronomical determinations
- Instruments for diffusion ordered by the fundamental clock: synchronized clocks, radio-electric time signal transmitters, and receptors (see Fig. 2.13)

2.5 The Official Foundation of the BIH Under the Auspices of the IAU (July 1919)

After World War I (1914–1918), former scientific associations were reconstituted on new bases.

In July 1919 (18–28), during the third session of the "Inter-Allied Conference of Scientific Academies" in Brussels, the "International Research Council" (IRC) was officially created (see Pelseneer et al. 1919, Blaauw 1994), one of its goal being to bring about the creation of Associations, known as "International Unions" (see Fauque & Fox 2022) setup for definite purposes as international, non-governmental, non-profit organizations. Several Unions were created during this Constituent Assembly of the IRC.

Among the Unions created were:

- The International Union for Geodesy and Geophysics (IUGG) in which geodesy set up here its own section, known since 1932 as the "International Association of Geodesy" (IAG)
- The International Astronomical Union (IAU), divided into thirty-two commissions

The statutes of the IAU commissions were adopted at the plenary session of July 26, 1919 of the Constituent Assembly of the IRC. The names and composition of these commissions were adopted at the same plenary session.

2.5.1 The "Commission de l'heure" (Time Commission) of the IAU

Commission number 31 of the IAU was the "Commission de l'heure" (Time Commission), also called "Commission Internationale de l'heure" (International Time Commission). Its composition is given below:

President: M. R.-A. Sampson.

Members: B. Baillaud, G. Bigourdan, G. Celoria, F. W. Dyson, G. Ferrié, J.A. Hoogewerff, S. S. Hough, G. Lecointe, and H. Philippot.

This commission occupied a peculiar position in the IAU due to its specific mission for unifying time worldwide and therefore needed specific regulations.

2.5.2 Regulations of the International Time Commission

The regulations of the Time Commission were adopted at the plenary session of July 26, 1919 of the Constituent Assembly of the International Research Council. Below is the English translation (see [International Astronomical Union] 1922) of the original text in French.

These regulations were clearly inspired by the 1913 draft statutes for the International Time Association (see Sect. 2.3.3.3), which was never created because of World War I. The purpose and domicile are the same and the constitutional regulations have many similarities.

1. An International Time Commission shall be formed for the purpose of unifying time by means of radiotelegraphy and other signals, whether these are scientific signals of great precision or ordinary signals adapted to the needs of navigation, meteorology, seismology, railways, posts and telegraphic services, public administrations, clockmakers, individuals, etc.
2. The domicile of the Commission shall be located at Paris.
3. The Bureau shall consist of a President, Vice President, and a General Secretary; the latter two shall be elected by the Commission itself. The General Secretary, whose tenure of office will terminate on December 31, 1931, shall fulfill the function of Director of the International Time Bureau.
4. The purpose of the International Time Bureau, located at the National Observatory of Paris, is as follows:

   a. As regards ordinary signals, to harmonize the determinations of universal time expressed in Greenwich time, which shall be transmitted to them by the National Time Centres. These shall be responsible for calculating, with the highest obtainable precision, the meantime, as deduced from determinations made by the Observatories of their respective countries. These results shall be sent as quickly as possible to the emitting stations and to the National Centres.
   b. As regards scientific signals of precision, to harmonize the time determinations made by the Associated Observatories and deduce from them the most exact time.
   The International Time Bureau shall publish the results of its comparisons. Any result whose publication is delayed shall be communicated in detail to all scientific associations and official institutions which may send in a request to that effect.

5. In addition to the Director, the International Time Bureau shall consist of:

   a. Scientific collaborators, nominated and dismissed by the Commission, on the proposal of the Director of the International Bureau. They shall be charged, with or without indemnisation, with the conduct of special studies. Their term of office shall not exceed two years, but they shall be re-eligible.
   b. Scientific assistants, nominated and dismissed by the Director of the International Bureau. They shall be responsible for the carrying out of the program drawn up by the Commission. Their term of office shall not exceed four years, but they shall be re-eligible.
   The budget shall determine the sums allocated to each of these classes.

6. The Director of the International Bureau shall publish the results of the work undertaken in accordance with the decisions of the Commission. He shall present to the President an annual report covering the whole field of activity of the Bureau, together with the program of work for the following year. This report and program shall be published.
   In his capacity of General Secretary, he shall, in consultation of the President, conduct the correspondence and administer the funds. He shall be responsible for the custody of documents.
7. The Director of the International Time Bureau shall prepare the estimate for the budget, which, after having been approved by the President and the Commission, shall be submitted to the International Astronomical Union.
8. The income of the Commission shall be devoted to:

   a. Covering the expenses of administration and publication
   b. Honorarium to the General Secretary as Director of the International Time Bureau
   c. Grants or remunerations for calculations, observations, or experiments undertaken at the request of the Commission
   d. The purchase and upkeep of material of the International Time Bureau
   The Director of the Bureau, under the control of the International Astronomical Union, shall be responsible for the expenditure allocated to the above objects.

9. The above regulations shall remain in force until December 31, 1931.

The Bureau international de l'heure (BIH) was thus officially created as the executive organ of the newly founded IAU Commission 31. This Bureau was substituted for another body with the same name and the same purpose, headed by B. Baillaud, which was established by the 1913 International Conference of Time and had operated since 1912 without having a legal status.

In fact, due to the actual date when the statutes of the IRC came into force,[5] the legal status of the BIH was only reached in 2021 (see Bigourdan 1920). Despite this, B. Baillaud, then President of the IAU, judged that the new organization should begin to function on January 1, 1920. He abandoned the effective management of the BIH and handed over the services and instruments of the old Bureau to G. Bigourdan who became the first official director of the BIH, a position he held until 1928. An official character being given to the BIH it was to take on greater importance from year to year in response to new requirements (see Lambert 1940; Guinot 2000).

[5] the statutes of the IRC stipulated that the Convention would come into force on January 1, 1920, on condition that at least three of the admitted countries have acceded to it and this condition was not fulfilled until late.

2.5.3 *Benjamin Baillaud and the* Bureau International de l'Heure

Benjamin Baillaud (1848–1934), eminent astronomer and wise administrator, was throughout his career an architect of international cooperation and the origin of important initiatives. As Director of Paris Observatory (1908–1927), member of the Académie des sciences, and member of the Bureau des longitudes, he played a leading role in the organization of the unification of time at the international level. He was, with G. Ferrié, at the origin of the operational time signal service, installed at the Eiffel Tower in conjunction with the Paris Observatory, and also of the two International Time Conferences in Paris. He then organized the Bureau international de l'Heure even before the signing of the Convention on Time. The history of this Bureau, from 1912 to 1919, reflects the sequence of efforts made by him, as its first director, to ensure its functioning even during the war-torn period that the Bureau went through in its early years. Being elected as the first President of the IAU (1919–1922), he had a prominent role in giving to the BIH a legal status under the auspices of this Union, with regulations largely inspired by the draft Time Convention that was never ratified.

2.5.4 *The Funding of the BIH*

After having received most of its resources from the Paris Observatory (premises, staff, and operating means) during the period 1912–1919, the BIH became in 1919 the executive organ of IAU Commission 31. Part of its funding then came from this Union (see Sect. 2.5.2) with an annual funding of 35,000 francs for the period 1920–1921. The report of the IAU Finance Commission at the subsequent 1922 and 1925 General Assemblies shows that the BIH received at that time the largest part of the budget allocated by the Executive Committee to scientific activities under the auspices of the IAU. The IAU Executive Committee also suggested in 1925 that IUGG should contribute to the financial support of the Bureau.

Extracts of the Reports of the IAU Finance Committee (see [IAU] 1922 and [IAU] 1925)

> **1922**
> The Finance Committee received from the Executive Committee the following estimates for each of the next three years 1923–1925 (see Table 2.1). This should be compared to the amount of 1500 Frs (Francs) for the unit of contribution for 80 units of actual contributions to the Union at that time.
> The effective number of units of contribution is about 80 according to the present constitution of the Union, and the Finance Committee recommends the continuation of the sum of 1500 francs as the unit of contribution, with a view to building up a reserve fund which may be allocated to special work.

Table 2.1 Budget attributed by the IAU to various committees—1923–1925

Committee	Amount
Bureau de l'Heure	50,000 Frs.
Variation of latitudes	10,000 Frs.
Carte du Ciel	5000 Frs.
Comets	1000 Frs.
Double stars	2000 Frs.
Variable stars	1200 Frs.
Printing (Report of Conference, etc.)	4800 Frs.
Secretariat	10,000 Frs.
Total	84,000 Frs.

1925
The Finance Committee suggests that the possibility of obtaining a contribution from the Geodetic Union toward the expenses of the *Bureau de l'Heure* should be considered by the Executive Committee.

2.6 Conclusion

The unification of time on a worldwide scale by means of wireless telegraphy, which appeared at the beginning of the twentieth century, was an exciting undertaking. However, the foundation of an international Bureau in charge of this unification was a very long and difficult process.

Thanks to a series of leading scientific personalities, with extensive experience in international cooperation, a very sophisticated organization has been prepared for the calculation and dissemination of international time. The initial organization prepared in 1912 and 1913 by two successive International Time Conferences, with both technical and diplomatic aspects, was very similar to that of the Metre Convention (1875). It would have created an intergovernmental organization, the *International Time Association*, whose executive body would have been the *Bureau International de l'Heure* (BIH) and whose funding would have been shared among the participating States. But, due to World War I, this project could not be carried out, and the BIH had to work for seven years without an official status at the Paris Observatory, maintained by the French government.

The creation of the International Astronomical Union in 1919 was the occasion for an official creation of the BIH on an international basis.

References

André, C. & Rayet, G. (1874). *L'Astronomie pratique et les Observatoires en Europe et en Amérique depuis le début du XVII^{è} siècle jusqu'à nos jours*. Paris: Gauthier-Villars.

Baillaud, B. (1913). *Rapport annuel sur l'état de l'Observatoire de Paris* pour l'année 1912. Paris, Imprimerie Nationale. Source gallica.bnf.fr/BnF.

Baillaud, B. & Chandon, E. (1922). Rapport relatif aux signaux horaires émis de l'Observatoire de Paris par le poste radiographique de la Tour Eiffel, depuis 1910 jusqu'à la fin de 1919. *Bulletin Astronomique*, 2^{è} *série, 1*, 129–167. Paris: Gauthier-Villars.

Barthalot, R. (2009). *Histoire de l'Observatoire de Paris, Chapitre 9*. New York: La Coopération Internationale. https://sites.google.com/site/histoireobsparis/.

Baüer, P. (2020). Contributions du Bureau des longitudes à cent ans de coopération internationale dans le domaine des sciences de l'univers (UAI, UGGI et URSI): Gustave Ferrié et l'opération "Longitudes mondiales". *Collection du Bureau des longitudes, 7*, under publication.

Bencker, H. (1945). Standardization of Time and accurate determination of exact time. *International Hydrographic Review, 22–24*, 41–51.

Bigourdan, G. (1920). Rapport annuel sur les travaux effectués par le Bureau International de l'Heure. *Bulletin Astronomique, 1*, 189–195. Paris: Gauthier-Villars.

Bigourdan, G. (1923). Les services horaires de l'Observatoire de Paris et le Bureau International de l'Heure (BIH), Histoire, description et fonctionnement. *Bulletin astronomique*, 2^{è} (série, 3), 294–329. Paris: Gauthier-Villars.

Blaauw, A. (1994). *History of the IAU, The Birth and First Half-Century of the International Astronomical Union*. Amsterdam: Kluwer Academic Press.

Bougeret, J.-L. (2019). The first President of the IAU, Benjamin Baillaud. In C. Sterken, J. Hearnshaw, & D. Valls-Gabaud (Eds.), *Under One Sky: The IAU Centenary Symposium* (pp. 139–146). Cambridge: Cambridge University Press.

Bureau des longitudes. (1897). Conférence Internationale des Étoiles Fondamentales de 1896. Procès-verbaux. *Annales du Bureau des longitudes, 5*, D1–D90. Paris: Gauthier-Villars.

Bureau des longitudes. (1912). *Conférence Internationale de l'Heure (Paris, Octobre 1912)*. Paris: Gauthier-Villars. Imprimeur-Libraire du Bureau des longitudes, de I'École Polytechnique.

Bureau des longitudes. (1913). Congrès International des éphémérides Astronomiques, & Conférence Internationale de l'Heure (Paris, Octobre 1912). *Annales du Bureau des longitudes, 10*, D1–284. Paris: Gauthier-Villars.

Bureau des longitudes. (1914). *Connaissance des Temps ou des mouvements célestes à l'usage des astronomes ou des navigateurs pour l'an 1916*. Paris: Gauthier-Villars.

Bureau des longitudes. (1933). Conférence Internationale de l'Heure (Paris, Octobre 1913). *Annales du Bureau des longitudes, 10*, A1–A114. Paris: Gauthier-Villars.

Bureau des longitudes-Observatoire de Paris. (1992). *Le Bureau International de l'Heure, 75 ans au service de l'Heure Universelle*. Paris: Imprimerie Observatoire de Paris.

Bureau International de l'Union Télégraphique. (1912). *Convention Radiotélégraphique Internationale*. Berne: Protocole Final et Règlement de service.

Bureau International des Poids et Mesures. (1921). *The Metre Convention and Annexed Regulations*.

Chandler, S. (1891). On the variation of latitude. *Astronomical Journal, 11*, 65–70.

Chinnici, I. (2022). *Precursors to IAU: Paris Observatory and the Carte du Ciel project*, this volume (Chap. 1), and refs. therein.

Débarbat, S. (2006). Public and scientific time at Paris Observatory: evolution over three centuries. In A. Brzezinski, N. Capitaine, & B. Kolaczek (Eds.). *Proceedings of the Journées 2005 Systèmes de référence spatio-temporels* (pp. 54–58). Warsaw, Poland: Space Research Centre PAS.

Débarbat, S. (2009). On the history of the astronomical constants. In M. Soffel & N. Capitaine (Eds.). *Proceedings of the Journées 2008 Systèmes de référence spatio-temporels* (pp. 75–76). Dresden, Germany: Lohrmann-Observatorium.

Elbert, W. (1908). Sur la variation des latitudes. *The Journal of Physical Theory and Application, 7*, 773–786.

Fauque, D., & Fox, R. (2022). A hundred years ago: The birth of the IAU as a model for others scientific unions (1919). In *This Volume*.

Guinot, B. (2000). History of the Bureau international de l'Heure. In S. Dick, D. McCarthy, & B. Luzum (Eds.). *Polar motion: historical and scientific Problems*. ASP Conferences (vol 208, pp. 175–184).

Höpfner J. (2000). The International Latitude Service—A Historical review, from the beginning to its foundation in 1899 and the period until 1922. *Surveys in Geophysics, 21*, 521–566.

International Astronomical Union. (1922). Resolutions of the IAU Finance Commission & Provisional Reports and Recommandations of Standings Committees. 31. Commission de l'Heure. In A. Fowler (Ed.). *Transactions of the IAU Vol. I—Proceedings of the 1st General Assembly*, Rome, Italy. *New Zealand Journal of Science, 110–114*, 175–181 and 203.

International Astronomical Union. (1925). Resolutions of the IAU Finance Commission & Provisional Reports and Recommandations of Standings Committees. 31. Commission de l'Heure. In A. Fowler (Ed.). *Transactions of the IAU Vol. I—Proceedings of the 2nd General Assembly*, Cambridge, UK. *New Zealand Journal of Science, 110–114*, 175–181 and 203.

Lambert, A. (1935). Benjamin Baillaud et les problèmes horaires. *Annales Françaises de Chronométrie, 5*, 15–22.

Lambert, A. (1940). Le Bureau international de l'heure, son rôle, son fonctionnement. *Annuaire du Bureau des longitudes pour 1940*, B2_B23. Paris: Gauthier Villars. Source gallica.bnf.fr / BnF.

Le Guet-Tully, F., & Davoigneau, J. (2012). L'inventaire et le patrimoine de l'astronomie: l'exemple des cercles méridiens et de leurs abris. *In Situ* 6 | 2005. http://journals.openedition.org/insitu/9177. https://doi.org/10.4000/insitu.9177.

Procès-verbaux du Bureau des longitudes. http://purl.oclc.org/net/bdl/.

Levallois, J.-J. (1980). The history of the International Association of Geodesy. *Bulletin Géodésique, 54*, 249–313.

Pelseneer, P., Swarts, F., & Lecointe, G. (1919). Compte rendu présenté à l'Académie royale des sciences, des lettres et des beaux-arts de Belgique. In *Conseil International de Recherches*. Assemblée constitutive tenue au Palais des Académies à Bruxelles, du 18 au 28 juillet 1919 (pp. 1–178). Bruxelles, Hayez: Imprimeur de l'Académie royale de Belgique. Source gallica.bnf.fr / BnF.

Royal Astronomical Society. (1885). Sixty-fifth Annual General Meeting. The International Meridian Conference. *Monthly Notices of the Royal Astronomical Society, 45*, 261–265.

Torge, W. (2005). The International Association of Geodesy 1862 to 1922: from a regional project to an international organization. *Journal of Geodesy, 78*, 558–568.

Nicole Capitaine is an astronomer emeritus at Paris Observatory where she was Director of the Astronomie Fondamentale Department and part of the Bureau international de l'heure for eight years. She is a correspondent of the Académie des Sciences and a member of the Bureau des Longitudes. Her scientific activity covers high-precision astrometry, time scales and space geodesy. She chairs the *Committee of the International Scientific Unions* of the Academy and is a member of its Science and Metrology Committee. She has served as President of the *Fundamental Astronomy* Division of the International Astronomical Union (IAU) and was a member of the Working Group that developed the IAU 2009 System of Astronomical Constants and the 2012 IAU Redefinition of the Astronomical Unit of Length. She is the IAU representative on the *Consultative Committee for Units of the International Committee for Weights and Measures* and has served as President of the ICSU *Federation of Astronomical and Geophysical Data Analysis Services*.

Chapter 3
A Hundred Years Ago: The Birth of the IAU as a Model for Other Scientific Unions (1919)

Danielle Fauque and Robert Fox

Abstract The date of 28 July 1919 is a familiar landmark in the history of the International Astronomical Union. This chapter sets the foundation of the Union in Brussels on that day in the context of the major reordering of both science and international politics that took place after the Great War. Mindful of the powerful position of German science before 1914, the victorious Allies fashioned a structure, based on a new International Research Council and unions for the individual sciences, from which the Central Powers, Germany in particular, were excluded. Founded at the same time as the International Union of Geodesy and Geophysics, the IAU served, in its structure and statutes, as a model for other disciplinary unions, even for the IRC itself. Although 12 Allied countries were present in Brussels, prominent figures in the scientific academies of just three nations—France, Britain, and the USA—were the main architects of both the IAU and the IRC. The task, in which purely scientific priorities of efficiency and coordination of effort interacted with diplomatic objectives, was fraught with complexity. Even the Allied position was not uniform. Behind an outward show of consensus, correspondence between Emile Picard, Arthur Schuster, and George Ellery Hale (all of them centrally involved in the founding of the union) and debates in their respective national academies reveal significant differences of perspective, notably on the pressing question of the possible admission of nations that had been neutral during the war.

3.1 Introduction

Formally, the foundation of the International Astronomical Union took place in Brussels on 28 July 1919. It did so as one of the concluding acts of the last of

D. Fauque (✉)
Université Paris-Saclay, Gif-sur-Yvette, France
e-mail: danielle.fauque@universite-paris-saclay.fr

R. Fox
University of Oxford, Oxford, UK
e-mail: robert.fox@history.ox.ac.uk

© The Author(s), under exclusive license to Springer Nature Switzerland AG 2022

T. Montmerle, D. Fauque (eds.), *Astronomers as Diplomats*, Historical & Cultural Astronomy, https://doi.org/10.1007/978-3-030-98625-4_3

three interallied conferences called to fashion the order of world science after the Great War. It was at this conference, following two earlier ones, in London (October 1918) and Paris (November 1918), that the representatives of the academies of a small group of Allied nations agreed to found a new organization, the International Research Council, as the parent body for disciplinary unions that would in due course extend across the whole range of the sciences. The IAU was among three unions that the IRC agreed to establish at the Brussels conference, along with the International Union of Geodesy and Geophysics (IUGG) and the International Union of Pure and Applied Chemistry (IUPAC). Other unions, including those for radiotelegraphy, physics, biological sciences, geography, and mathematics, followed quickly, bringing the number of unions to eight by 1925.

While the date of 28 July 1919 remains a landmark in the history of the IAU, the union, like the IRC itself, had roots that reached back into the war years and even before. The roots were complex, and an examination of them opens perspectives not only on the far from straightforward path that led to the creation of the IAU but also on the shifting interface at which science interacted with international relations and national interests on a far broader front and continued to do so for long afterwards. As we argue, in fact, the early history of the IAU can only be understood in the context of the profound reordering of world politics that occurred after the divisions engendered by the war.

3.2 Science in the Shadow of War

The creation of the initial group of three disciplinary unions in 1919 was by no means the first venture in the promotion of science on an international scale. Since 1899, the International Association of Academies (IAA) had brought together more than 20 leading academies, all but one of them (the US National Academy of Sciences—NAS) European, in a federal structure that fostered transnational collaborations in both the sciences and the humanities (Schroeder-Gudehus, 1978: 42–48, 1982).[1] The IAA's activities were constrained from the start by underfunding and a skeletal administrative structure, and the concrete results were few and of mixed quality. Far more effective in the half-century or so before the Great War were independent activities, many of them pursued through the growing number of international disciplinary congresses. Often, these initiatives responded to a mixture of scientific and economic imperatives that made cooperation in such areas as chemical nomenclature and physical units and standards essential. But astronomy, too, lent itself to cooperation, exemplified in the projects for the photographic star map (then as now referred to as the Carte du Ciel) and the Astrographic Catalogue, launched with participating observatories in more than a dozen countries in 1887, the International Union for Cooperation in Solar Research (ISU), founded

[1] For a more recent discussion and bibliography, see Fauque and Fox (2021).

in 1904, and the International Time Bureau (Bureau International de l'Heure—BIH), the result of two international conferences in 1912–1913, though not ratified diplomatically until after the war (Capitaine, 2022).

The state of war that began in early August 1914 was enough to curtail virtually all such activities. And within weeks any sense that scientists belonged to a seamless community that transcended national boundaries took another turn when 93 German intellectuals (including 12 past or future Nobel Prize winners in science) published a patriotic manifesto under the title "A call to the civilized world" in a Berlin newspaper on 4 October.[2] The manifesto, which protested against the "lies and calumnies" of the Allies and sought to justify German military actions in occupied Belgium, aroused a tide of mutual recrimination and hostility inflamed by scientific and scholarly communities on both sides of the conflict. The debates about the possible exclusion of enemy nationals from the ranks of their foreign members were just one sign of an animosity that still casts a shadow over relations long after the war.

It was against this background of what many saw as an irreparable rupture that the mathematician Gaston Darboux, one of the two permanent secretaries of the Paris Académie des Sciences (ADS), wrote to the secretary of the Royal Society (RS), Arthur Schuster, on 10 November 1916 (Fig. 3.1).[3] As part of a plan to promote closer Franco-British cooperation during the war, Darboux broached the possibility of a conference of the Allied nations in Paris at which, among other affairs of common interest, the future of scientific relations with the enemy would be addressed.[4] Darboux's unexpected death in February 1917 curtailed discussion. But the matter was not forgotten, and it resurfaced during a visit to London by Joseph Renaud, head of the French navy's hydrographic service, between 21 June and 3 July. Renaud's mission was primarily a naval one of liaising with the British Admiralty on timekeeping at sea. While in England, however, he took the opportunity of informal conversations with Schuster and Oxford's Savilian professor of astronomy, Herbert Hall Turner, to ascertain British opinion on the form that contacts with German scientists should have during and, more particularly, after the war. Renaud's impressions, conveyed to a meeting of the Bureau des Longitudes (BDL) on 11 July after his return to Paris, were unequivocal: as he reported, with support from a letter he had received from the British astronomer

[2] On the composition, diffusion, and impact of the manifesto, see von Ungern-Sternberg and von Ungern-Sternberg (1996), and MacLeod (2018).

[3] Darboux's letter, like several others in the exchanges we describe, no longer exists. Where the relevant correspondence is not available, we draw on a typed "Summary of correspondence relating to conference on international conventions after the war", written by Arthur Schuster, probably in late 1917, and now in Schuster's files concerning the International Research Council, Library of the Royal Society, London. The files, with a call number MS 540 and currently in course of classification, are cited hereafter as Schuster Papers. The typed summary is cited as Schuster Papers: "Summary".

[4] Schuster Papers: "Summary". Darboux's letter and Schuster's reply are missing. See also RS Minutes of Council, vol. 11 (14 February 1918), 288.

Fig. 3.1 Portrait of Arthur Schuster (1851–1934) by William Orpen, 1912 (P/0116). ©The Royal Society, London, United Kingdom

and geographer Arthur Robert Hinks, the British scientific community wished to see Germany excluded from international science and to discuss measures to that end with their French colleagues.[5]

Among the members of the BDL and in all probability present on 11 July was the distinguished mathematician Émile Picard, who had now succeeded Darboux as the ADS's permanent secretary for the mathematical sciences (Hadamard, 1942) (Fig. 3.2). A week after Renaud's report, Picard took up the matter of British attitudes towards the enemy in a letter to Schuster. The letter was cast around a blunt question: "Do we or do we not wish to resume personal relations with our enemies?" Picard's answer to the question, an unqualified no, was not in doubt. And, encouraged by a "formal reply" from Schuster on 22 July, he was soon set on concrete action, specifically a meeting of French and British scientists on the

[5] BDL, Procès-verbaux, 11 July 1917. Charles Lallemand had read Hinks's letter to Renaud at the previous meeting of the BDL, on 4 July. Hinks was adamant in his view that British scientists were resolved to break links with Germany immediately and to limit future international cooperation to Allied nations alone.

Fig. 3.2 Émile Picard (1856–1941). Photo D. Fauque/©Archives de l'Académie des sciences, Paris, France

subject, which he proposed to Schuster for late September or early October.[6] At this point, Picard's zeal seems to have caused disquiet as well as sympathy among senior figures in the Royal Society. This was evident in the advice of the Society's president Joseph J. Thomson to Schuster. While a meeting was desirable, Thomson argued, it should be delayed until the USA, an ally since entering the war in April 1917, could be represented.[7] Schuster duly put the case for postponement to Picard and, in a decisive broadening of the discussion, to George Ellery Hale, foreign secretary of the National Academy of Sciences and head of the Mount Wilson Observatory in California (Fig. 3.3).[8] What Schuster now proposed was a tripartite meeting in London in the spring of 1918 at which an American delegation would be present.

[6] Schuster Papers: "Summary". Picard to Schuster, 18 July, Schuster's reply, 22 July, then Picard's reply, dated 26 July 1917, all missing.

[7] Schuster Papers: "Summary", where Schuster refers to Thomson's intervention in a letter to Picard, 12 August 1917, missing.

[8] Hale Papers: Schuster to Hale, 27 August 1917, quoted in Blaauw (1994, 25).

Fig. 3.3 George Ellery Hale (1868–1938). Courtesy Caltech Archives, Pasadena, CA

Hale's was a receptive ear. Though well-travelled in Europe and a committed participant in international ventures, he had long felt that the IAA had fallen short of its objectives, and his experience of trying to work with the Prussian and other German academies in the IAA and in the ISU had often been unhappy. Nevertheless, he had reason to feel uneasy about what he was hearing from Europe. A characteristically forthright letter that Picard had written to him in July, immediately after Renaud's report and the exchange between Picard and Schuster that followed, was disturbing in the vehemence of its anti-German tone.[9] But, as foreign secretary of the NAS and with Schuster anxious to move ahead with a preliminary consultation in London to prepare for the following year's conference, Hale had no choice but to sound opinion within the Academy.

In sending members of the NAS council a copy of Picard's letter, along with a detailed printed memorandum and covering letter of his own, on 18 September, Hale

[9] Hale Papers: Picard to Hale, 22 July 1917, translation in Blaauw (1994, 24).

knew he was taking a risk.[10] In a letter to Picard on the following day, he expressed his "full agreement" with Picard's position, though with a clear hint of the distance between his own preference for "a liberal attitude towards neutral countries" and the view, which he knew to be Picard's, that the place of the neutrals should only be addressed once an Allied structure for international science was in place, and even then on a strictly "case by case" basis.[11] To Picard, as to Schuster soon afterwards, Hale was realistic, not to say pessimistic, about the chances of achieving consensus on the German question within the NAS, and his pessimism proved to be fully justified.[12] The terse message of a telegram that he sent a month later to Schuster, "Opinion divided regarding conference", conveyed what had been his worst fear.[13] Having launched a consultation, he found himself mediating within the NAS between the strong anti-German feelings of W. W. Campbell and the contrary view of colleagues of the stature of Robert Millikan, Edward Charles Pickering, and Frank Schlesinger, who favoured postponing further discussion until after the conflict; at all costs, as Millikan put it, nothing should be done that might "feed the flames of animosity, which are already sufficiently supplied with fuel" (Fig. 3.4).[14]

Such profound diversity of opinion weighed heavily on preparations for the conference scheduled for the following spring in London. So too did the unremittingly hard line that Picard maintained in a French newspaper article in October 1917 and an obituary notice on Darboux that he laced with rhetoric against a nation he portrayed as having brought on itself the penalty of exclusion, at least for some time, from scientific exchanges between "peoples who share the culture of humanity "(*peuples de culture humaine*) (Picard 1917a, 1917b). Marked though the differences of emphasis were, they did not undermine Allied support for some kind of boycott of the German scientific community, to be implemented once the war was over. By the autumn of 1917, the broad unanimity, which had taken shape in exchanges between the Allies since the beginning of the year, could draw further strength from the voice of Belgium, now increasingly insistent despite the shackles of occupation by German forces and the closure of the Belgian Académie des Sciences, des Lettres, et des Beaux-Arts from the earliest days of the conflict. In June 1917, the botanist and vice-president of the Academy, Jean Massart, conveyed

---

10 Hale to members of council of the NAS, and enclosures, 18 September 1917; quoted in Blaauw (1994, 29). The memorandum in question was the "Memorandum on the organization of international science", Hale Papers; identified in Blaauw (1994, 29 and 271), and Kikuchi (2011, 142, note 16).

11 Hale Papers: Hale to Picard, 19 September 1917; quoted in Blaauw (1994, 29–30). For Picard's attitude to the neutral nations, see Picard to Hale, 22 July 1917, cited above, note 9. See also our further comment, below, p. 000.

12 Hale Papers: Hale to Schuster, 22 September 1917; quoted in Blaauw (1994, 30–31).

13 Schuster Papers: Hale to Schuster, telegram dated 25 October 1917. Followed by letter from Hale to Schuster, 29 October 1917, Hale Papers; quoted in Blaauw (1994, 33).

14 Hale Papers: Millikan to Hale, 27 September 1917; quoted in Blaauw (1994, 31). For Campbell's view, expressed in a widely circulated statement on "International cooperation in science", dated 15 September 1917, see Blaauw (1994, 28–29).

Fig. 3.4 William Wallace Campbell (1862–1938). https://commons.wikimedia.org/wiki/File:Campbell_William_Wallace.jpg. Hector Macpherson (1905). Public domain. Via Wikimedia Commons

the strength of Belgian feeling when he spoke out, in the French *Revue scientifique*, with a plea for the Allies to move quickly to fashion a post-war structure for international science that would prevent the Germany's "scientific barbarians" (*barbares savants*) from re-asserting their position as a major force in science (Massart, 1917).

The solidarity of the Franco-Belgian position and the divisions within both the NAS and the Royal Society left Schuster facing a difficult winter. But he never wavered in his determination to implement the plan for the conference in the following spring. On 5 November 1917, with the approval of the RS council, he moved to maintain momentum with a proposal that plans for the conference should be discussed at an "informal and confidential" preparatory discussion between the

Académie des Sciences and the Royal Society.[15] On into the New Year, things moved fitfully. It was not until 17 December that Schuster's letter came before the executive committee ("comité secret") of the Académie. While the Académie's response on the principle of a conference of Allied nations to establish international bodies from which the enemy would be excluded was positive, it came with a caution. As Picard reported to Schuster, diplomatic conventions pertaining to certain existing organizations would necessitate a "conversation" with the Ministry of Foreign Affairs before even the preliminary Franco-British consultation could take place.[16]

The injection of an explicitly diplomatic dimension threatened administrative complications and a departure from Schuster's conception of a preparatory meeting that, as he insisted to Hale as well as to Picard, was to be a purely informal encounter with no authority to make binding decisions.[17] Another threat came from a renewed intervention by J. J. Thomson. Writing on 11 January 1918, Thomson again urged Schuster to consider postponement of the bilateral Franco-British discussion.[18] Thomson was clearly disturbed not only by the action of the ADS in passing what he described as its "very drastic resolution" of 17 December but also, and more immediately, by a call from "several" fellows for the Royal Society to pass a similar resolution to exclude the enemy nations from international scientific bodies after the war. For Thomson, such precipitate action risked harming the Society and would, at the very least, require approval at a special general meeting of the fellows. In the circumstances, he argued, a visit by a French delegation headed by Picard would "probably do more harm than good"; in any case he did "not think, from Picard's letter, that the French regard it as very urgent".

Despite Thomson's unease, the plan for a Franco-British consultation to begin preparing for the following spring's conference was still in place when the RS Council met on 24 January.[19] But it soon encountered another snag. This followed a quite unanticipated episode that began, also in January, with the discovery by the Italian Accademia dei Lincei that it had not been invited. The sense of being left out had led to a friendly but firm protest from Rodolfo Lanciani, a member of the humanities section of the Lincei, to the British Ambassador in Rome, and caused palpable embarrassment in the RS when news reached London. A specially summoned meeting of the RS Council on 14 February approved a lengthy placatory letter to the Ambassador, to be signed by the RS's foreign secretary William Abbott

[15] RS Minutes of Council, vol. 11 (1 November 1917), 257: Schuster to Picard (addressed as one of the two Académie's permanent secretaries), draft letter, 5 November 1917.

[16] ADS, Comité secret, register 7 (17 December 1917), 388–390: Picard's letter to Schuster is missing but referred to in Schuster's report to the Royal Society on 24 January 1918. See RS Minutes of Council, vol. 11 (24 January 1918), 280.

[17] For Schuster's conception of the meeting, see his draft letter to Picard, 5 November 1917, above, note 15. Hale, unable to attend, fully endorsed Schuster's proposal that the French and British alone should attend the preparatory meeting. See Schuster Papers, Hale to Schuster, 17 November 1917.

[18] Schuster Papers: Thomson to Schuster, 11 January 1918.

[19] RS Minutes of Council, vol. 11 (24 January 1918), 277–280.

Fig. 3.5 Vito Volterra (1860–1940). Photo D. Fauque/© Archives de l'Académie des sciences, Paris, France

Herdman.[20] And Schuster followed with a personal one to his fellow physicist and mathematician Vito Volterra (Fig. 3.5).[21] The letters did what they could to calm Italian indignation by stressing the intended informality of the meeting. But the harm was done. It was obvious that planning that had hitherto been largely the responsibility of the circle of Schuster, Picard, and Hale now had to become at once more formal and open.

[20] For the letter from the Ambassador, Sir Rennell Rodd, to the RS, 28 January 1918, and Council's approval of Herdman's draft reply (finally sent on 16 February), see RS Minutes of Council, vol. 11 (14 February 1918), 286–288.

[21] Schuster Papers: Schuster to Volterra, 16 February 1918 (copy).

3.3 An Interallied Affair

The establishment of a four-man Royal Society working party, the "Committee of International Scientific Organizations" (hereafter CISO), at the meeting on 14 February signalled the change in the Society's thinking.[22] The committee's brief was "to consider and report to Council upon the arrangements to be made for and the proposals to be brought before the Conference". Although the conference in question was still conceived as an informal preparatory one, the days of an exclusively Franco-British meeting were numbered. A week later, a French request for postponement of the meeting until after Easter further opened the door to a review and the decision, on 14 March, that invitations to a "preliminary consultation on questions relating to the future of International Scientific Organizations", now to be held on 8 May, should go not only to the ADS and the NAS but also to the Accademia dei Lincei and the Belgian and Japanese legations in London.[23]

Soon, though, external circumstances converged menacingly. On 25 April, with only two weeks to go before the planned consultation, the Council of the Royal Society heard (in a letter from Picard) that the ADS felt obliged to withdraw, almost certainly in response to the German spring offensives that had brought enemy troops to within 50 miles of Paris and exposed the city to long-range bombardments.[24] By now, Hale too had pulled out, in his case under pressure of work at Mount Wilson.[25] Anxious to see the meeting go ahead, he had arranged to be represented by three senior colleagues: the physicist Henry Bumstead (currently a scientific attaché at the US Embassy in London), the engineer William Frederick Durand (fulfilling a similar role at the Paris Embassy), and the physiological chemist Russell Henry Chittenden (resident in London from February to June 1918 as an American delegate to the Inter-Allied Scientific Food Commission).[26] But Hale's and the French withdrawal left the May meeting unviable.

The Royal Society's response to the disappointment was to shift attention immediately to preparations for the larger conference later in the year. Essential to the sustained momentum was the receipt, also on 25 April, of the report of the CISO.[27] Following a brief review of existing international organizations in science, the report proposed a number of subjects for discussion at the forthcoming conference. Of particular importance, and a clear priority, was the future of

[22] RS Minutes of Council, vol. 11 (14 February 1918), 287. The members of the committee were Schuster, the physiologist Hugh Kerr Anderson, the geologist and future director of London's Science Museum Henry George Lyons, and the electrical engineer James Swinburne.

[23] RS Minutes of Council, vol. 11 (21 February 1918), 289, and (14 March 1918), 294–295. Also Schuster to Volterra, 17 March 1918, Volterra archives, Accademia dei Lincei, Rome.

[24] RS Minutes of Council, vol. 11 (25 April 1918), 302.

[25] Schuster Papers: Hale to Schuster, 18 April 1918.

[26] On Chittenden's time in London and Paris as a member of the Commission, see Vickery (1944, 84–87).

[27] RS Minutes of Council, vol. 11 (25 April 1918), 297–302.

organizations subject to a diplomatic convention, such as the International Geodetic Association (IGA) and the International Association of Seismology (IAS). But there were other bodies, too, in which cooperation across national boundaries depended on a clear division of labour. Among them were the International Solar Union, the International Committee on Atomic Weights, and the International Association of Academies, whose coverage of both the sciences and the humanities raised particular problems.

In laying down what was effectively an agenda for the autumn conference, the CISO highlighted challenges and questions rather than solutions. What, for example, was to be done about the IGA, with its headquarters in Potsdam? The diplomatic convention governing the IGA had remained de facto in force, albeit at a reduced level, despite having formally expired on 31 December 1916. In the words of a clearly concerned CISO, "It would appear ... that the Geodetic Association presents in itself almost all the questions which are of importance at the present moment, and it may serve as a typical example for discussion".[28] On at least one important principle, however, the CISO could assume unqualified agreement. This was that the way forward depended on consultation between all the Allies and not just those in the limited circle invited for 8 May. In the new spirit of openness, the CISO report would be sent to the academies of every Allied nation, with requests for proposals concerning the future of individual scientific activities and administrative bodies. Formal invitations duly went out to the academies already involved, followed in July by invitations to academies or legations in Japan and, at Picard's request, Brazil, Portugal, and Serbia.[29] There was much to do, and the new target date for the autumn conference, October, allowed welcome time for further consultation and crucially the presence of an increasingly engaged Hale.[30]

The letter in which Hale had broken the news of his absence to Schuster on 18 April had borne no sign of diminished commitment. Quite the contrary. It came with a printed memorandum in which Hale presented a "scheme" for the future organization of international science that added a distinctive new dimension to the debate.[31] Central to the scheme was the proposal for an international body in which each nation would be represented by a multidisciplinary research council that would bring together representatives not only of the forms of research traditionally associated with academies but also of research in industry, agriculture, and medicine

[28] RS Minutes of Council, vol. 11 (25 April 1918), 297–302: "Report of Committee on International Scientific Organizations", esp. p. 300. At the same meeting James Jeans was appointed to the committee (p. 302).

[29] RS Minutes of Council, vol. 11 (4 July 1918), 320.

[30] Schuster Papers: Hale to Schuster, 17 June 1918.

[31] Which of two surviving documents Hale enclosed with his letter is unclear. It could have been either the "Memorandum on the organization of international science", cited above, note 10, or (as Blaauw, 1994, 33–34 seems to indicate) the 7-page "Suggestion[s] for the international organization of science and research". Both documents, which convey broadly similar ideas, are available in the Schuster papers at the Royal Society and the papers of B. Baillaud, Bibliothèque de l'Observatoire de Paris.

applied to both civil and military purposes. The model was the National Research Council (NRC) in Washington, a body, recently approved by President Wilson, that Hale had been instrumental in founding in 1916 and that he now chaired.[32] The very novelty of the proposal, which had its roots in the American push towards military readiness in the event of the country's entry into the war, exposed it to misunderstanding, not least on the function that academies, hitherto the main focus of discussions of the new post-war order, might have in such a structure.[33] Here, Hale's own view was clear: it was that each national research council should be "organically a part" of the relevant national academy and that the academy in question should take a lead role, as the NAS did in its oversight of the American NRC. In European eyes, though, this was just one unresolved question in a scheme that bristled with them. How, in practice, would "pure science" fare in a national body for research in which applied interests were so strongly represented? And, as Schuster asked Hale, was there not a danger that such a council might be seen as too controlling and hence as an impediment to independent initiatives in research, including new departures at both the national level and in individual disciplines?[34]

Bumstead, who lobbied hard through the spring of 1918 to promote Hale's ideas from his base in the American Embassy, was disappointed by Schuster's reaction, which he interpreted as a sign that Schuster was becoming "a little sluggish, as men of his age not infrequently do", in contrast to the more enthusiastic and younger new member of the CISO, James Jeans, who, perhaps not irrelevantly, had spent some years teaching at Princeton.[35] Exchanges during the summer of 1918 did little to diminish Schuster's reservations. Picard's view that it would be difficult to envisage the establishment of research councils in all countries was widely shared, not least within the Royal Society, where Henry Lyons stressed the difficulty of implementing Hale's plan in different national contexts.[36] In contrast with the unenthusiastic responses, however, the idea of representation by a national research council found support in a number of countries, notably in Italy, where Volterra saw it as helpful to his plans for a Consiglio Nazionale delle Ricerche (eventually created in 1923) (Guerraggio & Paoloni, 2012), and Japan, where the chemist Jōji Sakurai led the campaign for a similar institution, founded in 1920 (Kikuchi, 2011, 2019).

Despite the favour in certain quarters, the fact remains that Schuster had put his finger on objections that he knew would emerge in discussions involving other European partners. Far better, as he put it to Hale on 27 July, to aim for agreement on "a few main questions of principle ... and not to start with too ambitious a programme".[37] Among the questions of principle he had in mind was one on which

---

[32] Woodrow Wilson gave presidential approval to the NRC on 11 May 1918; see below, note 89.

[33] See Schuster Papers: Bumstead to Schuster, 13 June 1918.

[34] On Schuster's reservations, see Blaauw (1994, 36 and 41–42). Schuster raised the question in a letter of 28 April 1918, to which Hale responded on 17 June 1918; both in Schuster Papers.

[35] Hale Papers: Bumstead to Hale, 25 and 27 July 1918; quoted in Blaauw (1994, 41).

[36] Schuster Papers: Picard to Schuster, 20 August 1918; Lyons to Schuster, 7 June 1918.

[37] Schuster Papers: Schuster to Hale, 27 July 1918 (copy), responding to Hale's letter of 17 June.

he particularly insisted: the avoidance of "the diplomatic control which has hitherto been exercised over international scientific organizations". In this, Schuster was positioning himself as the champion of as loose a rein as possible. The goal, as he saw it, should be a structure that allowed "great latitude to different countries" and left them free to choose whether to be represented by a national research council, academy, government, or some other competent body. It was an approach that deliberately avoided pressing countries too strongly, at least at the start. Faced with such contrasting reactions, Schuster had no choice but to tread carefully and entrust discussion of Hale's proposals to an interallied committee constituted to represent the diversity of international opinion.[38]

It was all very well for the Allies to lay their plans. Never far from the surface, however, was the nagging matter of the place of other nations in the proposed post-war order. With regard to the Central Powers, Allied opinion stood firm on a period of exclusion or cessation of contact. But should there nevertheless remain a route to eventual reintegration, as Hale consistently believed, rather than a boycott of indefinite length? Even trickier questions turned on policy towards nations that had remained neutral during the war. On this, Hale's support for the early admission of the neutrals, to avoid their being driven "into the arms of the Germans", was at odds with the fear, commonly expressed in French and Belgian circles, that their presence within the new international body risked creating an internal lobby favourable to the admission of Germany.[39] The balance was delicate, but prevarication also had its dangers. Georges Lecointe, the scientific director of the Belgian Royal Observatory's astronomical service at Uccle, had already made the point forcefully in an 11-page printed report on "International intellectual relations after the war", written for the Belgian Ministry of Foreign Affairs in December 1917 but only now, in the summer of 1918, freely available (Fig. 3.6).[40] Lecointe's recommendation of a complete break with the Central Powers that might last for as long as twenty years identified him as a hardliner. His recommendation, though, came with the warning, similar to Hale's, that whatever post-war settlement the Allies adopted, self-interest decreed that it must offer a pathway to membership for such neutral nations as might wish to apply.

Lecointe's text left the Astronomer Royal Frank Dyson, who sent it on to Thomson on 5 August 1918, with mixed feelings (see Figs. 3.7, 3.8, and 3.9). Dyson went only so far as to state that he would "not wish to take part in any scientific conferences with Germans" and that he was under no illusion about the Bureau of Astronomical Telegrams, which "though nominally at Copenhagen is really

38 Schuster Papers: Schuster to Hale, 27 July 1918.

39 For Hale's advocacy of a flexible position on the future admission of non-Allied nations, see Schuster Papers: Hale to Schuster, 17 and 19 June 1918.

40 Schuster Papers: Lecointe and Herman De Guchtenaere, "International intellectual relations after the war", dated Paris, 4 December 1917, translated by Lady Lilian Rothermere from the French original. De Guchtenaere was a young literary figure, later better known by his pen name Paul Méral, who gave Lecointe secretarial help. Lecointe was an army officer, astronomer, and hydrographer, attached to the wartime Belgian government in exile at Le Havre.

Fig. 3.6 Georges Lecointe (1869–1929). Photo D. Fauque/© Archives de l'Académie des sciences, Paris, France

German".[41] But the firmness of Lecointe's intervention struck a chord by feeding into a wider sense of urgency for a decision about future attitudes to countries beyond the closed circle of the main Allies. At a meeting of the BDL on 3 July, Charles Lallemand, general inspector of mines and director of the French national survey, and a seasoned internationalist, delivered a particularly stark warning. As an unnamed German member of the currently inactive International Bureau of Weights and Measures (Bureau International des Poids et Mesures, BIPM) had apparently made clear to him, the Central Powers had no wish to enter into a collaboration with Allied scientists whose expressions of hatred for them remained, to all appearances, as uncompromising as ever; unless the Allies changed their rhetoric, Germany

[41] Schuster Papers: Dyson to Thomson, 5 August 1918.

Fig. 3.7 Sir Frank Watson Dyson (1868–1939). https://commons.wikimedia.org/wiki/File:Frank_Watson_Dyson.jpg. Bain News Service. Public domain. Via Wikimedia Commons

would take a lead in fashioning an alternative structure for international science in which the neutral nations would be welcome. It was a stark reminder that, on both sides of the conflict, the position of the neutrals would be central to whatever post-war restructuring took shape. Even Picard showed signs of heeding the message. In the same meeting, he acknowledged that the neutrals could be considered for admission, though only very selectively and once the Allied structure was firmly in place; significantly, there was no talk of a prearranged pathway.[42]

In broaching such questions, Allied scientists were entering a diplomatic minefield. At its heart lay the question not just of the wisdom but also of the legality of the proposed exclusion of the Central Powers from the new order. The British were quick to address the implications. The biologist and secretary of the Royal Society, Sir William Hardy, put a key question to his Cambridge colleague A.

[42] BDL, Procès-verbaux, 3 July 1918.

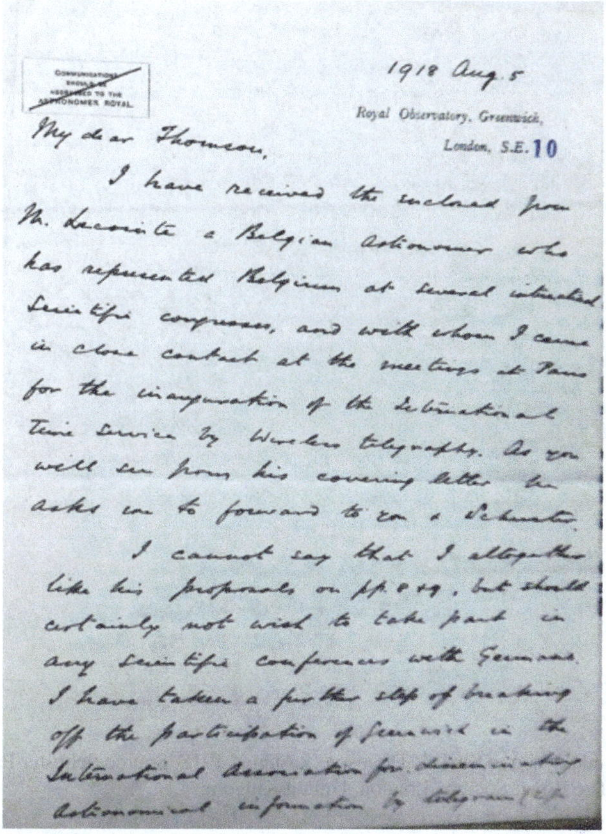
Communications should be addressed to the Astronomer Royal.

1918 Aug. 5

Royal Observatory, Greenwich,
London, S.E. 10

My dear Thomson,

I have received the enclosed from M. Lecointe a Belgian astronomer who has represented Belgium at several international scientific congresses, and with whom I came in close contact at the meetings at Paris for the inauguration of the International Time Service by Wireless telegraphy. As you will see from his covering letter he asks me to forward to you & Schuster.

I cannot say that I altogether like his proposals on pp. 8 & 9, but should certainly not wish to take part in any scientific conferences with Germans. I have taken a further step of breaking off the participation of Greenwich in the International Association for disseminating astronomical information by telegram [illegible]

Fig. 3.8 Sir Frank Dyson to Sir J. J. Thomson, 5 August 1918. Source Schuster Papers (MS/540). Photo R Fox/©The Royal Society, London, United Kingdom

Pearce Higgins, reader in international law: "Are agreements concluded between Governments annulled by a state of war or are they merely suspended?".[43] Higgins's reply was nuanced. As he saw it, the "general rule" was that war definitively annulled all treaties to which only the belligerents were parties. But treaties that also involved non-belligerent countries probably fell into the same category as, for example, the International Postal Union (IPU), always seen as an essential vehicle for communication to which Germany would be expected to belong, regardless of the conflict. Even that cautious opinion, though, left unanswered questions, not least in the always delicate case of the IGA. As Higgins observed, should the IGA (by analogy with the IPU) not be subject to automatic annulment, there remained the major complicating factor that it had its Central Bureau in the

[43] Schuster Papers: London Conference file: Higgins to Hardy, 5 October 1918 (copy), item 11.

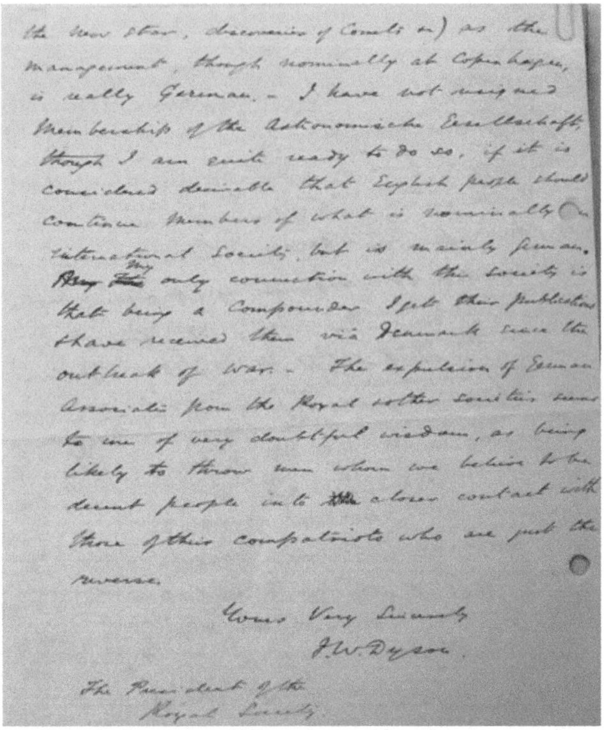
the new star, discoveries of Comets &c) as the
management, though nominally at Copenhagen,
is really German. – I have not resigned
Membership of the Astronomische Gesellschaft
though I am quite ready to do so, if it is
considered desirable that English people should
continue members of what is nominally an
International Society, but is mainly German.
My only connection with the Society is
that being a Compounder I get their publications
& have received them via Denmark since the
outbreak of war. – The expulsion of German
Associates from the Royal & other Societies seems
to me of very doubtful wisdom, as being
likely to throw men whom we believe to be
decent people into closer contact with
those of their compatriots who are just the
reverse.

Yours very sincerely
F.W. Dyson.

The President of the
Royal Society

Fig. 3.9 Sir Frank Dyson to Sir J. J. Thomson, 5 August 1918. Source Schuster Papers (MS/540). Photo R Fox/©The Royal Society, London, United Kingdom

distinguished Geodetic Institute in Potsdam on the outskirts of Berlin. If the IGA continued to exist, what hope was there of preventing Germany from re-emerging as the international leader in geodesy? One solution, favoured by the French and advocated by Picard before the BDL, would be to enforce Germany's withdrawal from international organizations and impose its relinquishment of such rights and property as it might have in them as part of the anticipated peace agreement.[44] Diplomacy at that level, however, lay far beyond the competence of the scientific community.

Higgins's advice served to highlight the complexity of the situation and hence to stoke rather than diminish Allied fears. In bringing geodesy to centre stage, it also raised questions for the disciplinary demarcations between geodesy and adjacent areas of astronomy. In correspondence between April and September 1918, Hale and Schuster exchanged a number of letters on this and wider issues concerning astronomy's proper scope. It was easy enough for the discipline to embrace astrophysics, as both Hale and Schuster believed it should. But might it

[44] BDL, Procès-verbaux, 18 September 1918.

not also include some or all of geophysics, in which geodesy, terrestrial magnetism, seismology, vulcanology, and meteorology all had a place?[45] Hale's preference was that geodesy should be separated from the other earth sciences and grouped in association with astronomy, on the grounds that the observational methods of geodesy were unlike those of the rest of geophysics: they were, as he put it to Schuster, "purely astronomical".[46] Tempting though it may have been for Hale and Schuster to try to settle the disposition of their favoured disciplines of geophysics and astronomy between themselves, private gentlemanly correspondence was no foundation for a decision, and the matter was shelved pending further consideration within the Allied community.

3.4 The First Interallied Conference. London, 9–11 October 1918

After a summer of intense but largely open-ended exchanges, and with the end of the war in sight, it was time to turn to detailed plans for the London conference, now scheduled for 9–11 October. As the date of the conference approached, the pace of the preparations increased, especially within the Royal Society, where Schuster assumed a heavy burden as the main local organizer. At a special general meeting of the Society on 26 September, fellows discussed the resolutions that they wished the British delegates to bring to the conference table, and a week later council appointed its delegates: 12 in all, headed by the President, J. J. Thomson, and including not only some of the Society's most senior fellows but also representatives of the Royal Irish Academy and the Royal Society of Edinburgh. The resolutions themselves, just two as recorded in the minutes of the RS's council, were restrained.[47] One expressed the desirability of establishing "organizations for scientific cooperation". The other called for the conference to consider the possible expulsion of enemy foreign members (a question that had limped along inconclusively in the RS since June) and to seek a common Allied policy that would then be referred back to the Society for discussion.[48]

Far less straightforward were the proposed resolutions Schuster received from the ADS and the NAS. Those from the ADS rehearsed the views that he had heard from Picard during the summer. But they had only been officially formulated late in the day, at a meeting of the ADS on 30 September, and then transmitted for consideration by the Royal Society, where the council discussed them on 3

45 Schuster Papers: see, for example, Hale to Schuster, 17 June and 8 August 1918.

46 Schuster Papers: Hale to Schuster, 8 August 1918.

47 RS Minutes of Council, vol. 11 (3 October 1918), 326–327.

48 The resolution's non-committal tone reflected the profound disagreements on expulsion within the society, with leading figures, including Thomson and Dyson (see above), known to oppose any such move, as was also the case in the NRC.

October.[49] They articulated the familiar French wish that the Central Powers should be obliged, as part of any peace treaty, to withdraw from existing international bodies in science, except for those governing essential public services, such as the railway networks and telegraphy. They also stipulated that the existing bodies should then be laid down and replaced by new ones, initially open only to the Allies, though with the possibility of their being opened at some future date to the neutrals as well. The French goal was an arena of international science incorporating a complete separation between the Allies and their enemies, to the point that the governments of all Allied countries would be called upon to send no delegates to any meeting or other activity in which the Central Powers were represented (see Appendix 2).[50]

The nine resolutions proposed by the NAS would similarly have caused Schuster no surprise, following his exchanges with Hale during the summer.[51] The proposal that an interallied committee should be formed to draw up a plan for the international organization of science was assured of an easy passage. But potentially more controversial and troublesome for Schuster were two points that bore the stamp of Hale's own priorities. One was that the plan should embrace the applications of science as well as pure science. The other was that the international organization which the NAS wished to see established in due course should be made up of representatives of the national research councils that each member country would be asked to create, on the model of the USA's NRC. Schuster, always anxious to achieve agreement, summarized his own position in a carefully crafted statement that attempted the balancing act of winning support both in the RS Council, which on 3 October duly approved submission of the statement to the conference, and from the authors of the resolutions submitted by the ADS and the NAS.

Schuster's statement conveyed the fierce personal commitment of a patriot devoted (despite his German family background) to the Allied cause.[52] Indeed, some within the Royal Society, as in the NAS, may well have been unnerved by the vehemence of his condemnation of the Central Powers, Germany in particular, for breaking "the ordinances of civilization, disregarding all conventions, and unbridling the worst passions which the ferocity of war engenders".[53] But while the belligerent tone conveyed genuine conviction, it also had strategic intent. With only days to go before the London conference, Schuster knew that his main task was to build bridges between Allies whose positions on the German question were far from unanimous. To that end, he made sure that the most important documents were

---

49 RS Minutes of Council, vol. 11 (3 October 1918), 327–328.

50 ADS: Comité secret, register 7, 1912–1918 (30 September 1918), 431–432.

51 Schuster Papers: IRC documents, London, "Resolutions proposed by the delegates of the NAS to interallied conference", item 7. Idem, Memoranda, items 6 and 6a.

52 Schuster was born in 1851 into a wealthy commercial family trading in textiles in Frankfurt am Main. Following the Austro-Prussian war of 1866 and the occupation of the town by the Prussian army, the family emigrated to Manchester, where Schuster took British citizenship in 1875. Early in the Great War, he and his family suffered from public attacks, notably from some academic colleagues who cast doubt on his patriotism (Schuster, 1918).

53 RS Minutes of Council, vol. 11 (3 October 1918), 328–329.

available in both French and English and that all participants received copies of key supplementary materials. Among the documents were the article by Jean Massart and Higgins's letter to Hardy, both referred to above[54]; a proposal by the French astronomer Guillaume Bigourdan for a regularization of bibliographical norms in scientific citations, with a view to replacing them with current German practices; a copy of Hale's printed submissions on the American NRC and suggestions for future NRCs; and the report of the RS's CISO.[55]

The 33 delegates from eight nations who gathered on 9 October did not want for information or, for that matter, a measure of luxury (see Appendix 1). The conference took place in the Royal Society's headquarters in Burlington House on Piccadilly, and delegates were housed nearby in one of London's most elegant hotels, the Ritz, complete with a private sitting room for discussions outside the formal sessions. Comfort, however, seems not to have yielded quite the calm path to consensus for which Schuster hoped. Daniel Kevles describes the meeting as having at one point "erupted in a cacophony of indecorous debate", and the impression of rancorous disorder is certainly one that Hale presented in private correspondence and the report he sent back to President Wilson, whom he had consulted before leaving for London (Kevles, 1971: 56). In the absence of minutes and with only brief published reports to go on, we cannot know how acrimonious the debates really were. But Kevles is probably correct in suggesting that the most heated exchanges were provoked by Picard's attempt to win over the British and American delegations from their more conciliatory positions. On this score, Picard was unsuccessful. Hale remained loyal to an understanding, established in discussions with President Wilson, that whatever decisions were taken in London would be moderate in their impact on relations with German scientists.[56] And his voice seems to have prevailed. The seven resolutions that were put to a concluding vote on 11 October, and approved unanimously, addressed the treatment of the Central Powers as a whole; there was no specific mention of punitive measures against Germany.[57]

It may be that outward unanimity concealed the dominance exercised by a combined Anglo-American majority of 18 out of the 33 delegates.[58] Nevertheless, there was much about the conference that would have pleased the French and Belgian contingents. Picard, who chaired most of the sessions, had his share of

54 An offprint of Massart's paper is in Schuster Papers: London Conference file. For the exchange between Hardy and Higgins, see above, note 43.

55 Schuster Papers: London Conference file, items 12–13. Baillaud Papers: "Conférence de Londres".

56 On Hale's exchanges with Wilson and his alignment with Wilsonian foreign policy before his departure for London, see Kevles (1971, 54–56).

57 For the resolutions and reports on the conference, see Picard and Lacroix (1918a, 566–570, and 1918c, 674–675), and *Year Book of the Royal Society*, 1918, 184.

58 The three largest delegations—of Britain (12), France (7), and the USA (6)—must have dominated proceedings. The other delegations were far smaller: Belgium (3), Japan (2), and one each from Italy, Serbia, and Brazil. See Appendix 1. The delegate from Portugal did not arrive in time to participate (Picard & Lacroix, 1918c: 673).

victories, not least in the resolutions themselves, which contained passages about future actions taken almost verbatim from those proposed by the ADS. Hale too had cause for satisfaction. A key survivor was the NAS's proposal that the future "international council" should be made up of representatives of national research councils which the Allied academies would now be "invited" to set up. To start the process, a Committee of Inquiry would be established. It would be left for this committee, under Schuster's chairmanship and probably with Picard, Hale, Lecointe, and Volterra among its members, to draw up a "general scheme" of international bodies that would meet the needs of scientific and industrial (including military) research.[59] With so many expectations satisfied, the keenest feeling of all, one of profound relief, would probably have been Schuster's. A summer of preparation and negotiation had taken a toll on his always fragile health, but he worked tirelessly to the end. Indeed, the speech he delivered just before the vote may even have been crucial in winning over any last waverers. Nothing could have done more to sustain his claim to understand the particular pain of the French and Belgian victims of enemy occupation than his account of his German childhood and his family's emigration to Britain in the 1860s (Schuster, 1918). As a declaration of his unqualified commitment to the Allied cause, it reinforced the message of his statement of 3 October, which was now reproduced (after the removal of its specific mention of Germany) as a preamble to the conference report.

While the resolutions dealt effectively enough with the broad principles of the new order for international science, their adoption inevitably papered over a number of cracks. Among the cracks was the very status of the London conference and the decisions taken there (see Figs. 3.10 and 3.11). As Brigitte Schroeder-Gudehus has observed, it had always been stated that the conference would have a consultative rather than a decision-making function (Schroeder-Gudehus, 1978: 109). That, in particular, was the position of the ADS, which had insisted that its delegates were not attending in an official capacity; their function was to report back to the Académie with proposals that would be discussed before submission for approval by the French government.[60] It must be said that the distinction between deliberative and consultative representation was something of an illusion. Consultations between academies and governments had taken place before the conference, to the point that several Allied governments had already expressed their unofficial approval of the general thrust, if not the details, of the resolutions that were later adopted and given quasi-official status at the conference.[61]

On the last evening, with the voting over, delegates to the conference enjoyed a banquet hosted in the name of the British government by the foreign secretary Arthur J. Balfour. As Picard put it in a banquet speech, the three days of debate in

[59] *Year Book of the Royal Society*, 1918, 183, where the preliminary report on the London conference mentions Schuster's appointment as "acting president" of the Commission of Inquiry but gives no details of other members.

[60] ADS: Comité secret, register 7 (30 September 1918), 431.

[61] Schuster Papers: London Conference file, item 10: "Resolutions passed by the conference".

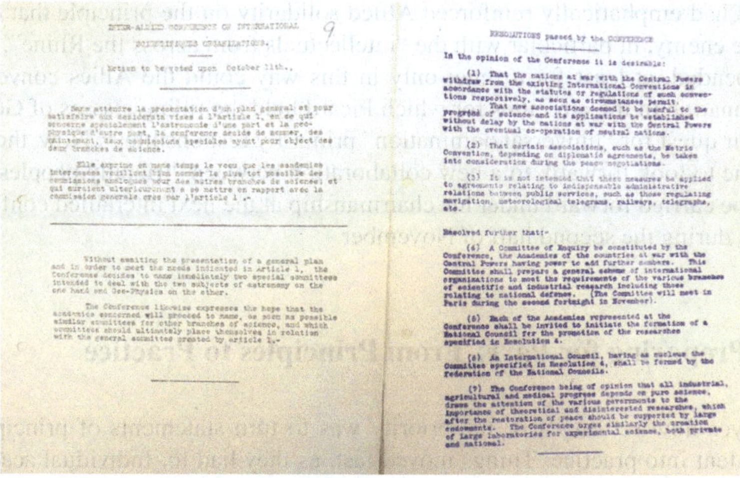

INTER-ALLIED CONFERENCE ON INTERNATIONAL

SCIENTIFIQUES ORGANIZATIONS.

Motion to be voted upon October 11th.,

Sans attendre la presentation d'un plan general et pour satisfaire aux desiderata vises a l'article 1, en ce qui concerne specialement l'astronomie d'une part et la geophysique d'autre part, la conference decide de nommer, des maintenant, deux commissions speciales ayant pour object ces deux branches de science.

Elle exprime en meme temps le voeu que les academies interessees veuillent bien nommer le plus tot possible des commissions analogues pour des autres branches de science; et qui auraient ulterieurement a se mettre en rapport avec la commission generale cree par l'article 1.-

Without awaiting the presentation of a general plan and in order to meet the needs indicated in Article 1, the Conference decides to name immediately two special committees intended to deal with the two subjects of astronomy on the one hand and Geo-Physics on the other.

The Conference likewise expresses the hope that the academies concerned will proceed to name, as soon as possible similar committees for other branches of science, and which committees should ultimately place themselves in relation with the general committee created by article 1.-

RESOLUTIONS passed by the CONFERENCE

In the opinion of the Conference it is desirable:

(1) That the nations at war with the Central Powers withdraw from the existing International Conventions in accordance with the statutes or regulations of such conventions respectively, as soon as circumstances permit.
That new associations deemed to be useful to the progress of science and its applications be established without delay by the nations at war with the Central Powers with the eventual co-operation of neutral nations.

(2) That certain associations, such as the Metric Convention, depending on diplomatic agreements, be taken into consideration during the peace negotiations.

(3) It is not intended that these measures be applied to agreements relating to indispensable administrative relations between public services, such as those regulating navigation, meteorological telegrams, railways, telegraphs, etc.

Resolved further that:-

(4) A Committee of Enquiry be constituted by the Conference, the Academies of the countries at war with the Central Powers having power to add further members. This Committee shall prepare a general scheme of international organizations to meet the requirements of the various branches of scientific and industrial research including those relating to national defence. (The Committee will meet in Paris during the second fortnight in November).

(5) Each of the Academies represented at the Conference shall be invited to initiate the formation of a National Council for the promotion of the researches specified in Resolution 4.

(6) An International Council, having as nucleus the Committee specified in Resolution 4, shall be formed by the federation of the National Councils.

(7) The Conference being of opinion that all industrial, agricultural and medical progress depends on pure science, draws the attention of the various governments to the importance of theoretical and disinterested researches, which after the restoration of peace should be supported by large endowments. The Conference urges similarly the creation of large laboratories for experimental science, both private and national.

Fig. 3.10 London Conference October 1918. Resolutions. Source Schuster Papers (MS/540). Photo R. Fox/©The Royal Society, London, United Kingdom

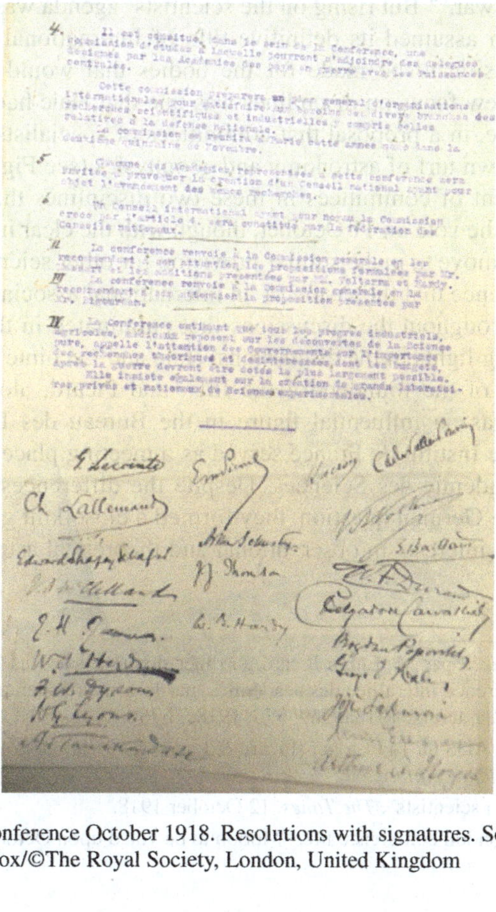

4\. Il est constitue, dans le sein de la Conference, une commission d'etudes a laquelle pourront s'adjoindre des delegues designes par les Academies des pays en guerre avec les Puissances centrales.

Cette commission preparera un plan general d'organisations internationales pour satisfaire aux besoins des divers branches des recherches scientifiques et industrielles y compris celles relatives a la defense nationale.

La commission se reunira a Paris cette annee dans la deuxième quinzaine de Novembre.

5\. Chacune des Academies representees a cette conference sera invitée a provoquer la creation d'un Conseil national ayant pour objet l'avancement des memes recherches.

6\. Un Conseil international ayant pour noyau la Commission creee par l'article 4, sera constitue par la fédération des Conseils nationaux.

II La conférence renvoie à la Commission generale en les recommandant à son attention les propositions formulees par Mr. [illegible] et les additions presentées par MM. Volterra et Hardy.

IV La Conference estimant que tous les progres industriels, agricoles, medicaux reposent sur les decouvertes de la Science pure, appelle l'attention des Gouvernements sur l'importance des recherches theoriques et desinteressees, dont les budgets, après la guerre devront être dotés le plus largement possible.

Elle insiste egalement sur la creation de grands laboratoires, privés et nationaux de Sciences experimentales.

Fig. 3.11 London Conference October 1918. Resolutions with signatures. Source Schuster Papers (MS/540). Photo R. Fox/©The Royal Society, London, United Kingdom

London had emphatically reinforced Allied solidarity on the principle that contact with the enemy, in particular with the "intellectuals from across the Rhine", should be suspended, at least for a time; only in this way could the Allies convey their condemnation of the atrocities for which Picard held the ruling classes of Germany and their quest for "universal domination" primarily responsible.[62] Now, though, it was time to look forward, to a new collaboration between "friendly peoples". This would be carried forward under his chairmanship at the next interallied conference, in Paris during the second half of November.

3.5 Preparing for Paris. From Principles to Practice

With eyes now set on Paris, the priority was to turn statements of principle and good intent into practice. Things moved fast, as they had to. Individual academies quickly endorsed the London resolutions and published them in their own and other scientific journals.[63] The press too published reports, typically (as in *The Times*) presenting the resolutions as a "boycott of German scientists" and part of the prosecution of the war.[64] But rising on the scientists' agenda was not just the move towards what soon assumed its definitive title of International Research Council but also the provision to be made for the bodies that would govern individual disciplines. This new focus on disciplines was one that Hale had urged, before the London conference, in a proposal that committees of specialists should be set up, beginning on his own turf of astronomy and geophysics (see Fig. 3.10). And it was to the establishment of committees in these two disciplines that attention turned immediately after the vote on 11 October, though with the clear intention that Allied academies should move to make similar provision for other sciences as well.[65]

It is no coincidence that the vision of an international association for astronomy was a presence throughout the discussions that culminated in the founding of the IRC. Of the leading lights in the discussions, Hale and Lecointe were astronomers; Schuster was one of the founders of the ISU; and Picard, along with his many other activities, was an influential figure in the Bureau des Longitudes, whose headquarters in the Institut de France served as a meeting place for the astronomy section of the Académie des Sciences. Despite the differences between them on the handling of the German question, they formed a close-knit group with a shared concern for Allied interests in observational and theoretical astronomy that would

[62] "Allocution prononcée par M. Émile Picard au dîner donné à Londres le 11 octobre 1918, à l'occasion de la conférence interalliée des académies, par le gouvernement anglais, et présidé par M. Balfour" (Picard & Lacroix, 1918a, 569–570; 1918c, 676).

[63] For example, in the ADS on 14 October (Picard & Lacroix, 1918a); also *Year Book of the Royal Society*, *Revue scientifique*, *Nature*, etc.

[64] "Boycott of German scientists", *The Times*, 12 October 1918.

[65] Schuster Papers: London conference file, "Motion to be voted upon October 11th".

necessarily come to the fore at the second interallied conference in November. In important ways, these practical priorities formed a bridge between the questions of general principle that were the business of the London conference and the emphasis on building new structures that was to give the gathering in Paris a different and distinctive tone.

When Picard returned to Paris, it was to the BDL, on 16 October, that he first reported on the decisions taken in London. Presenting them as evidence of a successful negotiation, he took pride in "resolutions consistent with the views of the French delegation" and the fact that they had been "adopted unanimously". The priority now was for organizations involved in each area of science to explore ways of forging closer links, even a complete merger of interests, within their discipline. This made it a matter of urgency for the BDL to consider what should be done in astronomy. If there was to be a merger, what form should it take, and what precisely should be merged? On what existing foundations might it be based? Was economy in monetary and material resources to be a major objective, as Lallemand, for example, suggested? The questions came before the BDL at its next meeting, on 23 October, along with the report of the CISO and other documents that had circulated in London.

The exchanges on 23 October soon laid bare tensions between Picard and Benjamin Baillaud, the director of the Paris Observatory (Bougeret, 2019). In answer to Baillaud's statement that discussions in London had envisaged an international body that would embrace applied as well as pure science, Picard was uncompromising: as he saw it, the concept that had emerged from the conference was of an organization that should be primarily devoted to pure science. There was also the question of membership, to be restricted in the first place, as Picard insisted, to the Allies and only subsequently extended, and then with caution, to the neutral nations. Baillaud's very different view was that the new structure should be underpinned by diplomatic agreements and, subject to such agreements, be as open as possible. As both men agreed, however, the priority was for action to be taken without delay. For Picard, that meant moving quickly to draw up statutes for the various bodies that would be established, perhaps using the statutes of existing associations as a starting point. And the sense of urgency prevailed. The BDL's council appointed a working party of Baillaud, Lallemand, now heavily involved in promoting the cause of an international association for geodesy, and Bigourdan to prepare a report for the next meeting, a week later (Figs. 3.12 and 3.13).

Discussions on 30 October showed how determinedly the working party had fulfilled its task. Bigourdan had drawn up a list of international bodies in astronomy, with a summary of their statutes and publications and an account of their meetings. He also underlined the importance of a programme of observations of the minor planets and the determination of their ephemerides in which the new body would be able to facilitate collaboration between participating universities and observatories. Most importantly, however, it was at this meeting that Baillaud introduced a first draft for an international astronomical union that would replace existing associations in the field and do so, as he insisted, without entailing the need for diplomatic ratification. At the following weekly meeting, on 6 November, members

Fig. 3.12 Charles Lallemand (1857–1938). Photo D. Fauque/© Archives de l'Académie des sciences, Paris, France

embarked on a point-by-point examination of the plan, presented in an annotated typescript entitled "Union astronomique—Projet de statuts. 1ère discussion" ("first discussion") (see Fig. 3.14).[66] The plan contained 28 articles distributed under seven headings: I. Organization (articles 1–6); II. Resources of the "society", amended by hand to "association", then "union" (articles 7–8); III. Without a heading, but probably bringing together articles 9–16 on national groups and their federation as an international council; IV. General assemblies (articles 17–20); V. Activities and scientific projects (articles 21–22); VI. Publications (articles 23–26); VII. Procedures for the revision of statutes and dissolution of the society (articles 27–28).

What the plan offered was a comprehensive set of statutes, with articles founded partly on ones in force before the war and partly on the proposals from Hale, though

66 BOP Archives. CNA Papers: File "Comité national d'astronomie".

Fig. 3.13 Benjamin Baillaud (1848–1934). Photo D. Fauque/© Archives de l'Académie des sciences, Paris, France

with no mention of Hale's principle that an academy should have oversight of a national group. Baillaud's view was simply that each nation's adhering body should have legal status and a title that conveyed the nationality in question. The national groups, all from belligerent countries on the Allied side (Baillaud listed 27 of them), would then constitute the union, with a General Council at its head. Conspicuously, no consideration was given to the neutral countries and little to the criteria governing the make-up of the national groups. An elaborate legal form, in fact, went hand in hand with a measure of freedom that left key questions unanswered. In particular, was membership to be open, closed, or a compromise between the two? Lallemand and the mathematician Paul Appell advocated a mixed association in which election would follow the presentation of the candidacy of a national group by two existing members. Such a loosely structured procedure had more in common with Baillaud's conception than with Picard's.

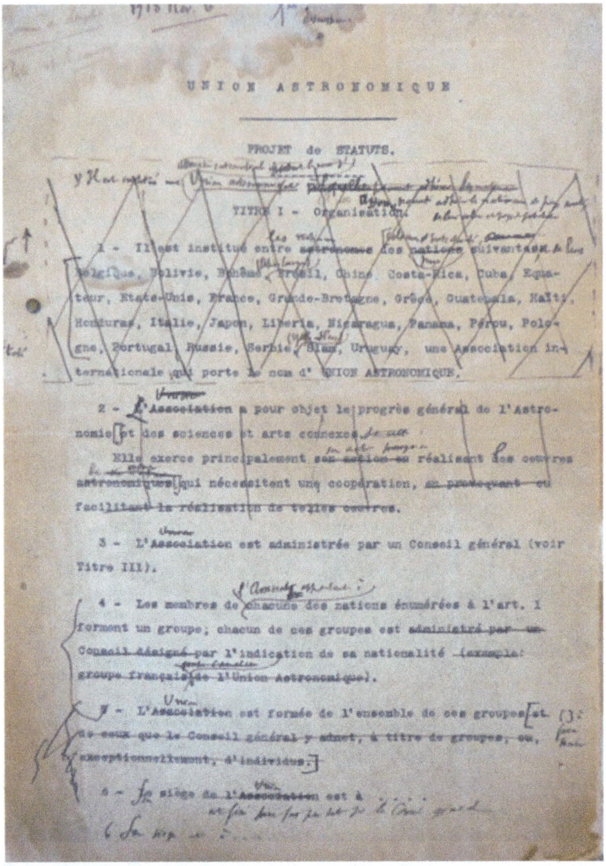

UNION ASTRONOMIQUE

PROJET de STATUTS.

TITRE I - Organisation.

1 - Il est institué entre astronomes des nations suivantes: Belgique, Bolivie, Bohême, Brésil, Chine, Costa-Rica, Cuba, Equateur, Etats-Unis, France, Grande-Bretagne, Grèce, Guatemala, Haïti, Honduras, Italie, Japon, Liberia, Nicaragua, Panama, Pérou, Pologne, Portugal, Russie, Serbie, Siam, Uruguay, une Association internationale qui porte le nom d' UNION ASTRONOMIQUE.

2 - L'Association a pour objet le progrès général de l'Astronomie et des sciences et arts connexes.

Elle exerce principalement son action en réalisant des oeuvres astronomiques qui nécessitent une coopération, en provoquant ou facilitant la réalisation de telles oeuvres.

3 - L'Association est administrée par un Conseil général (voir Titre III).

4 - Les membres de chacune des nations énumérées à l'art. 1 forment un groupe; chacun de ces groupes est administré par un Conseil désigné par l'indication de sa nationalité (exemple: groupe français de l'Union Astronomique).

5 - L'Association est formée de l'ensemble de ces groupes et de ceux que le Conseil général y admet, à titre de groupes, ou, exceptionnellement, d'individus.

6 - Le siège de l'Association est à

Fig. 3.14 Draft statutes of IAU, annotated, 6 November 1918. Source Comité national d'astronomie Papers. First page. Photo D. Fauque/© Bibliothèque de l'Observatoire de Paris

With so many fundamental decisions still to make, it was agreed that discussion would continue in a special meeting of the BDL, to take place exceptionally on a Saturday, 9 November. The first item was the establishment of a French national committee for astronomy. Picard, opposed to Baillaud's principle of openness, insisted that it was for the Académie des Sciences to set up such a committee and that the BDL could only be involved "in so far as it represents the members of the Académie's astronomy section"; his concern here was that governmental support would only be forthcoming if the ADS was centrally involved. The exchange left Picard and Baillaud far apart. Picard, fearful of any idea that membership might be available to any national group that paid a subscription, was not mollified by Baillaud's qualification that while the association he advocated would be open, the council governing it would be elected internally and hence be closed. On the activities that the new association might undertake, however, Baillaud and Picard did agree. The list included publications, the translation of abstracts, and research projects in such areas as solar spectra and the solar constant. It provoked differing opinions, not least on the financial implications. But, finally, the BDL members settled on the principle of a closed association, for which Picard agreed to prepare a new plan, limiting himself as usual to general questions and matters of principle.

Four days later, Picard laid a revised draft constitution before the BDL. His dramatically streamlined plan reduced the number of articles to nine (see Appendix 3). While some articles were taken unchanged from Baillaud's 28, the draft omitted those concerning financial arrangements and internal administrative procedures, leaving them as matters for astronomers to deal with at a later date. Significantly, though, it returned to Hale's idea of representation by national research councils that would operate in each country under the oversight of an academy and subject to precise criteria of eligibility for NRC membership. And, broaching matters that Baillaud had ignored or left imprecise, it tied the number of admissible delegates to a country's population and laid down strict conditions governing the future admission of neutrals.

Although Picard had prepared his constitution specifically for astronomy, he regarded it as a model for both the IRC and any of its associated disciplinary bodies. Recognizing that it could be adapted for the planned international geophysics association as well, Lallemand gave his warm support, and Picard's proposal was adopted unanimously and appended to the minutes of the meeting.[67] A week later, it was Lallemand himself who presented the BDL with draft statutes for a union that would bring together all the international associations with interests in the many constituent areas of geophysics. After minor modifications, these statutes too were approved.[68] On 25 November, on the eve of the Paris conference, the two sets of draft statutes, Picard's and Lallemand's, came before the "Comité secret" of the Académie des Sciences. There too, following some discussion, they were voted through without opposition.

For Schuster, as chairman of the working party set up during the London conference, the task of preparing for Paris began as soon as the London resolutions were approved by the Royal Society on 24 October. On the same day he wrote to Picard, inviting him to open the conference in Paris on Tuesday (26 November), by which time he would himself have been in the city over the weekend "so as to have a day or two to devote to preliminary discussions".[69] He hoped soon to be able to send a summary of the main points to be tackled. Among the more delicate of these was the position of Greece. No invitation to the London conference had been issued to the Greek scientific community; the explanation was, in Schuster's words, that "at the time the invitations went out, we were somewhat afraid that the majority of Greek savants were Germanophile". Now Schuster suggested that he should contact Greece's prime minister Eleftherios Venizelos, currently in London and known to

[67] BDL, Procès-verbaux, 13 November 1918.

[68] BDL, Procès-verbaux, 20 November 1918.

[69] Schuster Papers: Schuster to Picard, 24 October 1918 (handwritten copy).

be sympathetic to the Allies. If he received a satisfactory response, he would issue an invitation for the Greek Academy to be represented in Paris.

3.6 Second Interallied Conference. Paris, 26–29 November 1918

The conference began on Tuesday 26 November at 10 a.m. in the headquarters of the Académie des Sciences on the quai de Conti. Delegates from ten countries, 46 in all, were present (see Appendix 1).[70] Chairing the opening ceremony, Paul Painlevé, a member of the Académie and former president of the Council of Ministers, outlined the purpose of the conference, which was to create new associations, always with unity as the target in whatever decisions were taken. With an eye on his continuing political duties, now as a member of parliament, Painlevé retreated quickly to the role of honorary president, leaving Schuster to preside in his place, with Picard as vice-president and Durand and Lallemand as secretaries. Schuster's first act was to put forward for discussion two proposals that were to do much to structure the proceedings. One came from the Royal Society as a series of "Suggestions" for the essential first step of appointing an executive committee (EC) that would manage the transition to the proposed International Research Council; the EC's core task would be to prepare an outline plan for the creation of international disciplinary organizations, made up of national councils in the various disciplines, which would then be federated under the general international council.[71] The other proposal, from the Académie des Sciences, was for the discussion of Picard's statutes for the international association for astronomy and Lallemand's for the similar body in geophysics.

Drawing heavily on articles 4, 5, and 6 of the London resolutions and on Hale's ideas, the RS's "Suggestions" were presented on the first morning of the conference by Jeans; with Picard in support, they were quickly agreed to, with the additional decision that the EC would serve as a provisional IRC pending the council's constitutive assembly. The focus then turned to the two sets of statutes. The fact that geophysics in Lallemand's plan brought together geodesy and seismology (both with their leading research centres in Germany) immediately highlighted the implications for the wider world of international politics. Picard's proposed statutes similarly had a political dimension, the more so as he saw them as relevant not only to astronomy but also to "the associations that we shall in all probability go on to

[70] Belgium (3), Brazil (1), France (13), Great Britain (9), Italy (5), Japan (2), Poland (1), Portugal (1), Romania (4), Serbia (2), the USA (6). The delegates of Greece and Portugal did not arrive in time for the beginning of the conference, although the Portuguese representative arrived before the end, raising the final number of delegates to 47. See typescript in Baillaud Papers and (Picard & Lacroix, 1918b, 934–937).

[71] Schuster Papers: IRC documents, Paris Conference file, item 19.

form".[72] As Picard had anticipated, most of the articles passed without difficulty.[73] Among the sticking points, however, were two that he chose to highlight: the number of votes to be granted to each country and the always sensitive matter of the admission of the neutrals, both questions that applied to associations in all the sciences, including the IRC itself.[74] The discussion that ensued was emphatically not about astronomy alone.

The afternoon's exchanges began with Picard's advocacy of a merger of the existing associations active in astronomy. As he argued, these individual bodies, 19 of them by his count, could be "fused together so as to form a single association with the title of, for example, 'Astronomical union'".[75] The result would be a new type of association, described as "mixed", in the sense that it would be eligible to receive governmental support but without being subject to a diplomatic convention. In maintaining this balance, the principle that membership of a country's national committee in a discipline could not be had simply by the payment of a subscription was crucial in winning the confidence of governments. So long as delegates came from academies and other learned societies or the public services, Picard believed that some governmental representatives might well be tempted to join and so provide an appropriate measure of oversight.

The core of Picard's proposal was the set of nine articles headed "Union astronomique" that we reproduce in Appendix 3. The first article defined the nature of the association (the term "association" being used to describe a union or other similar group) and stipulated that the new body would be open only to "nations currently at war with Germany". In pursuit of its objective of advancing astronomy and related sciences and practices, the association would be governed by an international committee composed of representatives of national committees in the discipline (article 2). Articles 3, 4, and 5 set out the composition and functions of these national committees, with article 5 broaching the delicate question of the number of delegates they would be allowed to send to the international committee; this would be determined by the population of the nation in question. Articles 6 and 7 concerned the responsibilities and activities of the committee, followed by article 8, which outlined the procedure for voting on the admission of "non-belligerent" nations, and article 9, governing any future modification of the statutes.

Discussion of article 1 soon exposed the complex ramifications of any move to act against Germany. Which, as some asked, were the countries at war with Germany, and why did Picard's proposal only mention Germany, when the resolutions approved at the London conference spoke rather of the Central Powers as a whole? There were complications, too, with regard to Eastern Europe, where Romanians

[72] ADS: register 1 J25, second session, 26 November 1918, 14. Register 1 J25 is currently being reclassified.

[73] ADS: register 1 J25, second session, 26 November 1918, 8–15.

[74] With the original French text not yet entirely available in English, Jeans proposed that article 5, on the allocation of votes, should be translated immediately.

[75] ADS: register 1 J25, second session, 26 November 1918, 9.

were fighting Hungarians, Serbia was still at war with Bulgaria, and the territory that might be ceded from Germany to Poland had yet to be negotiated. In the same region also, the collapse of the Austro-Hungarian Empire had resulted in an Austria more closely tied to Germany, while Greece, neutral at the beginning of the war, had ended it on the side of the Allies. And all this was to say nothing of the still unresolved fate of the Ottoman Empire and the wishes of delegates with a variety of special interests, from the Balkan countries and Poland for example, who expected to have their say. The German question had become far more than a matter for the preponderant voices of the core founding nations.

Article 8, stipulating that "non-belligerent" nations could only be admitted if they received three-quarters of the votes, was another that provoked keen debate. Which were these nations? Were they simply those, loosely described as "neutrals", that had taken no part in the war, or perhaps only those that had broken off diplomatic relations with Germany? The case of several Latin American countries, far removed from the conflict, raised just such issues. And what of Greece, or of Poland and Czechoslovakia, both newly created with Allied support? Was Picard correct to regard these three nations as "belligerents"?

Allied attitudes to the scientifically more consequential neutrals were similarly far from straightforward. Sweden, Norway, Denmark, and the Netherlands were regarded with particular suspicion for the links they had maintained with Germany throughout the war, with the attendant danger, as Picard often warned, that they provided a route by which Germany might re-enter the international community. Hence his insistence that their admission to the Allied associations should be strictly by invitation; there should be no mechanism for application, and any future decision to admit them would have to receive the stipulated three-quarters of the votes, rather than the two-thirds favoured by certain of the more accommodating Allies. In the end, after two sessions dominated inconclusively by questions bearing more or less directly on policy towards Germany and the neutrals, it was decided to avoid direct engagement with the issues by simply listing, as in London, the 16 countries that would have the right to join the IRC.[76]

The voting regulations specified under article 5 also divided opinion. On many matters, the principle of "one country, one vote" would plainly not work. The astronomical community in the USA, for example, deserved greater voting power than those of small nations with little input into astronomy. But if a country's population was a determining factor, did it follow that China should receive a much larger number of votes than, say, the Netherlands, with its small population but high standing in world astronomy? Or was it enough to say that in practice the Netherlands would gain extra votes by being able to count on those of its colonies? Likewise for other colonial powers (Britain and France in particular), might the worst anomalies be avoided through the votes granted to dominions and protectorates, which it was finally agreed to treat as sovereign states on the grounds

[76] The 16 countries were Australia, Belgium, Brazil, Canada, France, Greece, Italy, Japan, New Zealand, Poland, Portugal, Romania, Serbia, South Africa, UK, and USA.

that they had parliaments of their own? Such questions were not new; they had been raised in discussions of conventions governing international bodies before the war, and article 5 only partially resolved them. In drafting the article in terms of numbers of delegates, rather than votes, and tying those numbers to national populations, Picard was avoiding engagement with issues that risked delaying a decision.[77] It was a strategy that bore the mark of Picard's determination to get things done. Faced in the debate with the procrastination of certain delegates who tended to refer delicate questions back to the executive committee, he took a firm stance. They were there, he reminded them, to get involved, specifically in the creation of new associations, beginning with the astronomical union. And time was short: action had to be taken before the Germans took the lead once again.

On 28 November, it was Lallemand's turn to present the draft statutes for the international union of geophysics (IUG). These were not the first statutes that he had put before the geophysical community. In 1917, following the lapse of the diplomatic convention governing the IGA on 31 December 1916, he had already proposed a reorganization involving the Allies alone and in December 1917 had even prepared a draft convention for consideration by the Ministry of Foreign Affairs and subsequently by Allied geodesists.[78] Lallemand's priority, then and throughout, was that any new association should have no links with "Prussia" and, to that end, should exclude the IGA, now reduced to an association of the neutral countries led by Raoul Gautier of the Geneva observatory.[79] Gautier's response was predictable and firm (Gautier, 1918). He saw the move as premature and flawed by its disregard for the neutrals. His view that any reorganization of the IGA should wait until two years after the war was one that Lallemand, who shared Picard's anxiety about the threat of renewed German dominance, could not countenance.

The statutes that Lallemand presented to the Paris conference were closely aligned with Picard's, and they conveyed the same priorities. But they embodied one major departure. This was that the new union would be divided into two sections: one for geodesy, the other for the rest of geophysics. Each section would have its own president, one of whom would be designated president for the union as a whole by a procedure yet to be determined. The unusual structure invited debate. Why create a single union, with two sections, rather than two separate unions? And was there not a case for attaching geophysics to astronomy? Or might that risk creating a union that would dominate the others, simply by virtue of its size? There was discussion, too, of the allocation of the budgets for each of the sections and the distribution between them of the sub-disciplines for which they would be responsi-

77 ADS: register 1 J25, second session, 26 November 1918, 22. See (Levallois, 1980, 259–264).

78 See, for example, RS Minutes of Council, vol. 11 (24 January 1918), 280. The draft was published in the *Bulletin de la Société astronomique de France*, 3 (March 1918), 96–97.

79 The neutrals in question—Switzerland, Netherlands, Sweden, Spain, USA (not yet involved in the war)—had agreed on a provisional convention that would expire two years after the end of hostilities, in which transitional period observations would continue to be sent to Potsdam. (Levallois, 1980: 268).

ble, including seismology,[80] vulcanology, terrestrial magnetism, meteorology, tidal research, and oceanography. As with the astronomical union, however, unresolved questions were not allowed to impede a decision, and Lallemand's statutes for an IUG were approved.

Picard and Lallemand had set a remarkable pace in pursuit of their aims for astronomy and geophysics, and their success in securing agreement on their respective sets of statutes had a galvanizing effect on the broader Allied project. A committee to consider the founding of a union for chemistry was one sign of the determination to push ahead in other disciplines (Fauque, 2020). With the plans for the first two unions in place, the focus then turned to outstanding points of debate concerning the International Research Council. These included the suggestion that the staffs of embassies should include permanent scientific and technical attachés; the plan for a bibliography free from domination by German sources; and a proposal for a system of exchanges of researchers between different countries, extending even to the establishment of international laboratories financed by the Allies. Finally, the members of the IRC's executive committee (EC) were elected: Picard as president, Hale, Volterra, and Lecointe as vice-presidents, and Schuster as secretary. It was a choice that forcefully reaffirmed the primacy of the five nations that had taken the lead from the start.

At the concluding session, on the morning of 29 November, the delegates put their signatures to the resolutions that had been approved, in preparation for consultation with authorities in their own countries.[81] Academies and governments, as appropriate, were to be asked to send representatives to meetings of the International Research Council, the representatives being chosen from members of national committees of the kind envisaged in London or a similar body. It would be the duty of the IRC's EC to examine scientific and technical questions that came before it and encourage the appointment of specialized disciplinary committees such as the one for chemistry in preparing proposals for future unions. And the whole process was to be administered from a provisional headquarters at the Royal Society, ready for a third interallied conference that would proceed to the definitive constitution of the IRC as soon as enough national councils had been formed.

Despite the differences of emphasis, the Paris conference delivered at least the appearance of unity that Painlevé had expected of it in his opening address. The statutes of two disciplinary unions had been agreed, and there was broad agreement on the way ahead to other similar bodies, including crucially the IRC itself. Minor amendments apart, the RS's "Suggestions" and the statutes proposed by Picard and Lallemand had provided important foundations for the consensus. This is unsurprising, given that the most prominent voices in the debates were those of

[80] The International Association of Seismology (IAS) had its central bureau in Strasbourg, where many of its instruments were housed, under German jurisdiction, until the armistice.

[81] The minutes and other reports on the Paris conference give an impression of confusion with regard to terminology. The IRC is sometimes taken to represent all the countries invited, sometimes the EC alone. As the ""voice" of the IRC, the EC is often confused with the IRC itself.

Picard, Lallemand, Schuster, and, also on the British side, the physiologist and veteran of wartime poison gas research Ernest H. Starling. By comparison, interventions by Bigourdan, Lecointe, Volterra, and Annibale Riccò, though forceful, were intermittent, as were the occasional contributions by other delegates, notably from Brazil and Romania.

Among the voices that weighed in the decisions taken in Paris, Hale's remains the most difficult to evaluate. While elements from his initial plan, in particular its proposal for countries to be represented on the IRC by members of national research councils, did survive, his idea of integrating applied science for civil or military purposes did not. It is striking, in fact, that while he spoke in all the sessions, he generally did so briefly and only came out of his shell in the closing session, when he spoke to a document entitled "The constitution of the national councils for developing scientific and industrial research", printed and pre-circulated in English though with copies of a French translation available. His intervention, as recorded (with linguistic idiosyncrasies) in the conference report, highlighted the centrality of astronomy as the model discipline for what had been achieved and the tasks that lay ahead:

> In the proposed constitution of an astronomical Union the composition of a national committee to represent astronomy in a given country was suggested, and it was proposed that such a national committee should consist of representatives of the astronomical societies of that country, of representatives of such bodies, for example, as the Bureau des longitudes here in Paris, of public services, which may be concerned, such as those relating to the determination of time or other questions directly relating to astronomy, of government departments, and in general of such organisations or institutions as are most directly interested in the promotion of astronomical research, and in the United States of astronomical observations for scientific or practical purposes. This plan of organizing a National Committee, or for astronomy is precisely the plan which, from the point of view of our National Research Council, would be used in organizing a committee on astronomy of a National Research Council. In other words, our idea of forming a committee relating to a given subject in the National Research Council is precisely similar to that which our French colleagues have proposed for the formation of a national committee on astronomy to take part in an international organization.[82]

Hale's words have the air of a gentle reminder of his personal role and that of his American colleagues in a project that had come to bear the strong imprint of the French. The course of debate, first in London and now in Paris, had owed much to Picard's powerful personality and the natural authority and clarity of expression that characterized all his contributions; he was equally skilled in advancing his own views and in manoeuvring to ensure their acceptance by others. Although the draft statutes for astronomy that he brought to the table in Paris had many loose ends, he succeeded in establishing them as the foundation not only for the statutes of what was soon to become the IAU but also for those of other unions, even for the IRC itself. In the process, the importance of Hale's original contribution was subtly lost from view. And it was all done without even a passing mention of the documents

[82] ADS: register 1 J25, seventh session, 29 November 1918, 2–3. Schuster followed with a translation into French of the essentials of Hale's speech.

that Hale had submitted during the summer. This was no accidental oversight. In the BDL as in the ADS, Picard's aim had been quite explicitly to promote the presence of France on the international stage. To that end, he had arranged for a large and likeminded French delegation to be present at the interallied conferences and articulated what he saw as two overriding missions. One was to ensure the long-term exclusion of Germany, the other was to delay for as long as possible the entry of the neutral nations. Both missions bore precisely the fruit for which he had hoped.

3.7 Preparing for Brussels

On 29 November, after the closing session, the astronomers and geodesists still present in Paris met to agree on a path that would lead as quickly as possible to the founding of the astronomical union (Schuster, 1920: 105).[83] In the interests of efficiency and speed, their aim was to arrange for the constitutive assembly of the union to take place in Paris on the same day, in the second half of May 1919, as the corresponding assemblies for the geophysical union and the IRC. They also decided that the agenda should include the presentation of a number of reports (See Appendix 4). These were to include the work of the central bureaus for time (BIH) and astronomical telegrams (Baillaud), the publishing of a regular digest of the most recent observational data (Bigourdan), and arrangements for determining ephemerides (Dyson) and the administration of the Carte du Ciel (Baillaud) and the Solar Union (Schuster). It was also decided that the Time Bureau and the Bureau of Astronomical Telegrams would function under the direction of the Paris Observatory until further notice (see Appendix 4).

The resolutions agreed at the Paris conference were approved by the ADS on 9 December and published in its weekly *Comptes rendus* (Picard & Lacroix, 1918b). On 4 January 1919, the members of the Académie royale de Belgique (ARB), followed suit with an endorsement of the report on the conference from its delegates. And two months later, on 1 March, it took an important further step by setting up a national research council. In this, it was responding to a recommendation by Lecointe, who announced that, in accordance with a suggestion by the IRC's provisional EC, the constitutive assembly of the Council would now take place in July, in all probability in Brussels, rather than in Paris as had originally been intended. It was a suggestion that the Académie accepted enthusiastically on 5 April.[84]

In London, the Royal Society also took swift action. Following a brief report by Schuster on 12 December, the fellows met in a special session on 16 January at which they insisted that the IRC should move quickly to admit neutral nations. They

[83] Astronomers present: Baillaud, Bigourdan, Deslandres, Dyson, Hale, Lecointe, Riccò, and Schuster (Schuster, 1920: 105; La Vallée Poussin et al., 1919: 72).

[84] ARB 1919: session 1 March, 159, and session 5 April, 215.

also appointed two committees, one of which was to examine the ADS's statutes for astronomy, while the other would deal with the resolutions concerning geophysics. In both cases, it was expected that the deliberations would result in the creation of national committees for the disciplines concerned.[85] Schuster wrote to Picard on 20 January, with the news that the committees had been formed, and to Turner on 8 February, confirming that the statutes proposed by the ADS would be discussed by the committee for astronomy before being passed on for consideration by the Royal Astronomical Society (RAS). The amendments to a surviving draft of Schuster's letter to Turner reveal his mixed feelings about the French proposals. While the draft shows him deleting his criticisms, his original comment that it was a plan "full of obscurities" conveyed what appears to have been a truer sentiment.

In writing to Lecointe on 10 February, however, Schuster did not hold back. He asked Lecointe for his opinion on the French proposal for astronomy, contrasting it with the IUG statutes, which he found better conceived. He was especially troubled by the French idea of a structure based on largely autonomous associations that would only come together in the context of a general assembly administered by a permanent council whose powers and procedures remained undefined. What followed was an unveiled criticism of Picard. As Schuster saw it, Picard had sought to centralize everything. Yet, he had stopped short of centralizing administrative correspondence, to the point that Schuster did not know which communications should go to all the delegates and which to the members of the EC alone.[86] Any reply from Lecointe has not survived. But by 27 February Schuster was ready to propose that the RS Council should authorize the Society's astronomy committee to submit its own proposals for the statutes and for a national committee to the RAS for comment, ready for the RAS to report back to Council.[87] The RAS Council, four of whose members belonged to the RS's astronomy committee, considered the proposals on 14 March and expressed its approval.[88]

As all this was happening, the USA's NRC was undergoing radical reorganization, to adjust to its transition from an element in the national war effort, as defined by Woodrow Wilson's presidential act of 11 May 1918, to a new profile of functions in time of peace.[89] The board of the NRC was preoccupied with the change through

---

[85] RS Minutes of Council, vol. 11 (12 December 1918), 347, and (16 January 1919), 360–361. The members of the committee for astronomy were F. Dyson, A. Eddington, A. Fowler, S. Hough, J. Jeans, H. Newall, R. Sampson, A. Schuster, H. Turner, and the Astronomer Royal for Ireland, Henry Crozier Plummer.

[86] Schuster Papers: Schuster to Picard, 20 January 1919; Schuster to Turner, 8 February 1919, draft typescript; Schuster to Lecointe, 10 February 1919, handwritten copy (in French).

[87] RS Minutes of Council, vol. 11 (27 February 1919), 380.

[88] Minutes of RAS Council meeting, 14 March 1919. We are grateful to Dr. Sian Prosser, Librarian and Archivist of the Royal Astronomical Society, for consulting the minute book on our behalf during the recent restrictions on access.

[89] *Proceedings of the National Academy of Sciences* [hereafter *PNAS*], 5/5 (1919), 188; see note 32. For the reorganization of the NRC, see the minutes of the joint meeting of its Executive Board and the Council of the NAS on 11 February 1919; ibid., 181–87.

the early weeks of 1919, and it was not until 11 March that it turned to the question of the future unions, responding now to a cable from Schuster (*PNAS*, 1919: 189–193). At a joint meeting of the Council of the NAS and the Executive Board of the NRC, Hale reported that the NAS had taken a lead by approving the creation of an "American Section of the proposed International Astronomical Union" (hereafter AmS) in accordance with the decisions taken in London and Paris and the wishes of Hale himself (whose encouragement led to the organization of American sections in other disciplines as well). The new AmS was given temporary status, under the physical sciences division of the NRC, pending the third interallied conference, planned for July.

The AmS had set to work quickly and held its first meeting in Washington three days earlier under the chairmanship of Hale, with Joel Stebbins (the director of the University of Illinois Observatory and representative of the American Astronomical Society) as secretary (*PNAS,* 1919: 193–196). At the meeting, it had appointed an interim executive committee to "consider the general matter of business, appoint all committees, and add additional members to the Section". It had also approved the London resolutions, asked for the neutrals to be admitted as soon as the peace was signed, and endorsed the role of a future union that would "take the place of previous international bodies in astronomy". As Frank Schlesinger, the director of the University of Pittsburgh's Allegheny Observatory, insisted, the kind of astronomical tasks that required international cooperation remained to be determined. He divided these into four categories: (1) Work too extensive to be undertaken by any one nation, the Carte du Ciel being an obvious example; (2) Observations calling for a wide diversity of locations, such as latitude variation, solar phenomena, and variable stars; (3) Collaborations involving international conventions on such matters as nomenclature, standards, and units; (4) Liaison activities aimed at avoiding a duplication of effort, for example in the preparation of ephemerides, data for almanacs, and abstracts. The result of Schlesinger's perception of the way ahead was the creation of 18 specialized committees (a 19th was to be created at a later date) that were asked to draw up a report for passing on to the US astronomers appointed as delegates to the third interallied conference (*PNAS*, 1919: 189–196).[90]

In France, the IRC and the unions were rarely mentioned at the BDL after Picard's report on the Paris conference on 4 December, and this remained the case during the first half of 1919. Even so, the determination to establish the unions of astronomy and geophysics without delay was undiminished, sustained as always by the fear that Germany might make a similar move itself.[91] In the BDL as in the ADS, however, the situation in Strasbourg, now occupied by French troops, remained an even higher priority. On this, Picard went so far as to approach the

[90] For the nineteenth committee, see (Campbell & Stebbins, 1920, 351). The Section voted for Hale, as its chairman, to "appoint a Committee on Committees, to act temporarily as an Executive Committee" (id.: 194); the new committee was composed by W. W. Campbell (chairman), C. G. Abbot, E. W. Brown, F. Schlesinger, and J. Stebbins. See also (Lattis & Lattis, 2019: 410).

[91] BDL, Procès-verbaux, 4 December 1918.

French Ministry of Foreign Affairs, which authorized the sending of a delegation to inspect the Central Bureau of Seismology and the currently sequestered instruments belonging to the IAS. The French preference was for the Central Bureau to stay in Strasbourg under French control. Lecointe, a corresponding member of the BDL, backed the French position, urging that the Bureau should assume responsibility for the Strasbourg Institute of Seismology and that a statement to that effect should be sent to the Allies and neutral nations. It was also intended that the BDL should remain the main driving force in the telegraphic determination of longitudes, a point that the Bureau made clear to the IRC, through which it hoped to persuade Hale, Dyson, and Schuster to keep it informed about the measures to be taken in their own countries (Schiavon & Rollet, 2021b). Finally, the BDL prepared questions that would be put to the representatives of the countries beginning to gather in Paris for the Peace conference. These detailed exchanges between Hale, Dyson, Shuster, and the BDL continued through to the Brussels conference in July.

When the IRC's EC met in Paris between 20 and 24 May 1919, it did so without Hale, whose duties at Mount Wilson made it impossible for him to attend. Returning to the measures agreed in Paris, the EC decided that the draft statutes for the IRC, IAU, and IUG were sufficiently advanced for them to go forward for approval at the Brussels conference (see Fig. 3.15). But during the meeting, on 21 May, Picard alerted the BDL that the EC had received the alternative British statutes for the IAU. While he insisted that the British and French texts only differed in matters of detail and that agreement would present no difficulty, he recommended that the French delegation in Brussels should be sufficiently large to "cope with any eventuality".[92] The statutes for the geophysical union also encountered a snag, following an intervention by Volterra, who insisted that the word "geodesy" should appear in the title, which now became Union géodésique et géophysique internationale (UGGI). The problem with regard to the chemical union was even trickier. Here the EC had to face two quite distinct proposals. One had been submitted by the International Committee for Cooperation in Chemistry, appointed under the chairmanship of Albin Haller at the Paris Conference in November 1918 (Fauque, 2011, 2020). The other had been drafted at an international conference of chemists chaired by Charles Moureu in April 1919, also in Paris but independently of the ADS. The plan for a further meeting of these chemists to discuss the matter, in London between 15 and 18 July, cut across the schedule for the Brussels conference, the first day of which had to be put back from 16 to 18 July, in the hope that by then only one proposal would come forward for consideration (Schuster, 1920: 76). In response to demands from several countries, the EC had by now also agreed a list of the neutral countries that might be eligible for invitations to join at a later date (Schuster, 1920: 77).[93]

[92] BDL, Procès-verbaux, 21 and 28 May 1919.

[93] The countries listed were Denmark, Spain, Monaco, Norway, Netherlands, Sweden, Switzerland, Czechoslovakia, and Finland. In July, Chile, China, Siam, Argentina, and Mexico were added (Schuster, 1920: 8).

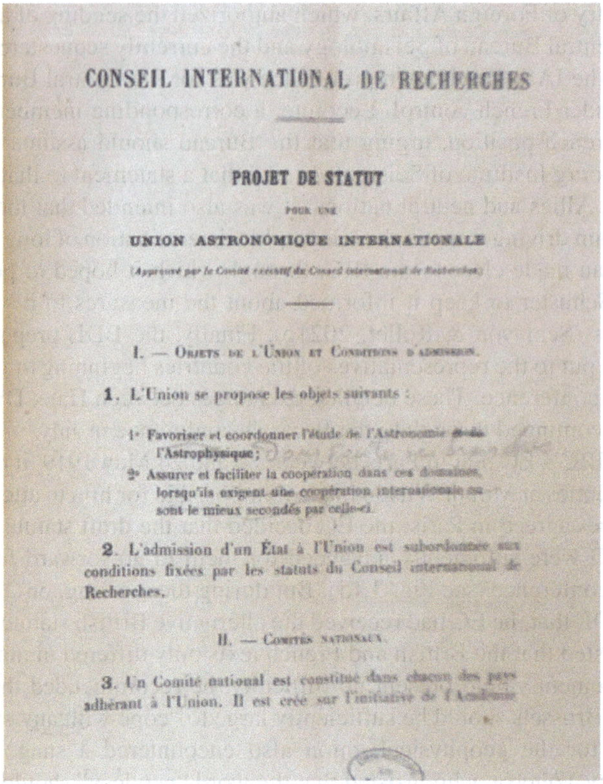

CONSEIL INTERNATIONAL DE RECHERCHES

PROJET DE STATUT

POUR UNE

UNION ASTRONOMIQUE INTERNATIONALE

(Approuvé par le Comité exécutif du Conseil international de Recherches)

I. — Objets de l'Union et Conditions d'admission.

1. L'Union se propose les objets suivants :

1° Favoriser et coordonner l'étude de l'Astronomie et de l'Astrophysique ;

2° Assurer et faciliter la coopération dans ces domaines, lorsqu'ils exigent une coopération internationale ou sont le mieux secondés par celle-ci.

2. L'admission d'un État à l'Union est subordonnée aux conditions fixées par les statuts du Conseil international de Recherches.

II. — Comités nationaux.

3. Un Comité national est constitué dans chacun des pays adhérant à l'Union. Il est créé sur l'initiative de l'Académie

Fig. 3.15 IAU proposed statutes. May 1919. First page. Source Baillaud Papers, "Documents préliminaires". Photo D. Fauque/©Bibliothèque de l'Observatoire de Paris

These were typical of the uncertainties that accumulated in the weeks preceding the conference, to say nothing of the absence of proposals as polished as those for astronomy and geophysics. The one exception was a proposal for physics, presented by the RS using the model of the statutes of the IAU (Schuster, 1920: EC report, 76). The French, in particular, responded anxiously, and Picard's comments to the BDL about the RS's alternative statutes for astronomy on 21 May only heightened the unease. Lallemand's worries were evident on 11 June in his request for Schuster to send him the statutes for both astronomy and geophysics, so that he could discuss them with colleagues well before the conference.[94] Despite the sense of unreadiness, on 12 June Schuster informed academies and other relevant bodies that the IRC, IAU, and UGGI would be formally established, and their inaugural meetings held,

[94] Schuster Papers: Lallemand to Schuster, 11 June 1919, received on 16 June 1919.

at the Brussels conference.[95] The Americans too were resolute. On 23–24 June the AmS for astronomy gathered in Washington for a last meeting before its departure for Europe. At this, it examined the committee reports it had received and agreed on the lines it would take in Brussels, notably in support of the exclusion of the Central Powers and the speedy admission of the neutrals (Campbell & Stebbins, 1920: 349–350). Leaving New York a week later, the American delegation reached London on 7 July (Lattis & Lattis, 2019: 410).

3.8 Third Interallied Conference. Brussels, 18–28 July 1919

By the time the US delegation moved on to Brussels on 17 July, it had had ten days in London, in which time it would have discussed with British colleagues the key issues that were likely to come up at the conference. Although the recent difficulty of access to archival sources, including the papers of certain key individuals, has prevented our pursuing such contacts as there must have been, we can be sure that the Americans, like the other leading delegations, arrived in Brussels well primed. They were in for a lengthy stay, during which, as they knew, international politics, as well as science, would be relentlessly to the fore. That was immediately evident in the theatricality of the formal proceedings. The opening session, on 18 July, took place in the presence of Albert I, king of the Belgians, and the Minister for Arts and Letters, and in a setting, in the Palais des Académies, which conveyed the pain of Belgium's wartime suffering. The building, badly damaged during the German occupation, was now available for occupation once again. But photographs, knowingly displayed, brought home to the more than 200 delegates from 12 nations (Greece having been unable to send a representative) the obstacles that the country had had to overcome in mounting the conference (see Appendix 1).

It was one of the ironies of the Brussels conference that illness prevented Picard, who had played such a central role in London and Paris, from attending; Alfred Lacroix, the ADS's other permanent secretary, presided in his place. With the formalities of protocol and welcome out of the way, Schuster presented the report of the EC, and Lacroix went on to guide discussions, not all of them easy, to a successful conclusion on 25 July. It was then, in a plenary session, that the statutes of the IRC were finally ratified, after only minor modifications. If any delegates had lingering reservations about the exclusion of the Central Powers and the cautious policy towards the neutrals, these could scarcely have survived organized visits to the sites of conflict and destruction in Louvain and Dinant: on these two fundamental points of principle, unanimity reigned. With the IRC's statutes endorsed, it remained for as few as three of the five "founder" nations to join in order for the convention,

[95] RS Minutes of Council, vol. 11 (12 June 1919), 395–396. See also *PNAS*, 1919, *5/10* (17 June 1919), 472–473.

scheduled to last for twelve years from 1 January 1920 to 31 December 1931, to come into force.

In the intervals between the conference's five general plenary sessions, specialists working to establish unions in such disciplines as scientific radiotelegraphy, mathematics, biological sciences, and physics were busy. The astronomers, too, were active. Under the chairmanship of Lecointe, they met, in more or less large groups, on at least five occasions to discuss the IAU's statutes, its internal structure, and the make-up of the Bureau. In the printed text distributed to them in Brussels, the provisional statutes were arranged under eight headings.[96] The 20 articles followed the outline plan, common to all the unions though conceived with special reference to astronomy, that had been agreed at the Paris conference a few months earlier; to this, administrative clauses appropriate to the discipline concerned were added. If we compare these provisional statutes with the ones approved at the end of the Brussels conference, however, certain revisions do appear important. Heading IV, "Sections of the Union", was reduced to "Standing Committees" (in the French version: "Commissions"), and the first two articles under this heading were substantially altered. The original article 7 had read: "The General Assembly may entrust the direction of work in certain branches to sections whose powers it has the right to determine". In the definitive statutes, however, this became: "The Union appoints Standing Committees for the study of special branches of astronomy, the encouragement of collective investigations, and the discussion of questions relating to international agreements or to standardization. The Standing Committees shall present reports on their work to the Union" (Schuster, 1920: 228).

The notes taken by Alfred Fowler as the IAU's secretary general bring out the import of the amendments (Fig. 3.16). As his personal record of the meetings shows, each section was initially conceived as embracing "a group of subjects" (Fowler's underlining). Yet finally, again in Fowler's words, "it was decided to carry on the work of the Union through a number of committees, one for each subject".[97] The thrust of the change was significant, in that that the *possibility* of action by the Union was replaced by an *expectation* of action. An alteration to article 8 pointed to a similar tightening of control: the original right of each committee to appoint its own bureau gave way to a requirement for each of them to submit a proposed list of members to the IAU's EC, which would then pass the list on for approval by the General Assembly. The adjustments signalled a clear move from the freedom of action embodied in the draft proposals (strongly influenced by Baillaud's original text of November 1918) to a more hierarchical structure in which power was centralized in an EC committed to promoting the Union as a decisive force in the discipline. Giving the sections undue freedom ran the inevitable risk of a dispersion of effort, and it was the goal of efficiency and coherence that prevailed. A later report to the US NRC by Campbell reveals how challenging such questions had proved to be:

---

[96] Baillaud Papers: draft statutes for the IAU, in preliminary documents for Brussels Conference, 13–19.

[97] IAU Archives: Fowler, folder I-1.A.

Fig. 3.16 Alfred Fowler (1868–1940). https://commons.wikimedia.org/wiki/File:Alfred_Fowler.jpg. Bain News Service. Public domain. Via Wikimedia Commons

> Should it [the union] be organized as a great astronomical society, comprehending all divisions of astronomical science in a somewhat general or indefinite manner, or should it consist of working committees, each committee concerned with some definite line of astronomical research? Should there, for example, be a great division of the union relating to the solar system, or, on the contrary, several comparatively small committees occupying themselves responsibly with such definite subjects as solar rotation, eclipses, solar radiation, etc.?... (Campbell & Stebbins, 1920: 352).

Discussions leading to minor changes on 21 July and endorsement on the following day by a meeting of the astronomers present culminated, on 24 July, in approval of the IAU statutes by the IRC's EC. The model of specialized committees, as proposed by the AmS and probably supported by the British contingent, had won out, and it now only remained to establish the membership of what it became increasingly common to refer to as "commissions", though the term "standing committees" was used in the English translation of statutes (Lattis & Lattis, 2019: 410). This was the main business of a separate special session on

24 July. It was decided that there would be 32 commissions in all, 19 of them with titles corresponding to the committees of the AmS. Among the other 13 were commissions for key enterprises that had functioned before the war—astronomical telegrams, the Carte du Ciel, and the International Time Bureau (BIH, in its French abbreviation)—as well as a new commission for relativity, designated "commission 1" (Schuster, 1920: 165–168). On the following day (25 July), delegates proceeded to the appointment of the commissions' bureaus. The names of those appointed (only four of them women) to a total of 284 positions underline the dominance of the USA (with 29% of the positions), Great Britain (27%), and France (26%). Italy (7.4%), Belgium (4.9%), and Japan (1.8%) lagged far behind, while all the remaining nations together accounted for only 2.8%. Even the small number of positions allocated to these nations had the effect of reinforcing the influence of the "big three". Two went to Poland and one to Czechoslovakia, both of them countries firmly in the French sphere of interest; three to the British dominions of Canada (one) and Australia (two); and one to India, which was administered by Britain. Only the single position for Russia, a country that the Allies hoped to retain in their orbit of influence, broke the mould.

The presidencies of the commissions followed the same pattern: eleven (34%) went to the USA, nine to Britain (28%), and eight to France (25%), a total of 87%. Just two went to Belgium (though with the additional presence of Cardinal Désiré-Joseph Mercier as honorary president of commission 32, for calendar reform) and one each to Italy and Japan.[98] The presidencies, like all the appointments, were sufficiently prized to attract leading astronomers. Among notable names were Eddington and Hale, presidents of the commissions on relativity (n°1) and the solar atmosphere (n°12), respectively; Baillaud, president of commission 5 on the analysis of publications and bibliography; and Lecointe, president of commission 6 on astronomical telegrams. There were some, too, who occupied more than one presidency. These included Bigourdan, president of commission 28 on nebulae in addition to 32 on calendar reform, and Turner, president of commission 17 on lunar nomenclature, who was persuaded by Baillaud to take on the presidency of commission 23 on the Carte du Ciel as well. The result was a strikingly large measure of continuity, with several members of the AmS, including the presidents of some of its committees who moved on to similar positions in the corresponding commissions of the IAU.[99]

On Saturday 26 July, at a final plenary meeting of the astronomers present, the composition of the commissions and their bureaus was formally approved. So, too, was the membership of the bureau of the IAU itself, with Baillaud as president, Campbell (head of the American delegation), Dyson, Lecointe, and Riccò (replaced, after his death in September, by Antonio Abetti of the Arcetri Observatory) as its

[98] Although the appointment of Mercier, a leading champion of Belgian resistance to German occupation during the war, had profound political significance, the work of the commission was led by Bigourdan as president.

[99] For further details, see (Montmerle, 2019).

four vice-presidents, and Fowler as secretary general. Again, the positions were shared between the five "founder" nations, though with the proviso that a fifth vice-presidency was to be kept vacant for an astronomer from a neutral country as soon as one of those countries joined the Union.

The continuity that was formalized in these decisions did much to ensure a coherence of purpose in the transition to a new order for astronomy in which important prewar functions had to be maintained, though now in a disciplinary community from which the Central Powers were excluded. The challenge was addressed in a series of reports on specific areas of the discipline that delegates in Brussels were asked to consider (see list, Appendix 4). Behind these, there was no mistaking the hand of Baillaud, who had been prominent throughout the war in trying to sustain existing international activities. As someone whose involvement in such activities went back many years, Baillaud took it upon himself to submit reports on three areas in which he had taken a particular interest: the Carte du Ciel, the determination of time and longitudes, and astronomical ephemerides. In all three activities, the changed circumstances made review unavoidable. The prewar list of observatories contributing to the Carte du Ciel, for example, included Potsdam, now excluded (Chinicci, 2022). Facing the same problem in the determination of longitude by telegraphy, Baillaud seized the opportunity of supporting the efforts of Gustave Ferrié to have the Eiffel Tower used for the transmission of signals. Likewise in the French interest, his coordinating action, not least in launching an inquiry into the transmitters used by the Allies during the war, had the strategic effect of giving the Paris Observatory a central position in the discussions (Capitaine, 2022; Schiavon & Rollet, 2021b: 267–317).

In his insistence on the importance of continuity between past and future initiatives, Baillaud was not alone. His ephemerides report, which made much of decisions promoting international collaboration that had been taken at conferences in 1896 and especially 1911, was complemented with one on wartime ephemerides work by his friend Dyson. The new material that Dyson added included information about a proposal to substitute "the Civil day, beginning at Midnight, for the Astronomical day, beginning at Noon, in Astronomical publications" that had been raised and warmly supported at a conference on timekeeping at sea held in London in 1917 (Schuster, 1920: 103). Now, with the German national bureau excluded, it had become a matter of urgency to organize a conference at which this and other pre-1918 conventions would be reviewed. A report on the ISU by Schuster similarly stressed the need for continuity adjusted to the new political realities, while in his report Bigourdan, too, evoked prewar precedents in his plan for an international publication, in English, French, and Italian (not German), which would circulate astronomical news rapidly and lend itself to being collected as an annual volume, complemented with bibliographical information in the form of a card catalogue. The model was unmissably the bibliographical initiative that Paul Otlet, the Belgian founder of the Institut international de bibliographie, had established in Brussels shortly before the war, although Otlet's name was not mentioned.[100]

Together, the reports presented a compelling account of prewar international collaborations that had continued to function more or less effectively during the war and that it was now essential to bring under the authority of the new union. The astronomers in Brussels responded positively and turned to the task of determining the way forward with the two thorny questions of time measurement and astronomical telegrams, both of which were the subject of formal international conventions. The future of the International Time Convention (Convention Internationale de l'Heure) raised especially perplexing problems, as Baillaud's historical retrospect made clear. The fact that the convention, agreed in 1913, had never been ratified diplomatically left the BIH in Paris without legal status; even though the work of the Bureau had continued throughout the war, with Baillaud as director, it was now an urgent matter for the convention to be amended in a way compatible with present circumstances. In Brussels, a Commission Internationale de l'Heure was substituted for the Association Internationale de l'Heure that had been established at the founding conferences of 1912 and 1913 in Paris. The upshot was that the new body's headquarters would remain at the Paris Observatory, where the BIH would pursue its work and publish its results under the authority of the Commission, with its director reporting to the Commission and its budget coming from the IAU (see Capitaine, 2022)

No less pressing was the future of the Central Bureau for Astronomical Telegrams, located before the war in Germany. The purpose of the Bureau was "to receive, centralize, and dispatch information concerning astronomical discoveries, observations, and calculations" (notably ephemerides of the minor planets) (Schuster, 1920: 234). Its status had come up in London and Paris, and again when the astronomers who met immediately after the Paris conference (November 1918) had suggested that, for the time being, the Bureau should be administered by the French. In fact, the war itself had already seen a number of changes. One had been the move of the Bureau from Kiel to Copenhagen, which had brought the work under the control of Elis Strömgren, the director of the Copenhagen observatory; this had particularly troubled the French and the British, who accused Strömgren of underhand cooperation with the Germans.[101] There had been changes, too, in the preparation of the ephemerides, largely a prewar responsibility of the Kiel Bureau. Baillaud's arrangement for some of the calculations to be undertaken by provincial

[100] Otlet, however, did attend the Brussels conference, chairing meetings to consider the plan for an international union for bibliography and documentation (Schuster, 1920: 40–42; 217–221; 277–281). On his bibliographical institute, see (Wright, 2014) and the website of the Mundaneum in Mons, which perpetuates the legacy of Otlet's original Mundaneum in Brussels; https://artsandculture.google.com/partner/mundaneum

[101] See BDL, Procès-verbaux, 19 and 25 June 1918, also Laguens (2019, 2021, and 2022). It was a sign of a tentative easing of tensions that Strömgren resumed the directorship of the Bureau in 1922, despite having been elected president of the German Astronomische Gesellschaft in the previous year, a position he was to hold until 1930 (Wielen, 2019 and 2022).

observatories in France had been a way of allowing the Allies to break free from the German stranglehold. But the work had proved a heavy burden and reinforced the need for a conference at which the Allies would distribute the tasks among themselves.[102] In Brussels, it was decided that the Bureau would be moved to the Belgian Royal Observatory at Uccle, where it would be supervised by Lecointe, newly appointed as the observatory's director general. The decisions went on to the final plenary meeting of astronomers on 26 July, where they were voted on and adopted, in the form of two sets of internal regulations cast as statutes: one for the International Time Convention, formally confirming Paris as its headquarters in association with the BIH (article 2), and the other for the Central Bureau for Astronomical Telegrams (Schuster, 1920: 169–172, 232–235). Both were quickly endorsed by the IRC's EC immediately afterwards.

While the published *Reports of Proceedings* of the Brussels conference summarize the decisions concerning astronomy well enough, they convey nothing of what was said. But Baillaud's speech to astronomers at their plenary session on 26 July does offer suggestive glimpses of the exchanges over the preceding few days (Schuster, 1920: 16–19). It was, in effect, a closing address. After an historical sketch of international relations in the years between the transits of Venus in 1874 and 1882 and the war, Baillaud stressed the feelings that would prevent a renewal of contact with the scientists of the Central Powers even though the hostilities were over. A significant thread in the speech was his affirmation that the idea of forming the IRC was Hale's, and that it was the IRC that had set in train the next step of founding the new disciplinary associations. He also saw it as important to insist that the astronomical union was the first of these associations and that it had grown from a proposal by Hale at the London conference in October 1918. Like any source of this kind, Baillaud's speech has to be read with caution. In the common manner of the genre, it was a considered reconstruction of past events, structured around the role of a founder, in this case Hale, whose vision of a collective, unifying project was driven equally by "a similar passion for science and a similar desire for union" (Schuster, 1920: 18). The idea of a union was driven home by Baillaud's repeated use of "you" when addressing his fellow astronomers. It was "you", he stressed, who set up commissions, chose their presidents and members, and selected those who were to serve on the executive committee, and it was by sharing the results of "your" labours that further advances would be achieved. Among those he thanked in his closing acknowledgements, he made special mention of Lecointe, whose clarity of thought, high intelligence, and unfailing cordiality had allowed so much to be achieved in so short a time. Strikingly, and significantly, there was no mention of Picard.

---

[102] See Appendix 4: Baillaud, Annexe X. See also BDL, Procès-verbaux, 10 October 1917.

When all the delegates to the whole conference met for a final plenary session at 10 a.m. on Monday 28 July, they endorsed the creation of three unions: the IAU, UGGI, and IUPAC, subject to subsequent approval by the relevant authorities in countries wishing to join. Although the draft statutes of some other unions were approved, the limited number of national delegates present and the complexity of ongoing negotiations within disciplinary groups meant that there were too many unresolved questions for these statutes to be formally endorsed: the statutes for mathematics, scientific telegraphy, biological sciences, and physics had this interim status, all of them modelled on those of the IAU. Despite the unfinished business, there was a sense of a job well done. It remained only to re-elect the IRC's executive committee (Schuster, 1920: 8), to rehearse the pleasantries of thanks (mainly to the Belgian hosts) and farewell, and in an act of special homage to recognize the particular contribution of the IRC's absent vice-president, Hale. A telegram that Alfred Lacroix agreed to send to Hale expressed the conference's widely shared gratitude: "To the illustrious astronomer, instigator of the International Research Council, the congress gathered in Brussels expresses its admiration and gratitude" (Schuster, 1920: 9). A manuscript note in Baillaud's papers suggests (unsurprisingly) that it was he who composed the warm words (see Fig. 3.17).[103]

Amid the general satisfaction, there was no discipline for which the conference had been a more decisive event than astronomy. Astronomers had taken part in a constitutive general assembly for their union, agreed on statutes and a structure for the IAU that would last for several years to come, and seen astronomy occupying centre stage in much of the wider debate. It was telling that 59 of the 286 pages of the *Reports of Proceedings* of the conference (21%) concerned astronomy, compared with only 14 pages (5%) recording the general plenary sessions. With the IAU bureau and commissions in place, attention turned quickly to the future. At 12.30, straight after the conference's concluding plenary session, the union's Executive Committee met in the Palais des Académies to finalize the make-up of the commissions. Fowler was deputed to convey instructions to the chairmen, who would keep members informed about the allocation of tasks to be done. In the anti-German spirit of the day, it was decided that English or French would be the union's languages, with no requirement for translation.[104] Looking further ahead, there were thoughts, too, of the next general assembly, scheduled to take place in 1922 at the same time as the general assembly of the UGGI, whose formation (albeit with its different, two-section structure) had closely paralleled that of the IAU during the Brussels conference. The venue, proposed by Riccò, would be Rome, with Cambridge as an alternative if the plan for Rome did not materialize.

[103] Baillaud Papers: File Conférence de Bruxelles.

[104] IAU Archives: Fowler, folder I-1.A.

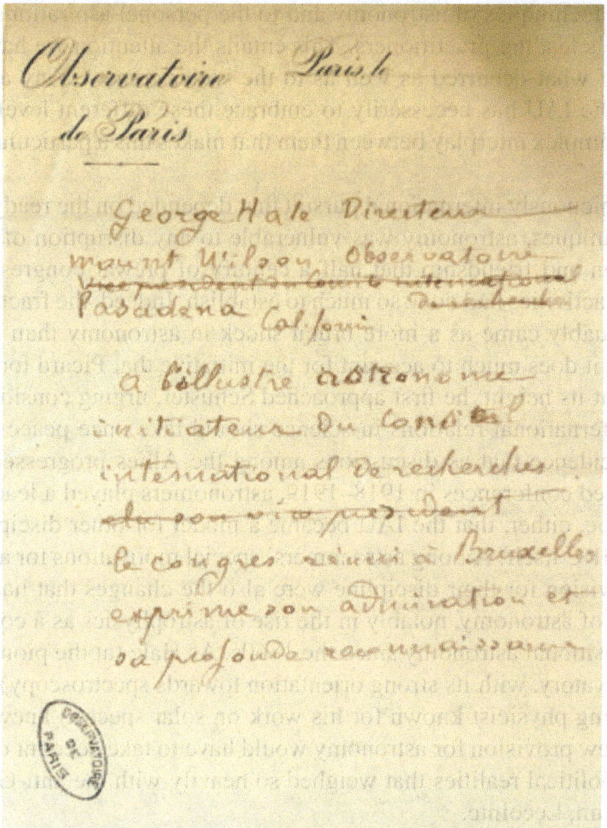
Observatoire de Paris

Paris, le

George Hale Directeur
Mount Wilson Observatoire
~~Vice président Conseil international~~
Pasadena Californie

A l'illustre astronome
initiateur du Conseil
international de recherches
~~et son vice président~~
le congrès réuni à Bruxelles
exprime son admiration et
sa profonde reconnaissance

Observatoire de Paris

Fig. 3.17 Baillaud's draft of telegram sent to Hale, Brussels, 28 July 1919. Source Baillaud Papers. Photo D. Fauque/© Bibliothèque de l'Observatoire de Paris

3.9 Conclusion

The fact that, a century on, the IAU looks to 28 July 1919 as the date of its foundation can easily obscure the tortuous train of events that culminated in the decisive vote in Brussels. Like other early unions, the IAU was fashioned within a context profoundly marked by the war and amid a determination, shared by all the Allies, that the Central Powers, and Germany in particular, should be excluded from international science for a period that remained to be determined. But the course that led to the founding of the IAU was also subject to considerations specific to the practices and techniques of astronomy and to the personal aspirations of certain of the discipline's leading practitioners. This entails the attention we have paid to the fine details of what occurred as well as to the wider context. Any account of the founding of the IAU has necessarily to embrace these different levels of analysis, and it is the complex interplay between them that makes this a particularly rewarding case history.

As a conspicuously international pursuit that depended on the ready exchange of data and techniques, astronomy was vulnerable to any disruption of the networks of cooperation and friendship that half a century of prewar congresses and other transnational activities had done so much to establish. Indeed, the fracturing of those networks arguably came as a more brutal shock in astronomy than in most other sciences, and it does much to account for the initiative that Picard took when, with the war still at its height, he first approached Schuster, urging consideration of the shape that international relations in science should take once peace was restored. It is no coincidence that as discussions among the Allies progressed through the three interallied conferences in 1918–1919, astronomers played a leading role, it is no coincidence, either, that the IAU became a model for other disciplinary unions and even the IRC itself. Among astronomers' special motivations for a review of the post-war provision for their discipline were also the changes that had come about in the profile of astronomy, notably in the rise of astrophysics as a competitor with traditional positional astronomy since the 1860s. As Hale (at the pioneering Mount Wilson observatory, with its strong orientation towards spectroscopy) and Schuster (a wide-ranging physicist known for his work on solar spectra) knew with special clarity, any new provision for astronomy would have to take account of this, as well as the more political realities that weighed so heavily with the anti-German Picard and, in Belgium, Lecointe.

Although the path taken by the IAU, in particular the form of its statutes, was such a powerful presence in discussions leading to the formation of other unions (both the first wave of unions, in geophysics and chemistry, and the later ones in mathematics, scientific radiotelegraphy, and physics) and even the IRC itself, transferring the model pioneered by the IAU's founders was not a straightforward matter. Hale's proposal for a structure in which national disciplinary committees, acting under a national research council, would be the representative bodies in the various unions was widely seen as unduly hierarchical and insensitive to local circumstances. And there were profound differences with regard to Germany. Was exclusion to end with the conclusion of peace (as the Americans and many in the British and Italian camps believed) or was it to be extended for a much longer period (the clear preference of the French and the Belgians)? And what of the neutral nations? Could a union in astronomy be seen as truly international when a leading astronomical nation like the Netherlands, neutral during the war, was not a member? The election of Willem De Sitter as a vice-president of the IAU in 1922 suggests how anomalous the international astronomical community saw the situation to be. But, despite a drift towards more inclusive policies during the 1920s, anomalies persisted, and it was only with the dissolution of the IRC in 1931 and its replacement by the International Council of Scientific Unions, conceived as a much freer federation of unions, that the IAU could truly live up to its title of international.

It has to be observed, in conclusion, that the processes leading to the foundation of most other disciplinary unions still await their historians. Throughout this study, we have stressed the specificity of astronomy, alongside the broader circumstances relevant to the genesis of all unions. We are also conscious of the exceptional richness of the sources at our disposal for this particular case. Mainly in French and British archives, and in large measure unexplored, these have done much to fashion the focus and profile of our account, which might well have been different if we had worked primarily from, say, American or German sources. The materials available to historians of other unions will necessarily have their own character and offer their own opportunities, and the resulting emphases will differ from ours. But where sources exist, a study of the founding of any of the eight unions that had come to exist by 1925 is likely to reveal the same complex interweaving of science, politics, and personal and disciplinary motivations that was so evident in the emergence of the IAU.

Acknowledgements We are grateful for the help we have received in the libraries and archives we were able to visit before the Coronavirus pandemic made access difficult. Our thanks go especially to the Royal Society Library, the Archives of the Académie des sciences, the Library of the Observatoire de Paris, the IAU Secretariat at the Institut d'astrophysique de Paris (IAU archives), and the Library of the Accademia dei Lincei, Rome. We are also indebted to Dr. Sian Prosser, Librarian and Archivist of the Royal Astronomical Society, London, who kindly consulted sources in her care at a time when the RAS library was closed.

Appendix 1

Countries Represented at the Interallied Conferences (1918–1919)

Country	London 9–11 Oct. 1918	Paris 26–29 Nov. 1918	Brussels 18–28 July 1919
Belgium	X	X	X
Brazil	X	X	0
Canada	0	0	X
France	X	X	X
Great Britain	X	X	X
Italy	X	X	X
Japan	X	X	X
New Zealand	0	0	X
Poland	0	X	X
Portugal	0	0	X
Romania	0	X	X
Serbia	X	X	X
USA	X	X	X
Total present	8	10	12

X = present 0 = absent

Sources

October 1918. *Nature*, *102*, 133 (17 October 1918); Picard and Lacroix (1918c).

November 1918. Report in the Benjamin Baillaud papers, Bibliothèque de l'Observatoire de Paris.

July 1919. Schuster (1920: 49–62).

Appendix 2

The following are the relevant passages from Articles 1 and 3 of the statutes of the IRC, as approved at the constitutive assembly in Brussels in July 1919, after their original formulation as resolutions at the interallied conference of October 1918 in London.

From article 1 ("Objects of International Council, and conditions of admission"), footnote to subsection 2.

> That it is desirable that the nations at war with the Central Powers withdraw from the existing Conventions relating to International Scientific Associations in accordance with the Statutes or Regulations of such Conventions respectively, as soon as circumstances permit; and.
>
> That new associations, deemed to be useful to the progress of science and its applications, be established without delay by the nations at war with the Central Powers with the eventual co-operation of neutral nations (Schuster, 1920: 222).

From *article 3 ("Conditions of admission").*

> After a Union has been formed, nations not included in the above list [of the 16 nations invited to become founder-members], but fulfilling the conditions of Article 1 of the resolutions of the Conference of London, may be admitted, either at their own request, or on the proposal of one of the countries already belonging to the Union.
>
> Such requests or proposals shall be submitted to a vote of the Union concerned. A favourable majority of not less than three-quarters of the countries already forming part of the Union shall be required for admission (Schuster, 1920: 223).

Note: Membership of the IRC was formalized as a precondition of membership of any of the unions at the IRC's second general assembly, on 26 July 1922. See (Schuster, 1923: 9).

Appendix 3

Picard's proposal, as approved by the BDL on 13 November 1918

UNION ASTRONOMIQUE

1. Il est institué, sous le nom d'Union astronomique, une association internationale entre les pays actuellement en guerre avec l'Allemagne. Cette association a pour objet le progrès général de l'Astronomie et des sciences et arts connexes. Elle exerce principalement son action en provoquant et facilitant la réalisation des œuvres de cet ordre, qui nécessitent une coopération.
2. L'association est régie par un Comité international formé des délégués de Comités nationaux relatifs à chacun de ces pays, formés comme il est dit à l'art. 3.
3. Chaque Comité national est composé de représentants des Sociétés savantes et des services publics compétents, ainsi que de représentants du Gouvernement. Il est formé par les soins de l'Académie nationale des sciences du pays, ou à son défaut, par les soins du Gouvernement.
4. Les ressources de chaque Comité national proviennent des subventions de l'Etat et des Sociétés savantes, ainsi que des dons.

5. Chaque Comité national envoie des délégués au Comité international. Le nombre des délégués de chaque pays est fixé d'après le chiffre de la population sur le taux suivant: un pour une population inférieure à cinq millions, deux entre cinq et dix millions, trois entre dix et quinze; quatre entre quinze et vingt, cinq au dessus de vingt millions. Le Comité international est nommé pour six ans.
6. Le Comité international, de sa propre initiative ou sur la demande de un ou plusieurs Comités nationaux, étudie l'opportunité et le mode de réalisation des entreprises nationales.

   Les dépenses relatives à chacune des entreprises de l'Association sont réparties entre les divers Comités nationaux, proportionnellement aux nombres de leurs délégués au Comité international.
7. Une publication spéciale est consacrée aux procès verbaux des séances du Comité international.

   Dès que les ressources générales le permettront, l'Union entreprendra le calcul des éphémérides des planètes et s'occupera d'en assurer les observations. Elle publiera un recueil donnant le plus rapidement possible les résumés des travaux astronomiques récents. La publication sera faite, sous la direction du Comité international, en trois éditions: en langue française, en langue anglaise, en langue italienne.

   La publication des résultats des entreprises faites sous les auspices de l'Union, si elle n'est pas assurée autrement, sera faite par le Comité international. Les dépenses de ces publications sont payées dans les conditions indiquées à l'article 6.
8. L'association une fois constituée, les nations non belligérantes dans la présente guerre pourront demander à faire partie de l'Association. Dans ce cas, leurs demandes seront soumises au Comité international qui décidera, sur l'admission à la majorité des trois quarts des voix.
9. Les statuts ne peuvent être révisés qu'à la demande de la moitié au moins des Comités nationaux. Les modifications sont votées par le Comité international à la majorité des deux tiers des voix.

Appendix 4

List of documents concerning astronomy (Brussels, 1919)
Published in (Schuster, 1920).

Special Meetings

Union astronomique internationale, 14–19.

Reports

Baillaud, Benjamin: Union internationale pour la détermination de l'heure et des longitudes, Annexe VIII, 78–83.
Baillaud, Benjamin: Entreprise internationale de la carte photographique du ciel, Annexe IX, 84–94.
Baillaud, Benjamin: Rapport présenté au nom du Bureau des longitudes concluant à une nouvelle réunion de la Conférence des éphémérides, Annexe X, 95–101.
Dyson, Frank W.: Astronomical ephemerides, Annexe XI, 102–103.
Schuster, Arthur: Report of the International Union for Co-operation in Solar Research, Annexe XII, 104.
Bigourdan, Guillaume: Rapport sur la publication d'un recueil donnant connaissance le plus rapidement possible des travaux d'ordre astronomique, Annexe XIII, 105–109.

Committees and Regulations (Approved 26 July 1919)

Union astronomique internationale: Composition des commissions, Annexe XXII, 165–168.
Union astronomique internationale: Commission Internationale de l'Heure. Règlement intérieur, Annexe XXIII. A1, 169–170.
Union astronomique internationale: Bureau Central International des Télégrammes Astronomiques. Règlement intérieur, Annexe XXIV, A2, 171–172.
International Astronomical Union: International Time Commission, régulations, Annexe 23, 232–233.
International Astronomical Union: International Central Bureau for Astronomical Telegrams, Regulations, Annexe 24, 234–235.

Statutes

Statuts de l'Union astronomique internationale, Annexe XXI, 160–164. (French version).
Statutes of the international Astronomical Union, Annexe 21, 227–231. (English version).

Appendix 5

Acronyms used in the text

Name	Acronym
Académie des sciences	ADS
American section (of the proposed IAU)	AmS
Académie royale des sciences, des lettres et des beaux-arts de Belgique	ARB
Bureau des longitudes	BDL
Bureau international de l'Heure	BIH
Bureau international des Poids et Mesures	BIPM
Conférence Internationale de l'Heure	CIH
Committee of international scientific organizations	CISO
Comité national français d'astronomie/French National Committee of astronomy	CNA
Executive committee	EC
International Association of Academies Association internationale des académies	IAA
International Association of Seismology Association internationale de sismologie	IAS
International geodetic association Association géodésique internationale	IGA
International postal union	IPU
International research council Conseil international de recherche	IRC
International solar union International Union for Cooperation in solar research	ISU
International Union of Geophysics	IUG
International Union of Geodesy and Geophysics Union géodésique et géophysique internationale	IUGG/UGGI
International Union of Pure and Applied Chemistry	IUPAC
National Academy of Sciences	NAS
National Research Council	NRC
Royal Astronomical Society	RAS
Royal Society	RS

Bibliography

Archives

Académie des Sciences (cited as "ADS"): Comité secret, registres des séances. Register 1J25, "Conférence interalliée des Académies scientifiques", is currently being reclassified. This

bound volume contains a typescript of notes taken at the interallied conference in Paris, 26–29 November 1918. App. 300 pages.
Accademia dei Lincei: Archives, Correspondence to and from Volterra. Cited as "Volterra Papers".
Bibliothèque de l'Observatoire de Paris (cited as "BOP"), Archives: Dossiers Baillaud (MS/1060-IV-C-1), cited as "Baillaud Papers"; Bigourdan (MS/1062-IV-1-A); BIH Archives (MS/1062-IV); Comité national français d'astronomie, CNA (MS-1062/IV-1-A. Box 115), cited as CAN Papers.
Bureau des Longitudes. Procès-verbaux des séances, available online at http://bdl.ahp-numerique.fr. Cited as "BDL, Procès-verbaux".
Caltech Archives, Pasadena, CA: Papers of George Ellery Hale, cited as "Hale Papers".
Royal Society Archives: IRC papers of Arthur Schuster, MS/540, IRC Papers. Cited as "Schuster Papers".

Printed Sources

Académie royale des sciences, des lettres et des beaux-arts de Belgique, *Bulletin de la classe des sciences*, 1919. (Cited as ARB).
Alter, P. (1980). The Royal Society and the International Association of Academies. *Notes and Records of the Royal Society, 34*, 241–264.
Bernhard, C. G., Crawford, E., & Sörbom, P. (Eds.). (1982). *Science and technology in the time of Alfred Nobel [Nobel symposium, 52]*. Pergamon Press.
Bigourdan, G. (2020). *Le Bureau des longitudes, son histoire et ses travaux, de l'origine à ce jour* in N. Capitaine & C. Le Lay (Eds.), in collaboration with G. Boistel. http://bdl.ahp-numerique.fr/source-histoire.
Blaauw, A. (1994). *History of the IAU. The birth and first half-century of the International Astronomical Union*. Kluwer Academic Publishers.
Bougeret, J.-L. (2019). The first president of the IAU. Benjamin Baillaud, in Ch. Sterken et al., *op. cit.*, 139-146.
Campbell, W. W. (1920). Report of the meetings of the International Research Council and of the affiliated unions held at Brussels, July 18–28, 1919. *PNAS, 6*(6), 340–348.
Campbell, W. W., & Stebbins, J. (1920). Report on the organization of the International Astronomical Union, presented for the American section, IAU. *PNAS, 6*(6), 349–396.
Capitaine, N., Bauër, P., & Débarbat, S. (2022). The foundation of the *Bureau international de l'heure*. In Th. Montmerle & D. Fauque, *op. cit.*, ch. 2.
Chinicci, I. (2022). Precursors to IAU: Paris Observatory and the Carte du Ciel project. In Th. Montmerle & D. Fauque, *op. cit.*, ch. 1.
Cock, A. G. (1983). Chauvinism and internationalism in science: The International Research Council, 1919-1926. *Notes and Records of the Royal Society of London, 37*, 249–288.
Conférence Internationale de l'Heure. (1913). *Convention*. Gauthier-Villars.
de Broglie, L. (1942). *La vie et l'œuvre de M. Émile Picard*. Gauthier-Villars.
de La Vallée Poussin, Ch, Lecointe, G., & Massart, J. (1919). Compte rendu de la conférence des Académies des sciences interalliées. *Bull. de la classe des sciences*, Académie royale de Belgique, séance du 4 janvier 1919, 63-81.
De Vorkin, D. (2019). George Ellery Hale's internationalism. In Sterken, Ch. et al, *op. cit.*, 153-160.
Eckart, W. U., & Fox, R. (Eds.) (2021). *Blockades of the mind – Science, academies, and the aftermath of the great war* [*Acta Historica Leopoldina, 78*]. Halle.
Fauque, D. (2011). French chemists and the international reorganisation of chemistry after world war I. *Ambix, 58*, 116–135.
Fauque, D. (2020). Reorganizing chemistry after World War I: The birth of the International Union of Pure and Applied Chemistry (IUPAC), in *Rendiconti dell'Accademia Nazionale delle Scienze detta dei XL*, Memorie di Scienze Fisiche e Naturali, serie V, vol. XLIII, parte II, t. II,

75-86. Atti del XVIII Convegno di Storia e Fondamenti della Chimica, a cura di M. Taddia, Roma 8-10 ottobre 2019. https://www.accademiaxl.it/pubblicazioni-2/rendiconti-on-line/[2].

Fauque, D., & Fox, R. (2021). Binding the wounds of war. Internationalism, national interests, and the order of world science, 1919–1931. In W. U. Eckart & R. Fox, *op. cit.,* 41-68.

Fowler, A. (Ed.). (1923). *Transactions of the International Astronomical Union. Vol.I* (*Proceedings of the first general assembly, Rome May 2–10, 1922*). Imperial College Bookstall.

Fox, R. (2016). *Science without frontiers. Cosmopolitanism and national interests in the world of learning, 1870–1940*. Oregon State University Press.

Friedman, R. M. (2001). *The politics of excellence. Behind the Nobel Prize in science*. Times Books.

Gautier, R. (1918). À propos de l'Association géodésique internationale. *Bulletin de la Société astronomique de France*, *6* (June), 212. (On line on Gallica.fr).

Greenaway, F. (1996). *Science international. A history of the International Council of Scientific Unions*. Cambridge University Press.

Guerraggio, A., & Paoloni, G. (2012). *Vito Volterra*. Springer.

Hadamard, J. (1942). Émile Picard, 1856–1941. *Obituary Notices of the Royal Society, 4*(11), 129–150.

Kevles, D. J. (1971). Into hostile camps: The reorganization of international science after world war I. *Isis, 62*, 47–69.

Kikuchi, Y. (2011). World war I, international participation and reorganisation of the Japanese chemical community. *Ambix, 58*, 136–149.

Kikuchi, Y. (2019). Pioneers of Japanese participation in IUPAC. *Chemistry International, 41*(3), 16–19.

Lacroix, A. (1919). Report to the Académie des sciences, 25 August 1919. *Comptes rendus hebdomadaires des séances de l'Académie des sciences, 169*, 345–349.

Laguens, F. (2019). "The sky seems heavy with clouds": Arthur Eddington and the IAU. In Ch. Sterken et al., *op. cit.*, 178–183.

Laguens, F. (2021). "International science is bound to win": Eddington, Strömgren et la coopération scientifique (1919–1922). In W.U. Eckart & R. Fox, *op. cit.*, 27–40.

Laguens, F. (2022). Crossing forbidden borders after WWI: Arthur S. Eddington's network. In Th. Montmerle & D. Fauque, *op. cit*, ch.4.

Lallemand, Ch. (1918). Association géodésique internationale. *Bulletin de la Société astronomique de France, 3* (mars), 95–98. https://gallica.bnf.fr/ark:/12148/bpt6k96404104/f109.item.

Lattis, J. M., & Lattis A. J. (2019). American astronomers in Belgium, 1919: Snapshots from the founding of the IAU. In Ch. Sterken et al, *op. cit.*, 406–418.

Leppan, P.-V. A. (1984). A note on the history of the International Association of Geodesy. *Journal of Geodesy, 58*, 224–229. Also on the International Association of Geodesy's site at https://www.iag-aig.org/iag-history/136

Lettevall, R., Somsen, G., & Widmalm, S. (Eds.). (2012). *Neutrality in twentieth-century Europe. Intersections of science, culture, and politics after the first world war*. Routledge.

Levallois, J.-J. (1980). The history of the International Association of Geodesy. *Bulletin géodésique, 54*, 249–313.

Macleod, R. (2018). The mobilisation of minds and the crisis in international science: The "Krieg der Geister" and the "manifesto of the 93". *Journal of War & Culture Studies, 11*, 58–78.

Massart, J. (1917). Une organisation scientifique interalliée. *Revue scientifique, 55/11*(26 May–2/9 June), 334–337.

Montmerle, Th. (2019). The IAU, from commissions to divisions… to commissions. In Ch. Sterken et al., *op. cit.*, 289–324.

Montmerle, T., & Fauque, D. (Eds.). (2022). *Astronomers as diplomats. When the IAU builds bridges between nations*. Springer.

Ninhos, C., & Clara, F. (Eds.). (2017). *Closing the door on globalization. Internationalism, nationalism, culture and science in the nineteenth and twentieth centuries*. Routledge.

Pelseneer, P., Swarts, F., & Lecointe, G. (1919). *Conseil international de recherches, Assemblée constitutive tenue au Palais des Académies, à Bruxelles, du 18 au 28 juillet 1919. Compte rendu présenté à l'Académie royale des Sciences, des Lettres et des Beaux-Arts de Belgique*. Hayez.

Picard, É. (1917a). Les relations scientifiques internationales après la guerre, *Journal des débats*, 17 October 1917. https://gallica.bnf.fr/ark:/12148/bpt6k4884086.item

Picard, É. (1917b). *Notice historique sur Gaston Darboux*. Gauthier-Villars. Read on 10 December 1917.

Picard, É., & Lacroix, A. (1918a). La Conférence interalliée des académies scientifiques à Londres. *Comptes rendus des séances hebdomadaires de l'Académie des sciences, 167*, 566–570.

Picard, É., & Lacroix, A. (1918b). La deuxième session de la Conférence interalliée des académies scientifiques. *Comptes rendus des séances hebdomadaires de l'Académie des sciences, 167*, 933–941.

Picard, É., & Lacroix, A. (1918c). La Conférence interalliée des académies scientifiques à Londres, *Revue scientifique*, 56e année (16-23 November 1918), 673–675.

PNAS. (1919). *Proceedings of the National Academy of Sciences*. 5/5, session 11 March, 189–196. https://www.pnas.org/content/by/year.

Reinbothe, R. (2006). *Deutsche als internationale Wissenschaftssprache und der Boykott nach dem Ersten Weltkrieg*. Peter Lang.

Reinbothe, R. (2010). L'exclusion des scientifiques allemands et de la langue allemande des congrès scientifiques après la Première Guerre mondiale. *Revue germanique internationale, 12*, 193–208.

Royal Society. (1945). *Minutes of Council Meetings*, vol. 11 (printed version of the original minutes). (Cited as RS Minutes of Council).

Schiavon, M., & Rollet, L. (Eds.). (2021a). *Le Bureau des longitudes au prisme de ses procès-verbaux (1795-1932)*. PUN - Éditions universitaires de Lorraine.

Schiavon, M., & Rollet, L. (Eds.) (2021b). Organiser la reprise en main des institutions scientifiques internationales: le Bureau des longitudes en 1919. In M. Schiavon & L. Rollet, *op. cit.*, 263–313.

Schroeder-Gudehus, B. (1978). *Les scientifiques et la paix. La communauté scientifique internationale au cours des années 20*. Presses de l'Université de Montréal.

Schroeder-Gudehus, B. (1982). Division of labour and the common good: The International Association of Academies, 1899–1914. In C.G. Bernhard et al., *op. cit.*, 3–20.

Schroeder-Gudehus, B. (1986). Pas de Locarno pour la science: la coopération scientifique internationale et la politique étrangère des États pendant l'entre-deux-guerres. *Relations internationales, 46*, 173–194.

Schuster, A. (1918). Allocution prononcée par M. Schuster, secrétaire de la conférence avant le vote de la déclaration et des résolutions. *Revue scientifique, 56*(16-23 November), 675–676.

Schuster, A. (Ed.). (1920). *Constitutive assembly held at Brussels, July 18th to July 28th, 1919. Reports of proceedings*. Harrison & Sons.

Schuster, A. (Ed.). (1923). *Second assembly held at Brussels, July 25th to July 29th, 1922. Reports of proceedings*. Harrison & Sons.

Smit, J. P. (2017). The politics of interwar chemistry: Neutrality and nationalism in the rhetoric and actions of internationalist chemists. In Ninhos and Clara, 2017, *op. cit.*, 7-29.

Smit, J. P. (2019). Ernst Cohen and the challenge of a truly international union. *Chemistry International, 41/3*(July), 9–11.

Somsen, G. (2012). "Holland's calling": Dutch scientists' self-fashioning as international mediators. In R. Lettevall et al., *op. cit.*, 45-64.

Sterken, C., Hearnshaw, J., & Valls-Gabaud, D. (Eds.). (2019). *Under one sky. The IAU centenary symposium. Proceedings of the IAU, symposium 349, Vienna, Austria, 2019*. Cambridge University Press and IAU.

Torge, W. (2005). The International Association of Geodesy 1862 to 1922: From a regional project to an international organization. *Journal of Geodesy, 78*, 558–568.

Vickery, H. B. (1944). Biographical memoir of Russell Henry Chittenden 1856–1943. *National Academy of Sciences Biographical Memoirs, 24*, 57–104.

von Ungern-Sternberg, J., & von Ungern-Sternberg, W. (1996). *Der Aufruf "An die Kulturwelt!". Das Manifest der 93 und die Anfänge der Kriegspropaganda im Ersten Weltkrieg: mit einer Dokumentation*. Franz Steiner.

Widmalm, S. (2012). "A superior type of civilisation": Science as politics in Sweden, 1917-1926. In R. Lettevall et al., *op. cit.*, 65-89.

Wielen, R. (2019). Germany's difficulties in becoming a member of the International Astronomical Union. In Ch. Sterken Ch. et al., *op. cit.*, 205-213.

Wielen, R. (2022). Germany's difficulties in becoming a member of the International Astronomical Union. In Th. Montmerle & D. Fauque, *op. cit.*, ch. 11.

Wright, A. (2014). *Cataloging the world. Paul Otlet and the birth of the information age*. Oxford University Press.

Danielle Fauque holds a Master's in the physical sciences and a PhD in the history of science for a thesis on the history of the heliometer (1748–1826). A member of the Groupe d'Histoire et de Diffusion des Sciences (GHDSO) at Orsay, University of Paris-Saclay, France, she was elected to the International Academy of the History of Science in 2005. She has written extensively on the history of astronomical navigation, instruments, and scientific voyages in the eighteenth century and is currently working on international relations in science in the twentieth century. Among her recent publications on astronomy are: "Testing Longitude Methods in Mid-Eighteenth Century France", in R. Dunn and R. Higgitt (eds.) *Navigational Enterprises in Europe and its Empires, 1730-1850* (Basingstoke: Palgrave Macmillan, 2016), 159–179 and "A thematic study, Observatoire de Paris", in Clive Ruggles and Michel Cotte (eds.), *Heritage Sites of Astronomy and Archeoastronomy in the Context of the UNESCO World Heritage Convention* (2016).

Robert Fox read physics at Oxford before going on to doctoral work in the history of science (D.Phil., 1967) and teaching for 20 years at the University of Lancaster. He became the founding director of the Centre de Recherche en Histoire des Sciences et des Techniques at the Cité des Sciences et de l'Industrie (La Villette) in Paris in 1986 and was appointed to the chair of the history of science at Oxford in 1988. He served as president of the International Union of History and Philosophy of Science from 1995 to 1997 and president of the newly created European Society for the History of Science from 2003 to 2006. Since retirement from Oxford in 2006, he has held visiting professorships at Johns Hopkins University, East Carolina University, and the Czech National Technical University, Prague. Among his recent books are *The Savant and the State. Science and Cultural politics in Nineteenth-Century France* (2012) and *Science without Frontiers. Cosmopolitanism and National Interests in the World of Learning, 1870–1940* (2016).

Chapter 4
Crossing Forbidden Borders After WWI
Arthur S. Eddington's Network

Florian Laguens

Abstract Arthur S. Eddington was the Director of Cambridge Observatory when the Armistice was signed on November 11th, 1918. He was also involved in several scientific organisations, especially the Royal Astronomical Society and the Royal Society. This paper presents his actions in favour of cooperation within the scientific community between 1914 and 1928. He had multiple contacts both in England and abroad, and he did his best to promote the renewal of scientific relationships in the aftermath of WWI. Thus, the reader is to encounter, among others, the Dutch Willem De Sitter, the French Emile Picard, the English Arthur Schuster, the American George H. Hale, the Australian Henry Brose, the Swedish Elis Strömgren, the Germans Max Wolf, and of course, Albert Einstein.

4.1 Introduction

Where to begin with someone like Eddington? Although he is so well known, there is so much to discover about him. Indeed, a lot is still to be done regarding his various activities and commitments, as so much archive material and published works remain unexplored. Hopefully, this chapter may be an opportunity to elaborate about Eddington's actions in favour of international cooperation, between 1914 and 1928.[1] During this period, he took part in many different events and had several occasions to promote international cooperation, if not acting seemingly as a diplomat—on a scientific level at least.

[1] The purpose of this paper being to examine the aftermath of WWI, it does not intend to present an exhaustive account of Eddington's action in favour of scientific cooperation. The interested reader may refer to Laguens, 2019, which gives some information up to 1944, when Eddington died.

F. Laguens (✉)
IPC-Facultés Libres, Paris, France
e-mail: flaguens@ipc-paris.fr

© The Author(s), under exclusive license to Springer Nature Switzerland AG 2022
T. Montmerle, D. Fauque (eds.), *Astronomers as Diplomats*, Historical & Cultural Astronomy, https://doi.org/10.1007/978-3-030-98625-4_4

Arthur Stanley Eddington was born on 28 December 1882. Both his parents belonged to the Society of Friends, that is to say a religious group founded around 1650 by George Fox (1624–1691). The members of the Society are usually named "Quakers". Eddington was taught since his earlier childhood the grounding claims of Quakerism (Trueblood, 1993; Dandelion, 2008):

- God enlightens each individual equally.
- Anyone can have a direct experience of His "Inward Light".
- Peace is to be defended in any circumstances.

Hence Eddington's pacifist tendencies.[2]

He entered Owens College in Manchester in 1898, and then Cambridge University in 1902. In January 1906, he became Chief Assistant at the Royal Greenwich Observatory and he was immediately elected a Fellow of the Royal Astronomical Society. Eventually, he was chosen to succeed George H. Darwin—a son of the great Charles Darwin—as Plumian Professor of Astronomy and Experimental Philosophy at Trinity College, Cambridge, in 1913. The following year he was appointed Director of the Cambridge Observatory.

4.2 Background: England Between 1916 and 1918

The first opportunity for Eddington to be involved in political matters seems to have been provided by Bertrand Russell. Great Britain had entered the war on 4 August 1914. Up to 2 March 1916, the British army was relying upon volunteer enrolment, but after that date military service became compulsory for men of age 18 to 41. At Trinity College, the tension was increasing within the Academic community. Some older Fellows were avowedly favourable to the conflict despite their inability to serve, and others were explicitly opposed to it. For fear that some Fellows might organise pacifist reunions, the University Council forbade meetings in private apartments in 1915. Eddington signed along with twenty other Fellows a request protesting against that decision and sent it to the Council. But the interdiction of meetings was confirmed on 22 January 1916. In April, a conscientious objector named Ernest Everett was sentenced to two years of hard labour for disobedience to the military authorities. A few days later, an anonymous leaflet was printed expressing a strong opposition against Everett's condemnation. When some people distributing the pamphlet undercover got to be arrested, Bertrand Russell came out as the author to be blamed for it. He did so thanks to a public letter issued in *The Times* on 17 May 1916. It was not two months before Russell was fined a hundred

[2] This implication (Quaker *ergo* pacifist) may be nuanced to a certain extent (Laguens, 2018, 193–196). The impressive historical research by Thomas C. Kennedy help realise the difficulties of the Society of Friends at the time regarding pacifism (Kennedy, 2001, 312–356). On that matter, see also Rae, 1970.

pounds and expelled from Trinity. Once again, Eddington and two dozen Fellows petitioned against that decision. In vain (Hardy, 1970; Monk, 1996, 462–466).

This Russell story conveys an impression of the heavy atmosphere in England at that time, even in the intellectual *milieu*. A famous victim of the somewhat generalised anti-German feelings was the physicist Arthur Schuster. While not an astronomer, his diplomatic manoeuvres are not to be neglected for the foundation of the International Astronomical Union. A conspiracy was already going on against him since his election as Secretary of the Royal Society in 1912, but when he became President of the British Association for the Advancement of Science in 1915 things went rogue. He had to face the death of his son at war (August 1915) and in the meantime the protests by his colleagues. Among his most aggressive opponents, there seems to have been Henry Edward Armstrong, a chemist and senior Fellow of the Royal Society, and Herbert Hall Turner. Turner had been Chief Assistant of the Royal Greenwich Observatory from 1884 to 1893, and then the director of the Radcliffe Observatory (Oxford). Both positions were at the time fairly prominent regarding the British astronomical community (Hutchins, 2008).

Schuster's name was mocked publicly because of his German consonance (indeed the Schuster family had emigrated from Germany in 1870), and some colleagues even attempted to ruin the 1915 BAAS general assembly scheduled to take place in Manchester. Sir Henry Roscoe, a then-renowned chemist, wrote to Joseph J. Thomson, President of the Royal Society: "I hear a well-founded rumour that certain persons are making a movement against Schuster as P. B. Assoc. [President of the British Association]. That is endeavouring to dissuade scientific men from attending the Manchester meeting. This, if true as I believe, is the most abominable intrigue and can only be stopped by men determined to attend—at least for the Address—at Manchester".[3] Eventually, the presidential address was a plain success, and 1200 BAAS members attended. Unsurprisingly, Schuster lectured about internationalism. Here are his closing words: "Only through victory shall we achieve a peace in which once more Science can hold up her head, proud of her strength to preserve the intellectual freedom which is worth more than material prosperity, to defeat the spirit of evil that destroys the sense of brotherhood among nations, and to spread the love of truth" (Schuster, 1916, 23).

How could Eddington not have been moved? How could he not have recalled these words when he wrote a few months later a letter to be published in *The Observatory* about "The Future of International Science"? The key sentence reads: "The welfare of astronomy urges a continuance of co-operation" (Eddington, 1916, 271). It should be noticed that Eddington and Schuster knew each other quite well, as they had first met in 1898 when Eddington had studied physics at the Owens College in Manchester under Schuster. In 1909 Eddington's former teacher had invited him to come back to Manchester on the chair for theoretical physics. But Eddington declined as he preferred to pursue his astronomical work at the Royal Greenwich Observatory. The year 1917 demonstrated once again the non-

[3] Roscoe to Thomson, 30 July 1915, Cambridge, University Library, MS Add. 7654/R53.

Fig. 4.1 A. Einstein, P. Ehrenfest and W. De Sitter standing, A.S. Eddington and H. Lorentz sitting, in front of the office of W. de Sitter in Leiden (late September 1923) (credit: Leiden Archives, Creative Commons)

hostility of Eddington towards German scientists. He paid homage to the late Karl Schwarzschild in the *Monthly Notices* of the Royal Astronomical Society. As he feared that the obituary be not properly circulated, he undertook to distribute twenty-five copies himself.[4]

Let us recall that in November 1915 Einstein published in Berlin four papers associated with a new theory he called the "general theory of relativity", followed by a synthesis in May 1916. Immediately, the theory was studied in depth and taught in Leiden, thanks to Ehrenfest (who was Austrian), Lorentz and De Sitter, who, as Dutch, were knowledgeable in German (Kox, 1992) (Fig. 4.1).

In England, the introduction of the theory was far from being as easy as it was in the Netherlands. The history of the general theory of relativity began, as far as Eddington was concerned in June 1916. The exact date cannot be traced back but the fact is certain: he ended up about the end of June receiving a letter from Willem De Sitter announcing the publication of Einstein's theory in November 1915. Here is Eddington's reply: "I am immensely interested in what you tell me about Einstein's

[4] Eddington (1917), Eddington to Wesley, 23 March 1917, London, Royal Astronomical Society, Library.

theory. [. . .] Hitherto I had only heard vague rumours of Einstein's new work. I do not think anyone in England knows the details of his paper".[5] Eddington, apart from his exceptional abilities in mathematics, had a great advantage over most of his British colleagues: he could read German too. No doubt he studied in depth Einstein's famous paper "Die Grundlage der allgemeinen Relativitätstheorie" which he received in early July 1916 (Einstein, 1996, 284). To get around the problem of physicists being unfamiliar with the German language, De Sitter suggested he could himself write a paper in English to give an overview of Einstein's theory in English. Eventually, he wrote four papers. Three were published in the *Monthly Notices of the Royal Astronomical Society* in October 1916, December 1916 and June 1917 (De Sitter, 1916a, 1916b, 1917). The fourth one was a brief summary for *The Observatory*, issued under Eddington's auspices in October 1916 (De Sitter, 1916c; Guichelaar, 2018).

Because of the First World War, the exchanges were restrained between the British Isles and the continent. There is no reason why the Eddington-De Sitter relationship should not have suffered from this situation. From time to time, it appears that their letters were returned by the Censor, in particular when, in June 1918, Eddington tried to send the proof-sheets for his 100-page account on Einstein's theory: "I have just completed a Report for the Physical Society on Einstein's Theory of Gravitation—a pretty full account with mathematical details, running perhaps to 100 pages. I venture to ask if you would kindly read the proof-sheets. [. . .] I should be greatly indebted to you if you would; it is a subject in which it is very easy to go astray. [. . .] I tried to send you some reprints some time ago but they were returned by the Censor".[6] And a couple of months later: "Thanks very much for consenting to read my proofs. I am sending the complete set under separate cover; it should reach you alright (*sic.*), as I have obtained a permit to send it. Perhaps you would send me a card when you receive them" (Fig. 4.2).[7]

In 1917, both the Royal Society and the Royal Astronomical Society decided to prepare for the solar total eclipse expected for 29 May 1919. This celestial event would represent a most favourable opportunity to test Einstein's theory by measuring the deflection of a ray of light grazing the Sun's limb. It meant for English astronomers to prepare for testing the theory of a German scientist. Erwin Freundlich, then director of the Potsdam Solar Observatory, near Berlin, who was a close working associate of Einstein, had attempted an analogous measurement in 1914. Freundlich had travelled to Crimea to take photographs of the eclipse but it all happened at the outbreak of WWI and he was briefly imprisoned by the Russian regime (Hentschel, 1994). As for his instruments, they were confiscated and stored at the Poulkovo National Observatory for several years (Eisenstaedt, 2002). Politics again.

---

[5] Eddington to De Sitter, 11 June 1916, Leiden, Universiteit Leiden, Archief Sterrewacht Leiden, WdS 14.

[6] Eddington to De Sitter, 11 June 1916 and 28 April 1918.

[7] Eddington to De Sitter, 24 June 1918.

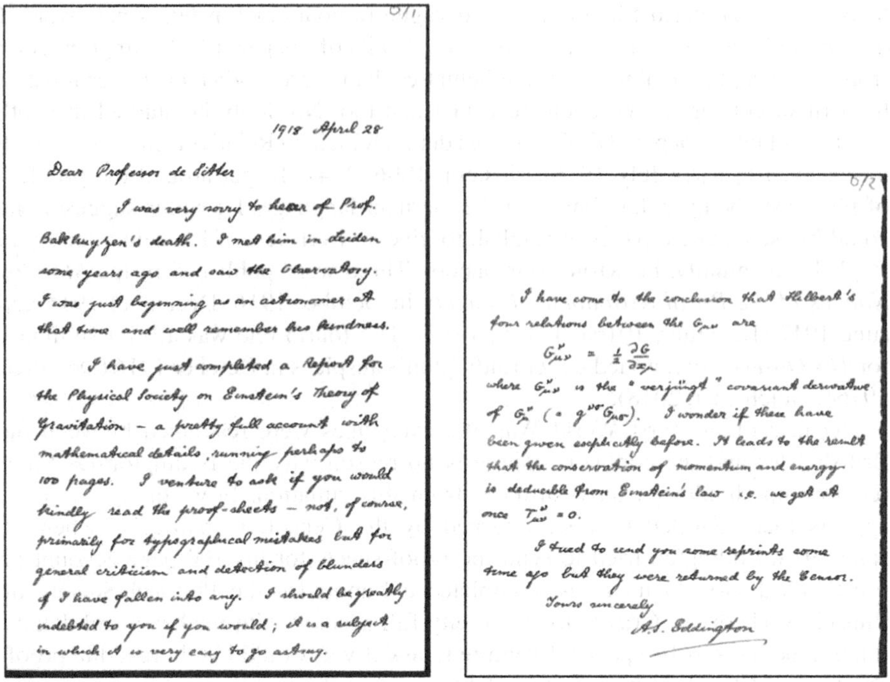

1918 April 28

Dear Professor de Sitter

I was very sorry to hear of Prof. Bakhuyzen's death. I met him in Leiden some years ago and saw the Observatory. I was just beginning as an astronomer at that time and well remember his kindness.

I have just completed a Report for the Physical Society on Einstein's Theory of Gravitation – a pretty full account with mathematical details, running perhaps to 100 pages. I venture to ask if you would kindly read the proof-sheets – not, of course, primarily for typographical mistakes but for general criticism and detection of blunders if I have fallen into any. I should be greatly indebted to you if you would; it is a subject in which it is very easy to go astray.

I have come to the conclusion that Hilbert's four relations between the $G_{\mu\nu}$ are

$$G_{\mu\nu}^{\nu} = \frac{1}{2}\frac{\partial G}{\partial x_{\mu}}$$

where $G_{\mu\nu}^{\nu}$ is the "verjüngt" covariant derivative of G_{μ}^{ν} ($= g^{\nu\sigma}G_{\mu\sigma}$). I wonder if these have been given explicitly before. It leads to the result that the conservation of momentum and energy is deducible from Einstein's law i.e. we get at once $T_{\mu\nu}^{\nu} = 0$.

I tried to send you some reprints some time ago but they were returned by the Censor.

Yours sincerely

A.S. Eddington

Fig. 4.2 Letter from A.S. Eddington to W. De Sitter, 28 April 1918. By permission of Leiden University Library

4.3 The International Research Council and the International Astronomical Union (1918–1919)

Apart from Arthur Schuster, two main characters were involved in the beginnings of the International Astronomical Union, namely Émile Picard and George Ellery Hale. Picard was a notorious French mathematician with an exceedingly brilliant academic career. From 1917 he was the *Secrétaire perpétuel* for mathematics at the Académie des sciences. Needless to insist on the prominence of his position among French scientific scholars. But some remarks have to be made on his personal life. He had lost two of his five children during World War I. Especially demanding had been the death of his older son in January 1915. A daughter of his also died in 1916 from a disease she had contracted while being a nurse in a military hospital. In the general opinion, he never completely coped with the mourning process and his temper changed durably (Broglie (de), 1942, 41).[8] As for George E. Hale, he had been the director of the Mount Wilson observatory since 1904. He was involved in international cooperation even before WWI as he took the leading role in the

[8] On Picard, see *infra* the paper by Fauque and Fox.

creation of the International Union for Cooperation in Solar Research, the fifth and last meeting of which was held in August 1913 in Bonn, Germany (Solar Union, 1914; Wright, 1966, 300–304; Kevles, 1971). When President Wilson asked for the constitution of a National Research Council in 1916, Hale was chosen as chairman. The NRC provided Hale with an institutional basis for consulting the Royal Society and the Académie des Sciences. Schuster represented the British side, and Picard the French one.

All three came in September 1917 to an agreement about a further international meeting. Hence the Inter-Allied Conference held in London on 9–12 October 1918. Thirty-four delegates were then around the table, representing Great Britain, Belgium, France, Italy, Japan, Portugal, Serbia, and the United States. They unanimously adopted a one-page common statement. Here is an extract: "If today the representatives of the scientific academies of the Allied nations are forced to declare that they will not be able to resume personal relations in scientific matters with their enemies until the Central Powers can be readmitted into the concert of civilized nations, they do so with a full sense responsibility" (RAS, 1918, 415).

This London conference was followed by another one in Paris on 26–29 November 1918. There, the decision was made to establish an "International Research Council" run by Picard (president), Hale, Vito Volterra, Georges Lecointe (vice-presidents) and Schuster (general secretary). Picard was willing to terminate the former international organisations and to create new ones regrouping the Allied Powers, with a definitive banishment of the Central Powers and a temporary exclusion of the neutral countries. The executive committee was to prepare for the first IRC assembly in Brussels in July 1919. The International Astronomical Union was officially born on 28 July 1919 that is on the last day of the Constitutive Assembly of the IRC. The first article of its statutes read: "The purpose of the Union [...] is to facilitate the relations between astronomers of different countries where international co-operation is necessary and useful" (IAU, 1922, 3).

In July 1920, the first countries to have officially joined the Union were Belgium, Canada, France, Greece, Japan, the United Kingdom, and the United States.[9] It may nonetheless be recalled that the members of the IAU have never been, strictly speaking, countries, but that each country is represented by a scientific body such as the Royal Society or the Académie des sciences. The following months were devoted to the reception of the subscription dues from the adhering countries. Some hesitation persisted regarding the venue for the first General Assembly. From the very beginning, it was planned to take place in Rome in 1922, but Italy took a long time to pay for its dues. Hence the hesitation of Benjamin Baillaud, the IAU President, to maintain this location. The IAU would not meet in a country that was not a formal member of the Union.[10] Once the dues paid, the matter was

[9] Schuster to Fowler, 23 June 1920, Paris, International Astronomical Union, IAU archives, file I.1.B.

[10] Baillaud to Fowler, 27 May 1921, Paris, International Astronomical Union, IAU archives, file I.1.B.

ipso facto settled and the time had arrived for sending out the invitations for the congress. It was decided to issue one to each country that had definitely joined, namely Belgium, Canada, France, Greece, Italy, Japan, Mexico, United Kingdom, and United States, but also to all the others that had also been invited to join. Those were at that time Australia, Brazil, Poland, Portugal, Serbia, South Africa, and New Zealand.[11] Alfred Fowler, the IAU General Secretary, raised another question: should the other countries having adhered to the IRC be also invited? In this category were the former neutral countries, in particular, Denmark and Holland. Schuster's answer arrived soon: "I have had a good deal of discussion with M. Picard about inviting representatives of Neutral Countries who have not joined the Unions to send representatives (without power to vote) to the meeting at Rome. Picard strongly objects on the ground that it is not in our power to do this according to the Statutes". The letter ends with a brief additional sentence: "German astronomers are, of course, excluded".[12]

The situation slightly evolved between 1921 and 1922, leading to the admission of some "neutrals" in 1922 and especially Denmark and the Netherlands. The first General Assembly of the IAU concluded with 19 members. Meanwhile, the General Assembly of the IRC approved on 26 July 1922 a resolution stating that: "Only countries which have adhered to the International Research Council are entitled to be members of the Unions connected with it" (IAU, 1925, 251). The statutes of the IAU mentioning that "The admission of countries to the Union shall be subject to the Regulations of the International Research Council" (IAU, 1922, 219; IAU, 1925, 252), this additional resolution was soon forwarded to Fowler, still the General Secretary of the IAU.[13] Obviously, the IRC meant to reaffirm its grip on the IAU admission process.

4.4 The General Theory of Relativity (1919–1920)

In 1919, Eddington's thoughts were mostly occupied by the 29 May total eclipse of the Sun. A lot has already been told about this decisive event (Kennefick, 2019; Stanley, 2019). But quite unknown is the fact that the British expeditions were not initiated by Eddington. Astronomer royal Frank Dyson was indeed the prime mover. As early as March 1917 he published a paper entitled: "On the Opportunity Afforded by the Eclipse of 1919 May 29 of Verifying Einstein's Theory of Gravitation" (Dyson, 1917). The decision was to be made by the Joint Permanent Eclipse

[11] See Fowler to Baillaud, 16 September 1921 and 20 September 1921, Paris, International Astronomical Union, IAU archives, file I.1.B.

[12] Schuster to Fowler, 21 September 1921, Paris, International Astronomical Union, IAU archives, file I.1.B.

[13] See Schuster to Fowler, 4 September 1922, Paris, International Astronomical Union, IAU archives, file I.1.B.

Committee, composed of eleven members from the Royal Society and eleven from the Royal Astronomical Society. At its 10 November 1917 meeting the Eclipse Committee agreed on Dyson's recommendation and decided that two expeditions be prepared to take pictures of the 1919 total eclipse, one on the island of Principe, off the coast of Equatorial Guinea, and the other in Sobral, near Fortaleza, Brazil.[14] The importance of Dyson's advocacy has been stated by Eddington in a private letter to Hermann Weyl: "It was Dyson's enthusiasm that got the eclipse expeditions ready to start in spite of very great difficulties in making the preparations. He was at that time very sceptical about the theory though deeply interested in it; and he realised its very great importance".[15] No doubt Eddington was primarily committed to these expeditions for scientific reasons. In December 1918, he wrote to Weyl: "You may be interested to know that in England we are preparing two expeditions for the eclipse next May, for the sole purpose of measuring the deflection of light. I shall go with one of them".[16] However, Eddington had also in mind the political implications. At least he realised how powerful the message sent to the scientific community would be—the theory by a German physicist being confirmed by British astronomers—after the official announcement of the positive results on 6 November 1919 (Eddington, 1919; Dyson et al. 1920). To a fellow astronomer, he wrote on 18 November: "International science is bound to win, and recent events—the verification of Einstein's theory—have made a tremendous difference in the last month. I actually have hopes that next February [1920] we shall award the R.A.S. Gold Medal to Einstein. You can imagine what a splendid step forward that will be"[17] . Things did not go as smoothly as expected, though.

Four days earlier, at the RAS council meeting, three names had been proposed for the Gold Medal: Annie J. Cannon, Henry Norris Russell and Albert Einstein.[18] At the next monthly meeting in December the majority cast their vote in favour of Einstein. But on 9 January 1920 the decision was not confirmed. Eventually, the council chose to award no Medal rather than giving it to Einstein. Writing to him, Eddington was desperate: "You were selected by an overwhelming majority in December. Meanwhile the 'irreconciliables' took alarm, mustered up their full forces in January, and managed to defeat the confirmation of the award in January. So for the first time for about thirty years no Gold Medal will be awarded this year! I confess I was very much surprised when the motion was proposed and carried

[14] The minutes of the Joint Permanent Eclipse Committee are kept among the Royal Society archives (CMP/109) and at the Royal Astronomical Society (RAS Papers 54/1).

[15] Eddington to Weyl, 18 August 1920, Zurich, Bibliothek der Eidgenössischen Technischen Hochschule, Nachlass Hermann Weyl, Ms. 91/523. See also Jeans (1925, 677), Stachel (1986, 229), Kennefick (2012). Weyl taught mathematics in Zurich (Switzerland) from 1913 to 1930.

[16] Eddington to Weyl, 16 December 1918, Zurich, Bibliothek der ETH, Nachlass Hermann Weyl, Ms. 91/522. See also Eddington (1919).

[17] Eddington to Strömgren, 16 November 1919, Lund, Lund University Library, Elis Strömgren's Archive. For a transcription of the correspondence between Eddington and Strömgren, see Laguens (2021).

[18] See the RAS Council Minute Book, vol. XI (1915–1928), p. 101 (London, RAS Library).

originally (it was proposed by two men who during the war have been violently 'patriotic'); but until a day of two before the January meeting we all regarded the confirmation as a matter of course".[19]

What a disappointment after his letter of 1 December 1919: "Our results were announced on Nov. 6, and you probably know that since then all England has been talking about your theory. [...] It is the best possible thing that could have happened for scientific relations between England and Germany. I do not anticipate rapid progress towards official reunion, but there is a big advance towards a more reasonable frame of mind among scientific men, and that is even more important than the renewal of formal associations".[20]

Just before the official announcement of the eclipse results, Eddington was writing to Cambridge University Press, so as to encourage the publication of a booklet devoted to the theory of general relativity in spite of his author being German. Erwin Freundlich had published *Die Grundlagen der Einsteinschen Gravitationstheorie* in 1916, and it had been translated into English by a young Australian physicist named Henry Brose. Brose had been staying in Hamburg in the summer of 1914, visiting some relatives. He had been arrested as a civilian prisoner and eventually interned at the Ruhleben camp in Berlin surroundings. He was not released until 1917, but he did his best to take advantage of his imprisonment by translating several scientific books. In October 1919, Eddington wrote to the University Press at Cambridge to introduce Brose: "The booklet translated is one which has had a great success in Germany. I find that there is a most amazing amount of interest taken in the subject (Einstein's theory of gravitation) not only among mathematicians & physicists but all classes in England. And apart from my own Report of the Physical Society (which is mainly mathematical) there is nothing in English on the subject. [...] I think there should be no doubt that the book would pay—in fact I think just now a book on this subject however bad would pay—and this book is distinctly good".[21]

Immediately afterwards, Brose wrote to promote his translation: "Professor Eddington has just informed me that he has sent in my translation of Dr. Erwin Freundlich booklet on 'The Foundations of Einstein's Theory of Gravitation', the sole English rights of which are in my hands. The book which is recommended by Professor Einstein has run into the third German edition and will, I believe, be alike appreciated by the more general reader here, who merely wishes to gain an insight into the development of the ideas which led to the enunciation of the theory, as by physicists who are anxious to avoid the heavy mathematics; the essentials are, however, mentioned and partly worked out in the appendix. No similar book has as yet appeared in English and I believe that the present moment is particularly favourable, as everyone interested is anxiously awaiting the result of the eclipse

[19] Eddington to Einstein, 21 January 1920, in Einstein (2004, 370). The "two men" Eddington is referring to are Turner and James Jeans.

[20] Eddington to Einstein, 1 December 1919, in Einstein (2004, 263).

[21] Eddington to Cambridge University Press, 20 October 1919, Cambridge, University Library, Pr.A.E.34(11–12).

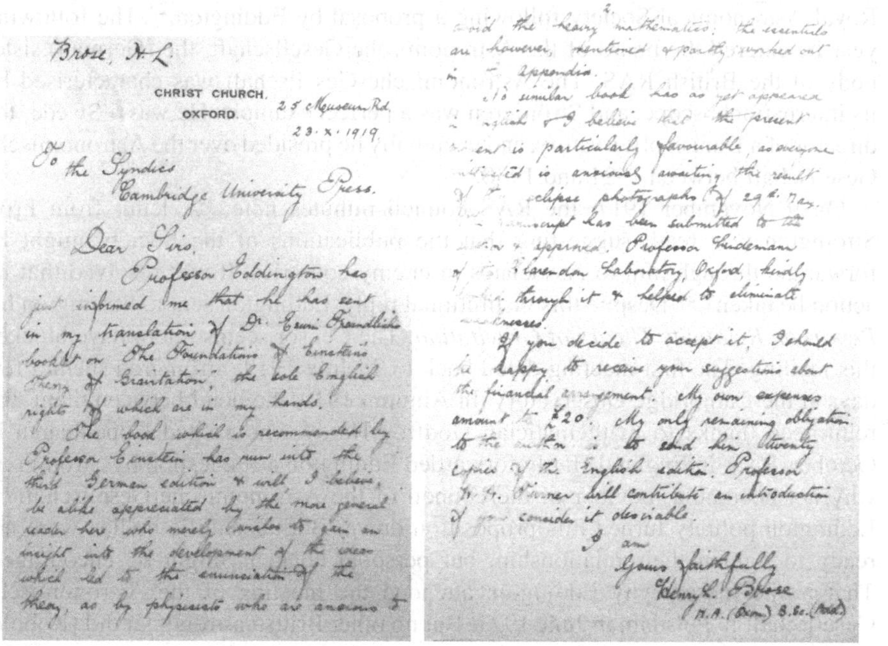
Brose, H.L.

CHRIST CHURCH,
OXFORD.
25 Museum Rd
23.x.1919

To the Syndics
Cambridge University Press.

Dear Sirs,
Professor Eddington has just informed me that he has sent in my translation of Dr. Erwin Freundlich's booklet on "The Foundations of Einstein's Theory of Gravitation", the sole English rights of which are in my hands.
The book which is recommended by Professor Einstein has run into the third German edition & will, I believe, be alike appreciated by the more general reader here who merely wishes to gain an insight into the development of the ideas which led to the enunciation of the theory, as by physicists who are anxious to

2
avoid the heavy mathematics; the essentials are however, mentioned & partly worked out in the appendix.
No similar book has as yet appeared in English & I believe that the present moment is particularly favourable, as everyone interested is anxiously awaiting the result of the eclipse photographs of 29th May. The manuscript has been submitted to the author & approved. Professor Lindemann of the Clarendon Laboratory, Oxford, kindly looked through it & helped to eliminate weaknesses.
If you decide to accept it, I should be happy to receive your suggestions about the financial arrangement. My own expenses amount to £20. My only remaining obligation to the author is to send him twenty copies of the English edition. Professor H. H. Turner will contribute an introduction if you consider it desirable.
I am
Yours faithfully
Henry L. Brose
(M.A. (Oxon) B.Sc. (Adel))

Fig. 4.3 Letter from H. Brose to the Syndics of CUP, 23 October 1919. By permission of the Syndics of Cambridge University Library

photographs of 29th May. The manuscript has been submitted to the author and approved. Professor Lindemann of the Clarendon Laboratory, Oxford, kindly looked through it and helped to eliminate weaknesses. [. . .] Professor H. H. Turner will contribute an introduction if you consider it desirable" (Fig. 4.3).[22]

Eventually, 1250 copies were published in early 1920, Eddington's intervention being mentioned in the "Translator's note" (Brose, 1920).

4.5 The IAU General Assembly in Rome and the RAS Centenary (1922)

An additional character in this story is Elis Strömgren. A Swedish astronomer, he taught at the German University of Kiel from 1904 onwards. In 1907, he was chosen to become the director of the Copenhagen observatory, a position he was to keep until 1940. In 1916, he was elected a foreign member of the

[22] Brose to Cambridge University Press, 23 October 1919, Cambridge, University Library, Pr.A.B/1073(1–2).

Royal Astronomical Society following a proposal by Eddington.[23] The following year he entered the board of the Astronomische Gesellschaft, the (German) sister body of the British RAS. The Astronomische Gesellschaft was characterised by its international scope, and Strömgren was a perfect example. He was a Swede, the director of a Danish observatory, and eventually he presided over the Astronomische Gesellschaft between 1921 and 1930.

On 9 November 1917, the RAS council minutes note: "A letter from Prof. Strömgren was read, suggesting that the publications of the Society might be forwarded through him to Associates in enemy countries. It was resolved that no action be taken".[24] Despite this institutional reply, Eddington sent to Strömgren his *Report on Relativity Theory of Gravitation*. The Censor seems not to have blocked this mailing. The first meeting dated back to 28 July 1919. Strömgren spent a few days at the Cambridge Observatory. In Autumn 1919, the bond between them was reinforced thanks to mathematician Godfrey Hardy who visited Copenhagen in October. Back in England Hardy forwarded Eddington a suggestion by Strömgren: why not become a member of the council of the Astronomische Gesellschaft?[25] Eddington politely turned this proposition down. Both scientific bodies were not ready to tighten their relationship, but personal contacts might be encouraged. That is the reason why Eddington attended the meeting of the Astronomische Gesellschaft in Potsdam in June 1921. But no other British astronomer did (Kobold, 1921).

The Rome General Assembly of the IAU in 1922 was also very important from Strömgren's point of view. First, Denmark entered the Union.[26] Second, he was once again entrusted with the *Bureau des télégrammes astronomiques*. This Bureau had been created by the Astronomische Gesellschaft in 1884 and had in the first place been settled in Kiel. When the First World War broke out, several organisations—and notably the Royal Astronomical Society—refused to keep sending their messages to Kiel and asked for the Bureau to be transferred to a neutral country. Strömgren put forward the name of Copenhagen, and ran the Bureau until 1920. Suddenly, the Bureau was moved to Belgium under the authority of Georges Lecointe, the director of the Uccle observatory and a vice-president of the IAU. A couple of years later Strömgren took advantage of Lecointe stepping out and brought back the Bureau to Copenhagen (IAU, 1922, 232; RAS, 1922a, 180).

Eddington also undertook to revive international relations on the occasion of the Royal Astronomical Society centenary. As it had been created in 1820—by William Herschel—the Society should have celebrated its 100th anniversary in 1920, obviously. But the event could not be properly prepared until January 1922, presumably due to the aftermath of WWI. The date of May 30th was chosen, close

---

[23] See RAS Council Minute Book, vol. XI (1915–1928), p. 23 (London, RAS Library).

[24] RAS Council Minute Book, vol. XI (1915–1928), p. 53 (London, RAS Library).

[25] Strömgren to Eddington, 2 October 1919, Lund, Lund University Library, Elis Strömgren's Archive.

[26] For a complete list of countries adhering to the IAU in 1922, see IAU 1922, 4.

enough to the Rome General Assembly to encourage the attendance of the American delegates. In February, the secretaries of the Society sent an invitation to each fellow and to every associate. Including German associates. Strömgren wrote to Eddington: "It was a great pleasure for me to see that the R.A.S. has invited all Associates to the Centenary in June. I write to you not in your quality of President of the R.A.S. nor in your quality of a Cambridge-Professor but confidentially to a good Colleague to ask you whether you think it convenient and useful that German Astronomers come and that the Astronomische Gesellschaft sends an Address (probably to be presented by me)".[27] The only problem was that the secretaries were mistaken in thinking that the invitation was to be extended to *all* associates. They even were not far from being fired.[28] Eddington replied to Strömgren: "My frank opinion (not as President but as an individual) is that the presence of German Associates at the Centenary Celebration would not be desirable. It is with great reluctance I say this, remembering the very kind reception I had in Germany, but I am afraid they would not have such a satisfactory time here. Of course my own view is that there could be no better way of celebrating our centenary than by a complete reunion after the war, and if a question of this kind could be settled by a majority vote I have no doubt that this view would be endorsed. [...] It has been decided not to have any presentation of addresses from other scientific societies, so that I cannot make an exception for the Astronomische Gesellschaft; otherwise I should welcome your proposal as a very happy idea. But I see no objection to the Astronomische Gesellschaft sending a brief message of congratulation which could be read or mentioned from the chair at the meeting. Also if any German Associate sent a message of goodwill such as members unable to be present often do send, I would gladly read it along with others. (I might mention that Max Wolf is the best-known and most popular of our Associates and a word or two from him would go down well.)"[29] Wolf had been the director of Heidelberg observatory since 1902 and had been awarded the RAS Gold Medal in 1914. Moreover, he had not signed the "Manifesto of the Ninety-Three" in October 1914. This document, published under the title "A Call to the Civilized World", was, following Robert Fox's account, "a vibrant expression of national pride and resentment against the Allied powers, all cast in an unyielding tone that was to colour perceptions of Germany long after the Armistice of 1918" (Fox, 2016, 47). The list of signatories included several first-class scientists such as Max Planck, Wilhelm Ostwald, Fritz Haber and Walther Nernst.

The centenary meeting started at 11 am on Tuesday 30 May 1922. Eddington being then the president of the Royal Astronomical Society, he read the congratulations sent by King George V, followed by those from Baillaud (representing France), Kapteyn (The Netherlands) and Wolf (Germany). By doing so, he deliberately

---

[27] Strömgren to Eddington, 1 March 1922, Lund, Lund University Library, Elis Strömgren's Archive.

[28] See RAS Council Minute Book, vol. XI (1915–1928), p. 156 (London, RAS Library).

[29] Eddington to Strömgren, 8 March 1922, Lund, Lund University Library, Elis Strömgren's Archive.

placed the messages from a neutral country and a former enemy before those sent by American institutions. No doubt this was done on purpose. Eddington's own opening address closed with a humorous remark: "We celebrate today our centenary, and by a simple subtraction sum (subtracting 100 years from May 30th, 1922) you arrive at the date of our foundation, January 12th, 1820 according to the most accredited calculations—which I trust no mathematician here will venture to criticise. The great event caused a ripple in the ether which has spread out ever since in widening circles. [. . .] We shall look out on the heavens again tonight with renewed enthusiasm and joy; and if from the majority of the stars we can expect no more than an unrecognising stare, there are half a dozen old favourites which—we may fairly be persuaded—will give us an answering twinkle" (RAS, 1922b, 204).

Later on, Eddington wrote to Wolf: "It was a great pleasure to me to receive your kind letter to the Royal Astronomical Society and to read it at the opening of our proceedings on Tuesday [May 30th]. Your message was received most cordially by the Fellows; and if you could have heard the warm applause I am sure you would have been delighted. I had a considerable number of messages to read representing all the chief countries; these gave an excellent start to the meeting emphasising your own words that it was not an English festival only but a feast of all the citizens of truth".[30]

The celebration went on with an exposé by John Dreyer about the history of the Royal Astronomical Society. In the afternoon, six talks were given by some associates members, namely Robert Aitken, Ejnar Hertzsprung, Frederick Seares, Harlow Shapley, Elis Strömgren and Charles St. John: four Americans and one Dane from the Leiden observatory (Hertzsprung), not to mention the Swedish-born Danish Elis Strömgren.

4.6 From the Cambridge General Assembly to the Leiden General Assembly (1925–1928)

From 1922 onwards, Willem De Sitter was the only "non-Allied" vice-president of the IAU. In the winter of 1925, he took part in a series of letters written as a pressure onto IRC officials, so that the conditions of admission may be loosened in favour of the former Central Powers (Blaauw, 1994, 72–74). Several proposals were indeed made, some referring to the League of Nations membership, others allowing for a country to become an IAU member without having previously adhered to the IRC. All these were examined at the third IRC meeting held in early July 1925. And all were rejected as the statutory number of two-thirds of the votes of the adhering countries could not be reached (Fauque & Fox, 2021).

[30] Eddington to Wolf, 2 June 1922, Heidelberg, Universitätsbibliothek Heidelberg, Nachlass Max Wolf, Hs. 3695E.

Shortly after the IRC meeting occurred the second IAU General Assembly, held in Cambridge from 14 to 22 July. Disappointment and resentment seem to have been the shared feelings among the IAU delegates. During the Assembly, Vincenzo Cerulli first wrote to Frederick Stratton (the incoming General Secretary) in order to express the thoughts of his fellow astronomers: "The Italian delegation wishes to state that it is its opinion that no further meeting of the IAU should be held until it can be completely international. This is also quite in conformity with an address delivered by Prof. Strömgren to the President and the executive committee and signed by 59 attendants of the Cambridge meeting".[31] Among these signatories were such prestigious names as Giorgio Abetti, Frank Dyson, A.S. Eddington, John Evershed, Ejnar Hertzsprung, Bertil Lindblad, Hugh Newall, Jan Oort, Harold Spencer Jones, and Pieter Johannes Van Rhijn. American astronomers were no less direct: "The delegation from the United States is strongly of the opinion that the time has come when the removal of restrictions as to nationality is essential to the success of the Union as a truly international organization, and the delegation is not prepared to recommend the participation of astronomers from the United States in a future congress until the statutes permit astronomers of all countries to be present and to share in its deliberations".

The Belgium delegation wrote afterwards, explicitly following a motion adopted in February 1925 by the British National Committee for Astronomy, to request the executive committee of the IAU to ask the IRC to authorise the admission of the Central Powers when they become members of the League of Nations. Messages of support were sent by eight delegations, namely Canada, Czecho-Slovakia, France, Japan, Poland, Portugal, Romania, and Spain. On the last day of the General Assembly (i.e. 22 July) Strömgren was invited to express his views officially before the other delegates: "The Delegation of Denmark is of the opinion that the time has come when the removal of restrictions as to nationality is essential to the success of the Union as a completely international organisation" (IAU, 1925, 223).

At the end of this last meeting, Willem De Sitter was elected President of the Union. Among the vice-presidents were V. Cerulli, A.S. Eddington and the Japanese astronomer Shin Hirayama, all of whom had signed the Strömgren manifesto (IAU, 1925, 226). Needless to say, they all acted—and De Sitter most of all—towards an amendment of the IRC statutes and subsequently of the IAU, with the 1928 General Assembly in mind. These efforts proved fruitful, eventually. In 1926, under the pressure of several members including astronomers, the IRC voted a change in the rules of admission, and in parallel Germany was admitted within the League of Nations. All this led to the invitation of German astronomers in Leiden in 1928, and fourteen did participate (IAU, 1928; Wielen, 2019). Once again, Strömgren played an important role as an intermediary between the IAU and the German community. He was the one to forward the invitations. De Sitter even planned for an honorary doctorate to be awarded jointly to Karl Küstner and Henri Deslandres. The former

[31] Cerulli to Fowler, 20 July 1925, Paris, International Astronomical Union, IAU archives, file I.3.A. In comparison, 189 delegates attended the Cambridge GA.

had been the director of the Bonn Observatory until 1925, and the latter was at that time the director of the Paris Observatory. Both had been on active duty during World War I. What a sign of reconciliation!

References

Blaauw, A. (1994). *History of the IAU: The birth and first half-century of the international astronomical union*. Kluwer Academic Publishers.

Broglie, L. d. (1942). *La Vie et l'œuvre de M. Émile Picard*. Institut de France.

Brose, Henry. (1920). Translator's note. In Erwin Freundlich, *The foundations of Einstein's theory of gravitation*, viii. : Cambridge University Press. Reprinted 2011.

Dandelion, P. (2008). *The Quakers*. Oxford University Press.

De Sitter, W. (1916a). On Einstein's Theory of Gravitation, and its Astronomical Consequences. First Paper. *Monthly Notices of the Royal Astronomical Society, 76*, 699–728.

De Sitter, W. (1916b). On Einstein's theory of gravitation, and its astronomical consequences. Second Paper. *Monthly Notices of the Royal Astronomical Society, 77*, 155–184.

De Sitter, W. (1916c). Space, time, and gravitation. *The Observatory, 39*, 412–419.

De Sitter, W. (1917). On Einstein's theory of gravitation, and its astronomical consequences. Third Paper. *Monthly Notices of the Royal Astronomical Society, 78*, 3–28.

Dyson, F. (1917). On the opportunity afforded by the Eclipse of 1919 May 29 of verifying Einstein's theory of gravitation. *Monthly Notices of the Royal Astronomical Society, 77*, 445–447.

Dyson, F., Eddington, A. S., & Davidson, C. (1920). A determination of the deflection of light by the Sun's gravitational field from observations made at the total Eclipse of May 29, 1919. *Philosophical Transactions of the Royal Society of London A, 220*, 291–333.

Eddington, A. (1916). The future of international science. *The Observatory, 39*, 270–272.

Eddington, A. (1917). Karl Schwarzschild. *Monthly Notices of the Royal Astronomical Society, 77*, 314–319.

Eddington, A. (1919). The Total Eclipse of 1919 May 29 and the influence of gravitation on light. *The Observatory, 42*, 119–122.

Einstein, A. (1996). *The collected papers of Albert Einstein* (Vol. 6). Princeton University Press.

Einstein, A. (2004). *The collected papers of Albert Einstein* (Vol. 9). Princeton University Press.

Eisenstaedt, J. (2002). *Einstein et la relativité générale*. CNRS Editions.

Fauque, Danielle, & Fox, Robert. (2021). Binding the Wounds of War. Internationalism, National Interests, and the Order of World Science, 1919–1931. In Eckart, W. U., & Fox, R. (Eds.), Blockades of the mind – Science, academies, and the aftermath of the great war, *Acta Historica Leopoldina, 78*. Halle: Deutsche Akademie der Naturforscher Leopoldina.

Fox, R. (2016). *Science without Frontiers*. Oregon State University Press.

Guichelaar. (Jan, 2018). *Willem De Sitter: Einstein's Friend and Opponent*. Springer.

Hardy, G. (1970). *Bertrand Russell and Trinity*. Cambridge University Press.

Hentschel, K. (1994). Erwin Finlay Freundlich and testing Einstein's theory of relativity. *Archive for History of Exact Sciences, 47*, 143–201.

Hutchins, R. (2008). *British University Observatories: 1772–1939*. Ashgate.

IAU. (1922). *Transactions of the IAU*, vol. 1. London: Imperial College Bookstall.

IAU. (1925). *Transactions of the IAU*, vol. 2. London: Imperial College Bookstall.

IAU. (1928). *Transactions of the IAU*, vol. 3. Cambridge: Cambridge University Press.

Jeans, J. (1925). Address on presenting the Gold Medal of the Society to Sir Frank Watson Dyson. *Monthly Notices of the Royal Astronomical Society, 85*, 672–677.

Kennedy, T. C. (2001). *British Quakerism: 1860–1920*. Oxford University Press.

Kennefick, D. (2012). Not Only Because of Theory. In C. Lehner, J. Renn, & M. Schemmel (Eds.), *Einstein and the changing worldviews of physics* (pp. 201–232). Birkhäuser.

Kennefick, D. (2019). *No Shadow of a Doubt*. Princeton University Press.

Kevles, D. J. (1971). 'Into Hostile Political Camps': The reorganization of international science in World War I. *Isis, 62*, 47–60.

Kobold, H. (1921). Generalversammlung der Astronomischen Gesellschaft in Potsdam. *Astronomische Nachrichten, 214*, 97–104.

Kox, A. (1992). General relativity in the Netherlands, 1915–1920. In J. Eisenstaedt & A. J. Kox (Eds.), *Studies in the history of general relativity* (pp. 39–56). Birkhäuser.

Laguens, Florian. (2018). Eddington philosophe. *La nature et la portée de la science physique d'après Arthur S. Eddington*. PhD dissertation. Paris: Université Panthéon-Sorbonne.

Laguens, F. (2019). 'The sky seems heavy with clouds.' Arthur S. Eddington & the IAU. In C. Sterken, J. Hearnshaw, & D. Valls-Gabaud (Eds.), *Under one sky: the IAU Centenary Symposium* (pp. 178–183). Cambridge University Press.

Laguens, F. (2021). 'International science is bound to win.' Eddington, Strömgren et la coopération scientifique (1919–1922). In W. Eckart & R. Fox (Eds.), *Blockades of the Mind – Science, Academies, and the Aftermath of the Great War* (pp. 27–39). Deutsche Akademie der Naturforscher Leopoldina.

Monk, R. (1996). *Bertrand Russell*. Jonathan Cape.

Rae, J. (1970). *Conscience & Politics*. Oxford University Press.

RAS. (1918). Inter-allied conference on international organisations. *The Observatory, 41*, 414–416.

RAS. (1922a). Meeting of the International Astronomical Union at Rome. *The Observatory, 45*, 176–190.

RAS. (1922b). Royal Astronomical Society Centenary Celebrations. *The Observatory, 45*, 201–212.

Schuster, A. (1916). The common aims of science and humanity. In *British Association for the Advancement of Science, Report of the eighty-fifth meeting*. John Murray.

Solar Union. (1914). *Transactions of the international union for co-operation in solar research* (Vol. 4). Manchester University Press.

Stachel, J. (1986). Eddington and Einstein. In E. Ullmann-Margalit (Ed.), *The Prism of Science* (Vol. 2, pp. 225–250). D. Reidel.

Stanley, M. (2019). *Einstein's War*. Dutton.

Trueblood, D. E. (1993). *The People Called Quakers*. Friends United Press.

Wielen, R. (2019). Germany's difficulties in becoming a member of the International Astronomical Union. In C. Sterken, J. Hearnshaw, & D. Valls-Gabaud (Eds.), *Under One Sky: The IAU Centenary Symposium* (pp. 205–213). Cambridge University Press.

Wright, H. (1966). *Explorer of the Universe: A biography of George Ellery Hale*. Dutton.

Florian Laguens defended his PhD dissertation devoted to Arthur S. Eddington's philosophy at Panthéon-Sorbonne University (Paris) in 2018. He organised in May 2019 an international conference to celebrate the centenary of the 29 May 1919 eclipse, that was held at the Institut d'Astrophysique de Paris and brought together thirty physicists, philosophers and historians from a dozen countries. He is now a lecturer in Philosophy at IPC-Facultés Libres (Paris), where he teaches mainly Modern Philosophy and History of Science. He also teaches History of Physics at the Université de Paris.

Part II
After WWII: Divided Countries

Chapter 5
The China Crisis

Xiaowei Liu

Abstract The so-called *China crisis*, well documented in *History of the IAU* by Adriaan Blaauw and in *Under the Same Starry Sky: History of the IAU* by Chengqi Fu and Shuhua Ye, refers to the withdrawal in 1960 of the People's Republic of China (PRC) from the Union. The crisis stemmed from the admission by the IAU, amidst strong protest from PRC and some other member countries, of the Republic of China (ROC) to the Union, creating the so-called *"Two Chinas"*—or *"One China, one Taiwan"* problem. The crisis directly led to the absence of mainland Chinese astronomers from the stage of international collaborations and exchanges and was only solved two decades later. The solution, accepted by all the parties involved, is that China is to have two adhering organizations, with mainland China astronomers represented by the Chinese Astronomical Society located in Nanjing (China Nanjing) and China Taiwan astronomers represented by the Academia Sinica located in Taipei (China Taipei). The denominations "*China Nanjing*" and "*China Taipei*" represent the IAU official resolution and should be used in all IAU events.

The China crisis, probably the most serious one in IAU history, was a painful lesson in the 100-year development of the Union. Yet, with its eventual solution, the Union has emerged stronger, upholding its spirit of promoting astronomical development through international collaboration of astronomers from all regions and countries, regardless of the political systems, religion, ethnicity, gender, or level of astronomical development.

Keywords IAU · National membership · P.R. China

This chapter is republished by permission from the IAU from "Liu, X. (2018). The China crisis. Proceedings of the International Astronomical Union, 13 (S349), 222–227. https://doi.org/10.1017/S1743921319000346".

X. Liu (✉)
South-Western Institute for Astronomy Research, Yunnan University, Kunming, P.R. China
e-mail: x.liu@ynu.edu.cn

© The Author(s), under exclusive license to Springer Nature Switzerland AG 2022

T. Montmerle, D. Fauque (eds.), *Astronomers as Diplomats*, Historical & Cultural Astronomy, https://doi.org/10.1007/978-3-030-98625-4_5

5.1 The Entry and Early Involvement of China with the Union

China has a rich tradition in astronomical observations and studies and boasts the longest and most comprehensive ancient records of astronomical phenomena dating back to the Shang dynasty. In its 5000-year continuous civilization, China has nurtured world-class astronomers including Heng Zhang (78–139 AD), Shoujing Guo (1231–1316 AD), and Guangqi Xu (1562–1633 AD).

The works of Euclid of Alexandria, Claudius Ptolemy, Nicolas Copernicus, and Tycho Brahé were translated and introduced to and practised by Chinese astronomers in the late Ming dynasty, following the arrival of Jesuit priests (Matteo Ricci, Nicolaus Longobardi, Giacomo Rho, Johann Schreck, and Johann Adam Schall von Bell). In 1644, Johann Adam Schall von Bell (1592–1666 AD) was appointed as Director of the Imperial Observatory, the first time the position was held by a foreigner.

From the *First Opium War* in 1840, China was forced to open up ports under the weapons of the rising Western powers and was reduced gradually to a semi-colonial, semi-feudal country which ushered in a warlord era. In 1912, the last feudal dynasty Qing fell and the Republic of China (ROC) was founded. The Lugou (Marco Polo) Bridge Incident on July 7, 1937 marked the beginning of full-out invasion of China by Imperial Japan and the Eight-Year Chinese War of Resistance which followed and which ended in 1945 along with the World War II. This was followed by a bitter period of civil war between the ruling Nationalist Party (Kuomingtang—KMT) of Jieshi Jiang (Kai-shek Chiang) and the Communist Party of China (CPC) led by Zedong Mao (Tse-tung Mao). The war ended in 1949 with the founding of the People's Republic of China (PRC) and the retreat of KMT from mainland China to the Chinese island of Taiwan.

The World War II ended with the emergence of two political blocs, the Western Bloc led by the United States of America (to which the provincial KMT government in Taipei belonged) and the Eastern Bloc (with which the PRC was associated) led by the Soviet Union, which fought between them the Cold War that only ended decades later in 1992. In 1950, the Korean War broke out between the Eastern Bloc North Korea (the Democratic People's Republic of Korea—DPRK) and the Western Bloc South Korea (the Republic of Korea—ROK). The war evolved rapidly into one involving PRC forces on the DPRK side and US forces fighting for the ROK. The war ended with the *Korean Armistice Agreement* signed in 1953.

Several astronomical observatories were built in China by foreign Jesuit priests in the late nineteenth century, including Shanghai Xujiahui (1873) and Sheshan (1901) Observatories by the French, Hongkong Observatory (1882) by the British, and Qingdao Observatory (1898) by Germans. In 1922, Qingdao Observatory was reclaimed by China. Chinese meteorologist and astronomer returning from the Université libre de Bruxelles, Bingran Jiang (1883–1966) became director, commencing modern Chinese astronomical observation and research.

Meanwhile, more Chinese intellectuals studying abroad returned back to China. These included Lu Gao (1877–1947), inspired by French astronomer Camille Flammarion (1842–1925), Qingsong Yu (1897–1978) returning from the Lick Observatory, University of California, and Yuzhe Zhang (Yu-Cheh Chang; 1902–1986) from the Yerkes Observatory, University of Chicago. In 1926, the Mathematics Department of Sun Yat-sen University (SYSU) was renamed the Mathematics and Astronomy Department and began to offer astronomical courses, marking the beginning of astronomical higher education in China. In 1929, Yun Zhang (1896–1958) who studied astronomy at the University of Lyon built the SYSU Observatory and became its first Director. In 1928, the Institute of Astronomy, Academia Sinica was founded. On October 30, 1922, the Chinese Astronomical Society (CAS) was founded with 47 members at the Beijing Ancient Observatory (constructed in 1442 during the Ming Dynasty on the original site first built by Shoujing Guo in 1279 of the Yuan Dynasty). Lu Gao became its first President. By 1947, the CAS had 688 individual and 6 institutional members.

Initiated by Yuanpei Cai (1868–1940), the first Director (1928–1940) of Academia Sinica and President (1916–1927) of Peking University, Purple Mountain Observatory (PMO), the best astronomical observatory at the time in the Far East, was inaugurated in 1934. Since its establishment, the CAS took liaison with the Union and collaboration with the international community was one of its vital tasks. CAS member Yun Zhang attended in 1925 the 2nd IAU General Assembly (GA) in Cambridge as an observer, and members Qingsong Yu and Jinyi Zhao participated in 1928 in the 3rd GA in Leiden. All these individuals paved the way for China joining the IAU. In 1935, at the 5th GA in Paris, China was formally admitted to the Union as its 26th National Member, with the CAS located at PMO in Nanjing as the adhering organization. Four initial individual members, Qingsong Yu, Lu Gao, Bingran Jiang, and a Japanese astronomer Shinjo Shinzo (1873–1938), were also admitted. At the 6th GA in 1938 in Stockholm, China's individual members increased to 11.

In spite of the war, Chinese astronomers did their best to engage with the Union and paid the arrears in 1947. After the civil war and the establishment of PRC in 1949, a delegation of four astronomers from mainland China, invited by President Otto Struve (1897–1963), including Yuzhe Zhang, attended the Dublin GA in 1955. The PRC resumed its legitimate National Membership in the Union and cleared the arrears.

5.2 The Conflict and the Withdrawal of China from the Union

The "application" for Union membership from the ROC, submitted in late April and early May 1958, just two months before the Moscow GA, was not initiated by astronomers from the island (there were few if any at the time) but was orchestrated by the US government in order to further isolate the PRC after the Korean War, at the height of the Cold War.[1] It came at a time when the US National Committee for Astronomy (NCA) was about to submit their invitation to the EC to host the 1961 GA in Berkeley, after the 1958 GA in Moscow. The plot by the US Department of State, using visas to the US as a threat, was deliberately designed to (one stone, three birds) (1) block astronomers from the "communist" mainland China, (2) promote the status of "Free China (Taiwan)," and (3) create the so-called "*Two Chinas*" or "*One China, one Taiwan*" problem.

The plot was opposed by Leo Goldberg (US NCA), Otto Struve (former IAU President and still consultant to the EC), and Detlev Wulf Bronk (President, National Academy of Sciences), worrying that this would tarnish the US scientific reputation. In spite of the pressure from the US government, the EC decided to postpone any decision until after the Moscow GA, in recognition of "*the serious implications*" that "*acceptance of the Taiwan application during the Moscow meetings might have had: immediate withdrawal of mainland China from the IAU, and possibly also that of USSR, the host*" (Blaauw 1994, p. 193).

Goldberg contacted his representative in the US Congress, George Meader, a conservative and fair-minded Republican, who presented the case to John Foster Dulles (the US Secretary of State), who referred it to his science advisor, Wallace Brode. Brode promptly demanded that Taiwan be invited to the IAU.

> The fact that Taiwan then had no astronomers and would have to qualify for IAU membership in the approved way meant nothing to the militant anti-communist Brode. Brode wanted Goldberg to go to the 1958 Moscow meeting and submit the 1961 invitation but with the condition that Taiwan be admitted at once. Such a demand could well wreck the IAU. From Brode's point of view, if the astronomers would not go along with his orders, so much the worse for them. [Biographical Memoirs, US National Academy of Sciences, 1997; (cited from Fu and Ye 2009, p. 154)].

Unfortunately, after the Moscow GA, the attitude of the EC made a U-turn. In spite of opposition by the IAU Vice Presidents from the USSR and Czechoslovakia, who stated that admission should be judged solely on scientific grounds, that the astronomical activity in this applicant's country was too low, and that the admission of Taiwan might risk the withdrawal of mainland China, the EC pushed through the admission by the ballot, with five votes for and two against.

Yuzhe Zhang, President of the CAS in Nanjing, in his letter to J.H. Oort, serving President of the IAU, expressed surprise and indignation:

> ...Taiwan is an inseparable part of Chinese territory, it is a province of China...Should the report be authentic, I, on behalf of the Astronomical Society of the People's Republic China, hereby lodge our strong protest with you and insist that the Executive Council of IAU rescind the illegal decision...Otherwise, the Astronomical Society...will resolutely and definitely withdraw from the IAU (Blaauw 1994, p. 193).

---

[1] The "Chinese Astronomical Society" in Taipei was only established in July 1958, two months after the application submission.

IAU *Information Bulletin* (IB) No. 2 of November 1959 announced the adherence of Taiwan as a member of the IAU. The withdrawal of mainland China was announced in IB No. 3 of May 1960. The rapid announcements underscored the acceptance as a "*fait accompli.*"

The EC further brushed aside letters of protest from the Polish and Bulgarian Academies sent in March 1960, as well as concerns raised by Vice President O. Heckmann during the 1961 GA. Also during the Assembly, President Oort, in an unusual move, before the vote, asked representatives to vote against a combined motion submitted by the USSR and Czechoslovakia Academies of Sciences requesting the decision of the EC to admit Taiwan be revoked.

The decision of the EC, succumbed to the political pressure from the US government, led to the two-decade absence of astronomers of mainland China from the international stage of astronomical collaboration and exchange. Squeezed between the demands of the then two superpowers, the United States of America and the Soviet Union, the IAU survived but unfortunately was clearly damaged.

5.3 The Return of China

In the following two decades, in spite of the growing dissatisfaction and concern, restrained by the then prevailing political environment, both international and domestic, little progress was made on the restoration of the legitimate position of China in the IAU until the 1979 Montreal Assembly. Several events before then eased the way forward from the prevailing deadlock:

1. In 1971, UN Resolution No. 2758 restored the membership of the PRC and expelled the ROC from the UN.
2. In 1972, Richard Nixon visited China. The joint communiqué recognized Taiwan as a part of China, thereby ushering in a new era of Sino-American relations.
3. In 1976, the ten-year Cultural Revolution in mainland China ended.
4. In 1978, China started the economic reforms and opening up.

All these paved the political way for China rejoining the Union.

The key was to find "*a way acceptable to the Chinese of reinstating their membership in the IAU without expelling Taiwan, an action that would violate the statutes of the Union*" (Goldberg 1977, see p.20).

Initiated by President Adriaan Blaauw, under the invitation of the EC, a six-member delegation from mainland China, including Yuzhe Zhang, Shuhua Ye (Vice President of the IAU 1988–1994), and Zhaohua Yi, and a single-member delegation from Taiwan represented by C.S. Shen, President of the Taiwan NCA, arrived in Montreal to discuss the matter prior to the GA on August 13–24, 1979 and to explore the possibility of "dual membership," that, on the one hand, expresses the indivisibility of China (emphasized and agreed by both parties across the Straight), and, on the other hand, reinstates the membership of PRC without blocking China Taiwan's further adherence.

reiterating that the Chinese Astronomical Society opposes any solutions which can suggest the existence of "two Chinas" or "one China, one Taiwan";

the Chinese Astronomical Society proposes to the IAU Executive Committee

a) that for the time being dual adherence should be introduced, the adherence and the adhering bodies to be specified as:

China — Chinese Astronomical Society (Purple Mountain Observatory, Nanking)
— (a name for the adhering organization representing the astronomical community of Taiwan; for this name we propose "Chinese Taiwan Astronomical Society" or, provisionally, as reference "Astronomy Taiwan, China"; the name to be established by the IAU Executive Committee in consultation with the two adhering organizations);

Fig. 5.1 Parts of the letter from Yuzhe Zhang, President of the Chinese Astronomical Society (1943–1982), to Adriaan Blaauw, President of the IAU (Zhang & Blauuw 1979, pp. 525–527)

The negotiations resulted in a proposal communicated to the GA at its closing session, which was presented in the form of an exchange of letters dated Montreal, August 22, 1979, between the President of the CAS, Yuzhe Zhang, and the outgoing President of the IAU, Adriaan Blaauw. The two letters were reproduced in full in IAU *Transactions*, Vol. XVII (Bappu 1979, pp 48–50). The only problem left was the name of the adhering organization in Taiwan (Fig. 5.1).

The problem was shortly solved in 1980. The arrangements were ratified at the 1982 Patras GA XVIII. The solution, accepted by all the parties involved, is that China is to have two adhering organizations, with mainland Chinese astronomers represented by the Chinese Astronomical Society located in Nanjing (*China Nanjing*, since 1935) and *China Taiwan* astronomers represented by the Academia Sinica located in Taipei (China Taipei, since 1959). The denominations "*China Nanjing*" and "*China Taipei*" represent the IAU official resolution and should be used in all IAU events. At the time of this writing (August, 2018), "*China Nanjing*" and "*China Taipei*" have, respectively, 666 and 74 IAU individual members.[2]

> The Union had succeeded in overcoming political schism and in restoring harmony again among its membership for its prime purpose: the unhampered pursuit of scientific research and intercourse (Blaauw 1994, p. 204).

Shouguan Wang (President of the CAS, 1985–1989), on behalf of the CAS, gave a warm speech after the ratification of the membership of China by the GA (Wang 1983, p. 26):

> Dear friends & colleagues,
>
> The Chinese Astronomical Society celebrates its 60th anniversary this year. Its reunion with this international community today is an event that is highly appreciated by all its 900 members. I and my colleagues here are very glad to have this opportunity of speaking on behalf of our Society and its members to express our most cordial greetings and most sincere thanks to you all. Thank you!

[2] www.iau.org/administration/membership/national.

Zdeněk Kopal, writing on *The IAU—the first 60 years. Reminiscences and Reflections*, published in *Astrocosmos*, the newspaper of the Patras GA, wrote

> ... the Union has really never been free of political interference from many directions ever since. Perhaps the most conspicuous example of such an interference in recent years was the technical expulsion of the (People's) Republic of China, which was eased out of our midst in 1955 (1960) by the United States (during the enlightened era of John Foster Dulles), in collaboration with certain astronomers from Western Europe. Only God knows what good should have come to the science of astronomy and to the International Astronomical Union from severing (albeit temporarily) its official ties with the most populous nation of the Earth; but such acts did happen, and will continue to happen as long as the present structure of the IAU remains unchanged (Kopal 1982).

The resolution of the China crisis, combined with the continued improvement of relation across the Taiwan Straight, has benefited astronomers on both sides, stimulated and facilitated collaborations amongst them as well as with the international community at large. Astronomy in both mainland China and the Island has entered an era of rapid development. Four decades after the crisis, China is poised to make major contributions to the world astronomy research and education development.

In 2012, the IAU GA XXVIII was held in Beijing, being the first time in China in the Union's nearly 100-year history. In his opening ceremony speech, Vice President Jinping Xi remarked:

> The development of science and technology requires extensive international cooperation. Science and technology have no nationality! The vast expanse of space is the common home of all humankind; to explore this vast universe is the common goal of all humankind; astronomy in fast development is the shared fortune of all humankind (Xi 2012).

References

Bappu, M. K. V. (1979). Report of the XVIIth General Assembly of the IAU. *Transactions of the IAU, 17B*, 15. Published by Association of University for Research in Astronomy.

Blaauw, A. (1994). *History of the IAU: The Birth and First Half-Century of the International Astronomical Union* (296 pp). Dordrecht: Kluwer.

Fu, C., & Ye, S. (2009). *Under the Same Starry Sky: History of the IAU*. Shanghai: Jiaotong University Press.

Goldberg, L. (1977). *Transactions of the IAU, 16B*, 15. Published by Association of University for Research in Astronomy.

Kopal, Z. (1982). *Astrocosmos*. Greece: Newspaper of the IAU GA in Patras, 2, 5.

Wang, S. G. (1983). *Transactions of the IAU, 18B*, 15. Published by Association of University for Research in Astronomy.

Xi, J. P. (2012). *Inquiries of Heaven*. (Newspaper of the IAU GA in Beijing, Wednesday, August 22, 2012) 3, 1.

Zhang, Y. Z., (Chang, Y. C.), & Blaauw, A. (1979). Appendix IV, Letters exchanged between the President of the IAU and the President of the Chinese Astronomical Society at Nanking. *Transactions of the IAU, 17B*, 525–527. Published by Association of Universty for Research in Astronomy.

Xiaowei Liu is a Chair Professor of Yunnan University. He completed in 1986 his college education in astrophysics at Peking University and received his PhD degree in 1992 from the Beijing Astronomical Observatory (now the National Astronomical Observatories), Chinese Academy of Sciences. He then joined the Department of Physics and Astronomy, University College London, first as a postdoctoral, then as a senior research fellow. He joined the faculty of Department of Astronomy of Peking University in 1999, as a Cheung Scholar and University Chair Professor. He was a major founder of the Kavli Institute for Astronomy and Astrophysics at Peking University, serving firstly as associate and then deputy director. He moved to Yunnan University in 2017 and founded the South-Western Institute for Astronomy Research. He was a vice president of the International Astronomical Union (IAU) between 2012 and 2018.

Chapter 6
When China Left the IAU: A Reappraisal

Thierry Montmerle

Abstract The most important crisis in the post-WWII history of the IAU is undoubtedly "the China crisis", in which China withdrew from the IAU in protest against the admission of Taiwan by the Executive Committee in 1959. This crisis lasted until 1982, when China officially re-joined the IAU. While the highlights of this story can be found in the reference book by Adrian Blaauw (IAU President, 1976–1979, and a major actor of the return of China) in his *History of the IAU*, published in 1994, many details remained unsettled or obscure (like the implication of the US State Department in pushing for the admission of Taiwan).

On the occasion of the IAU Centenary (2019), I took up the task of revisiting this painful episode with data that were unavailable (or not used) by Blaauw: going through the IAU Archives on China kept by the Secretariat in Paris, having access to Chinese documents (that were translated on this occasion), broadening the picture by revealing information about the role of ICSU (the International Council of Scientific Unions), and even bringing up analogies with the Olympic Games and the context of high military tension between China and the USA across the Taiwan Strait in 1958. My conclusion is that the admission of Taiwan (by the IAU but also simultaneously by all the scientific unions) was the inevitable result of the geopolitical situation at the time, only dramatized in the case of the IAU by the fact that, coincidentally, its following General Assembly was scheduled to take place in the USA (in Berkeley) in 1961.

6.1 Introduction

In 1958, while the IAU was happy to hold its General Assembly (GA) "behind the Iron Curtain" for the first time (in Moscow), a related crisis unfolded that would take twenty years to solve. This crisis had its roots at the other side of the world, in

T. Montmerle (✉)
Institut d'Astrophysique de Paris, Paris, France
e-mail: montmerle@iap.fr

© The Author(s), under exclusive license to Springer Nature Switzerland AG 2022
T. Montmerle, D. Fauque (eds.), *Astronomers as Diplomats*, Historical & Cultural Astronomy, https://doi.org/10.1007/978-3-030-98625-4_6

one of the three countries "divided" between the Western and Eastern political blocs in the aftermath of WWII: China. The other two were Germany and Korea.[1]

On Oct. 1, 1949, after a bloody civil war broke out between Communist and Guomintang (Nationalist) forces following the Japanese surrender in 1945, Mao Zedong seized power and established the People's Republic of China (PRC), with Peking (now Beijing)[2] as its new capital, and his historic rival, Chiang Kai-shek, head of the Republic of China (ROC) since 1928, retreated to the island of Taiwan. The island had been occupied by Imperial Japan from 1895 until 1945, and is separated from the mainland by a strait 130 km wide. As a result, the ROC had shrunk from being as large as the USA and having about 400 million inhabitants (in 1949) to an island a little smaller than the Netherlands, with a population of eight million (to which two million would be added, fleeing from the mainland). Yet, being a member of the United Nations since 1945 and one of the victors of WWII, the ROC even had a seat at the UN Security Council, and was officially recognized by most countries, except those of the Eastern bloc.

Thus, while Mao and Chiang both shared the dream of a "unified" China (with different political regimes, of course) in practice China was (and still is) divided into two "countries" (politically speaking), or two "territories" (geographically speaking). For the IAU, the "China crisis" started when Taiwan applied for membership to the IAU in 1958, a membership that its Executive Committee (EC) accepted in 1959, and which provoked the withdrawal of (mainland) China in 1960, followed by the ratification of the EC decision in 1961 at the next GA, which was held in Berkeley (USA). Eventually, China re-joined the IAU (side by side with Taiwan) at the Montreal GA in 1979, some arrangements being delayed until the following GA in Patras (Greece) finally ratified its renewed membership in 1982. This was certainly the longest, most important and painful, crisis in the post-WWII history of the IAU.

A nice summary of "the China crisis" is presented by my Chinese colleague Xiaowei Liu in the present volume,[3] and this crisis is also mentioned by my Dutch colleague George Miley,[4] but the most extended work to date on this period remains a 15-page chapter (#9) in the well-known book by Adriaan Blaauw, who was IAU President (1976–1979) and a key architect of the return of China to the IAU (Blaauw,

[1] Not to forget Vietnam, emerging from the French colonial rule in 1954. But Vietnam joined the IAU much later, in 2009, whereas China joined the IAU in 1935 (this Chapter), Germany in 1951 (Chap. 11), and (North) Korea in 1961 (Chap. 10 and Sect. 6.5 below).

[2] Throughout this chapter I will preferentially use the "old" western transcription for names, rather than the so-called pinyin modern transcription introduced by China in 1979. This is for consistency with the IAU archival documents which, for the period considered here, and including those of Chinese origin, all use the "old" transcription. When necessary I indicate the correspondence between the two, including the reverse order for family names vs. given names, e.g., "Y.C. Chang" (standing for "Yu-Che Chang"), vs. "Zhang Yuzhe" (see below).

[3] Chap. 5, and Liu (2019).

[4] Chap. 12.

1994: *History of the IAU*).[5] Yet, to cover such a long period of time (almost twenty-five years) in 15 pages is very short, and since its publication a lot of new material has become available, allowing a much deeper study to be conducted.

First and foremost, there are the documents and archives of the IAU. Many documents are now available online (minutes of EC meetings since 1950; proceedings of GAs, known as "Transactions of the IAU;" GA daily newspapers, etc.; see the list of References). For the Chap. 9 of his book (which he entitled "The China conflict"), Blaauw appears to have used mostly his personal archives, covering only the "late period" of the crisis, and some correspondence stored in the University of Berkeley Bancroft Library covering the period just preceding the admission of Taiwan (1957–1958), to which, for this reason, I did not need to consult. But on the other hand, the IAU archives in Paris cover the whole twenty-five-year period, and have been essentially untapped. They give a fascinating "insider's view" of the China crisis. For instance, I discovered an important handwritten and confidential document, in the form of a "Personal Account" concerning the circumstances of the admission of Taiwan, by Donald Sadler, General Secretary at that time. This five-page "account" had been written eighteen years later, in December 1976, at the request of Patrick Wayman, who was IAU Assistant General Secretary during the Blaauw presidency.

Also, in opening his Chap. 9 Blaauw wrote that "little will be said about the efforts towards reconciliation made by ICSU." But shortly thereafter Frank Greenaway published what is considered now as a reference book about ICSU (1996: *A history of the International Council of Scientific Unions*),[6] which helps putting the IAU decision to admit Taiwan and subsequent events in a much broader perspective.

Last, but not least, is the necessity to reflect the Chinese viewpoint. In addition to the (short) Chapter by Xiaowei Liu, and thanks to him, I had access to a book published in 2009 (in Chinese) by Fu Chengqi and Ye Shuhua (she had been IAU Vice-President from 1988 to 1994): *Under one starry sky.*[7] The relevant part of this book (its Chap. 5) has been translated especially for the present work by Yi Zhou, a Chinese journalist and translator who also kindly wrote a Chapter for this volume.[8]

The story of "the China crisis" that emerges is very rich and much more complex than I expected on the basis of published work (mainly Blaauw), and is the subject of a separate book[9] (hereafter quoted as "MZ22"). This book includes in particular a new approach (a "reappraisal") of the events having led to the admission of

[5] Referred to as "Blaauw's book" (Blaauw, 1994) for short in the present chapter. The full title is "*History of the IAU: The Birth and First Half-Century of the International Astronomical Union.*" It is important to note that this book was, in a way, commissioned by the IAU (see the book's Preface), while the present chapter is not. A more recent history of the IAU has been published on the occasion of its centenary, but the China crisis is mentioned only briefly (Andersen et al., 2019).

[6] Referred to as "Greenaway" (Greenaway, 1996) for short.

[7] Referred to as "Fu&Ye" (Fu & Ye 2009) for short.

[8] About the Chinese astronomer Y.C. Chang (Zhang Yuzhe): Chap. 7.

[9] Montmerle and Zhou (2022).

Taiwan and to the withdrawal of China, as well as a previously unpublished, detailed account of the long separation period between withdrawal and return (1960–1979). This period involves such major events as the Cultural Revolution, the admission of China to the UN in 1971 and its consequences, the roles of ICSU, UNESCO, and, surprisingly, even the role of the Olympic Games, as well as the non-IAU initiatives of individual astronomers on both sides to keep in contact in spite of the obstacles, etc.

Here I have opted to concentrate on the most contentious, and still somewhat controversial, episode of the admission of Taiwan and the withdrawal of China, which covers the period 1958–1961, in the light of the new documents mentioned above.

6.2 China and the IAU: Together Since 1935

Modern China starts with the establishment of the "Republic of China" in on Jan. 1st, 1912, after a revolution (called the Xinhua Revolution) overthrew the Manchu Qing dynasty, which had been ruling China since 1636. To separate itself ostensibly from the past, the new Republic adopted Nanking (now Nanjing) as its new capital, demoting Peking (now Beijing) and its Forbidden City, home to the Chinese Emperors since 1420. In spite of many difficulties, political as well as economic, an intellectual elite, often educated abroad (notably in the USA and in France) managed to establish in 1928 an Academy of Sciences, best known under its "Latin" name: the *Academia Sinica*. Several years earlier, in 1922 (the year of the first IAU General Assembly in Rome), a group of 47 astronomers had founded the Chinese Astronomical Society (CAS) in Peking, which subsequently moved to Nanking in 1932. The famous "Purple Mountain Observatory," near Nanking, was inaugurated in 1934.[10]

Thanks to the creation of the Academia Sinica, Chinese scientists could open to the outside world and join their colleagues abroad in various scientific unions: for instance, China joined IUPAP (the *International Union of Pure and Applied Physics*) in 1934, the IAU in 1935 (at its Vth General Assembly in Paris), and ICSU (the *International Council of Scientific Unions*, itself replacing in 1931 the International Research Council, created in the aftermath of WWI in 1919),[11] in 1937.

In the background, however, turmoil prevailed and the political situation in China became increasingly difficult. After having colonized Manchuria in 1931, Imperial Japan continued its expansionist policy (which had started with the occupation of Taiwan in 1895, and the annexation of Korea in 1910) and invaded China in 1937 (the so-called "Second Sino-Japanese war," some say the real start of WWII in Asia). The main cities of China, Peking, Nanking, Shanghai, soon fell, with a

[10] Xiaowei Liu, Chap. 5

[11] D. Fauque & R. Fox, Chap. 3. See also Greenaway (1996).

terrible human toll. The head of China (since 1928), Generalissimo Chiang Kai-shek, moved the nation's capital to Chungking (Chongqing), then the remote capital of the Sichuan province. This is where a group of astronomers from Nanking fled, under the leadership of Y.C. Chang (now Zhang Yuzhe),[12] whom we will meet often below. These astronomers were nevertheless able (quite heroically) to establish a new observatory at 2000 m altitude near Kunming in the Yunnan province, 650 km South-West of Chungking (now the Yunnan Observatories, comprising several institutes). This observatory became the wartime capital of astronomy in China: out of 11 astronomers registered as IAU members at the time of the Stockholm General Assembly (1938), six belonged to the Kunming Observatory (not including Y.C. Chang himself, based at his headquarters in Chungking).[13]

After 1945, the Guomintang government moved back to Peking, and Zhang and the Chinese Astronomical Society returned to Nanking. However, civil war raged. In 1948, the Nationalist troops were retreating and the government transferred the Academia Sinica in Taipei, the capital of the liberated "Taiwan province." As Communist troops approached Nanking, the government instructed astronomers to go to Canton or Taiwan via Shanghai; however, Zhang and most, but not all, of his colleagues chose to remain in Shanghai. They waited there for several months until the city was taken over without a fight by the People's Liberation Army on May 27, 1949. They moved back to Nanking and the Purple Mountain Observatory, which then hosted the CAS, since the Academia Sinica had left the city.[14]

For the IAU, this was not a problem: since the CAS was its adhering organization (and *not* the Chinese government itself), it was natural that its corresponding "National member" would be the PRC (where the CAS was geographically located) rather than the ROC (Taiwan). And in fact, Chinese astronomers (i.e., PRC nationals) attended the IAU GAs after the creation of the PRC: understandably none in Rome (1952), but four in Dublin (1955), and eight in Moscow (1958).

6.3 Enter Taiwan: *Coup de théâtre* in Moscow

The Xth General Assembly opened in Moscow on Aug. 12, 1958, the first to be held behind the "Iron Curtain." The idea was not new: as early as 1948, at the VIIth GA in Zürich (the first GA after WWII), A.A. Mikhailov, the first Soviet IAU Vice-President, had extended an invitation to the IAU to hold the next GA in Leningrad in 1951. The EC was divided: this was still the Stalinist USSR, the Berlin blockade had ended only two years before, the Korean War (mainly between the USA and China in the background) had just started (1950). Eventually, as recorded in the minutes of the EC meeting held in Paris (#13, 1951), the difficulties in visas, transportation,

[12] See his biography by Y. Zhou (Chap. 7).

[13] Transactions Vol.VI (1939), p. 502

[14] For more details on this episode, see Chap. 7

and accommodation prevailed, at least officially, and the invitation was turned down. Then Rome was presented as an alternative venue by Italy at the last moment, and the next GA took place there, but understandably with a one-year delay (1952), which triggered a new three-year cycle. The Soviet invitation was renewed at the Dublin GA in 1955, with Moscow as the venue, and this time it was accepted. As a show of good will, even though a firm decision could not be taken on the spot, the USA followed suit, with an invitation to hold the following GA (in 1961) either on the West Coast or on the East Coast.

For the USSR, the timing of the GA could not have been better: the previous year, on Oct. 4, the country successfully launched the Earth's first artificial satellite, *Sputnik-1,* defining the birth of the space age. The Soviets also invented the first "daily newspaper," featuring scientific news contributed by participants, as well as local information (including excellent cartoons), all in three languages (Russian, English, and French), an initiative which subsequent GAs all repeated in various forms (but never in three languages). Named "KOCMOC" (*Cosmos*) in Russian, its front page boasted "Science needs peace" (Fig. 6.1). A promising article written by Y.C. Chang, the head of the Chinese delegation, was entitled "Astronomy in the New China (Cosmos, Issue #4, p.3)." The Cold War seemed far away from Moscow.

As is the case at each GA, new IAU Executive Committee members would be elected, for one or two three-year terms. As per the Statutes in force at the time, the President had to change every three years, while changing the General Secretary was mandatory only after two terms (normally six years). At the end of their term(s), the past President and General Secretary become "Advisers": they participate in the discussions, but cannot vote. The EC also comprises six Vice-Presidents, renewed by half at each GA: a given Vice-President may thus accept to serve one or two terms (most agree to serve two terms). The "outgoing" EC (which has been responsible for supervising the GA) serves throughout the sessions of the GA as an "acting" EC, while the "incoming" EC takes over at the end of the GA; needless to say, there are many interactions between the two ECs (which in fact have three Vice-Presidents in common) during the GA to ensure a smooth transition.

In Moscow, the "outgoing" EC (1955–1958) was composed of André Danjon (President, France), Pieter Th. Oosterhoff (General Secretary, the Netherlands), Donald Sadler (Assistant General Secretary, UK), and six Vice-Presidents: André Couder (France), Otto Heckmann (Germany), Boris Kukarkin (USSR), Eugeniusz Rybka (Poland), Pol Swings (Belgium), and Richard van der Riet Woolley (UK), with Otto Struve (USA, past President, 1952–1955) as Adviser. The "incoming" EC (1958–1961), elected during the GA and taking over thereafter, was composed of Jan Oort (President, the Netherlands), D. Sadler (General Secretary),[15] and six Vice-Presidents: two would remain, in their second term (O. Heckmann, and B. Kukarkin), and four were newcomers (Leo Goldberg, USA, the Chair of the Local Organizing Committee for the next GA, to be held in Berkeley in 1961;

[15] There was no "Assistant General Secretary" in this EC because D. Sadler would remain General Secretary in 1961.

Сегодня в Москве открывается X съезд Международного Астрономического Союза

C'est aujourd'hui qu'aura lieu à Moscou l'inauguration de la X-me Assemblée Générale de l'U.A.I.

The Xth General Assembly of the International Astronomical Union Opens To-day

COSMOS КОСМОС

12 1958 АВГУСТ AUGUST AOUT

Газета Организационного комитета по подготовке X съезда Международного Астрономического Союза
Gazette du Comité d'Organisation de la X-me Assemblée Générale de l'Union Astronomique Internationale
Newspaper of the Organizing Committee for the Xth General Assembly of the International Astronomical Union

NOTRE ASSEMBLEE DOIT ENCORE MIEUX RAPPROCHER LES ASTRONOMES

A. DANJON, President de l' U. A. I.

Voici enfin cette journée impatiemment attendue où l'Union Astronomique Internationale, répondant à l'invitation de l'Académie des Sciences de l'Union des Républiques Socialistes Soviétiques, ouvre à Moscou sa X-e Assemblée Générale. Des astronomes sont venus nombreux de toutes les parties du monde, pour participer à nos réunions de travail. Mais quelque chose de plus les attirait à Moscou: la perspective de nouer avec leurs collègues de l'Union Soviétique des liens de confraternité scientifique plus étroits. Le rôle de l'Union Astronomique Internationale n'est pas seulement de constater les progrès récents de l'astronomie et d'élaborer des plans de recherches, mais encore de rapprocher ceux qui, à travers le monde, s'occupent de mêmes sujets. A cet égard, cette X-e Assemblée Générale qui réalise la conjonction des astronomes de l'Est et de l'Ouest, aura sans aucun doute les plus heureuses conséquences, pour le progrès de l'Astronomie.

НАШ СЪЕЗД ДОЛЖЕН ЕЩЕ БОЛЬШЕ СБЛИЗИТЬ АСТРОНОМОВ

Президент МАС проф. А. ДАНЖОН

Наконец наступил долгожданный день, когда Международный Астрономический Союз по приглашению Академии наук СССР открывает в Москве свою X Генеральную Ассамблею. Многочисленные астрономы съехались из ... Международного Астрономического Союза заключается не только в том, чтобы констатировать новые достижения астрономии и вырабатывать планы научных исследований, но также в том, чтобы сблизить людей, работающих в разных частях света над одними и теми ...

НАУКЕ НУЖЕН МИР

В. АМБАРЦУМЯН, председатель Организационного комитета по проведению X съезда МАС

За три года, протекших со времени Дублинского съезда МАС, астрономия сделала огромный шаг вперед. Достаточно посмотреть, какие изумительные средства исследования появились за это время. Гигантский радиотелескоп в Манчестере, несколько новых очень крупных оптических телескопов и, наконец, искусственные спутники Земли, открывшие эпоху освоения космоса человеком, помогают сегодня нам в решении астрономических проблем.

Мне кажется, что в последующие три года мы станем свидетелями еще более поразительных достижений как в области средств научного исследования, так и в деле раскрытия загадок Вселенной.

Но, чтобы эти перспективы стала реальностью, нужно, чтобы не было войны. Поэтому мы, астрономы, всем своим существом должны стремиться к миру.

LA SCIENCE AU SERVICE DE LA PAIX.

V. Ambartsoumian
President du Comité d'organisation de l'U.A.I.

Pendant les trois années écoulées depuis la réunion à Dublin de l'Union Astronomique Internationale l'astronomie a fait de grands progrès. Il suffit de voir les moyens nouveaux d'observation et de recherche qui ont été établis pendant ce temps. L'immense radiotélescope de Manchester, plusieurs nouveaux modèles de télescopes optiques et enfin les satellites artificiels qui marquent le début de la conquête de l'espace cosmique par l'homme, nous aident aujourd'hui à résoudre les problèmes astronomiques.

Il me semble que dans les trois années à venir nous serons les témoins de succès frappants, tant en ce qui concerne les moyens de recherche scientifique, que dans la résolution des énigmes de l'univers.

Mais pour que ces perspectives puissent se réaliser il faut qu'il n'y ait pas de guerre. C'est pourquoi nous devons, nous les astronomes, de toutes nos forces défendre la paix.

SCIENCE NEEDS PEACE

V. AMBARZUMJAN, Chairman of the Organizing Committee of the IAU

In the three years that have passed since the General Assembly of the IAU was held in Dublin astronomy made a great stride ahead. It suffices to see what wonderful means of research appeared in this period of time. A gigantic radio-telescope in Manchester, a few very large new optical telescopes and finally the artificial Earth satellites that have opened the epoch for the mastery of the Cosmos by man are now of aid to us in the solution of astronomical problems.

I presume that in the following three years we will witness still more astonishing achievements both in the field of the means for scientific research and in the task of solving the Riddles of the Universe.

These great perspectives will become a Reality only if there will be no war. Therefore, we, astronomers, are bound to strive with all our zeal for Peace.

Fig. 6.1 Front page of the Moscow GA daily newspaper *KOCMOC*, the first of its kind [https://www.iau.org/static/publications/ga_newspapers/19580812.pdf]. To the left, André Danjon, IAU President; to the right, Viktor Ambartsumian, Chair of the Local Organizing Committee, and future IAU President (1961–1964)

Robert Petrie, Canada; Bohumil Šternbek, Czechoslovakia; and Richard Stoy, South Africa). In addition, the (non-voting) Advisers were A. Danjon and Th. Oosterhoff.

In Moscow, in the backstage of a seemingly happy GA, this transition between ECs took an unusual turn. The story is told by Donald Sadler, in his "Personal Account" found in the IAU archives (see Sect. 6.1 above). Just before the start of the GA, André Danjon, the French outgoing President, summoned a last-minute confidential meeting at the famous Hotel Ukrainia (where most of the GA participants were accommodated), between selected members of the outgoing and incoming ECs, *"excluding nationals of the USSR and other eastern countries."* The reason: Danjon *"had received, by hand, a communication from a high authority in USA (which* [Sadler] *could not identify), to the effect that the U.S. government would withdraw all support for the promised invitation* [to hold its next GA in California in 1961] *if the IAU did not admit the Republic of China* [i.e., Taiwan] *to membership during G.A. X in Moscow."* Then Leo Goldberg, a new EC member and responsible for this GA, obviously already aware of the issue, confirmed that *"without government support (especially as regards visas) it would be impossible to hold G.A. XI in U.S.A."*

Actually, the EC had been notified of the application of Taiwan for membership on the previous Apr. 30, by a letter sent by Bing-Ran Jiang, the President of its Astronomical Society, just founded in Taipei under the auspices of the Academia Sinica,[16] i.e., long after the previous EC meeting (#17) had been held (Liège, Jul. 4–6, 1957). However, another one was not planned until the Moscow GA, so the EC could not be discuss the letter contents in advance, although Th. Oosterhoff, the General Secretary, did ask for complementary information. Thus the news brought by Danjon was not Taiwan's application, but a last-minute pressure from the USA (the exact nature of which I will return below).

The visa issue was indeed crucial for the IAU. Normally, citizens of countries having no diplomatic relations with the USA, for instance, countries of the Socialist bloc, and especially the PRC, had to apply for special visas (usually via other countries that were recognized by the USA). These applications were accepted at the discretion of the State Department (the American Foreign Ministry), and for the IAU it was essential that all astronomers, irrespective of their nationality, could be admitted on the American territory, in the same way that the Soviets had admitted them in Moscow.[17] In other words, a deal was proposed to the IAU Executive: the Chinese delegation would be granted US visas *only* if the IAU would admit Taiwan at the Moscow GA. Sadler then continues:

> During the meeting, the opinion grew that the original application from Taiwan must have been originated by the U.S. State Department. There could be no doubt that the Union was being blackmailed in the most blatant manner. Danjon, quite justifiably, was furious and would, I think, have been prepared to reject the application outright.

The EC members present were however somewhat cautious, probably warned by Goldberg: they decided to ignore the immediate request from the "high authority in USA" but to defer Taiwan's application *"until the information requested* [by Th. Oosterhoff] *have been received."* This was all the more reasonable that any new membership application must be submitted several months in advance, i.e., in time to be discussed at the EC meeting preceding the following GA, which, as mentioned above, was not the case for Taiwan.

In parallel, the Chinese delegation had also been notified of Taiwan's application, but only upon arriving in Moscow, on Aug. 9. They were informed by two Soviet astronomers: Boris Kukarkin, IAU Vice-President (who had not been invited at the Hotel Ukrainia meeting), and Viktor Ambartsumian, past Vice-President and Chair

[16] For more details, see Chap. 9

[17] There were 820 participants in Moscow, out of an IAU membership of 1127, a very high participation rate (75%). For comparison, at the XXXth GA in Vienna (2018), there were 3004 participants, out of 13,595 IAU individual members (see Andersen et al., 2019)

of the Moscow GA Organizing Committee. The delegation was caught off-guard, and had to consult their authorities about what to do. Fu&Ye tell the story:

> The deputy director of the Purple Mountain Observatory Zhu Renjun, who was also in charge of the Party affairs of the delegation, quickly reported this information to the Chinese ambassador Liao Xiao in Moscow. Zhu Renjun also served as the deputy secretary of the Communist Party branch at the Purple Mountain Observatory. Soon after, the embassy received orders from the Chinese Ministry of Foreign Affairs in Beijing. Beijing asked them to firmly oppose the invitation of Taiwan. (...) After having received orders from the Chinese embassy, the Chinese delegation made various preparations for the possible scenarios prior to the General Assembly. In the worst-case scenario, they were ready to completely withdraw from this meeting, if the Taiwan issue was ever raised.

Of course, should this have happened, it would have created an unprecedented scandal, and a major diplomatic problem for the relations between the USSR and the USA.

However, deferring the problem of the admission of Taiwan was by no means solving it. The two stories above clearly showed that the IAU would soon be caught between opposite political pressures: on the one hand, from "high-level authorities" in the USA, and on the other, from "the Communist Party" in China.

But for now, the IAU did not yield and, at its outgoing EC meeting (#18) ending the Moscow GA on Aug. 19, item #7 of the minutes simply mentions:

> Request from Taiwan for IAU membership: very little information had been given concerning the National Committee of Astronomy in Taiwan and about astronomical activity in this country, the Executive Committee decided to postpone a decision in this matter until the next meeting of the new Executive Committee.

At precisely that same moment, and in a totally different domain of international relations, another story unfolded, apparently unrelated with the IAU, but which, although the astronomers in Moscow could not realize it, set a precedent to the succession of events that would take place the following year.

The link appears in the Obituary of Leo Goldberg, written by Lawrence Aller.[18] On p. 129, Aller mentions a surprising connection with the Olympic Games: in 1960, the Winter Olympics were scheduled to take place in Squaw Valley, California. For the International Olympic Committee (IOC) and its President, the American Avery Brundage, the Olympic charter demanded that the participation of athletes from the PRC be ensured in the same way as those of any other country–a condition exactly analogous to the one required by the IAU. In 1957, Brundage had threatened the State Department to withdraw the Games from the USA and find another venue if this request was not satisfied. Eventually, so much publicity was given to this episode that the State Department yielded and promised to grant visas to PRC athletes due to compete in Squaw Valley. Apparently inspired by this analogy, Goldberg, who was in touch with the State Department about the Berkeley GA, was trying to obtain the same "favor" for the astronomers. He had not succeeded yet in Moscow, but, as reported by Blaauw in his book, the correspondence with the

[18] Aller (1997).

State Department found in Goldberg's archives (kept at the University of Berkeley) indicated that the threat on Berkeley was "gradually mellowing," in Blaauw's words. This was the good news.

But Goldberg, and presumably the EC, were not aware of the bad news. There was an important difference between the IOC and the IAU: the ROC (Taiwan) was already a member of the IOC, and in 1954 a vote had been taken at a short majority to admit the PRC as well. In other words, after that date *both* the ROC and the PRC coexisted, and they did in fact compete in a number of (non-olympic) sports events. But by recognizing two committees the IOC was in effect recognizing two countries –and the coexistence was not going to last. As told by the historian Richard Espy in his book *The Politics of the Olympic Games* (Espy, 1979: pp. 62–63):

> by 1957 and 1958 the PRC authorities had retrenched and were beginning the 'Great Leap Forward.' The retrenchment became immediately obvious in Olympic circles. After a relatively quiescent period following the Melbourne Games, the PRC began a barrage of demands for the expulsion of the Taiwan committee from the Olympic movement. Throughout that year IOC President Avery Brundage and Tung Shou-yi, the PRC member on the IOC, corresponded on the issue. Brundage emphasized that two committees existed, that political issues were not the province of the IOC. (. . .) Tung replied with demands for Taiwan ouster and stated that it was not he, but Brundage, who was introducing politics by the continued insistence on two committees. Finally, in a letter dated Aug. 19, 1958

[the very date of the EC meeting in Moscow], (. . .) Tung told Brundage:

> 'I will no longer cooperate with you or have any connection with the IOC while it is under your domination'. With that, the PRC withdrew from the IOC and, at the same time, from all the international federations with which they were affiliated.

It had thus become the new PRC policy in 1958 that belonging to the same international organization as Taiwan was no longer possible, foreshadowing the future relations between the IAU, China, and Taiwan in the aftermath of the Moscow GA.

6.4 After Moscow: Preparing for the Next Executive Committee Meeting

The leadership of the new EC was highly experienced in astronomy and IAU affairs. The President, Jan Hendrik Oort (1900–1992; Fig. 6.2, left), Director of the Leiden Observatory since 1945, already had a worldwide reputation for his pioneering work on the structure of the Milky Way, the origin of comets, and fostering the birth of radioastronomy in the Netherlands (see, e.g., MZ22, and obituaries such as Wilford, 1992, Blaauw & Schmidt, 1993, van de Hulst, 1994). Before becoming its President, Oort had served the IAU continuously as General Secretary since 1935 (the year China was admitted to the IAU) until 1952 (except during the German occupation of the Netherlands, 1940–1944, when he was temporarily replaced by Walter S. Adams, an American). The General Secretary, Donald Harry Sadler (1908–1987;

Fig. 6.2 Jan Oort (left) and Donald Sadler (right), both in 1961. (From the Berkeley General Assembly daily newspaper "IAU News Bulletin;" © IAU) [https://www.iau.org/static/publications/ga_newspapers/19610815.pdf]

Fig. 6.2, right), was a specialist in celestial mechanics. He became Superintendent of H.M. Nautical Almanac Office (HMNAO),[19] then located in the Greenwich Royal Observatory, in 1937, and was appointed Secretary of the Royal Astronomical Society from 1939 to 1947. He had also been Vice-President (1948–1952), and President (1952–1955) of the IAU Commission 4 "Ephemerides," then, as we have seen, IAU Assistant General Secretary from 1955 to 1958.

This is the reason why the next Executive Committee meeting (#19, Sep. 7–9, 1959, i.e., more than a year after the Moscow GA), took place, at the invitation of Sadler, in the beautiful castle of Herstmonceux (in East Sussex, England, near Brighton), where the Royal Observatory had been recently relocated. Although the agenda was, as usual, a heavy one (30 items), it was clear that the main item (#4) would be the decision about the admission of Taiwan, upholding the promise made in Moscow by Danjon.

By coincidence, in October 1958, just following the GA in Moscow, ICSU held its Eighth GA in Washington. As a member of ICSU, the IAU had to send representatives at its GA: the opportunity to discuss the Taiwan case with the "Union of Unions," as ICSU was sometimes dubbed, was not to be missed, so Oort and Sadler flew to Washington in preparation for the forthcoming EC meeting in Herstmonceux. Note here that ICSU does not have "individual members" like the IAU; in fact ICSU and all the other large Unions GAs gather only *representatives* of their member unions: "National members," topical committees, etc., representing their respective scientific communities (which can be very large, much larger than the astronomical community), with in general specific mandates for votes, discussions, etc.

In his "Personal Account," D. Sadler gives a disturbing testimony of what happened in Washington:

19 H.M. stands for "Her (or His) Majesty." This Office was established in 1832 under King William IV. For more details on D. Sadler, see the obituary published by his successor G.A. Wilkins (1991).

> The question of the two Chinas was, of course, brought up at the meeting of the Executive Board of ICSU at the Academy of Sciences, in Washington D.C. (almost in the shadow of the State Department) in October 1958. As far as I can remember, it was not discussed during the General Assembly. Jan Oort and I represented the IAU. We were immediately appalled by the list of adhering countries, boldly announcing the membership of China (The Republic of China) with adhering organization the Academia Sinica, Taipei. We discovered that several other Unions (to which China adhered) had recently received applications from Taiwan, and I was informed that the (*illegible*) letters of application had been drafted in the U.S. State Department by a scientist (who shall be nameless), who was actually a member of the ICSU Executive Board. I have never had any reason to doubt the accuracy of that information. During the meeting of the Board, Laclavère (IUGG),[20] Oort and I repeatedly attempted to raise the question of the representation of China, but were always faulted by the President of ICSU (L.V. Berkner); also there was a majority (mainly from the ranks of the non-Union representatives) who clearly did not want the issue discussed in Washington.

This was an odd situation, because (and it seems that neither Oort nor Sadler were aware of it during the meeting), ICSU itself was facing the same situation: as reported by Greenaway (p. 89), at the time of its Fifth GA in Copenhagen in October 1949, China, having the Academia Sinica as an adhering member of ICSU, was based in Nanjing, but it had moved to Beijing when the Sixth GA took place (Amsterdam, October 1952). In other words, ICSU, like the IAU, had *de facto* accepted the "new China" as a member, not the ROC. Yet, at least as seen from the seats of Sadler and Oort, its President appeared to have pushed ICSU towards admitting Taiwan, then logically done the same for its member Unions -including the IAU. At any rate, with its forthcoming EC the following year, the IAU did not receive any guidance from ICSU, except that admitting Taiwan was implicitly in conformity with the ICSU Statutes, implying that for the Unions to admit Taiwan was also in conformity with their own Statutes. Indeed, all these Statutes were essentially the same by construction (the general coherence of ICSU and all its Unions) since the creation of ICSU in 1931 (at its first GA in Brussels), especially as regards the adherence of its members.[21] As for the IAU, it had changed its Statutes accordingly at its GA immediately following that of ICSU (IVth GA in Cambridge, USA, 1932), and these Statutes had remained the same in 1958 as they were in 1932. An important consequence for our discussion here is that Taiwan (and North Korea) would become IAU members according to the same Statutes than all the other countries admitted since 1932.

So examining these Statutes is essential to understand why the IAU and China became irreconcilable as a result of the admission of Taiwan. The first articles of these Statutes are reproduced below (Fig. 6.3), as published in the IAU Transactions[22]:

The key article here is of course the second one: "*Adherence of countries.*" A detailed discussion of this article is presented elsewhere (MZ22), but its essence is fairly simple—or so it seemed until the case of Taiwan was brought up. For the IAU,

[20] IUGG: International Union of Geodesy and Geophysics.

[21] See Fauque & Fox, Chap. 3.

[22] Transactions Vol.X, p. 731

INTERNATIONAL ASTRONOMICAL UNION

STATUTES

I. OBJECTS OF THE UNION

1. The purpose of the Union is:
   (i) to facilitate the relations between astronomers of different countries where international co-operation is necessary or useful;
   (ii) to promote the study of astronomy in all its departments.
2. The Union adheres to the International Council of Scientific Unions.

II. ADHERENCE OF COUNTRIES

3. A country which adheres to the International Council of Scientific Unions may adhere to the Union (*a*) by the organization through which it adheres to the International Council, or (*b*) by a National Committee of Astronomy formed by or recognized by such organization.

A country not adhering to the International Council may adhere to the Union by a National Committee of Astronomy recognized by the Executive Committee of the Union.

The term 'country' is to be understood as including Dominions, diplomatic Protectorates, and any dependency in which independent astronomical activity has been developed.

Fig. 6.3 The IAU Statutes of 1958, unchanged since 1932

after declaring (Article I.2) that it "adheres to the International Council of Scientific Unions," its main criterion is whether or not the "country" (defined at the end of the article) is a member of ICSU. We see that adhering to ICSU is *not* a prerequisite to adhering to the IAU: if not, adherence to the IAU can be decided by its Executive Committee, provided it recognizes a National Committee of Astronomy.

A look at Table 6.1 below illustrates this. I list the countries having adhered to the IAU (in chronological order) since 1932 (thus according to the Statutes above), and in the last column the year of their adherence to ICSU: they are uncorrelated. For instance, China and the USSR both adhered to the IAU in 1935, but China adhered to ICSU in 1937, and the USSR only in 1954.[23] Conversely, Bulgaria adhered to ICSU in 1931 but joined in IAU only in 1957. So this distinction between the two ways of joining the IAU had been indeed put into effect many times previously. In spite of the way the succession of arguments was presented in Moscow about the application of Taiwan, this was not the problem at all.

The real problem was the definition of a "country," on which the applicability of the article was resting.

On purpose, the definition in the Statutes was rather loose, non-governmental, i.e., non-political (the interface with the IAU was, and still is, a "National Committee of Astronomy," not a government) but basically geographical. Interestingly (and keeping in mind that the Statutes were those adopted in 1932), it included entities

[23] This was in connection with the organization of the International Geophysical Year. See discussion in Greenaway, p. 88.

Table 6.1 Adherence of national members to the IAU (1932–1961: Statutes of 1932) and to ICSU

Year IAU	General Assembly (GA)	Country[a,b]	Ind. Members[c]	Year ICSU
1932	Cambridge, USA	Vatican State	1	1931
1935	Paris, France	*China*	11 [1938]	1937
		USSR	25 [1938]	1954
		Yugoslavia	6 [1938]	1931
1938	Stockholm, Sweden	South Africa[d]	12	1931
1939		Australia[d]	13 [e]	1931
1946	(Copenhagen conferente)[f]	India[g]	7	1931
1947		Hungary	1	1931
1948	Zürich, Switzerland	Finland	5	1931
1951		Germany	0[h]	1952
1952	Rome, Italy			
1953		Venezuela	0	1955
1954		Israel	1	1952
		Lebanon	1	1974
1955	Dublin, Ireland	Austria	1	1950
1957		Bulgaria	1	1931
1958	Moscow, USSR			
1959		*Taiwan*[i]	2	1958
1961	Berkeley, USA	*Taiwan*[j]	2	
		DPR Korea	0	1961
		Turkey	2	1968

Source: IAU Transactions or Transactions B for each General Assembly
[a] Japan re-admitted in 1952; Brazil re-joined in 1961
[b] Adherence (i.e., new members), as accepted by the Executive Committee
[c] Registered IAU individual members at date of adherence. Census given at the following General Assembly. Until 1935 the Transactions list "members and guests", so it's not clear who the members are; stricto sensu the numbers here are upper limits, but 1 or 2 doesn't really matter…
[d] Re-joined
[e] Census of 1948
[f] Executive Committee preparatory Conferente for the first post-war GA (March 6–13, 1946). J. Oort had been IAU General Secretary since 1935 (Paris GA), to step down in 1948 [Zürich GA)
[g] India became independent later (on Aug.15, 1947)
[h] A "zero" means that no individual member was recorded at the time of publication of the corresponding Transactions B, likely for purely administrative reasons. In the case of Germany for instance, the individual members started to be listed only at the Moscow GA in 1958
[i] Admission in 1959 by the Executive Committee (see Sect. 6.3)
[j] Ratification by National Members voting at the Berkeley GA (see text for details). The mandatory ratificati on of new adherences by a vote of the General Assembly was introduced in the Statutes only in 1970 (Brighton GA). The name "Taiwan" was the one used for its membership to the IAU, never the "Republic of China"

that had a colonial, post-WWI connotation ("Dominions," "Protectorates," "any dependency…"),[24] but more broadly defined as "in which independent astronomical activity has been developed," a flexibility that did leave some room for interpretation. The idea was to be as inclusive as possible, not to repeat the errors of the International Research Council (IRC; and its affiliated Unions, including the IAU) in 1919,[25] which imposed severe restrictions on national membership in the aftermath of WWI (excluding the defeated "Central Powers:" Germany, Austria, and their allies, also some neutral countries, suspected to have supported them).[26] These restrictions were soon felt in contradiction with the adjective "International," and this is what had led to an overhaul of the IRC, replacing it with ICSU, which got rid of them, in 1931. For the IAU, as of 1932, this meant being ready to accept any "country" (in the statutory sense of the word), whatever the size of its astronomical activity, especially after WWII and its terrible destructions, which had in some cases annihilated any past activity.

This is again illustrated in Table 6.1, where we see many "small" countries having adhered to the IAU, with a very small number of astronomers (e.g., Hungary, Venezuela, Israel, Lebanon, etc.), simply in the hope that their communities would grow and flourish, in accordance with Article Ia ("relations between astronomers") and Ib ("promote the study of astronomy").

The goal of Oort and Sadler was to admit Taiwan (after cross-checking its astronomical credentials), *and* keep China as a member. Already in Moscow, there were rumors that China might withdraw if Taiwan were admitted (and the remote example of the IOC given above showed that this was more than a rumor), and there was also (now more in the background, apparently) some pressure by the State Department to block visas for the GA in Berkeley two years later. Would the Statutes indeed legally allow to admit Taiwan *and* retain China?

In MZ22 I discuss the Statutes in detail, and also I quote the reaction of Chinese astronomers (Fu&Ye) to the issue of the amount and independence of "astronomical activity" in Taiwan (which will be discussed further below, during the EC meeting in Herstmonceux), but my conclusions can be summarized as follows:

1. Statutes, article 3: *The membership of China to the IAU is confirmed* by the fact that it was admitted in 1935 with the "Chinese Astronomical Society" as its adhering body, recognized by the EC as National Committee of Astronomy, then located in Nanjing, now in Beijing. This is irrespective of ICSU (exactly when China actually pulled out of ICSU after 1958 is not clear, but again, this is irrelevant).
2. Statutes, article 3(a): *The membership of Taiwan follows from the admission to ICSU* of the organization (the Academia Sinica, formerly located in Nanking, now in Taipei) which recognizes the "Astronomical Society of the Republic of

[24] This terminology would be discarded when the Statutes were revised at the Brighton GA in 1970 (Transactions Vol. XIVB, 1971).

[25] See Fauque & Fox, Chap. 3.

[26] On the case of Germany, see Chap. 11 by R. Wielen, and Wielen (2019).

China, Taipei." This Society is, perhaps temporarily but factually, independent of the "Chinese Astronomical Society."

On the second point, D. Sadler, in his "Personal Account," would later comment:

> My interpretation of Statute No.3 was that there was no valid reason, or cause, for rejection, bearing in mind that:
>
> (a) Taiwan was an adhering member of ICSU.
> (b) It fulfilled all the technical conditions, including having a National Committee of Astronomy, recognized by the Academia Sinica.
> (c) There was some independent astronomical activity.
> (d) It could not be excluded, on the grounds that it was a "country," within the terms of the Statute, especially as no other "country" had de facto control of astronomical activity within Taiwan.

Of course, there was no way of treating separately the case of Taiwan while leaving the case of China aside, so Oort and Sadler drew what would become twenty years later coined a "*dual adherence,*" clarifying the respective status of "China" and "Taiwan," in terms of separate geographical, not political entities (the reason why these two short names were used, rather than their political names, resp. PRC and ROC). This scheme was in the form of a Resolution to be submitted to the forthcoming EC (here re-typed as a fac-simile):

Whereas there is the possibility of confusion in the descriptions variously accorded to the two Republics of China, both of which adhere to the Union, and

in order to indicate unambiguously the two geographical areas in which independent astronomical activity has been developed,

it is resolved that short names, as indicated below, shall be used for all purposes of the Union, with the exception of formal correspondence and descriptions in which the use of the official titles of the two countries is desirable,

1.	Official title of the country:	The People's Republic of China
	Site of Government:	Peking, China
	Adhering Organization:	National Committee of Astronomy (Nanking)
	Date of Adherence:	1935
	Geographical area in which Astronomers are represented:	The mainland of China
	Short name:	China (in French: Chine)
2.	Official title of the country:	The Republic of China
	Site of Government:	Taipei, Taiwan
	Adhering Organization:	The Astronomical Society of the Republic of China (Taipei), recognized as a National Committee of Astronomy by the Academia Sinica, (Taipei), through which organization the Republic of China adheres to the International Council of Scientific Unions
	Date of Adherence:	1959
	Geographical area in which Astronomers are represented:	Taiwan (Formosa)
	Short name:	Taiwan (in French: Taiwan)

Note that here the "political" names are also mentioned (as well as the expression "the two Republics of China"), but for reference only, even though at that time the "Republic of China" (Taiwan) was the only representative of "China" at the United Nations (for instance). At the same time, albeit unknowingly, this resolution, while strictly in conformity with the IAU Statutes of 1958, could admittedly be interpreted as suggesting that there were "two Chinas" or "one China, one Taiwan," a political issue external to the IAU, and this was the worm that corrupted an otherwise beautiful fruit, until all reference to a "country" would be removed in a way that would politically satisfy both parties.

6.5 The Difficult Decision: Hertsmonceux, 1959

The EC meeting (#19) started in the Hertsmonceux Castle Library on Sep. 9, 1959. Blaauw's book (p. 192) gives an account of the discussion that led to the admission of Taiwan, based on the minutes written by the General Secretary (D. Sadler),[27] and I also analyze them in detail (MZ22) in the light of his "Personal Account." Here I give only the highlights, listed by topics as recalled by Sadler and quoted above.

(a) Taiwan was a member of ICSU. As the time of the meeting, it had become official, while, as mentioned previously, at the time of the Moscow GA the ICSU decision was pending because its own GA had not yet convened in Washington. It was a strong argument for admission: in his "Personal Account," Sadler adds this comment:

> The IAU adhered to ICSU and its statutes, especially in respect to National Members, were linked with those of ICSU. It would have created much confusion if the IAU had interpreted "country" in a manner contrary to that which ICSU had already adopted.

However, Sadler makes no mention of the fact that other unions were probably in the same situation. The "ICSU argument" was apparently not brought up at the EC meeting, although it was explicitly included in the draft resolution above.

(b) In spite of the fact that some doubts about the reality of the Taiwan National Committee for Astronomy had been raised in Moscow, the endorsement by the Academia Sinica in Taipei was sufficient, being a matter internal to Taiwan, so it had to be accepted by the IAU.

(c) Consistent with (b), evidence for astronomical activity, in the form of names of astronomers and publications, had been provided upon request to the General Secretary. The EC debated at length whether some "minimum" level of activity had to be considered, but it was clear that the inclusive policy of the IAU (i.e., of the previous ECs) meant that this level was left to the appreciation of the EC,

[27] https://www.iau.org/static/administration/ecminutes/ec19minutes.pdf

and, without precise criteria, could not be taken in any way against the admission of a country.[28] (As argued above, this is also quite clear from Table 6.1.)

(d) The topic most debated was the topic of "country," as defined by the Statutes, and on the correlated issue of the "independence" of the astronomical activity. No details are given in the minutes, but it was hard not to admit that there was no connection, not to mention support, of the Chinese Astronomical Society in Peking to the one in Taipei. In reality, the question of "independence in astronomy" was overshadowed by the issue of the "independence" of the island Taiwan from mainland China. This was a political issue internal to China, which obviously the IAU could not resolve.

As written in the minutes, *"the question* [of Taiwan] *was discussed at considerable length,"* the EC being fully aware of the consequences of admitting Taiwan. The President, J. Oort, following the General Secretary, concluded that, based on the Statutes, *there was no ground for refusing to admit Taiwan.* The minutes do not record the details of the arguments, except that there was no unanimity on this conclusion: two Vice-Presidents, the Soviet Boris Kukarkin, and the Czech Bohumil Šternbek, opposed admission. Then the President called for a formal vote: 5 in favor, 2 against.[29] Although the voting method (secret ballot or show of hands) is not mentioned, the discussion likely stayed within the polite "astronomical" limits, but the vote unquestionably reflected the Cold War divide...

On the basis of this "positive" vote, the President then presented the resolution he and Sadler had prepared in an attempt to retain China while admitting Taiwan (see above), and after another discussion, put it also to the ballot. Unsurprisingly, the same result came out: 5 in favor, 2 against.

Normally, if no consensus is reached on a given issue, and a vote is taken, the EC follows the majority, and this decision is final. However, this is not what happened: Kukarkin, supported by Šternbek, decided to oppose the majority vote, and requested to make a formal statement explaining his motivation, written (in French) in six, one-sentence sections, of which the key ones (translated) read as follows (the full version is given in the minutes)[30]:

- Considering that I categorically disapprove of the decision by the Executive Committee of the IAU, voted by 5 against 2, to admit Taiwan to the International Astronomical Union.
- ...
- Emphasizing that there is no independent scientific activity in Taiwan in the field of Astronomy, and thereby the decision of the Executive Committee is in contradiction with the precise meaning of the IAU Statute.

[28] For details brought up only recently, see Chap. 9 by Shiang-Yu Wang, Yun-Chun Wei, & Mei-Yin Chou.

[29] As Fu&Ye noted, one vote is missing, but this was simply because the Canadian Vice-President, R. Petrie, was absent due to illness. For the same reason, the two Advisers, the past President A. Danjon and the past General Secretary Th. Oosterhoff, were not attending the meeting either, but they had no right of vote anyway.

[30] https://www.iau.org/static/administration/ecminutes/ec19minutes.pdf

- ...
- I believe it is necessary and even obligatory for me to address the IAU General Assembly of 1961 to ask for a revision of the question of the admission of Taiwan to the IAU.
- In connection with this I propose that a *status quo* be observed until the General Assembly of 1961.

The President then stood by the majority vote, stating that

> the majority clearly considered that the astronomical activity in Taiwan was independent and that the Statutes of the I.A.U. did not permit rejection of the application, either on this ground or on the low level of astronomical activity. He said that Kukarkin was free to raise this matter again before the General Assembly in 1961, but that he could not accept the proposal to reverse the decision just taken; this decision to admit Taiwan must stand.

6.6 The Withdrawal of China

After the Hertsmonceux meeting, a daunting task was awaiting J. Oort: to announce the EC decision to admit Taiwan, and at the same time to convince China to remain in the IAU.

His contact person in China was Y.C. Chang, the President of the Chinese Astronomical Society, and also Director of the Purple Mountain Observatory. Using the modern pinyin transcription of his name, Zhang Yuzhe (1902–1986) had graduated from the prestigious Tsinghua College in Peking in 1923, and won a scholarship to study in the USA, first studying mechanics at Purdue University in Indiana. He then moved to Cornell to study architecture, then on to the University of Chicago to study physics and mathematics, where he met S. Chadrasekhar, and started to do graduate work at Yerkes Observatory (operated by the University of Chicago) in 1927 under G. Van Biesbrock, defending his PhD in 1929. He returned to China that same year, and became a Professor at the National Central University in Nanking. As mentioned above, after the war came with the Imperial Japan invasion of China in 1937, he moved to Chongking, to return to Nanking at the end of the war. He went back to the USA from 1946 to 1948 with a grant from the ROC government and stayed again at Yerkes Observatory, then headed by Otto Struve. There, he met many colleagues from around the world, including the USSR, and also in particular Jan Oort and Adrian Blaauw, in addition to his longtime friend S. Chandrasekhar. But Zhang missed China and his family, so he returned home for good in March 1948. As we have seen, he chose to stand by the "new China" in 1949, and became Director of the Purple Mountain Observatory in 1950 until the early 1980s. (He is widely considered today as "the father of modern Chinese astronomy.")[31]

Thus, strictly speaking Oort and Chang may not have been close friends, but they knew each other from their Yerkes years. Later, both were present, as well as

[31] For details, see his biography in this volume (Chap. 7).

Sadler, at the inauguration of the restored Pulkovo Observatory, near Leningrad, in 1954,[32] i.e., just four years before the Moscow GA, where they may have met again, although there is no record of such an encounter. So Oort presumably expected some difficult, but cordial, exchanges, given in particular that Chang of course wrote fluently in English. Unfortunately, this proved not to be the case.

For some reason, there is no trace in the IAU archives of a letter from Oort to Chang informing him personally of the EC decision to admit Taiwan, but the IAU announced it in the Nov. 1959 issue (No 21) of its *Information Bulletin.* It seems that Chang had received the news unofficially by another channel (Kukarkin, for instance?), but at any rate, in a letter to Oort dated Nov. 20, 1959, Chang writes:

> Dear Professor Oort,
>
> We learn that the Executive Council Meeting of the International Astronomical Union held in England on September 7, 1959, has adopted the decision of accepting the so-called "Chinese Astronomical Society" of the Chiang Kai-shek clique[33] for membership in the Union. We are greatly surprised by this decision.

Then, the following paragraph gives the origin of the problem, which turned out to have nothing to do with the so-called "level of astronomical activity" of Taiwan (which had been discussed so extensively by the EC), by going head-on into politics, and which, for this reason, I quote here in full for the first time:

> As all the people of the world know, Taiwan is an inseparable part of Chinese territory, it is a province of China. It is because of the armed occupation by the American imperialists that Taiwan has not been yet liberated up to the present. The Astronomical Society of the People's Republic of China is the only legal organization to represent China in joining the IAU. The acceptance of the so-called "Chinese Astronomical Society" of the Chiang Kai-shek clique by the Executive Council Meeting of the IAU is evidently in keeping with the hoax of the American imperialists of creating "two Chinas." It is illegal and wrong. Not only is it a trespass on the legal rights of China in an international organization but also an obstruction of the normal development of international scientific cooperation.

The logical conclusion follows:

> [we] insist that the Executive Council of IAU rescind the illegal decision about the acceptance of the so-called "Chinese Astronomical Society" of the Chiang Kai-shek clique. Otherwise, the Astronomical Society of the People's Republic of China will resolutely and definitely withdraw from the IAU.

In this letter, two key sentences summarize the roots of the crisis between the IAU (and other Unions) and China, even though it took different forms over the years:

- *Taiwan is an inseparable part of Chinese territory, it is a province of China.*
- *The Astronomical Society of the People's Republic of China is the only legal organization to represent China in joining the IAU.*

---

[32] Sadler, in his *Personal history of H.M. Nautical Almanac Office* (edited by Wilkins, 2008; p. 115).

[33] A word of French origin, frequently used in China to designate a despicable group, or gang, here standing for "government" (which included members of his in-law powerful family Song).

Actually, whereas the "extensive" discussions at the Herstmonceux obviously missed this point, it is in a way a pity that the Oort-Sadler resolution was not discussed more thoroughly, because it was much closer to the real issue. And this is what Oort tried to explain to Chang in his reply of Dec. 2, here summarized in a few quotes (the letter is reproduced in full in MZ22):

- China's representation in the IAU is a temporary, dual one, with one adhering organization located in Peking, and the other in Taipei.
- I must emphasize that the representation covered by the Astronomical Society of the Republic of China in Taipei is limited to those astronomical activities in Taiwan that are at present (and for whatever reason) independent of the guidance, supervision and representation of the Astronomical Society of the People's Republic of China in Peking.
- The application from the Astronomical Society of the Republic of China in Taipei formally conformed to the Statutes of the Union and to the Statutes and Directives of the International Council of Scientific Unions, to which the I.A.U. adheres.

Like Chang, Oort also set his stage for the future relations: *"The decisions of the Executive Committee cannot be rescinded. In this case the Astronomical Society of the Republic of China (Taipei) has already been informed of these decisions."*

The next question was to wait for Chang's reply, and since in all likelihood China would pull out of the IAU, to announce the painful news. Blaauw's book reports that the official announcement of China's withdrawal from the IAU was given in the May, 1960, issue of the *Information Bulletin*, but actually, as evidenced by the IAU archives, the news was spread publicly by China itself three months earlier, in a most unusual manner.

The full story appears in MZ22, but it can be summarized as follows. On Feb. 8, 1960, before any answer from Chang had reached Oort, Sadler writes (letter to L. Goldberg)[34] that he had been informed, "by an acquaintance," of a broadcast by an American radio station in Europe announcing the PRC withdrawal from the IAU. This was a typical Cold War situation: where was this information coming from? Was it reliable? Was it official? As it eventually turned out, the broadcast had originated in a press release by the official "New China News Agency" (now the Xinhua News Agency), sent around by "radioteletype" (i.e., telex), on Feb. 6. Later, upon at last receiving Chang's letter, Sadler realized that it had been sent to Oort on the previous day, so it must have been part of a plan to make China's response widely public even before the IAU itself was informed.

Actually, the move was not very successful, because the broadcast was apparently not taken up by newspapers, but the transcript I have found in the IAU archives draws heavily on Chang's letter, or, given its tone, more probably his letter edited by some Party official. (It is published in full in MZ22.) To quote excerpts (verbatim):

> Oort in his letter 'openly put on a par the great People's Republic of China and the Chiang Kai-shek clique which has long since been disowned by the Chinese people', and with ulterior

[34] There is a lot of correspondence between Sadler and Goldberg, in relation with the preparation of the Berkeley GA, where the issue of China and Taiwan would be raised again (see below, Sect. 6.7).

motives quoted article three of the statutes of the union as "grounds" for representing Taiwan as a country. Evidently, this is in line with the U.S. imperialist intrigue of creating "two China's."

(...)

The Taiwan question is China's internal affair and the Chinese people are determined to liberate Taiwan. We are firmly against the conspiracy of the American imperialists to create "two China's" and we will oppose it to the end.

(...)

It is an open interference in China's internal affairs on the part of the IAU to take this as a pretext to deny the right of our astronomical society to represent the astronomical workers in Taiwan. This is utterly intolerable to us.

The Transcript concludes:

The admittance of Taiwan to membership is a thing which goes far beyond the scope of "science," and reduces the IAU to a mere tool in the political intrigue of "two China's." This is not only an act of hostility to the Chinese people, but also runs counter to the will of fair-minded scientists in the IAU itself.

Notwithstanding, after having seen Chang's reply to Oort and the Xinhua press release, on Feb. 18 Sadler writes to Goldberg with some optimism:

The withdrawal of China from the Union is a sad blow, but it is by no means a fatal blow to the Union. Our correspondence with Chang has been (allowing for the enforced vituperation against the Americans and anti-communists) very friendly; and Chang has intimated that Chinese astronomers would wish to co-operate in international programmes as in the past.

In May, the IAU *Information Bulletin* tersely "confirms" the news:

Professor Yu-Che Chang, the President of the Astronomical Society of the People's Republic of China, has formally announced the withdrawal of the People's Republic of China from adherence to the Union, in protest against the decision of the Executive Committee to admit Taiwan (the Republic of China) to membership of the Union. As from this announcement astronomers in the People's Republic of China will cease to be members of the Union and therefore will cease to be members of Commissions.

6.7 Ratification at Berkeley

After China broke all ties with the IAU as a result of its internal political decisions, the question of a possible re-admission would not be put on the agenda of the EC for many years. However, one burning question remained to be settled: the formal admission of Taiwan.

In principle, the case was closed. According to the Statutes, the decision of the EC was final. No issue had ever been raised, even countries with obviously little astronomical activity had been admitted (Table 6.1). But in the past, the EC decisions had been unanimous. For Taiwan, there was this embarrassing dissent of two Vice-Presidents from the Socialist bloc, who did not accept the majority decision. Whatever the reasons they invoked (and we saw that in the end, they were not the relevant ones in the case of China), Kukarkin and Šternbek had a point: they wanted to withhold the decision until it could be publicly debated at the GA. Now that this decision had been taken, the next move was not to engage in a debate, but to seek ratification. Such a process had never taken place before, and would have to be exceptional (the necessity of a ratification by the GA of the admission of new "national members" would become mandatory only much later, in 1970, after a revision of the Statutes which was adopted at the GA in Brighton).[35] After all, Oort had said to Kukarkin in Herstmonceux that he was *"free to raise the matter again before the GA in 1961."*

For the time being, however, at the following EC meeting taking place almost a year after Herstmonceux, and two months after the withdrawal of China (#20, Prague, Jul. 6–8, 1960), the admission of Taiwan was not on the agenda.

The next EC meeting was due to take place a year later, just before the start of the General Assembly in Berkeley (#21: Monday, Aug. 14, 1961). Nothing had happened about Taiwan for several months, until, on March 20, Sadler received a letter from Šternbek asking him to put the admission of Taiwan on the agenda of the GA, and another from Kukarkin containing the same request. At this date, it was too late to introduce resolutions by individual members for a vote by the GA, it was only possible to present "statements," possibly followed by a discussion. However, Sadler replied that if resolutions were formally presented by their adhering bodies (the Czechoslovak and Soviet Academies of Sciences, respectively), they could be included in the Agenda provided they would reach the IAU Secretariat before Apr. 5, so that they could be circulated to the community in advance of the GA via the *Agenda and Draft Reports*.

And sure enough, the two resolutions reached the Secretariat in time. The Czechoslovak Academy of Sciences submitted a single sentence: *"That the General Assembly revoke the decision of the Executive Committee, in September 1959, to admit Taiwan as an adhering country;"* the Soviet Academy of Sciences was more detailed, paraphrasing the Kurkarkin-Šternbek statement at Herstmonceux, but now mentioning the withdrawal of China, and ending by: *"The General Assembly resolves to cancel the decision to admit Taiwan, as adopted by the Executive Committee on 8 September 1959."*

Eventually, for practical reasons, both resolutions were combined into one (the Soviet text). However, since the admission of a member "country" had never been subjected to a vote by the GA, a new procedure had to be invented. In addition, in this case, the Soviet resolution was not to *approve* an admission (then calling for a "positive" vote, as usual for resolutions), but to *repel* a decision by the EC on an admission (then a "negative" vote), a very unusual situation since this was *de facto* calling the GA for a vote of no-confidence of the EC, potentially triggering a major institutional crisis within the IAU. One condition was clear, however: in such a case, the voters were not the GA participants, but the "National representatives" of the 38 IAU member countries present at Berkeley. This was a much more predictable set of voters: one country, one vote. Indeed, Goldberg, as IAU Vice-President and Chair of the LOC, replying to O. Struve, who was expressing concerns about the outcome of the vote, wrote on July 2: *"In Sadler's opinion, the Soviet-Czech resolutions will*

[35] See the Transactions Vol.XIV B (1971).

be supported by about six countries and there may also be an equal number of abstentions, but the resolution will be easily defeated."

The EC (including Kukarkin and Šternbek) had agreed on the following steps to carry on the vote, which was public but took place during the Business Sessions (as is usual for National representatives, who vote on items like the IAU budget and other administrative issues):

- Presentation of the motion and its history, by the General Secretary.
- Statement read by the President, as prepared by the Executive Committee: *"On behalf of the Executive Committee, I formally move the rejection of the motion and ask representatives of adhering countries to vote against the motion."*
- Justification and explanations for this recommendation (based on the minutes of the Herstmonceux meeting).
- Open discussion.
- Ballot.

The Business Session discussing the resolutions (called "the motion" by Oort) took place as planned, on Aug. 24, 1961. The Transactions (Vol.XI B), edited by Sadler, give a very detailed account of the session (three dense pages),[36] so I hereby select the most relevant excerpts:

> The General Secretary (...) pointed out that it was quite possible to adopt different points of view on this matter and that some of these are incorporated in the resolution as proposed by the U.S.S.R. Academy of Sciences. However, the majority of the Executive Committee had taken a rather different view, which would be presented to the General Assembly in a statement by the President (...)

After reading the above statement asking the representatives to reject the motion, the President went on to read the text prepared by the Executive Committee:

> There can be no question that Taiwan qualifies for adherence to the Union under the terms of the Statutes of the Union." (...) In recent years, the Union has been glad to welcome the adherence of other countries in which astronomical activity is on no higher level. There is no evidence that this astronomical activity is not independent; it is directed by a properly constituted National Committee of Astronomy.

Then alluding to China's withdrawal:

> It must be made absolutely clear that the decision to withdraw was made by China alone, for reasons that have nothing whatsoever to do either the Union or with Astronomy. The Executive Committee was, however, informed the China might withdraw if Taiwan were admitted, and for that reason approved a resolution which it was hoped would enable China to maintain its adherence to the Union. This resolution reads:

[follows the "dual" scheme presented by D. Sadler at the Herstmonceux meeting, see above].

And concluding:

[36] It is most unfortunate that this volume of the Transactions cannot be downloaded at present (due to an unsolved copyright issue). See https://www.iau.org/publications/iau/transactions_b/. Hopefully this problem will be fixed in the future.

> The Executive Committee can only express the hope that circumstances will soon change in such a way as to enable China to resume its adherence to the Union, and so to play its full part, formally, in international co-operation in Astronomy. We have noted the great strides that are being made in China, and have learned with pleasure from Chinese astronomers that international co-operation will continue informally even though China no longer adheres to the Union.

Then the discussion was opened by the President. Kukarkin raised his hand, and commented that "*at this stage he wished to dissociate himself from the statement that had been made* [by the President] *on behalf of the Executive Committee.*" Rather surprisingly, in spite of the fact that the "motion" had been publicized long in advance of the GA, apparently no further comments or questions came from the audience, and it was put to the ballot without further ado.

Following the recommendation of the Executive Committee, among the National representatives, there were (by a show of hands) 24 votes against, 5 in favor, 3 abstentions and 1 absence because of illness: "*The President declared that the motion was lost.*"

Doing the math, and as predicted by Sadler, the result could be expected: there were five countries present belonging to the Socialist bloc (Czechoslovakia, Hungary, Poland, the USSR, and Yugoslavia; Bulgaria and Romania were absent); the three abstentions might have included Egypt (this was shortly after the Suez canal crisis) and Indonesia (a leader of the "non-aligned countries"). In any case, the outcome of the vote was clear: the resolution submitted by the USSR was a lost cause, as they probably anticipated themselves. In the same way as the vote at Herstmonceux, the result clearly followed the East-West division of the world . . .

The session on Taiwan, however, was not completely over. After giving the result of the vote, the President bade farewell to astronomers from mainland China, by "*regretting their absence from the General Assembly and sending every good wish for astronomical research in China.*" The proposal to send them a message "*was received with acclamation.*"

In parallel, North Korea had been unanimously admitted to the IAU by the EC (along with Turkey), but likely no one noticed among the participants of the GA.

In retrospect, the vote at Berkeley on the admission of Taiwan, which was triggered by the submission of resolutions that, potentially, could have overthrown the Executive Committee (at least its Officers, namely the President and the General Secretary, who would likely have had to resign) had the opposite result: the admission of Taiwan being ratified by the General Assembly (whereas, as noted before, this was not a statutory obligation) not unanimously but with an unquestionable majority, the legitimacy of the Executive Committee decision, as the main IAU institution, came out reinforced.

6.8 Conclusions: Science and Politics in the Time of the Cold War

To many rank-and-file astronomers, however, the admission of Taiwan by the Executive Committee in 1959 was a mistake, because it had forced the withdrawal of China, and overall because they felt the IAU had yielded to a direct political pressure by the US State Department. Long after Taiwan's admission, at the Grenoble GA in 1976, one could read in the meeting's daily newspaper *La Gazette d'Uranie* an article signed by a group of participants and entitled "China and the IAU," containing the following sentence:

> It is now generally admitted that the application of Taiwan for membership was politically motivated. It was part of a plan by the US State Department under Dulles to gain Taiwan's admission to as many international organizations as possible, thereby isolating the Peoples Republic of China internationally. By then admitting Taiwan (where there was a negligible amount of astronomical activity) to membership, we believe that the IAU unwittingly allowed itself to be used for political ends.[37]

After having examined the new documents available since Blaauw's book, I have come to the conclusion that in reality, the situation in 1958–1959 was much more complex than simply a "plan by the US State Department under Dulles," and had (in spite of a widespread opinion) nothing to do with "a negligible amount of astronomical activity" in Taiwan. The stakes were so much higher that *de facto* Oort, Sadler and the EC had no choice, in spite of the obviously politically motivated, but riskless, attempts by two Socialist bloc Vice-Presidents to overrule a majority vote, first by the EC at the Hertsmonceux meeting in 1959, and later by the National member representatives at the GA in Berkeley in 1961.

D. Sadler's "Personal Account" offers an entirely new perspective, even if it is just a "private" testimony. This is when he reports (quoted above) that, attending the ICSU GA with Oort in Washington in October 1958:

> We discovered that several other Unions (to which China adhered) had recently received applications from Taiwan, and I was informed that the *(illegible)* letters of application had been drafted in the U.S. State Department by a scientist (who shall be nameless), who was actually a member of the ICSU Executive Board.

In other words, the pressure was exerted on all the Unions at the same time, and it appeared to be *coming from ICSU* (itself admittedly pressured by the American policy, enforced by the State Department). As Sadler also noticed (without mentioning the context), it was therefore impossible for the IAU to adopt a position different from the one adopted by ICSU, i.e., not to admit Taiwan, and, as discussed above, its Statutes were quite appropriate to do so and did not imply at all that China should be expelled.

The difference, however, between the IAU and the other Unions at that particular time was that the IAU was going to hold its next GA in the USA. Because of the

[37] https://www.iau.org/static/publications/ga_newspapers/19760825.pdf (last issue, Aug. 31).

visa issue, there had to be a close contact with the State Department, which then appeared to be directly pressuring the IAU, thereby hiding the role of ICSU. In other words, seen now from the IAU perspective, the often-quoted direct role of the State Department in the admission of Taiwan was purely circumstantial: would the next GA have taken place in Hamburg, for instance (which was the venue for the GA following Berkeley in 1964), of course the IAU Executive would never have been in contact with the State Department, and the visa issue would simply not have existed . . . but Taiwan would certainly have been admitted nonetheless because the IAU was a member of ICSU, only perhaps in a less dramatic way because the USA would have had no role in it.

A further proof is the case of North Korea, which, as recalled below, had fought (with the help of China) the USA during the Korean war just a few years before. Of course, it was inconceivable that North Korean astronomers would be granted visas to come to Berkeley, but there was no obstacle whatsoever for North Korea to adhere to the IAU: the State Department simply did not have to be involved.

A decade after the Berkeley GA took place, in 1971, a major geopolitical change occurred when China was admitted at the UN. This event resulted in the expulsion of Taiwan, since according to the UN Statutes, one "country" (defined this time in governmental, political terms) could have only one seat. For China, a logical next step was to try and convince the Unions to do the same (forgetting that itself had been an IAU member, for instance, during the period 1949–1959 without being a member of the UN). Even UNESCO (this is an information I discovered in the IAU archives), being a UN organization which at the time contributed about 60% of its budget, literally ordered ICSU (in 1973 and 1975) to contact its Unions with explicit instructions to expel Taiwan (see the details in MZ22). But, as a kind of further confirmation of Taiwan's membership, all Unions (except two) refused to comply with UNESCO, so the re-admission of China was still in a deadlock—explaining, in the case of the IAU, reactions like the one at the Grenoble GA in 1976.

Going back to 1958, the year of Taiwan's application to ICSU and all scientific Unions, the next—and most important—question is: why at that time? Why were the USA suddenly so anxious, not to expel China from scientific Unions, but to support Taiwan so strongly?

The answer is provided by the Chinese themselves. Fu&Ye report:

> In December 1954, Washington signed the "Mutual Defense Treaty between the United States of America and the Republic of China" with Chiang Kai-shek. This development further aggravated the long-term standoff and tensions across the Taiwan Strait. Since then, the Taiwan issue had become a major dispute between the US and China.

As a matter of fact, a crisis erupted on Aug. 23, 1958 (i.e., right after the end of the Moscow General Assembly), over the status of two small archipelagos (the Quemoy and Matsu islands) close to the coast of mainland China but belonging to Taiwan, with fights between the Chinese People's Liberation Army artillery and the Taiwanese army supported by the US Seventh Fleet. For several weeks, with the Korean War just five years back, China and the USA were again almost at war. This crisis is also mentioned by R. Espy in his book (p. 63) as the reason for China withdrawing from the IOC after having coexisted with Taiwan for many years. (Fortunately, the political status quo about the Taiwan Strait still holds today.)

In view of this volatile geopolitical situation, and in spite of many efforts by ICSU, the IAU, and other Unions over the years, the key to resolving what was in reality the China-Taiwan issue could not lie in the hands of scientists. It is only at the end of the Cultural Revolution (which had started in 1966), with the death of Mao Zedong (in 1976), and also of Chiang Kai-shek (just a year before), that the relations between China and the USA started to improve for good (after the pioneering official visit of R. Nixon to China in 1972). Deng Xiaoping opened the economy and improved relations with Western countries, China and the USA opened diplomatic relations on Jan. 1, 1979 (Deng himself visited the USA later that month), and, most importantly for the IAU and the scientific Unions, started to normalize its relations with Taiwan, with which the USA severed its diplomatic relations (replacing them with a simple cooperation treaty).

While these radical geopolitical improvements are not mentioned by Blaauw and other authors, they were absolutely decisive in paving the way to "triangular agreements" between the Unions, China and Taiwan. These agreements were all similar in that they endorsed a "dual adherence" of China to the Unions, very comparable in spirit, if not in letter, to the "dual scheme" devised by Oort and Sadler and approved by the EC twenty years before at the Herstmonceux meeting: one adhering body geographically located in mainland China (Nanjing for the IAU), and the other one located in Taipei. For the IAU, this was called "the Montreal agreement," after the following GA which took place there in August 1979, just a few months after the "opening of China."

It took a few more months to resolve a number of details, and eventually, a final agreement was signed by the three parties in early 1980, allowing China to be ready to officially re-join the IAU, its astronomers being re-instated as IAU individual members. The re-admission of China side by side with Taiwan (which necessitated a slight change in the Statutes, which were now those of 1970) was ratified at the GA in Patras (Greece) in the summer of 1982, with the "short names" "China, Nanjing," and "China, Taipei," respectively.[38]

A painful episode in the history of the IAU, lasting over twenty years, was over at last.

Acknowledgments I wish to thank Xiaowei Liu, whom I got to know when he was IAU Vice-President during my term, for enlightening discussions and having given me access to the book by Fu Chengqi and Ye Shuhua ("*Under one starry sky,*" Fu & Ye, 2009, in Chinese; their Chap. 5), which tells for the first time "the China view" of this period, and to Yi Zhou, journalist and translator, who did a superb translation of this work (now in the IAU archives). I am also grateful to George Miley for exchanges during the early stages of this work.

[38] https://www.iau.org/administration/membership/national/

References

"IAU Transactions"[39]

Stockholm (GA VI, 1938): Transactions Vol.VI, Ed. J.H. Oort, Cambridge University Press (1939).
Moscow (GA X, 1958): Transactions Vol.X, Ed. D.H. Sadler, Cambridge University Press (1960).
Berkeley (GA XI, 1961): Transactions Vol.XI B, Ed. D.H. Sadler, Blackwell Scientific Publications (1962).[40]
Grenoble (GA XVI, 1976): Transactions Vol.XVI B, Ed. E.A. Müller & A. Jappel, Association of Universities for Research in Astronomy (AURA) (1977).
Montreal (GA XVII, 1979): Transactions Vol.XVII B, Ed. P. Wayman, Association of Universities for Research in Astronomy (AURA) (1980).
Patras (GA XVIII, 1982): Transactions Vol.XVIII B, Ed. R. West, Association of Universities for Research in Astronomy (AURA) (1983).

Books

Andersen, J., Baneke, D., & Madsen C. (2019). *The International Astronomical Union, uniting the community for 100 years.* Springer, 376 p.
Blaauw, A. (1994). *History of the IAU, The Birth and first half-century of the International Astronomical Union.* Kluwer Academic Publishers, 296 p.
Espy, R. (1979). *The politics of the Olympic Games.* University of California Press, 212 p.
Fu, C., & Ye, S. (2009). *Under One Starry Sky* 《同一个星空:国际天文学联合会史》 – *History of the International Astronomical Union.* (In Chinese: Shanghai Jiao Tong University Press, Shanghai), pp. 128–167. The English translation of Chapter 5 by Yi Zhou is available upon request from the Editors. It is referred to as "Fu&Ye 2009" in the text, and is reproduced as an Appendix in MZ22.
Greenaway, F. (1996). *A history of the International Council of Scientific Unions.* Cambridge University Press, 280 p.
Montmerle, T., & Zhou, Y. (2022). *[MZ22], China and the International Astronomical Union: Divorce, separation, and reconciliation (1959–1982).* Springer.

[39] *In full:* "IAU Transactions [Vol. x], Proceedings of the International Astronomical Union." Until and including the Moscow General Assembly in 1958, only one volume of the Transactions was published, edited by the General Secretary elected at that Assembly. As of the Berkeley General Assembly in 1961, two volumes were published, separating the Scientific part (vol. "A") from the Administrative part (vol. "B"). Vol. A is edited by the incoming General Secretary, vol. B by the outgoing General Secretary. In the present text, only the Transactions "B" are relevant after 1961. The list gives the volumes referred to in the text.

[40] Due to outstanding copyright issues, this volume is unfortunately not yet available online.

Articles

Aller, L. H. (1997). *Leo Goldberg (1913–1987).* Biographical Memoirs, Vol. 72, National Academy of Sciences.

Blaauw, A., & Schmidt, M. (1993). Jan Hendrik Oort (1900–1992). *Pub. Astr. Soc. Pacific, 105*, 681.

Liu, Xiaowei. (2019) *The China crisis,* IAU Symposium 349, Under One Sky: The IAU Centenary Symposium, Eds. C. Sterken, J. Hearnshaw, and D. Valls-Gabaud (Cambridge: Cambridge University Press), pp. 222–227.

van de Hulst, H. C. (1994). *Jan Hendrik Oort (28 April 1900–5 November 1992)*, Biographical Memoirs, The Royal Society, pp. 321–326.

Wielen, R. (2019). *Germany's difficulties in becoming a member of the International Astronomical Union,* IAU Symposium 349, Under One Sky: The IAU Centenary Symposium, Eds. C. Sterken, J. Hearnshaw, and D. Valls-Gabaud (Cambridge: Cambridge University Press) pp. 205–213.

Wilford, J. N. (1992, Jan). *H. Oort, Dutch Astronomer In Forefront of Field, Dies at 92.* The NewYork Times, Nov.12, 1992 [https://nyti.ms/29bhGYj].

Wilkins, G. A. (1991). *Donald Harry Sadler, O.B.E. (1908–1987),* Q. Jl. R. astr. Soc., vol. 32, pp. 59–65.

Wilkins, G. A. (Ed.). (2008). *A personal history of H.M. Nautical Almanac Office (1930–1972), by Donald H. Sadler.* HMNAO's website [http://www.hmnao.com/history].

Thierry Montmerle has been General Secretary of the IAU (2012–2015). This made him familiar with IAU history and procedures, as well as with « backstage » information about the work of successive Executive Committees and Commissions. He participated in the IAU Centenary celebrations in 2019 by co-organizing the « Astronomers as Diplomats » conference in Paris, on which this book is founded. His scientific domain is space research in high-energy astrophysics, specializing in the role of energetic particles and photons (X-rays, gamma-rays) on the interstellar medium, and the resulting star and planet formation and evolution processes. After doctoral and post-doctoral studies in Montreal, Canada, he held a position at the Saclay Center for Nuclear Studies near Paris, France (operated by CEA, the French Atomic Energy Commission), where he spent most of his career.

Chapter 7
The Prominent Chinese Astronomer and "Father of Modern Chinese Astronomy": Zhang Yuzhe

Yi Zhou

Abstract Zhang Yuzhe (16 February 1902–21 July 1986) is widely regarded as the father of modern Chinese astronomy. He was born in the Southeast of China in Fujian province. He earned his B.A. from Tsinghua College in Beijing, as well as his astronomy M.S. and Ph.D from the University of Chicago in the 1920s. He then returned to China to embark on his life's mission of bring the knowledge of modern astronomy to the Chinese people. Facing many arduous challenges and historical events such as WWII, the Chinese Civil War, the Cultural Revolution, Zhang still managed to groom generations of Chinese astronomers, establish astronomy research institutions nationwide and personally founded many astronomical research topics. Not only being the president of the Chinese Astronomical Society for more than forty years, but also as an individual member of the IAU since at least 1938, Zhang Yuzhe had played an integral role in the departing and later reconciliation between the PRC's side and the IAU in the second half of the twentieth century (1959–1979).

7.1 A Gifted Student and His Driven Objective to Study in the USA

Zhang Yuzhe (张钰哲; February 16, 1902–July 21, 1986), is also known as Yu-Che Chang or Y. C. Chang.[1] During his lifetime, Zhang Yuzhe had served as the longtime director of the Purple Mountain Observatory (director emeritus since Feb.

[1] His first name is Yuzhe (钰哲) and his last name is Zhang (张). His name (in the modern hanyu pinyin transcription) will henceforth be written according to the Chinese tradition as Zhang Yuzhe (张钰哲), as will other personal names quoted in the text, with a few exceptions. "Yu-Che Chang," or "Y.C. Chang," was the transcription (and name order) commonly used in the West until the 80s, although the reverse name order was also used, for instance in some archival documents of

Y. Zhou (✉)
Radio France Internationale (RFI), Sciences Po, Institut d'Etudes Politiques de Paris, Paris, France
e-mail: yi.zhou@rfi.fr

© The Author(s), under exclusive license to Springer Nature Switzerland AG 2022
T. Montmerle, D. Fauque (eds.), *Astronomers as Diplomats*, Historical & Cultural Astronomy, https://doi.org/10.1007/978-3-030-98625-4_7

Fig. 7.1 Zhang Yuzhe's statue unveiled at the Purple Mountain Observatory [credit: Purple Mountain Observatory]

1984) among other academic and political positions. He is widely regarded as the father of modern Chinese astronomy, especially after the founding of the People's Republic of China (PRC) in 1949 (Fig. 7.1).

The story of Zhang Yuzhe started on February 16, 1902, in a coastal county in the Southeast of China named Minhou (Fig. 7.2 shows his former residence in Fuzhou City). It is located in the province of Fujian, on the western (mainland) side of the Taiwan Strait. Zhang Yuzhe's father was a government official. Zhang had six other siblings and he was the youngest one among the two daughters and five sons. His oldest brother was sixteen years older than him. Unfortunately, his father passed away when he was only two years old. So from an early age, his widow mother had to look after the seven children in a rather impoverished lifestyle. Zhang Yuzhe thus developed an enduring bond and had abundant affection for his mother. When he was finally old enough to attend school, he was accepted into the Minglun Elementary School in the nearby Fuzhou (福州) city, the province capital. This school was free of tuition fees. Zhang Yuzhe studied diligently from an early age. He had read many classical Chinese books before entering middle school. He was thus able to obtain an advanced level in mastering classical Chinese literature.

Interestingly enough, his first personal experience with an astronomical phenomenon occurred when he was only eight years old. While living in his hometown, he was coincidentally able to witness the passing of the famous Halley's Comet with its long shining tail. This experience had left a profound impression in his

the IAU. For completeness, for proper names we add in the text the original (simplified) Chinese ideograms.

Fig. 7.2 His former residence in Fuzhou City [Photo source see (Fjshangyi.com, n.d.)]

young heart. After finishing his elementary school studies, Zhang Yuzhe followed his second eldest brother to Peking (now Beijing). Upon arrival, he first studied in Beijing JiFu Middle School, then advanced to the High School affiliated to Beijing Normal University. This school had been established in 1901 during the last years of the Qing imperial dynasty. It was the first Western tradition public secondary school in China, which still remains as one of the most elite schools in Beijing. Zhang Yuzhe was a gifted student, and considering his hard work and dedication, he obtained excellent grades on all-around subjects. He also excelled in learning English. He graduated in 1919 being the first of his class. Zhang Yuzhe was then admitted into Tsinghua College (清华学堂) after passing its entrance exams with ease. Tsinghua College was the former body of today's Tsinghua University. It was established in April 1911 using part of the American compensation from the Boxer Rebellion, called the Boxer Indemnity Scholarship.[2]

[2] The Boxer Indemnity Scholarship (庚子赔款奖学金) was a scholarship program for Chinese students to be educated in the USA, funded by the Boxer Indemnity. This indemnity was paid by the Qing government and its successor the Republic of China government to Western powers as

Tsinghua College was a college for Chinese students, whom the Qing government and its successor the Republic of China (ROC) government, after 1911, planned to send to the USA to study. Tsinghua College was also free, and due to its future graduates being potentially selected to further their studies in the USA, the competition for its entrance was particularly fierce. Once Zhang Yuzhe entered into Tsinghua College, as he would later divulge to his American interviewer Dr. David DeVorkin some sixty years later,[3] Zhang was "astonished" by the English level of some of his peers. That was because Tsinghua College also had its own Middle School and High School programs and they all had many American teachers. So when Zhang joined the college stage, some of his classmates had been studying in English for years by then. He mentioned at the beginning it was challenging for him to adjust to the English environment while at the College. Zhang said, "*Some teachers called the roll by number, you understand. Some called by name. I had to be very attentive, so as not to miss when my name was called.*"[4] Even so, he always had his aim at heart while studying and working towards this goal. It was to come to the USA and to study. As mentioned above, he moved to Beijing with his second eldest brother, who eventually studied banking in the USA. His studies were only cut short due to the Xinhai Revolution (known as the "Chinese" revolution in the West) that broke out in 1911. The Qing government was overthrown and followed by the establishment of the ROC. Still, with only two years of studying experience in the USA, his brother was highly sought after by Chinese employers at the time, so he was able to take a position with a good salary. Witnessing this, Zhang Yuzhe told himself at a young age "*oh, I must go to the United States*"[5] and follow his brother's footsteps. (Figure 7.3 Zhang Yuzhe approximately in the early 1920s).

7.2 Moving to the USA and First Steps in Astronomy

Following his brother's trail, so he did. In 1923, Zhang Yuzhe graduated from Tsinghua College and was subsequently selected to study in the USA, where he moved and began his studies in the same year. He later described this experience in a

requested war reparations for the damage and victims caused by the Boxer Rebellion between 1899 and 1901. In 1908, the US Congress passed a bill to return to China the excess of Boxer Indemnity, amounting to over 17 million dollars. President Theodore Roosevelt's administration decided to establish the Boxer Indemnity Scholarship Program to educate young Chinese generation. He recognized this program as a chance for "American-directed reform in China" that could maximize the USA's profit by improving the Sino-American relations, bridging China with American culture, and promoting the USA's international image.

[3] Zhang Yuzhe conducted an extensive interview with the astronomy historian Dr. David DeVorkin about his personal life, academic and professional career on 18 August 1979, during the XVIIth General Assembly of the IAU in Montreal (DeVorkin, 1979). This interview is also quoted several times in this chapter.

[4] The content of this section comes from Zhang Yuzhe's interview with Dr. David DeVorkin, see DeVorkin (1979).

[5] Same Reference as Note 4.

Fig. 7.3 Zhang Yuzhe approximately in the early 1920s [Photo source see Bai (2013)]

classical Chinese poem as *"being away in a foreign country for a long time not able to return, just like one can't help but have to change his clothes every season."*[6] It is noteworthy that among the 81 Tsinghua College students that came to the USA that year, Zhang Yuzhe was not the only one that would leave their footprints in modern Chinese history. His classmates included "Rommel of the East," the celebrated WWII General Sun Li-jen (孙立人), the renowned educator Liang Shiqiu (梁实秋), one of the first Chinese to earn a doctorate degree from MIT, Yu Hsiu Ku (顾毓琇), and other historical figures. There were five soon-to-be Chinese Academicians from this tiny class. While in the USA, Zhang Yuzhe first studied mechanical engineering in Purdue University. (Figure 7.4 shows Zhang Yuzhe in the USA in the 1920s.) He then transferred to Cornell University to study architecture. In spite of having

[6] The content of this section comes from Zhang's student Zhang Jiaxiang's memorial article, see Zhang (1995).

Fig. 7.4 Zhang Yuzhe in the USA in the 1920s [Photo source watch (Stories of scientists Zhang Yuzhe, 2020)]

witnessed Halley's Comet, and as he would confess to Dr. DeVorkin in 1979,[7] he was not interested in astronomy during the early days of his studies in the USA. After diving into the subjects mentioned above, he again changed his orientation at the University of Chicago to study physics and mathematics.

Zhang Yuzhe pointed out that this decision was made under the recommendation of one of his friends who was studying physics in the same university at the time. After having arrived at the University of Chicago, Zhang Yuzhe first studied mathematics and dynamics, optics, and electricity. One may ask how did this young man later become the father of modern Chinese astronomy? Zhang Yuzhe told DeVorkin that after thinking over his decision to study engineering and architecture, he read some popular astronomy written by the American astronomer Harold Jacoby one day. He found it really interesting and thought *"maybe I should go and continue my study in astronomy."*[8] Zhang explained that he was more interested in the mathematics side of astronomy, which included the computation of orbits and the prediction of position in the heavens. He said *"because at that moment that was my conception of astronomy."*[9] While at the University of Chicago, he took classes from Professor William Duncan MacMillan, Professor Forest Ray Moulton,

[7] Zhang told DeVorkin in the same interview: *"Well, in the beginning, I was not interested. I came to United States and for half a year I studied mechanical engineering, and then I studied architecture. Finally, I went to University of Chicago to study physics and mathematics."*

[8] The content of this section comes from Zhang Yuzhe's interview with Dr. David DeVorkin, see DeVorkin (1979).

[9] Same as Note 8.

Fig. 7.5 Yerkes Observatory Staff, August 1927, from left (back row): Nicholas T. Bobrovnikoff, Karel Hujer, *Yu-Che Chang,* William H. Garrett, Harriet Bigelow, Otto Struve, Storrs B. Barrett, Christian T. Elvey, Bertha Hirt, Edwin Brant Frost (Director), George W. Moffitt, Esther Searles, George Van Biesbroeck, Francis Easton Carr, Elisabeth Struve, William W. Morgan, Alexander Pogo, Frank R. Sullivan, George C. Blakslee. *(Credit: The University of Chicago Photographic Archive* [apf6–00449], *Hanna Holborn Gray Special Collections Research Center, University of Chicago Library).* Yerkes Observatory was famous for its 40-inch (102-cm) refractor (lens) telescope, once the largest in the world, to be modernized only in the late 1960s. It ceased operations in 2018

Professor Walter Bartky, and others. (Figure 7.5 shows Zhang Yuzhe among the Yerkes Observatory Staff on August 1927).

However, it was a professor named Laves of German origin that greatly influenced his interest in astronomy. Zhang Yuzhe stated that he did not think about whether he would obtain a job or not after studying astronomy. He rather recounted a fascinating story during his time at the University of Chicago. He befriended the astrophysicist and future Nobel Prize for Physics recipient Subrahmanyan Chandrasekhar.[10] Once, when the two were talking on a train together *[was not recalled by Zhang],* Chandrasekhar asked Zhang where he should work: should he stay in the USA, or should he go back to India? Zhang replied, *"I think I said, 'Of course you should stay in the United States because you can make a contribution to*

[10] Prize awarded in 1983.

astronomy while there. As for me, I am not as talented as you are. I will go back to China to teach astronomy.'" Zhang Yuzhe recalled *"that's what I told him."*[11]

Zhang Yuzhe graduated from the University of Chicago with a bachelor's degree in 1926. Based on his excellent grades, he was then sent to Yerkes Observatory in Williams Bay, Wisconsin, which was operated by the University of Chicago Department of Astronomy and Astrophysics. And it was there that Zhang Yuzhe furthered his graduate research under Professor George van Biesbroeck. (Figure 7.6 shows Zhang Yuzhe and Professor George van Biesbroeck). He did observations of minor planets and comets and computed orbits with him. Zhang Yuzhe earned his master's degree in 1927, with a thesis on the subject of latitude measurement using transit instruments. He further conducted his doctoral research at Yerkes Observatory on the topics of observational studies of asteroids, comets, and double stars. The topic of his Ph.D thesis was given by Professor Van Biesboeck, and he defended it in 1929. It was published in *The Astronomical Journal*, with the title "A study of the orientation of the orbit-planes of 16 visual binaries having determinate inclinations."[12] While undergoing his doctoral studies at Yerkes Observatory, on the night of November 22, 1928, Zhang discovered a new asteroid while making an observation with his adviser. This was the first asteroid discovery by a Chinese astronomer. Its international registration number is No. 1125. Since the discoverer has the right to name the newly discovered asteroid, which is a rule commonly practiced by the international astronomer community, he named it after his motherland as "1125 China."[13] Zhang Yuzhe was only twenty-six years old.

Having both his mother and country at heart, just like Zhang Yuzhe told Chandrasekhar that he would return to China to teach upon finishing his studies, it was exactly what he did. He made this decision also based on the fact that, to his knowledge, there were no observatories in China at the time. On his way home in 1928, he wrote a poem entitled "Upon nearly completing the studies in the USA, Sending a poem to Mother." He wrote: *"studying science and technology as needed by home and country, taking carriages and boats, carrying book bags in a foreign land; having romantically set an ambitious goal, not able to attend the parents day and night for a long time; did not repay the loving-kindness shown by mother who brought me up, not even pay back my gratitude just like when spring comes, the*

---

[11] The content of this section comes from Zhang Yuzhe's interview with Dr. David DeVorkin, see DeVorkin (1979).

[12] The Astronomical Journal, vol. 40, p. 11, Nov.1929.

[13] Actually, the story is a little bit more complicated. The object registered as "1125 China" discovered in 1928 at Yerkes was real but subsequently thought lost. Then in 1957 another object was discovered by Zhang at Purple Mountain Observatory, but not numbered at the time; in 1978, however, with Zhang's approval, the official decision was made to move the number and name from the object discovered in 1928 to the object discovered in 1957. But in early 1986, just before Zhang passed away, an asteroid was found which turned out to be the same as the one initially discovered by him -but the number and name had been already (and mistakenly) assigned in 1978. So the new/old asteroid was given (in 1988) a new number and name: "3789 Zhongguo," which is "China"... in Chinese (中国, the Middle Kingdom). So China has been twice honored, albeit for the *same* asteroid—the one initially discovered by Zhang at Yerkes.

Fig. 7.6 Zhang Yuzhe and Professor George van Biesbroeck in between 1927–1928 [Photo source see (Yscqcn.org.cn, n.d.)]

grass grows accordingly; writing back home with much joy to finally bring back the good news, that coming home next year with the books and Guqin (古琴, an ancient Chinese table music instrument) *across the ocean."*[14]

7.3 Back to the Motherland: Love and Science in Wartime

In fall 1929, Zhang Yuzhe finally returned to China. Before crossing again the Pacific Ocean, he purposely visited and examined in detail Lowell Observatory, Lick Observatory, Mount Wilson Observatory, as well as Victoria Observatory in

[14] The content of this section comes from Zhang Jiaxiang's memorial article, see Zhang (1995).

Fig. 7.7 Zhang Yuzhe and his future wife Tao Qiang in 1930 [Photo source see (Yscqcn.org.cn, n.d.)]

Canada. He wanted to use these experiences as learning materials, that would help him establish the first Chinese observatory one day.

After his return, Zhang Yuzhe was hired as a professor upon the recommendation of his friend K.C. Fong[15] at the Department of Physics of the National Central University in Nanjing. Nanjing was the capital of the ROC at the time and he knew K.C. Fong from their Chicago days studying together. Zhang Yuzhe then became the first astronomy teacher at the University. He taught general astronomy, theoretical mechanics, astrophysics, and celestial mechanics among other related subjects. He was also hired as a researcher at the Academia Sinica Institute of Astronomy.

As fate would have it, a nineteen-year-old young female student named Tao Qiang (陶强) would become one of Zhang's astronomy class students at the University. (Figure 7.7 Zhang Yuzhe and his future wife Tao Qiang in 1930.) Tao

[15] His name in Chinese cannot be found.

Qiang was born in the South of China in Jiangsu province, but she was brought up in Beijing, because her family had moved there. She was also a classmate of Wu Chien-Shiung (吴健雄) at the University, who would later immigrate to the USA and work on the Manhattan Project, and become the first female president of the American Physical Society. Tao Qiang was eight years younger than Zhang Yuzhe. A Chinese documentary later recounted that Zhang first fell in love with Tao Qiang. He pursued her with handwritten Chinese poems. But having read his handwriting, Tao half-jokingly suggested to Zhang that it was better for him to first practice his handwriting and then they would talk about romance. After hearing this, Zhang really put in the effort into practicing and finally captured Tao's love and obtained a polished handwriting. They married in February 1933, and gave birth to a daughter and a son. Their daughter was born in the same year and their son came to the world two years later. After the Second Sino-Japanese war broke out in 1937, when Japanese troops expanded their 1931 invasion of Manchuria and captured Beijing and Nanjing, Zhang Yuzhe moved to the wartime capital Chongqing, in Sichuan province, with his entire family. In early 1941, he was promoted as the director of the Academia Sinica Institute of Astronomy. Being the newly named director of the Institute, Zhang Yuzhe moved 650 kilometers southwards to the city of Kunming, in Yunnan province, with his Institute staff, leaving his mother, wife, and two children behind in Chongqing. Despite the many challenges and difficulties caused by the ongoing war and lacking material assistance, they established the Fenghuang Mountain Observatory or the "Phoenix Mountain" Observatory there. Zhang Yuzhe wrote in one of his letters to his family that *"the holy war is at its peak, how dare one to indulge the joys of family? Having the responsibility of computing the astrophysics, could not abandon the duty of traveling to a barren land."*[16]

One of the major achievements that Zhang Yuzhe carried out during this time was to lead an expeditionary trip into the North-West of China. They went there to observe a total solar eclipse in Gansu province, West of Inner Mongolia. One has to keep in mind that this happened during the height of the war against the Japanese invaders. On September 21, 1941, it was predicted that one could clearly observe the total solar eclipse in Gansu province. Zhang Yuzhe was determined to overcome all difficulties. He wanted to travel over 3000 kilometers to observe this event from the ground. (Figure 7.8 shows Zhang Yuzhe and his total solar eclipse expeditionary team in 1941). Although his resources were greatly limited and the necessary instruments were poorly scarce, he managed to collect enough funding through various means. Zhang Yuzhe also asked for assistance to gather enough suitable instruments through his local contacts. He then selected an observation and exportation team consisting of 11 members. After 42 days of travel under the constant threat of Japanese bombers, they were able to arrive in their destination Lintao county in Gansu province.

This was the first organized effort in observing a solar eclipse in modern Chinese history. The team had obtained great results in measuring the total brightness of the

[16] The content of this section comes from Zhang Jiaxiang's memorial article, see Zhang (1995).

Fig. 7.8 The tall Zhang Yuzhe (second row fourth from the left) and his total solar eclipse expeditionary team in 1941 [Photo source see (Ktli.sinica.edu.tw, 2020)]

stellar corona. They also took multicolor photos of it, as well as filmed the entire process of the total solar eclipse with colored filmstrips. After this adventure, Zhang Yuzhe wrote a paper entitled "Solar eclipse observed in China under the shadow of Japanese bombers." This paper was published in the "Popular Astronomy" magazine in the USA.[17] Great success also came with great sacrifices. While Zhang Yuzhe was away on this observation trip, his mother unfortunately died of old age. In a tumultuous wartime period like this, it was rather unusual to see a scientist daring to go on such a faraway journey, to carry out observational duties as Zhang Yuzhe and his team did (Fig. 7.8).

7.4 Returning to Yerkes and Befriending Famous Astronomers

After the end of WWII, Zhang Yuzhe went to the USA for a second time to further his exchange with leading international astronomers. He was sent there by the ROC government with a grant plus 300 dollars per month of living expenses. He visited and conducted research in the USA for one and half years from 1946 till 1948. (Figure 7.9 shows the visit of Soviet Astronomers to Yerkes Observatory.) While there, Zhang Yuzhe stayed at his former Professor George van Biesbroeck's boarding house. Coming back to Yerkes Observatory this time around, he met many well-known astronomers, such as Jan Oort, Bengt Strömgren, Kaj Aage Gunnar Strand, and of course his old pal Chandrasekhar among others. It is also compelling to mention that Zhang would work with and meet at least four future presidents of the International Astronomical Union (IAU). They were, as already mentioned, Jan

[17] Chang (1942).

Fig. 7.9 Visit of Soviet Astronomers to Yerkes Observatory, November 1946. *Left to Right: (Floor):* Henry Chun, Code, Anne B. Underhill, K. N. Rao, Marjorie Hall Harrison, Marvin L. White, Hall, Nancy G. Roman, Daniel E. Harris. *(Sitting):* Gerard P. Kuiper, Andrei B. Severny, Schalen, V. B. Nikhonov, George Van Biesbroeck, Oleg A. Melnikov, Gunnar Malmquist, Pelageya Sannikova Shajn, Subrahmanyan Chandrasekhar and Grigory A. Shajn. *(Standing):* William W. Morgan, Hendrik C. van de Hulst, William A. Hiltner, Jesse L. Greenstein, Louis G. Henyey, Kaj A. Strand, Daniel M. Popper, Gerhard Herzberg and *Yu-Che Chang*. (*Credit: Yerkes Observatory, University of Chicago, courtesy of AIP Emilio Segrè Visual Archives, Yerkes Observatory Collection*)

Oort (Netherlands; term: 1958–1961) and Bengt Strömgren (Denmark, 1970–1973), but also Otto Struve (USA, 1952–1955), and Adriaan Blaauw (Netherlands, 1976–1979), who all had something to do later with the leaving and returning of the PRC delegation in the IAU (see below).

While Zhang Yuzhe was still studying at the University of Chicago, Otto Struve was first an instructor and then an assistant professor at the University. They knew each other at least since 1927, as can be seen in the group photo of Fig. 7.5. Zhang himself would later recall after his return to the USA, *"Professor Struve asked me what kind of work I wanted to do. I said: 'This time I would like to do something in the line of astrophysics.'"*[18] And since McDonald Observatory in Texas was under the direction of Yerkes Observatory, Struve, as its Director, sent Zhang to observe with the 82-inch telescope equipped there. At the same time, we can also find a Yerkes Observatory Staff photograph dated in 1947 (Fig. 7.10), where Zhang Yuzhe was literally sitting right next to Adriaan Blaauw. Both Zhang and Blaauw would play a prominent role in the 70s in welcoming back the PRC delegation headed by Zhang to the IAU (see below).

[18] The content of this section comes from Zhang Yuzhe's interview with Dr. David DeVorkin, see DeVorkin (1979).

Fig. 7.10 Yerkes Observatory staff, November 1947. *From left (back row, standing):* Daniel E. Harris, Guido Münch, Henry Chun, Su-Shu Huang, Douglas Duke, Mrs. Marshall H. Wrubel, Marshall H. Wrubel, Mr. Clearman, Arne Slettebak, William Bidelman, Roy Wickham, Irene Hansen, Frances H. Breen, Marjorie Hall Harrison, Gertrude Peterson, Robert H. Hardie, Arthur D. Code, K. N. Rao, Henry G. Horak, Marvin L. White, Frank N. Edmonds, Anne B. Underhill, Nancy G. Roman, Mr. Robinson, John Vosatka, Margaret Phillips, John G. Phillips, Mr. Sinha. *(Front row):* Thornton L. Page, William W. Morgan, Charles Ridell, *Yu-Che Chang, Adriaan Blaauw,* Luise Oettinger Herzberg, Gerard P. Kuiper, *Jan H. Oort,* George Van Biesbroeck, William A. Hiltner, *Otto Struve,* Gerhard Herzberg, Marguerite Van Biesbroeck, Fred Pearson, Subrahmanyan Chandrasekhar. (Apart from Chang, italics here indicate the names of future Presidents of the IAU.) *(Credit: The University of Chicago Photographic Archive* [apf6–00496], *Hanna Holborn Gray Special Collections Research Center, University of Chicago Library).* See also (Montmerle & Zhou, 2022).

While in the USA, Zhang Yuzhe observed the spectrum of two spectroscopic binaries. Near the end of 1946, he read his paper entitled "Velocity-curve of BD-6°2376, a new eclipsing variable" at the annual meeting of the American Astronomical Society. Professor Struve asked him to present this paper during the event. And this paper was later published in *The Astrophysical Journal.*[19] His studies in regards to the spectrum observation of eclipsing binaries were also published in this journal and *The Astronomical Journal.*[20]

In early 1948, Zhang Yuzhe was scheduled to return to China after he had completed his tenure in the USA. However, due to the ongoing civil war between the Nationalist Kuomintang and the Communist troops, the ROC government was not only losing the war, but also facing serious financial troubles, Zhang was stranded in the USA. His wife Tao Qiang pleaded with different Chinese government officials asking for their assistance, but her efforts were futile. Thankfully, the American Geographical Society coincidentally decided to send a delegation to China in order to observe a solar eclipse. It was due to take place in May that year in Zhejiang

[19] Chang (1947).

[20] Chang (1948a, 1948b).

province. So with the help from Professor van Biesbroeck, Zhang Yuzhe was permitted to tag along this American delegation and return home.

7.5 Chinese Astronomy in the People's Republic: The Pioneering Days

In March 1948, Zhang Yuzhe finally returned to China. Due to the fact that the Kuomintang government was retreating, all organs of the Academia Sinica were transferred to Taiwan, just separated by a strait 130 kilometers wide from Zhang's native province. But Zhang Yuzhe and his family wanted to stay in the mainland. Many intellectuals and a large number of Chinese had become disillusioned with the Kuomintang Party's rampant corruption and the ROC government's ineptitude of governance in the late 1940s. Zhang himself would tell to DeVorkin, *"but when the victorious army of our Communists were near Nanking* [now Nanjing], *our government wanted us to go southward, either to Canton or to Formosa (Taiwan), but we didn't want to go. But we couldn't tell our government we wanted to stay there to wait for the People's Republic, so I said 'We go but we have first to go to Shanghai, then from Shanghai we go to Formosa or to Canton.' The fact is, we knew that all the institutes in Shanghai had agreed not to leave Shanghai. So when we went there, we acted in unison. We acted together in solidarity."*[21]

After staying in Shanghai for nine months due to safety reasons, Zhang and his family finally welcomed the Communist take-over of the city. He then went back to Nanjing and was appointed as the Director of the Institute of Astronomy on May 20, 1950. Previously, he had already been on the observatory staff. The Institute changed its name into the Purple Mountain Observatory[22] after 1949. He stayed as the Director there for over three decades. He also served as the President of the Chinese Astronomical Society, which had been founded in 1935, for more than forty years (1943–1985). Zhang Yuzhe was elected as a member of the Chinese Academy of Sciences (academician) in 1956. His research throughout the years had entered into various subjects such as celestial mechanics, planetary science, solar physics, astrophysics, almanac astronomy, practical astronomy, astronomical instruments, astronomical history, and other topics. In his lifetime, Zhang Yuzhe published over 100 academic papers. There were also 10 published books either written or translated by him. Among the tremendous volume of his research, when interviewed by DeVorkin in 1979 (DeVorkin, 1979), he considered the observation of the relation of the brightness of asteroids and the period of it as the most interesting or satisfying research that he had ever done. Zhang Yuzhe explained

[21] The content of this section comes from Zhang Yuzhe's interview with Dr. David DeVorkin, see DeVorkin (1979).

[22] Actually, the direct translation from Chinese (紫金山天文台) is "Purple Golden (or Gold) Mountain Observatory."

that *"(this is the) one thing, I should say, that was sort of new at that time we were working on it."*[23]

Politically and subsequently, he was elected as a member of the second CPPCC (Chinese People's Political Consultative Conference) in 1959, as well as a deputy to the Third, Fourth, Fifth, and Sixth National People's Congress (NPC). He also served as a central committee member of the legally recognized "Jiusan Society" (九三学社) Party. As the leading figure of the Cold War era of Chinese astronomy, Zhang Yuzhe not only shouldered the extraordinary responsibility of developing the Chinese capacities in the relevant fields, but also worked diligently in the study of astronomical science all at the same time. He actively participated in sketching out the blueprint for Chinese astronomical science, and helped to establish the Beijing Observatory. Zhang Yuzhe is highly remembered for having played a central role in supplying the countless astronomy talents, from the Purple Mountain Observatory to all the other observatories in China as well. Besides, he personally founded the study and observation of both minor planets and comets in China. He further paved the groundwork for the study of orbit determination in astrophysics, as well as the study of artificial satellite orbits in China. As the newly nominated director of the Purple Mountain Observatory, Zhang Yuzhe continued his research and observation of minor planets and comets in China. The 60 cm reflecting telescope possessed by the Observatory since 1930 (Fig. 7.11) was damaged during the war against the Japanese. After some repairing, a smaller 12-inch diameter refracting telescope stored there was used.

At a time where all resources were quite limited, starting from December 1949, he adamantly founded the photographic observation of fixed stations work for the minor planets and comets. He led the observation works, established the darkroom equipment, and guided his young team members. The Purple Mountain Observatory lacked even the basic coordinate measuring instrument, Zhang Yuzhe thus had to periodically bring stacks of glass negatives to the Physics Institute in Shanghai for measurement. They also gradually started to offer time service by a manual transit instrument. They were among the only two mainland observatories that were providing time service in China at the time. Later the PRC government welcomed engineers from Zeiss to come to the Observatory and repair the 60-centimeter reflecting telescope. Afterwards, they also received a twin astrograph of 40-centimeter diameter to observe comets and minor planets. In the beginning of the 1950s, while studying and observing the fixed positions and motions of asteroids, Zhang Yuzhe considered establishing the study of the physical photometry of asteroids. Along with his young researchers, Zhang carried out observations with great enthusiasm day in and night out. He would say that *"those who didn't understand would consider it as bitter, though those who*

[23] The content of this section comes from Zhang Yuzhe's interview with Dr. David DeVorkin, see DeVorkin (1979).

Fig. 7.11 The Purple Mountain Observatory 60-cm telescope under construction in the Zeiss factory in Jena, ca. 1930. *[Credit: Bundesarchiv, Deutsches Ausland-Institut]*

understand would appreciate the joy."[24] This work continued for many years and he was able to obtain several dozens of asteroids' light curves in their entirety. The series of papers published by Zhang Yuzhe and his colleagues were entitled "The Photoelectric Photometry of Asteroids." Since most asteroids have odd shapes, its level of brightness changes along with its reflecting dimensions that are exposed to the Sun. Thus, the period of the level of brightness of a given asteroid can also be interpreted as its rotation period. Zhang Yuzhe's work was a trailblazing landmark in the physical observation of asteroids in China. He assessed the different rotation periods of many asteroids. He also further dove into the study of the directions of certain asteroids' axis of rotation. Zhang Yuzhe's learnings have provided valuable materials for planetary science study in China.

7.6 Building a New Generation of Chinese Astronomers

While carrying out his scientific duties, Zhang started to build a generation of young staff for the future core of Chinese astronomers. He took students from the nearby Nanking University Astronomy Department and also absorbed students,

[24] The content of this section comes from Zhang Yuzhe's interview with Dr. David DeVorkin, see DeVorkin (1979).

Fig. 7.12 Zhang Yuzhe at the Purple Mountain Observatory in 1960, operating the Zeiss 60-cm telescope shown on Fig. 7.11 [Photo source see (Yscqcn.org.cn, n.d.)]

who had physics and engineering backgrounds. On January 20, 1955, Zhang and his assistants discovered a new asteroid for the first time in China by Chinese astronomers, which they temporarily named it as "Zi Jin Yi Hao" (Purple Golden No.1).[25] This is the field of study that he personally established and participated in until his death. He worked tirelessly for over forty years without interruption. Through these countless hours of investment, Zhang Yuzhe and his team collected over 8600 negatives of both minor planets and comets. They also obtained more than 9300 precise positions, as well as discovered over one thousand new asteroids. Among them over 100 asteroids were all given an international permanent registration number.[26] Additionally, Zhang and his team discovered three new comets as well. With the growth of his staff, Zhang also helped to establish various sections at the Purple Mountain Observatory. At first, they had the Department of Almanac and Ephemeris. And the planetary section, that came early. Afterward, they started an artificial satellite section for observing artificial satellites of the Soviet Union and the USA (Fig. 7.12).

Since the 1950s were the honeymoon period of the Sino-Soviet relations, Zhang Yuzhe made several trips to the Soviet Union for academic reasons. As mentioned above, he had already met with several Soviet astronomers in 1946, while they were visiting Yerkes Observatory. In 1953, he was a member of the Chinese Academy

[25] Later registered as 3960 Chaliubieju, after Cha Liubieju, the name of a female Chinese astronomer noted for her social work.

[26] See official biography provided by the Purple Mountain Observatory website in Chinese, "Zhang Yuzhe – the Star Seeking Kind in the immense heavens." (Pmo.cas.cn, n.d.-a).

Fig. 7.13 Zhang Zhang (first one on the right) attending the reopening ceremony of the Pulkovo Observatory of the Russian Academy of Sciences, near Saint Petersburg in May 1954 [Photo source see Bai (2013)]

of Sciences delegation to visit the Soviet Union. He visited many of the Soviet observatories at the time. During this trip, he was extremely impressed by the 13 planetariums that the Soviets had possessed. So much so, he wrote in his journal that China should build its first planetarium as planned without delay. This motion was put on schedule in 1954 and the first Chinese planetarium opened in Beijing in September 1957. There are also photographs of Zhang attending the reopening ceremony of the Pulkovo Observatory of the Russian Academy of Sciences, near St. Petersburg in May 1954. (See Fig. 7.13. above) However, these were not his first encounters with the Soviet Union. Zhang had previously been sent to Siberia with his colleague Li Yan to observe a solar eclipse in 1936.

Starting from 1956, Zhang Yuzhe participated in the astronomy section of the National Scientific Development Planning for the 1960s. He contributed immensely to the establishment of Shanghai, Beijing, Yunnan, Shaanxi, and other observatories in China, as well as the founding of the Nanjing National Astronomical Instruments Factory in 1958. Zhang recommended generations after generations of astronomers and engineers who were trained in the Purple Mountain Observatory to these newly founded astronomy stations. Academically, in early 1957, there was not yet a single country that had successfully launched an artificial satellite. Zhang Yuzhe used his scientific foresight and applied his knowledge of the fundamental theories of celestial mechanics and orbit determination, he then established the first orbit study of artificial satellites revolving around the Earth. This became the basis of the theory of satellites in China. Their consequential finding was finished in the first half of 1957, it was then written into an article entitled "An Investigation On The Orbital Motion Of An Artificial Satellite".[27] This paper was published in "Acta Astronomica Sinica" in late 1957.[28] It is worth noting that the theories coming from this study were later proven to be accurate, following the Soviet Union's first artificial satellite's launching experiment on Oct.4, 1957. This paper was published at a time where there were barely any studies on the subject of artificial satellite orbits in the world. Despite his countless achievements in the early days of modern Chinese astronomy, to what extent his academic and administration roles have been affected by the never-ending political movements under Chairman Mao's rule is unfortunately little known to the public. This is due to lack of available historical records and first-hand testimonies. But if we dig deeper, we can find that a pure scholar like Zhang Yuzhe was not able to be left entirely dry when facing historical waves, as we shall see in the next section.

In the early 1960s, Zhang Yuzhe once again led his team in the study of orbit for the Moon's rocket. They later published the paper titled "Fixed Point Strikes and Aerial Survey of Moon's Rocket Orbit." Together with his team of researchers from the Purple Mountain Observatory, he led them in the designing of the very first Chinese artificial satellite's orbit and orbit solution works. In January 1965, the famed Chinese physicist and mathematician Qian Xuesen (钱学森) proposed the idea of preparing and launching the first Chinese satellite to the central government, which was well received by the Chinese Premier Zhou Enlai (周恩来) and other senior level Party officials. This project was approved and code-named "Project 651." As one of his pride graduate students Lu Benkui (陆本魁) would later recall, Zhang Yuzhe was chosen to be the director of "Project 651," since he was the expert in orbit calculation and determination.

7.7 The Hard Times of the Cultural Revolution

All the same, Mao's Cultural Revolution broke out on May 16, 1966, and Lu suggested Zhang Yuzhe was left untouched in the beginning. And they were able to continue to work on this project, but soon enough Zhang was labeled as one of "the reactionary (bourgeois) academic authorities" just like the great majority of Chinese intellectuals and scientists at the time.[29] Zhang was politically persecuted by the institutions that he worked for and some of the colleagues that he worked

[27] See Acta Astronomica Sinica second edition 1957, the authors were Zhang Yuzhe and his student Zhang Jiaxiang or Chang Chia-Hsiang (张家祥).

[28] Acta Astronomica Sinica is the earliest astronomical journal (in Chinese), that was started in 1953 by Zhang. It is sponsored by the Chinese Astronomical Society and published by the Purple Mountain Observatory.

[29] Photo source watch (Stories of scientists Zhang Yuzhe, 2020).

with. He was asked to stop working, as told by Lu Benkui, who would later become the Director of the Purple Mountain Observatory (1996–2000). He stressed that Zhang Yuzhe still cared very much about the progress made on this project. During the early years of the Cultural Revolution, Zhang would secretly snuck to see Lu and ask him how far the first Chinese satellite project had been developed. As a young scientist Lu Benkui was not considered as an "authority," thus was able to continue to work on this project. Lu said in the same interview, as his student he could understand Zhang's passion for his country and his enthusiasm on this project, so he answered Zhang's questions in secrecy. Lu later[30] stated that Zhang Yuzhe had made a pioneering contribution in Chinese satellite development.[31]

The Cultural Revolution was a truly historical event, since seldom in the history of nation-state and humanity had the entire population of a given society been totally mobilized politically to this extent. Based on what we know, if Zhang Yuzhe was partially spared because he was a dedicated science scholar with international importance and name recognition, his wife Tao Qiang would suffer a terrible fate. During the Cultural Revolution, Tao Qiang was a math and geometry teacher at the Middle School Affiliated to Nanjing Normal University. As many of her former students would remember her in their memoirs, she was a respected and outstanding teacher prior to the Cultural Revolution.[32]She had much better people skills and sharpness in dealing with her students and colleagues than her husband. Tao would give out small gifts to other teachers and be extra helpful to the students in need on and off campus. On July 4, 1966, not even two months after the start of the Cultural Revolution, the housekeeper who was working for Zhang and Tao's household suddenly came to Tao's school. She told Tao's school that the latter had broken a plaster statue of Chairman Mao at their house. Such offense was deemed as a serious counter-revolutionary event, a "crime" that could lead to death penalty under such a lawless social context. Although Tao Qiang would later argue it was the housekeeper that broke Mao's statue, she helped her to clean up afterwards to not let anybody know, but when such an outrageous accusation had been made, the students at Tao's school were furious at her.

Tao Qiang was subsequently ganged up by the students and they put a waste paper basket over her head. They also dripped ink all over the basket and her head. The students apprehended her and marched her around the campus like a "counter-revolutionary." After a while, the mob stopped in front of a school building and held a public struggle session against her. She was openly disowned, criticized, and humiliated by the students and by some of her own fellow teachers. As remembered by the students who were present, there were at least two young teachers who threw

---

[30] Lu Benkui told this anecdote in the short documentary on Zhang's life entitled "Stories of scientists: Zhang Yuzhe."

[31] The first Chinese artificial satellite, Dong Fang Hong 1 ("the Orient is Red," 东方红一号) would eventually be launched on Apr.24, 1970.

[32] The contents of this section and below mainly come from Tao Qiang's former students' memoirs on the Cultural Revolution, see (Yi ke xiao cao, 2008; 303-1 Historic, 2020).

Fig. 7.14 Zhang Yuzhe and his wife Tao Qiang at Tiananmen Square in the late 1970s [Photo source watch (Stories of scientists Zhang Yuzhe, 2020)]

those gifts that Tao had previously given them under her feet and publicly drew the line between themselves and her. The gifts included textbooks, a watch, a violin, and others. Her colleagues publicly accused her of trying to buy their good will and denounced her in front of the mounting students. This happened while the students shouted out slogans like *"down with Tao Qiang"* and other attacking political phrases. Some students were remembered to have spat over Tao and her clothes. This all happened on the same day and Tao was only saved by the headmaster Wu, who tried to control the scene before it got out of hand. For the next several months this type of so-called struggle sessions were constantly held in Tao Qiang's school and elsewhere in China. She was one of the main targeted teachers that were persecuted by the Red Guards, who were made up by the students from her school. She experienced having her hair shaved by half, the popular style of humiliation called "Yin-Yang hairstyle" during the Cultural Revolution. Tao and others were physically and verbally abused by the Red Guards, who equipped themselves with belts and sticks as forms of punishment. One has to remember these Red Guards had been Tao's students just days before. Yet, they put her and other targeted teachers into forced labor as well. (Figure 7.14 shows Zhang Yuzhe and his wife Tao Qiang at Tiananmen Square in the late 1970s).

Tao Qiang later confessed to one of her students that, in the beginning of the Cultural Revolution, the amount of persecution, physical and psychological abuses that she endured made her think about committing suicide. She also chose not to share her sufferings with her husband Zhang, in fear that this would bear too much pressure on him. As remembered by one of Tao Qiang's students, Zhang dedicated his entire energy in astronomy-related works, every aspect of his life was reliant on his wife's assistance. No matter what Zhang Yuzhe would eat during the day or what he wore to work, all were arranged by Tao Qiang. When Zhang would travel afar, she even worried he would get lost. It was said that Tao Qiang looked after Zhang as her "big child." This is also why Tao stated that despite the enormous amount of pressure and persecution she did not choose to commit suicide, because *"she thought to herself, who would look after Zhang when she would be gone?"*[33]

After Mao's death and the arrest of the "Gang of Four" in 1976, the Chinese Communist Party was soon to be taken under the direction of Deng Xiaoping. By 1980, with Mao's appointed successor Hua Guofeng losing the internal Party struggle to Deng, Mao's overemphasis on social political struggle was soon to be diminished. Within these few years, Zhang would regain all of the fame, recognition, and positions at the twilight of his life. He started again to have a correspondence with the IAU in early 1973, and during the 1970s. In those times, the international astronomical community tried to reestablish the link with PRC astronomers as shown in the IAU membership withdrawal episode (see below). Some Australian and other foreign astronomers started to visit China and Zhang's Purple Mountain Observatory.[34]

Some foreign astronomers even asked to visit Zhang and his family at their home. This demand troubled the local authorities and those who were in charge at the Observatory. They thought the two-bedroom apartment in which Zhang and his family lived for decades was no longer suitable for an important figure like him. So they asked Zhang and his wife to move to a bigger place. However, because Zhang and his wife grew attached to their apartment and refused to move. The authorities then asked a neighbor of theirs to move and refurbished the neighbor's apartment into the salon of Zhang's family. It was there that he received many foreign visitors. With Zhang Yuzhe regaining his social status, many old friends and officials would constantly show up at his door once again.

One could say that Zhang and his family experienced a complete turn-around of fortune after the Cultural Revolution ended. Zhang Yuzhe himself would suggest that when Leo Goldberg and other American astronomers visited China, or when Chinese astronomers visited the USA in the 1970s, he personally was not involved in bringing the two sides together. He told DeVorkin in 1979 that *"well, I had nothing to do with that. If you read Goldberg's articles, you will see. There are two institutions, one in China and one in the United States. They made the arrangements between them."*[35] When asked by DeVorkin if, with his long-standing experience of contact with the USA, he wanted to participate in bridging the two sides, for which Zhang replied "Yes." But he insisted that *"well, it was so arranged. I couldn't go*

---

[33] The content of this section comes from a Tao Qiang's former student' memoirs on the Cultural Revolution (see Yi ke xiao cao, 2008).

[34] See the chapters by Wang Shouguan (Chap. 8) and George Miley (Chap. 12), this volume.

[35] The content of this section comes from Zhang Yuzhe's interview with Dr. David DeVorkin, see (DeVorkin, 1979).

there, take part in that. Couldn't do it." He also further stated he did not ask to take part in it.[36]

7.8 Leaving the IAU for Twenty Years

Nonetheless, based on the "Under One Starry Sky" book written by Ye Shuhua (叶叔华) who eventually became the first Chinese IAU Vice-President (1988–1994), and her colleague Fu Chengqi *(傅承启)*, Zhang, as the longtime president of the Chinese Astronomical Society, not only participated in its early activities, but also personally experienced the entire historical process when the Society left the IAU after the establishment of the PRC, as well as its renegotiation of reentering this international organization.[37] This series of events lasted for over two decades. Zhang Yuzhe had been an individual member of the IAU at least since 1938, he even served as a member of the IAU Finance Committee at the ninth General Assembly in Dublin in 1955. Due to various domestic and foreign obstacles, the Chinese Astronomical Society's membership at the IAU stopped before the establishment of the PRC. This was only resumed in 1953, after the Korean War was over, and the President of the IAU at the time, Zhang Yuzhe's friend, Professor Otto Struve sent an invitation to his former student in 1955.

Even so, the PRC delegation in the IAU would not stay long, with Taiwan's application to join the IAU coming soon after. In September 1959, the Executive Committee of the IAU at the time, under the Presidency of Jan Oort, held a meeting and its seven members voted on Taiwan's application for membership. The result was 5 vs. 2 in favor of Taiwan's adherence to the IAU. This decision quickly enraged many parties. Zhang Yuzhe who was representing the Mainland forwarded his letter of protest against the Executive Committee. In his letter on November 20, 1959, he clearly argued to Jan Oort that "*Taiwan is an inalienable part of the Chinese territory, it is a province of the PRC*".[38] He also urged the Executive Committee to renounce this decision. As mentioned above, Zhang had met Oort at Professor van Biesbroeck's boarding house more than ten years before. In the coming months, Zhang would write to Oort several times to protest the Taiwan decision. On February 5, 1960, he once again expressed the PRC side's opposition and protest to Oort, in the name of the president of the Chinese Astronomical Society. He stated that the mainland participation would be halted immediately and that China would formally withdraw from the IAU and all of its sub-committees.

---

[36] The content of this section comes from Zhang Yuzhe's interview with Dr. David DeVorkin, see DeVorkin (1979).

[37] See also the chapters by Xiaowei Liu, Thierry Montmerle and George Miley, this volume (resp. Chaps. 5, 6, and 12).

[38] The content of this section comes from Ye Shuhua's book, see Ye & Fu (2009).

Zhang Yuzhe also indicated that all relations with the IAU would therefore be broken completely. At the same time, all individual Chinese members issued their personal statements and declared their resignations from the IAU. Furthermore, the official Xinhua news agency of the PRC soon published an article titled "A protestation to the fabrication of 'two Chinas' plot." It officially announced the PRC's withdrawal from the IAU. From then on, the PRC had left the IAU and broken all ties with the latter body. Soon after, the third edition of the IAU Information Bulletin in 1960 printed the resignation of the Chinese mainland from this organization on its front page. It was written that the president of the PRC's Astronomical Society, Professor Zhang Yuzhe officially announced that the PRC left the IAU. This was a gesture of protest against the Executive Committee's decision in accepting the ROC (Taiwan) as one of its members... From this day and onwards, the PRC astronomers were no longer members of the IAU, nor would they still remain as members of its professional commissions.

After Nixon's visit to China in 1972, the IAU under the Presidency of Leo Goldberg had started to make efforts in mending the ties with Beijing. However, true breakthrough was not reached until the 17th IAU General Assembly in Montreal in 1979. As the successor of Leo Goldberg, Adriaan Blaauw, the other Dutch astronomer whom Zhang had also met (this time thrity years before) displayed his masterful diplomatic skills and political savviness in helping organize a meeting between the Taiwanese delegation and the Chinese mainland delegation at the Assembly. After days of challenge solving, the Chinese mainland delegation headed by Zhang, the Taiwanese delegation and the Executive Committee of the IAU reached "the Montreal Agreement." It is also meaningful to note that it is during this diplomatic mission that Zhang Yuzhe conducted his interview with DeVorkin in Montreal on August 18, 1979. Although the three sides still had some differences on the official name of the Taiwanese delegation at the IAU, the Chinese Astronomical Society was finally able to come back to the IAU, its return being ratified by the 18th General Assembly in Patras, Greece in August 1982.

7.9 Building the Future

Aside from his academic dedication and his love for his family and country, Zhang Yuzhe was also a man of many talents and strengths. His first foreign language was English, but he later also learned French and German. It was not until the 1950s that he started to learn Russian too. When the1960s came, he made the decision to translate the entire book of "Stellar Evolution" written by Otto Struve from Russian into Chinese.[39] Through his research and educational career, he helped to groom

[39] Otto Struve (1897–1963) was born in Ukraine, then belonging to Tsarist Russia. He had emigrated to the USA in 1921. Struve's book was first published by Princeton University Press in 1950, and translated into Russian in 1954. It is likely that at that time Zhang did not have access to the English edition, which would presumably have made his translation into Chinese easier.

countless young talents with great attentiveness and grit. In showing the ropes to the younger generations, Zhang Yuzhe always led by example not just handing out advice. He did not simply just focus on spreading out the tasks, rather he paid attention to incite curiosity and delegate responsibility to those who worked with him. He believed that in conducting research, one must be highly career-focused. He or she also has to have a deep level of interest and keep a steadfast mindset. Apart from being a leading scientist in his field, Zhang was also a Renaissance man when it came to the arts. He had substantial knowledge of classical Chinese literature. He could write ancient poems with ease, which are full of rhythm and profound meanings. Zhang Yuzhe further perfected himself on performing seal carving and paintings. "*...Hundreds of battles through blood and sweat, overseeing the results on the three destroyed mountains. Traveling the sky and measuring angles alone were non-beneficial, lying on the sick bed with disappointment because of not repaying the country's debt...*".[40]

This was part of the poem that he wrote while having a stomach surgery at the Beijing Hospital in 1963. Zhang Yuzhe sighed about being in his sick bed, watching the revolution go by, as well as not being able to conduct his astronomy activities. Zhang Yuzhe's love for his fellow countrymen was well demonstrated through his writings. Moreover, Zhang Yuzhe was a tenacious force of nature full of perseverance spirit. In 1977, at the advanced age of 75, he published a paper titled "The Tendency In Orbital Evolution Of Halley's Comet And Its Ancient History."[41] This study took into account our solar system's 9(8) planets' perturbation, as well as calculated the motion orbit of the Halley's Comet for over 3000 years (Zhang 1978). This paper further studied and analyzed the early records of Halley's Comet in Chinese history, it provided numerous clues regarding the solutions to the related chronology questions. This was highly regarded by historians. In 1980, when Zhang was already 78, he still led a team to the Chaidamu Basin in the mountainous Qinghai province, north of Tibet, to select the right location to build the first Large Millimeter Telescope in China. He made it to the top of Kunlun Mountains, which is 4800 meters above sea level. Three years after, he went further west to visit the artificial satellite tracking station based in Ürümqi in the Xinjiang Uyghur Autonomous Region, in Northwest China. This station had been built in 1958 based on his suggestion.

In 1984, when Zhang Yuzhe was already 82, he made a final trip to the USA and visited Harvard University. (Figure 7.15 shows Zhang Yuzhe, Tao Qiang and their two children in the USA in 1984.) He gave a speech entitled "Today's Chinese Observatories" at the Harvard-Smithsonian Center for Astrophysics. Previously, the Minor Planet Center had announced in 1978 that it had named the 2051st asteroid, discovered by the Harvard University Observatory two years before, as "2051Chang," in honor of his contribution to astronomy and as a sign of respect for

[40] The content of this section comes from Zhang Jiaxiang's memorial article on Zhang Yuzhe see Zhang (1995).

[41] Chang (1978).

Fig. 7.15 Zhang Yuzhe, Tao Qiang and their two children in the USA in 1984 [Photo source see (Yscqcn.org.cn, n.d.)]

his works. In 1985, not long before his death, Zhang Yuzhe once again witnessed the passing of the Halley's Comet at the Purple Mountain Observatory. It was the same exact experience that started his astronomy journey over seventy years before. (Figure 7.16 Zhang Yuzhe with Deng Xiaoping at the Purple Mountain Observatory in 1985.) Zhang passed away on July 21, 1986, when he was eighty-four years old. After he had passed, Zhang's wife Tao Qiang was greatly saddened, and she followed suit not long after. In October 1990, the Ministry of Posts and Telecommunications of the PRC issued its second edition of "Chinese Modern Scientists Memorial Stamps," Sohu.com (2018) Zhang Yuzhe's portrait was honored among other famous Chinese science figures. (See Fig. 7.17 below).

The highest award in Chinese astronomy, the "Yu-Che Chang Award" was also named after him Zhang (2018). It is issued by the Chinese Astronomical Society, for which Zhang is its longest-serving president to date.

Acknowledgments It is a pleasure to acknowledge the precious advice of Gareth Williams, past associate Director of the IAU Minor Planet Center (Cambridge, Mass., USA), on sorting out the intricacies of asteroid naming and the Chinese contributions to their discoveries. I want to also acknowledge the treasured advice given by Thierry Montmerle, past General Secretary of the IAU for his kind and tireless assistance in improving this article for better readership. Without his encouragement, this paper and the story of Zhang Yuzhe would not have come to the light for the English-speaking audiences.

Fig. 7.16 Zhang Yuzhe with Deng Xiaoping at the Purple Mountain Observatory in 1985 [Photo source see (Pmo.cas.cn, n.d.-b)]

Fig. 7.17 Zhang Yuzhe Post stamp (1990), featuring Comet Halley which Zhang had witnessed as a boy and which had reappeared in 1985, one year before his death [Photo source see (Yscqcn.org.cn, n.d.)]

References[42]

303-1 Historic account of the High School Affiliated to Nanjing Normal University during the Cultural Revolution part 1. (2020). In: 《南附中老三》. https://nsfz2.wordpress.com/2012/03/05/303-1-%E5%8D%97%E5%B8%88%E9%99%84%E4%B8%AD%E6%96%87%E5%8C%96%E5%A4%A7%E9%9D%A9%E5%91%BD%E9%A3%8E%E4%BA%91%E5%BD%95%EF%BC%88%E4%B8%8A%EF%BC%89/. Accessed 11 Sep 2020.

[42] The references include a personal interview with Zhang Yuzhe himself, his official biographies from the Purple Mountain Observatory website and the official Chinese news agency Xinhua, articles and interviews with his former students and colleagues, such as Ye Shuhua and Lu Benkui. Both Zhang Yuzhe and Tao Qiang's experience during the Cultural Revolution mentioned in the article were mainly based on personal memorial pieces and interviews by their former students, which were published on various Chinese blogs.

Bai, Y. (2013). *The Astronomer Zhang Yuzhe, observing the skys from Zhongshan for 60 years*, 天文学家张钰哲:钟山观天60载. [online] Uux.cn. Available at: <http://www.uux.cn/viewnews-54973.html> [Accessed 13 October 2020].

Chang, Y. C. (1942, Apr). Solar eclipse observed in China under the shadow of Japanese bombers. *Popular Astronomy, 50*, 198.

Chang, Y. C. (1947). Velocity-curve of BD – 6°2376, a new eclipsing variable. *Astrophysical Journal, 106*, 308.

Chang, Y. C. (1948a). Spectrographic observations of the eclipsing binaries of the W Ursae Majoris type: AH Virginis and TZ Bootis. *Astrophysical Journal, 107*, 96.

Chang, Y. C. (1948b). Spectrographic observations of the eclipsing binaries of the W Ursae Majoris type, AH Virginis and TZ Bootis. *Astronomical Journal, 53*, 111.

Chang, Y. C. (1978). *Acta Astronomica Sinica*, 1st ed.

Chinese Academy of Sciences, P. (n.d.). Yuzhe Zhang—-Purple Mountain Observatory, CAS. [online] English.pmo.cas.cn. Available at: <http://english.pmo.cas.cn/pe/cm/200907/t20090730_26495.html> [Accessed 11 September 2020].

DeVorkin, D. (1979). Oral History Interviews: Yu-Che Chang. [online] Aip.org. Available at: <https://www.aip.org/history-programs/niels-bohr-library/oral-histories/4553> [Accessed 11 September 2020].

Fjshangyi.com. (n.d.). *Zhang Yuzhe's former residence*. [online] Available at: <http://www.fjshangyi.com/case/65.html> [Accessed 13 October 2020].

Ktli.sinica.edu.tw. (2020). *Total eclipse of the sun*. [online] Available at: <http://ktli.sinica.edu.tw/residence/wp-content/uploads/Image 148.jpg> [Accessed 13 October 2020].

Montmerle, T., & Zhou, Y. (2022). *China and the IAU: Divorce, separation, and reconciliation*. Springer.

Pmo.cas.cn. (n.d.-a). *Zhang Yuzhe – The star seeking kind in the immense heavens*,《张钰哲:苍茫高天追星族–紫金山天文台》. [online] Available at: <http://www.pmo.cas.cn/ztjj2019/ysfc2019/201909/t20190915_5383354.html> [Accessed 5 October 2020].

Pmo.cas.cn. (n.d.-b). *Purple mountain observatory, the cradle of modern Chinese astronomy*紫金山天文台——中国现代天文学的摇篮—中国科学院紫金山天文台. [online] Available at: <http://www.pmo.cas.cn/zhxxpt_90872/gk/201305/t20130502_3830868.html> [Accessed 13 October 2020].

Sohu.com. (2018). *Using postage stamps to commemorate the one who discover stars*, 用邮票铭记那颗发现星星的您_张钰哲. [online] Available at: <https://www.sohu.com/a/278214528_100014089> [Accessed 13 October 2020].

Stories of scientists Zhang Yuzhe. (2020). 《河汉之星——张钰哲》 https://v.qq.com/x/page/f0839l6qah0.html . [Accessed 11 Sep 2020.]

Ye, Shuhua, & Fu, Chengqi. (2009). Under One Starry Sky 《同一个星空:国际天文学联合会史》 – History of the International Astronomical Union. Shanghai Jiao Tong University Press, Shanghai, pp. 128–167 [English translation of Chapter 5 by Yi Zhou available upon request from the IAU Secretariat https://www.iau.org/administration/secretariat/ , or from the Editors of this book].

Yi ke xiao cao. (2008). Me and Ms. Tao Qiang. 《我与陶强老师》, http://nsfz63.blog125.fc2blog.us/blog-entry-156.html [Accessed 11 September 2020].

Yscqcn.org.cn. (n.d.). History of Chinese Academicians – Zhang Yuzhe, 张钰哲. [online] Available at: <http://www.yscqcn.org.cn/yuanshi/info.aspx?id=2362&cid=1> [Accessed 13 October 2020].

Zhang, Y. (1978). *The tendency in orbital evolution of Halley's comet and its ancient history*. [online] Gb.oversea.cnki.net. Available at: <http://gb.oversea.cnki.net/KCMS/detail/detail.aspx?filename=TWXB197801010&dbcode=CJFQ&dbname=CJFD1979> [Accessed 5 October 2020].

Jiaxiang Zhang. (1995). The brief biographies of Chinese experts in science and technology, physics chapter: Astronomy Volume. 1,《中国科学技术专家传略》, 理学编:天文卷1. Beijing: Chinese Science and Technology Publishing House, Beijing, pp. 73–82.

Zhong Y. (2018). Zhang Yuzhe: The pioneer of heavens popular science-China. 《张钰哲:璀璨星空的探路者》 In: Xinhuanet.com. http://www.xinhuanet.com/science/2018-07/21/c_137337830.htm. [Accessed 11 Sep 2020].

Yi Zhou was born in China. He is a seasoned Journalist/Translator/Researcher based in Paris France and a curious observer of current world events. He is passionate about contemporary history and investigative journalism. He has worked for the Chinese service of the French public radio station, Radio France Internationale (RFI), since 2015. Previously, he finished high school in Columbus, Ohio, USA. He went on studying in France at the Institut d'Etudes Politiques de Paris ("Sciences Po") for both his Bachelor's and Master's degree, specializing in political and social sciences. He obtained a Master's degree in International Development from the Paris School of International Affairs (PSIA) of Sciences Po. Coming from China and having studied in the USA, France, and the Netherlands, has given him a special set of understanding, skills, and rare experience in dealing with cross-regional and international political and social subjects.

Chapter 8
Personal Recollections of W.N. Christiansen and the Early Days of Chinese Radio Astronomy

Wang Shouguan

Abstract Between 1963 and 1998, Prof. W.N. Christiansen visited China more than a dozen times, bringing valuable scientific information, expert guidance, and all possible help to the young Chinese radio astronomy team. Here, the writer presents his memories of two typical, deeply shared experiences, "The Shahe Experiment" and "The Making of the Miyun Meter Wave Aperture Synthesis Telescope," as expressions of the kind thoughts of a whole generation of Chinese researchers in astronomy.

Keywords W.N. Christiansen · Prof. Ke · Chinese radio astronomy · The Shahe experiment · The Miyun aperture synthesis telescope

Note From the Editors
Wang Shouguan (王绶琯) succeeded Zhang Yuzhe as President of the Chinese Astronomical Society in 1985. Born in 1923, he was one of the pioneers of radioastronomy in China in the 1960s. This chapter tells about the exchanges he had with W.N. Christiansen, who was himself a pioneer of radio interferometry in Australia, and with whom he developed a long-time collaboration and friendship during the difficult times when China had withdrawn from the IAU because of the admission of Taiwan in 1959. Wang wrote a more recent, but also more technical, article later (Shouguan Wang 2016: *Reminiscence of my sixty-five year voyage in Astronomy,* Research in Astronomy and Astrophysics, vol.16, No 6, 86 [12pp].)

The editors respectfully thank JinLin Han, from the National Astronomical Observatories in Beijing, and Wang Ying, Wang Shouguan's daughter, for facilitating

---

This chapter is republished by permission from the Journal of Astronomical History and Heritage from "Wang Shouguan, Journal of Astronomical History and Heritage, 12(1), 33–38 (2009)".

W. Shouguan, (deceased)
Shouguan wrote this chapter while at National Astronomical Observatory, Chinese Academy of Sciences, Beijing, China
e-mail: shouguanwang@aliyun.com

© The Author(s), under exclusive license to Springer Nature Switzerland AG 2022
T. Montmerle, D. Fauque (eds.), *Astronomers as Diplomats*, Historical & Cultural Astronomy, https://doi.org/10.1007/978-3-030-98625-4_8

their contacts with Wang Shouguan. This chapter is reproduced here with the kind permission of the author and of Wayne Orchiston, co-founder and editor of the *Journal of Astronomical History and Heritage* (JAHH).

Prof. Wang Shouguan, photo taken in 2011 at home. (Courtesy Zhang Yi)

Post-Scriptum: While we were preparing this book, we had several exchanges with Prof. Wang via Wang Ying. Unfortunately, Prof. Wang passed away on 28 January, 2021, in Beijing. He was 98. We thank him and his family again for their kind cooperation, and praise his contributions to Chinese astronomy and to the IAU in troubled times. His obituary is available on the IAU web site (https://www.iau.org/administration/membership/individual/2951/). We add below a Prologue describing Prof. Wang's early years in the UK, and how a naval officer became an astronomer.

8.1 Prologue: A Brief Biography of the Young Wang Shouguan (1923–1953)

Yi Zhou (周易) and Thierry Montmerle

Wang Shouguan was born on January 15, 1923, in Fuzhou, Fujian Province, the same city as Zhang Yuzhe, and home to the famous Mawei naval arsenal (尾造船厂). When he was about three years old, his father, a city official, died, and his mother, who came from a family of Chinese doctors, had a hard time raising him as her only son. He attended both traditional and newly reformed schools (one studying Confucius texts and the other modern science subjects), reading a lot and getting good grades, but there was financial pressure on the family. So, by the time he was 13, as his uncle was a naval officer for the Guomindang (the Nationalist government of the Republic of China [ROC], then under Chiang Kai-shek), he could be recommended to apply to the Fujian Mawei Naval Academy (formerly known as the Mawei Shipbuilding Academy), and went to Nanjing to successfully take the exam. He became one of the 100 accepted students.

At that time, the Academy students were trained as naval reserve officers for the ROC's Navy, learning to steer and command a ship, shipbuilding, etc.; the study and internship took a total of eight years. After graduation, the students could directly find a job, not needing to worry about employment—this is the reason why Wang Shouguan chose to enter this school. After all, his family was poor, and if this opportunity was not seized when finishing high school, there was no hope for him to work as soon as possible to support his mother and the family, and going to the university was only a distant dream.

The opportunity was really a good one, but the problem was, at that time, that he was unaware that his real interests were the stars, not the sea. Wang Shouguan began to learn how to command a crew, and to become the captain of a ship. However, probably because he read too much, by the age of 17, he found himself nearsighted. Now the situation became a bit worrisome - how can a nearsighted person become a captain? Having had already spent four years in training, was he going to have to interrupt his studies? After having considered the situation, the school decided to allow him to switch departments and study shipbuilding instead. "The reverie is useless," Wang said in an interview, complaining that he really did not like to do this: "Building a ship is a matter of life and death. Even when building a normal ship human life is at stake, everything is very rigid. People must have acquired a lot of experience before being able to use one of them. You can't get a drawing wrong, you have to calculate, you have to use the curved board for the hull, do all the detailed work, but you are not allowed to change anything, you are not allowed to test something yourself, even a little. Special tests could be conducted, but they weren't allowed at that stage, so I never really got used to shipbuilding."[1]

And at that time, Wang had already discovered astronomy as a hobby, but he just did not dare think about actually turning it into a life's work. In 1944, at the age of 21, he finally graduated from the Academy, and his first job was to build mines at a factory near the front line in Hunan Province (which was not occupied by Japanese forces) for use against enemy warships. At the end of the war, he applied for a publicly funded foreign exchange program to the UK to continue his studies at the Royal Naval College, in Greenwich near London. He probably did not expect that this voyage across the ocean would completely change the course of his life.

In 1945, at the age of 22, Wang Shouguan was thus admitted to the British Royal Naval College, where he studied at public expense. The Naval College had been built next to the Greenwich Observatory [resp. in 1696 and 1675], which led him to aspire to astronomy: "The fire of interest was burning in my heart, and I could not sit still . . . " He was admitted into the shipbuilding major class, where half of the students in the class were graduates from Cambridge and London University. There were also some graduates from workers' colleges, so the academic pressure was very high and the elimination rate was correspondingly high. At that time, there were six Chinese students in this group (Fig. 8.1); by 1949, only three of them graduated successfully,

[1] 微信公众平台. 2021. *星罗·棋布——自称"无所作为"的王绶琯老大爷，您还想咋？（一）*. [online] Available at: <https://mp.weixin.qq.com/s/lr2sngR8plszsJUTklID3g> [Accessed 10 June 2021].

Fig. 8.1 Wang during his time at the Royal Naval College in Greenwich. (Source: https://finance.sina.com.cn/tech/2021-01-29/doc-ikftssap1703310.shtml, photo credit: CCTV 10)

Fig. 8.2 1950 in London, Wang with classmates. He is standing in the middle. (Same source as Fig. 8.1)

and Wang was one of them (Fig. 8.2). The title of his memoir was "How to design an aircraft carrier" . . .

Starting in 1949, Wang wrote to several British astronomers to exchange information related to astronomy. Later on, he went by car from time to time to the University of London Observatory in Mill Hill [now the University College London Observatory], about 30 km to the North-East of Greenwich, and talked several times with Christopher C.L. Gregory, the Director of the Observatory, expressing his desire to work there. At that time, the Greenwich Observatory, the next-door neighbor of the Royal Naval College, was under the jurisdiction of the Royal Navy, and although it was near, Wang Shouguan, as a Chinese student, did not have easy access. Gregory, an old scientist in his 60s, expressed his appreciation and

Fig. 8.3 The caption in the picture (伦敦大学天文台) means “the University of London Observatory.” Wang Shouguan is standing on the right. Source: http://www.nao.cas.cn/wsg/tjsp/202101/P020210129422839803551.pdf

encouragement to the young man who had come all the way from China and was interested in astronomy, and not only gave him generous guidance, but also accepted him into the University of London Observatory in 1950. Could Wang become a real astronomer in two or three years? He wanted to try. “From then on, a pair of astronomer’s ‘big hands’ and a pair of untrained ‘small hands’ were held together,” he said. From 1950 to 1953, Wang spent endless nights stargazing with the 24-inch/18-inch twin refractor of the observatory (Fig. 8.3).

“When I first encountered this new discipline, I started from scratch.”[2] At that time, Wang Shouguan was mainly responsible for the nighttime actual testing from 8 pm to 4 am. During the long nights, he raised his eyes to look at the sky full of stars, and his thoughts were overwhelming. Many years later, he had recalled the

[2] Wang, Y., 2021. *王绶琯：科海勉相随，诗文天籁间-中国科普网*. [online] Kepu.gov.cn. Available at: <http://www.kepu.gov.cn/www/article/50dc7471baf046c7a758d4258b4dd59e> [Accessed 10 June 2021].

past in the article "*A small memory of a night outside London*"[3]: "At that time I was at the University of London Observatory, located in the northwest suburbs of London, surrounded by very flat and wide fields, a road stretched from London. The road is very wide and straight . . . after dusk, the night cover down, hazy, the road is like a straight canal, the bank on both sides of the pulse of thoughts sent to the other side of the sky . . ."

When it comes to participating in astronomy research in China, the clock goes back to 1952. At that time, the motherland had undergone radical changes, and overseas students were returning to China to serve, so Wang Shouguan decided to return to China without hesitation after receiving an invitation from Zhang Yuzhe, then Director of the Purple Mountain Observatory and his fellow countryman from Fuzhou. In 1952, Wang Shouguan wrote to the Purple Mountain Observatory to ask if astronomy was still being done in China. Wang said in an interview, "I learned that the Purple Mountain Observatory, as it turned out, had been in existence before 1949, and that Mr. Yu Qingsong was running it. Later, I learned that Mr. Yu had gone to the United States when Mr. Zhang Yuzhe was in charge, and as a result, I received a letter from Mr. Zhang Yuzhe, saying that I should come back quickly, and that now China was going to develop astronomy."[4]

Wang returned to China in 1953.

8.2 Foreword[5]

Childhood memory is precious. The "childhood"—or early days—of Chinese radio astronomy went through unusual trials and tribulations. The first team of Chinese researchers in radio astronomy, so very young at the time, and so deeply engulfed in stormy weather during their first steps, will hold dear in their memory the arrival of a senior colleague from the West. During this period, W.N. Christiansen (affectionately known by us as "Prof. Ke," "Ke" being the first of the five Chinese characters that transcribe his surname) made the long journey to us more than a dozen times (see Fig. 8.4), and brought to us valuable scientific information, specialist guidance, and assistance.

The memories of our discipline's infancy are still fresh with us, and our feelings cannot be expressed by ordinary words. Here, we recall a couple of the most typical events, as emissaries carrying the thoughts of an entire era.

---

[3] Wang, S., 2006. *Page 163*. [online] Nao.cas.cn. Available at: <http://www.nao.cas.cn/wsg/tjsp/202101/P020210129422839803551.pdf> [Accessed 10 June 2021].

[4] 微信公平台. 2021. *星罗·棋布——自称"无所作为"的王绶琯老大爷，您还想咋？（二）*. [online] Available at: <https://mp.weixin.qq.com/s/iFNxFCku-wKDqasoeFT91A> [Accessed 10 June 2021].

[5] This is where the original JAHH article by Wang Shouguan actually starts.

Fig. 8.4 Beijing 1987. Professor and Mrs. Christiansen with the then young Chinese team of radio astronomers

8.3 Recollections of the Shahe Experiment

Prof. Ke visited China for the first time in 1963. At that time, the Chinese Academy of Sciences' Beijing Observatory had a site in a Beijing suburb, Shahe, and the Radio Astronomy Section had installed there two cm-wave solar radiometers, copies of ones then in Soviet Russia.

In the early 1960s, our contact with Russia had fallen to a low ebb, and our contact with the West was nil. Members of our team at that time, with one or two exceptions, were all young people in their early 20s. In line with the whole of China, we subscribed to the slogans, "Self-Renewal Through Self-Effort" and "March Into Science." However, electronics in China was just being born, and if we were to pull ourselves up by our own bootstraps the only option that was technically feasible at all was working at meter wavelengths.

After going over the problem again and again, we opted to start with solar observations at meter wavelengths, and we considered constructing a copy of the Chris Cross (Fig. 8.5) which Prof. W.N. ("Chris") Christiansen had erected at Fleurs near Sydney (Australia) in 1957 (see Christiansen et al., 1961; Orchiston & Mathewson, 2009).[6] In 1963, we were working on the 32 antennas specified in the project, having basically decided on the site for the antenna array, but we had not solved the key technical problem of the transmission lines. At that time, China

[6] Christiansen was on the staff of the CSIRO's Division of Radiophysics in Sydney when the Chris Cross was built, but in 1960, he moved to a Chair in Electrical Engineering at the University of Sydney.

Fig. 8.5 View of the central part of the Chris Cross, looking west at sunset. This array comprised E-W and N-S arms each with 32 steerable parabolic antennas of 5.8 m diameter. This was the world's first crossed-grating interferometer, and it was used to produce daily isophote maps of solar emission at 1423 MHz (photography courtesy John Leahy)

still could not produce co-axial cables, and they were hard to import from Russia or from Eastern Europe. And, in particular, none of us had any idea of the overall technology of the antenna array. So, when we were told that Prof. Christiansen, the inventor of the Chris Cross, was coming to visit us, it was like happiness that had fallen from heaven.

Our joy was redoubled by a fact known to all: the Chinese scientific community at the time had been cut off from the West for more than a decade, as though we were sealed inside a hermetic wall. And this was the first time that a small door would open in this wall, and who should come through that door but the very man we most wanted to meet, Prof. W.N. Christiansen!

After a brief meeting, the professor invited the author to visit Australia, and in 1964, Dr Wu Huai-wei and I took up this invitation and went to Sydney. At that time there were no diplomatic relations between China and Australia, so we stayed in Prof. Christiansen's home and at the Hall of Residence at the University of Sydney.

During this trip, we made a point of visiting all of the radio astronomy establishments in Australia (but especially Fleurs), and we got an inkling of the scientific developments of the time and established many useful contacts. We discussed in detail with Prof. Ke ways of developing radio astronomy at the Beijing Observatory, and we arranged for Prof. Ke to visit us again in 1965 in order to help us with some of the technical challenges of the antenna array.

In 1965, we installed four 6-meter antennas at Shahe (Fig. 8.6), so that we could carry out "interim tests." As soon as Prof. Ke arrived, he guided us to start building the twin wire transmission line system that he himself had specially designed, using a material which we could easily obtain at that time, copper wire. Two parallel copper wires of 4 mm diameter made up the transmission lines, and the copper tubes that encased the wires were used to make an "adder" that added up two radio frequency signals, and a connector that served as a "matching transformer" between the cable and the copper tube. And there was the detector Prof. Ke himself

Fig. 8.6 The four-antenna Shahe test array

designed for testing stationary waves. By left-right sliding movements of the adder, the relative phase of the two signals was adjusted, thereby completing the "two-two addition."

When the Shahe Experiment was entering its final stage, China began sliding into a ten-year period of chaos, and all work was derailed, and even stopped completely at one stage. Nevertheless, by 1967 a meter wave "Christiansen array" of 16 east-west elements was installed at the new Miyun Observing Station, and test observations produced the first one-dimensional maps of the Sun. Originally the aim was to erect a 32-element crossed-grating interferometer (similar to the Fleurs one), but we had to revise this plan and ended up with the 16-element array instead.

The realization of the twin-wire transmission line system marked not only the completion of the new radio telescope, but equally important, it taught us (complete "new hands" at the game) a profound lesson: by making the best of the situation, one can successfully carry out scientific research under difficult conditions. And this is precisely what we understood by the term "Christiansen Style."

8.4 The Making of the Miyun Meter Wave Aperture Synthesis Telescope

After 1966, work at observatories throughout China stopped for a time, but the situation relaxed somewhat during the 1970s and from time to time we were able to carry out some work at the Miyun Observing Station. During this period, Prof. Ke came to China many times to bring us news of developments in radio astronomy abroad. And whatever little work we could do at the time always got support from him (on occasions he even brought us small electronics components that we needed). When in 1973 we heard the news that the Fleurs array was being converted into an aperture synthesis instrument,[7] we felt that once work resumed at Miyun this would also be the best goal for our endeavor. During one of Prof. Ke's visits we discussed this idea in detail, and the professor then made a move that was most extraordinary at the time: he proposed that China send two radio astronomers to Australia on a cultural exchange. As a result, in 1975 two members of the Miyun team, Drs Chen Hong-shen and Ren Fang-bin, spent eight months in the School of Electrical Engineering at the University of Sydney where they learnt about the hardware associated with the analogue receiver system of the Fleurs Aperture Synthesis. This provided, ahead of time, useful preparation for our later work.

The year 1976 saw the end of chaos in China and order was gradually restored at Miyun. Our first task was to convert the Miyun array into an Earth-rotation aperture synthesis instrument. Originally 32 antennas had been constructed when we began making the Mi-yun array, but many of these found their way to various locations in China during the years of unrest. Fortunately, we managed to track down most of these surplus antennas and bring them back to the Miyun Observing Station, ending up with a 28-element east-west array (Fig. 8.7), where the diameter of each aerial was increased from 6 to 9 m.

The making of the Miyun Aperture Synthesis Telescope was a gradual process. We started at a very low point, when the material conditions were difficult and the technological base was weak. But we had made preparations beforehand, and our target was clearly defined (as during the earlier Shahe period), so by making the best of a very difficult situation and bringing out our hidden potential we were able to forge ahead. The main problem at this time was the introduction of digital techniques and solving the complicated problem of data processing. Here, again, we had Prof. Ke's wholehearted support. In 1979, he sent his research student, Dr C.K. Kwong, to Beijing to help set up a digital receiver, and in 1980, he again invited two Miyun colleagues, Drs Chen Hong-shen and Zheng Yi-jia, to Sydney to familiarize themselves with the software and hardware of the Fleurs Synthesis Telescope.

[7] For details of the Fleurs Synthesis Telescope, see the various papers in the September 1973 special issue of the *Proceedings of the Institution of Radio and Electronics Engineers Australia* (Vol. 34, No. 8).

Fig. 8.7 The Miyun 28-antenna Meter Wave Aperture Synthesis Array

At this point, the blueprint of the Miyun Meter Wave Aperture Synthesis System was finalized, but it then took another four years to complete the experimentation, installation, and testing.

The Miyun Aperture Synthesis Telescope consists of 28 antennas each of 9 m diameter, divided into Array A (16 antennas) and Array B (12 antennas) arranged as shown in Fig. 8.8), making up 192 interferometer pairs, with baselines 3do, 4do, ... 194do (where do = 6 m). The system works at two frequencies, 232 and 408 MHz. At 232 MHz, each cycle of 12 hours' observation gives an overall resolution of 3.8×3.8 arc min. csc δ, and a "thermal noise limited" sensitivity of 0.05 Jy (SNR = 6) covering a field of 8×8 square degrees.

Now, when we recall this series of efforts, we recognize it to be precisely what enabled the Miyun team to pass the starting line of technological modernization, and throughout this eventful journey, which lasted ten years, we benefited from the guidance and concern of our good teacher and friend, Prof. Ke.

The Miyun Meter Wave Aperture Synthesis Tele-scope was formally commissioned in 1984 (see Wang, 1986), and Prof. Ke made a special journey to take part in the appraisal of the Facility. He and other Appraisal Committee members are shown in Fig. 8.9.

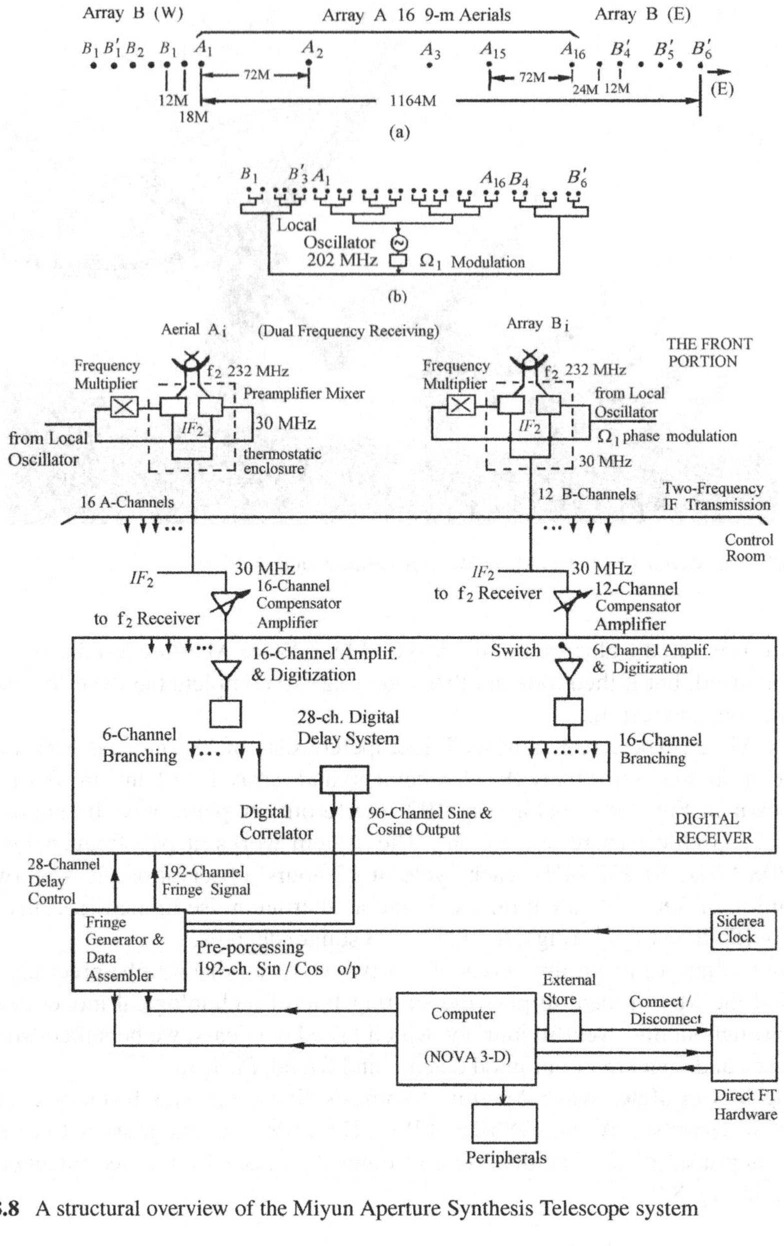

Fig. 8.8 A structural overview of the Miyun Aperture Synthesis Telescope system

Fig. 8.9 The Appraisal Committee of the Miyun Meter Wave Aperture Synthesis Telescope

8.5 Prof Ke and the Chinese Academy of Sciences (CAS)

Although just two examples may not do full justice to the history of "Prof. Ke and China," which spanned more than a third of a century, these two examples are typical, and are so deeply and indelibly impressed in our memories that they serve to illustrate the impact of his strong personality on our hearts and his contribution to the establishment and growth of radio astronomy in China. Over thirty-plus years, he became familiar with all of China's astronomical institutions and their radio astronomy divisions. Besides Beijing Observatory, the Purple Mountain Observatory, Shanghai Observatory, Yunnan Observatory, and the Urumqi Astronomical Station all kept records of the many lectures and visits by Prof. Ke, and of the numerous astronomers (and not just radio astronomers) from various places in China who visited Australia over the years and enjoyed kind hospitality and concrete help from Prof. and Mrs. Ke. Furthermore, Prof. Ke encouraged other Western radio astronomers to come to China, and many of them—including Rob Frater, Miller Goss, George Miley, Bruce Slee, and Richard Wielebinski—have established deep friendship with us. Like Prof. Ke, many of them made personal efforts to help China move out of isolation and rejoin the international astronomical community. Prof. Ke also, through his personal influence in international academic circles, did much to help return China to such international organizations as the IAU, URSI, and the ICSU.

Prof. Ke's deep friendship toward China was not confined to the Miyun radio astronomy team, or even to the Chinese astronomical community. His academic distinction and his sincerity elicited widespread respect and admiration in the wider scientific community. Apart from his aforementioned dealings with astronomers,

Fig. 8.10 Professor and Mrs. Christiansen with Go Moruo, the first President of the Chinese Academy of Sciences (sixth from left), Y.H. Woo, physicist and Vice-President of the Chinese Academy of Sciences (fourth from left), Chou Peiyuan, physicist and President of Peking University (third from left) and the astronomer Tcheng Mao-lin (second from left)

especially the younger ones, many scientists from our older generations, including Go Moruo, Y.H. Woo, and Chou Peiyuan, became close personal friends of his. Figure 8.10 is an historically significant photograph. It was taken in the 1960s and although rather "formal," it records some of these "older scientists"—who were not very old then—in company with Prof. and Mrs. Ke.

In recognition of his long and important contribution to Chinese astronomy, in 1996 Prof. Christiansen was elected a Foreign Member of the Chinese Academy of Sciences.

8.6 Concluding Remarks

I would like to conclude this short chapter with two photographs taken during the Ninth Assembly of Members of the CAS in 1998 (Figs. 8.11 and 8.12 on next page), for this was to be the last of the many occasions when Prof. Ke was with us (since 1963).

Fig. 8.11 Professor Christiansen (front row, second from left) at the Ninth Assembly of Members of the Chinese Academy of Sciences in 1998, during his last visit to China

Fig. 8.12 Professor Christiansen speaking at the Ninth Assembly of Members of the Chinese Academy of Sciences

Let these solemn records convey from all of us—friends in a country that he loved—our deepest feelings for him as we look on the passing of history.

Acknowledgements I am grateful to Dr John Leahy (through Dr Wayne Orchiston) for kindly supplying Fig. 8.5, and to Dr T. Kiang for translating this paper from Chinese into English.

References

Christiansen, W. N., Labrum, N. R., McAlister, K. R., & Mathewson, D. S. (1961). The crossed-grating interferometer: a new high-resolution radio telescope. *Proceedings of the Institution of Electrical Engineers, 108*, 48–58.

Orchiston, W., & Mathewson, D. (2009). Chris Christiansen and the Chris Cross. *Journal of Astronomical History and Heritage, 12*, 11–32.

Wang, S. G. (1986). Miyun Meter wave aperture synthesis radio telescope. *Chinese Astronomy and Astro-Physics, 10*, 3–11.

Wang Shouguan (1923–2021) was a member of the Chinese Academy of Sciences and a Research Professor at the National Observatories. He was formerly Director of the Beijing Astronomical Observatory and leader of the first team of Chinese researchers involved in radio astronomy.

Chapter 9
Birth and Development of Astronomy on the Island of Taiwan

Shiang-Yu Wang, Yun-Chun Wei, and Mei-Yin Chou

Abstract Taiwan, a small island on the west rim of the Pacific Ocean, has a short history with written records from around 400 years. Every several decades, different regimes controlled the island with very different purposes. During each transition, the official language and the education system also changed. Thus, it has been extremely difficult to educate the people with science. It was not until the early twentieth century that modern astronomy started to take root on the island. The efforts in the past century finally flourished in late 1990's. Taiwan's astronomical research now is visible worldwide. In this chapter, we review the history of this development in astronomy of Taiwan and provide some details regarding the IAU membership of Taiwan.

9.1 Astronomical Observation Records in Taiwan Before the Twentieth Century

Although Taiwan is separated from Mainland China by a strait just 150 km wide, the recorded history of Taiwan started only about 400 years ago when the Dutch colonized the island. Even so, there were already astronomical records in the seventeenth century. According to Han-Guang Chen's research on the astronomical

About family names : when transcripted into Western languages, the custom in Taiwan is now to adopt the Western order (the family name following, not preceding, the first name), contrary to the traditional Chinese order adopted in other chapters of this book, sometimes for the same persons, when referring to Chinese names.

S.-Y. Wang (✉) · Y.-C. Wei · M.-Y. Chou
Institute of Astronomy and Astrophysics, Academia Sinica, Taipei, Taiwan
e-mail: sywang@asiaa.sinica.edu.tw; emilywei@asiaa.sinica.edu.tw; cmy@asiaa.sinica.edu.tw

© The Author(s), under exclusive license to Springer Nature Switzerland AG 2022

T. Montmerle, D. Fauque (eds.), *Astronomers as Diplomats*, Historical & Cultural Astronomy, https://doi.org/10.1007/978-3-030-98625-4_9

data in Taiwan during the Ming Cheng and Ching dynasties,[1] there were records of observations of comets in 1680 and 1685, a lunar eclipse in 1761, and many meteors or comets appearing between 1786 and 1854 (Chen, 1969). These represent official records of astronomical observations during the Ching Dynasty in Taiwan. During this period, these astronomical phenomena were often regarded as the warning signals from the gods and linked to bad weather or warfare. For example, it was recorded in the official regional history 道光福建通志臺灣府 "(道光三年) 八月既望, 臺灣彗星起於東南, 而氣沖西北, 闇淡無光燄, 約二尺有奇, 至來年春尚有之, 未知何時消滅.是年遂有嘉、彰、淡、 閩、之釁。"[2] which links the observation of a comet in 1823 with the wars in various locations in Taiwan in the following year (Wang, 2016). Furthermore, these records are all of naked-eye observations, and there is no record of utilizing astronomical instruments at that time.

9.2 Astronomical Development in Taiwan During the Japanese Occupation Period (1895–1945)

The modern astronomical observation started during the Japanese occupation era. After Imperial Japan occupied Taiwan in 1895, as the state of governance gradually stabilized, more attention was focused on education and science activities, and Japanese became the official language, replacing the local Taiwanese dialect derived from Chinese ("helou"). The Taipei Weather Station under the Communications Bureau of the Governor's Office of Taiwan was established in 1896, and initially carried out only meteorological observations. In 1911, it set up a meridian room and purchased a British-made 3.5-inch meridian and a time chronometer. This was the first modern astronomical instrument and device in Taiwan (Chen, 1969; Liu, 2021). Science was formally added into the curriculum in Taiwanese public schools in 1912, and astronomy was also included in the teaching content. Akira Wada (和田彰), who came to Taiwan in 1903 and taught in the Japanese school at the Governor's Office, published articles in the *Journal of the Taiwan Education Association* to disseminate astronomical knowledge. The *Taiwan Daily News* published astronomical observation results in Japan and popular astronomical articles by Japanese. (Note: "Public schools" are equivalent to the elementary schools, exclusively for Taiwanese students.)

[1] The "Ming Cheng" dynasty relevant for Taiwan is not the same as the Ming dynasty of China. It is a remnant of Ming which has no blood relationship with Ming Emperors. However, the "Ching" dynasty is the same as the one known as the Chinese Qing dynasty (pinyin transcription).

[2] Translation: (The third year of Daoguang) *After August 15, a faint comet was observed with a tail extended toward Northwest direction. The visual length is around 60 cm. It was still visible until the next spring but no record about when it disappeared. Then, ChiaYi, ChangHua, TamShui, Fujian and Guangdong were infested with bandits*. See also the References.

Fig. 9.1 Issei Yamamoto (Asahigraph by *Asahi Shimbun* on 1953/9/9, 朝日新聞社-『アサヒグラフ』1953年9月9日号)

The Taiwanese society learned modern astronomy from school education, journals, and newspapers. After Issei Yamamoto (山本一清; Fig. 9.1) founded the Japanese astronomical society "Oriental Astronomical Association" in Japan in 1920, Ryo Mimoto (見元了) established its branch in Taiwan in 1926, becoming the first astronomical society in Taiwanese history. This association later developed into the "Astronomical Observation Association" (later renamed to the "Taipei Astronomical Association" which was the predecessor of the current "Taipei Amateur Astronomers Association") (Tsai, 2007; Chou, 2012; Wang, 2019). In 1938, the Governor's Office in Taiwan established a meteorological observatory and incorporated the Taipei Weather Station. The main astronomy equipment included a 4.5-inch refracting telescope, a 4-inch equatorial mount, and a 3.5-inch meridian. Its important tasks included timing, latitude/longitude observations by the meridian, sunspots, and planet observations with equatorial mount telescopes (Chen, 1969; Liu, 2021).

臺灣日日新報

本社寄贈の天體觀測望遠鏡

盛大な竣工受納式

けふ臺北市公會堂で

天體觀測同好會を組織

記念講演と映畫の夕

寫眞（1）公會堂屋上に常設された五藤式天體觀測室（2）内部の望遠鏡（3）寄贈並に受納式、河村社長の挨拶（4及び5）披露宴

Fig. 9.2 The *Taiwan Daily News* (in Japanese) reported on the opening ceremony of the newly gifted telescope by the newspaper company

To celebrate the 40th anniversary of the *Taiwan Daily News* and to promote popular science in 1938, a 4-inch refractor telescope made by Goto Optical MFG. Co. was gifted to the Taipei City Hall (the Taipei City Government at the time). The City Hall consulted the Meteorological Observatory of Governor's Office and the technician Kazuo Kubokawa (窪川一雄), to build a small observatory dome on the top floor of the south side of the Taipei (Taihoku) City Public Auditorium. The Taipei City Public Auditorium Observatory held a completion ceremony on May 1, 1939 (See Fig. 9.2), and Kazuo Kubokawa gave a popular science lecture on the topic of "The Universe" on that evening (Chen, 1969; Wang, 2019, 2020). The opening of the Taipei City Public Auditorium Observatory in 1939 triggered public interest in astronomy and the number of astronomical observers drastically increased. Kazuo Kubokawa and Masahisa Yoshimura (吉村昌久), an employee of the Taipei Imperial University Library, launched the "Astronomical Observation Association." Figure 9.3 shows the picture of the observatory.

The association was organized with four divisions: General Affairs, Research, Observation, and Data. In addition to sunspot studies, it also promoted astronomy

Fig. 9.3 Taipei City Public Auditorium Observatory, in 1940. (Courtesy of Taipei Astronomical Museum)

education for the general public. Its astronomical activities included: (1) 35 astronomical lectures (1939–1942, more than ten speakers that included Kazuo Kubokawa and Masahisa Yoshimura, with Kubokawa alone giving 16 lectures); (2) an exhibition of astronomical data (from March 28 to April 5, 1942), with 101 exhibition items; (3) 5 astronomical observations (held in 1938–1943). The observation targets included the Moon, Jupiter, Saturn, and solar eclipses (Chen, 1969; Wang, 2019). According to the *Taiwan Literature Quarterly* records, Taiwan's achievements in astronomical observation during the Japanese occupation period included Mercury transit (November 1940), a total solar eclipse (September 1941; see below), Martian surface (October–November 1941), positions and brightness of comets (March 1942), brightness of variable stars (April 1941 to August 1944) and sunspots (1940 to 1944). Most of the observations of sunspots were made by Chang-Hsien Tsai (蔡章獻), and most of the results were published in the *South Star* (南の星) magazine. In addition to the sunspot observation results, this magazine also published the observations and data of the total solar eclipse viewed from Taiwan on September 21, 1941, for which many Japanese astronomers came to

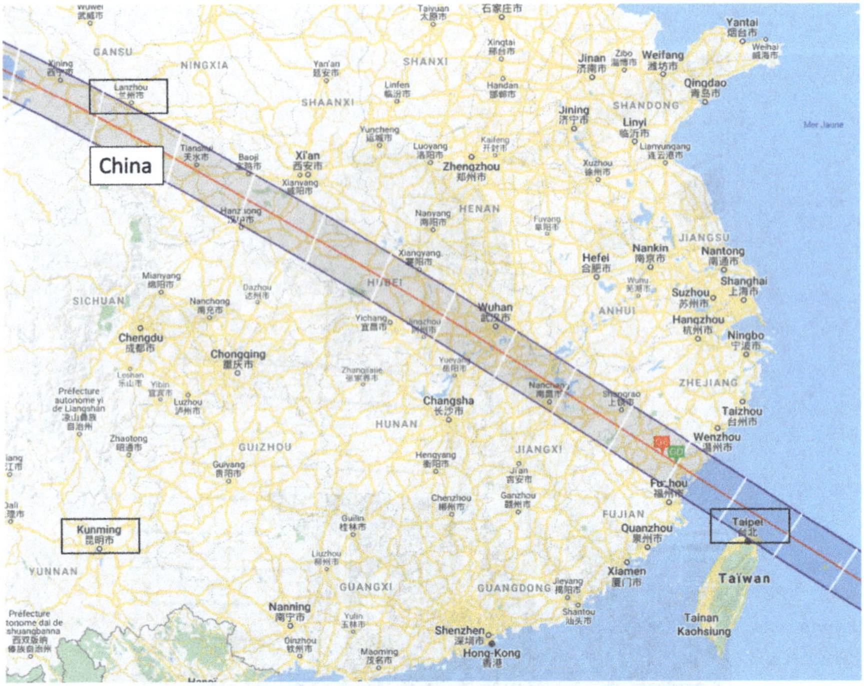

Fig. 9.4 (**a**) Path of the total solar eclipse of Sep. 21, 1941. This eclipse was observed both by Japanese astronomers off the Northern coast of Taiwan (almost longest duration: 3 min 22 s), and by a Chinese team led by Zhang Yuzhe (Y.C. Chang) near Langzhou, Gansou province (upper left corner), after having travelled about 3000 km from their base in Kunming, Yunnan province (lower left corner), "under the shadow of Japanese bombers" (Chang, 1942; see Chap. 7). The red marker labeled GE is the point of Greatest Eclipse. The green marker labeled GD is the point of Greatest Duration. (Map by F. Espenak, www.EclipseWise.com; enlargement of http://wwpw.eclipsewise.com/solar/SEgmapx/1901-2000/SE1941Sep21Tgmapx.html)

Taiwan, including Issei Yamamoto. In addition, Choushou Yoshida (吉田長祥) also published the book *Report of the solar eclipse in Taiwan* (Chen, 1969).[3] Fig. 4a shows the path of the eclipse over China, and Fig. 4b shows a picture of the observation led by Kazuo Kubokawa.

Naozou Ichinohe (一戸直蔵) was entrusted by the Governor Office to visit Taiwan twice in 1909 and 1911 to investigate the feasibility of setting up an observatory in the high mountains of Taiwan. After investigations, two plans was drawn up to build an observatory on the highest mountain (新高山, now玉山, Yu Shan; elev. 3950 m) in Taiwan and also on the Hsiaota mountain (小塔山, elev. 2291 m). Unfortunately, the project could not be executed due to high cost and World War II. The observatory at the highest mountain became a weather station that is still in operation nowadays. Director Issei Yamamoto of the Kwasan

[3] http://webcatplus.nii.ac.jp/webcatplus/details/creator/304205.html

Fig. 9.4 (**b**) The Solar eclipse observation preparation at Pengjia Islet (about 50 km to the North-East of the Northern coast of Taiwan, on the path of almost perfect totality) on September 21, 1941, organized by Kazuo Kubokawa. (*Taiwan Daily News*, 1941.09.18)

Observatory of Kyoto University, Japan, observed the transit of Mercury in Taichung on November 10, 1927, and observed a comet in Tainan the following year. In 1935, the Kwasan Observatory delivered a 10-inch equatorial mount telescope from Japan to Taiwan. It was set up in the Taichung Girls High School as a temporary observatory for southern sky photography. It was later moved to the Bagua Mountain Observatory in Changhua City in 1937. The Kwasan Observatory also sent Honda (本田氏) to carry out observations in Alishan (阿里山) in 1935 when it collaborated with an Australian astronomical organization to observe the zodiacal light (Chen, 1969; Liu, 2021).

9.3 The Development of Astronomy in Taiwan After World War II

After World War II and the surrender of Japan in 1945, the political situation in China and Taiwan changed drastically. Before the war, Japanese scientists dominated the development of astronomy in Taiwan, and afterwards they were replaced by local Taiwanese astronomers and scientists from mainland China. These scientists usually were not full-time astronomers, and were also engaged in research related to meteorology, surveys, physics, mathematics, and even geology. Chinese

Mandarin became the official language instead of Japanese, and the use of "helou" was forcefully banned in schools. As the Astronomy Institute of the Academia Sinica in Nanjing did not move to Taiwan, there was no academic institution for astronomy research in Taiwan.[4] The astronomy researchers were scattered in different institutes, such as Taiwan Provincial Weather Institution, Academia Sinica Institute of Mathematics, and Navy Surveying brigade and various universities, etc. After 1945, the Taiwan Provincial Weather Institution was officially approved to incorporate astronomy research. It established an astronomy department, but no observing facilities were available. The Weather Institution cooperated with the Zhongshan Hall Observatory (the original Taipei City Public Auditorium Observatory) in Taipei City and officially started operation in July 1947. The main tasks included sunspot observation (since July 1948), star observation for timing, solar radiation observation (since January 1948), astronomical phenomenon forecasts, and calculation of various astronomical data. In addition, there were observations of comets, total lunar eclipses, Jupiter, and Venus from 1948 to 1951. In March 1948, the "Taipei Astronomy Association" initiated by Jie-Chien Huang (黃介騫) and Chang-Hsien Tsai (Fig. 9.5) was formally established, and the journal of *Astronomical Communication* was published. In the meantime, the Astronomical Observatory of the Joint Logistics Geodetic Survey Team also started some activities since its relocation in Taiwan. Observations of Class 1 astronomy trigonometrical points and Class 2 astronomy trigonometrical points started in Beidou, Changhua, in January 1951 (Chen, 1969).

In 1957–1958, with the satellite tracking projects of the *International Geophysical Year*, the Taipei Astronomy Association and the Taiwan Shin Sheng Daily News jointly formed the "China Satellite Observation Committee," which received an approval letter from the Smithsonian Observatory on March 7, 1958, for establishing the "Taipei Observatory." Standardized visual observation stations (Moonwatch Station) were set up in Taipei (Zhongshan Hall Observatory) and Hsinchu (National Tsinghua University). Chang-Hsien Tsai was the team leader, and there were about 50 participants during the entire observation, including Chien-Ming He (何健民), Shao-Chin Chang (張紹進) (both later obtained Ph.D. and taught at universities in the USA), Li-Chen Wang (王理瑱) (formerly at the Stanford Linear Accelerator Center), and future Nobel laureate Yuan-Tseh Lee (李遠哲), who was only twenty years old at that time. The results of this observation were affirmed by the United States, and the Taiwanese contribution finally received a certificate of appreciation from the Smithsonian Astronomical Observatory as shown in Fig. 9.6 (Fan, 2007; Liu, 2007).

[4] Academia Sinica originally had around 20 institutes when it was operating in China before WWII, mostly in Nanjing. When Academia Sinica restarted in Taipei, there were only three institutes as most scientists did not move to Taiwan. On the other hand, the mainland Chinese side changed its academic structure, called the Chinese Academy of Science (CAS), with its headquarters in Beijing. In the beginning, the CAS only took over the institutes of Academia Sinica in Beijing. The original Academia Sinica Institute of Astronomy was integrated into the Purple Mountain Observatory.

Fig. 9.5 Mr. Chang-Hsien Tsai. (Courtesy of Taipei Astronomical Museum)

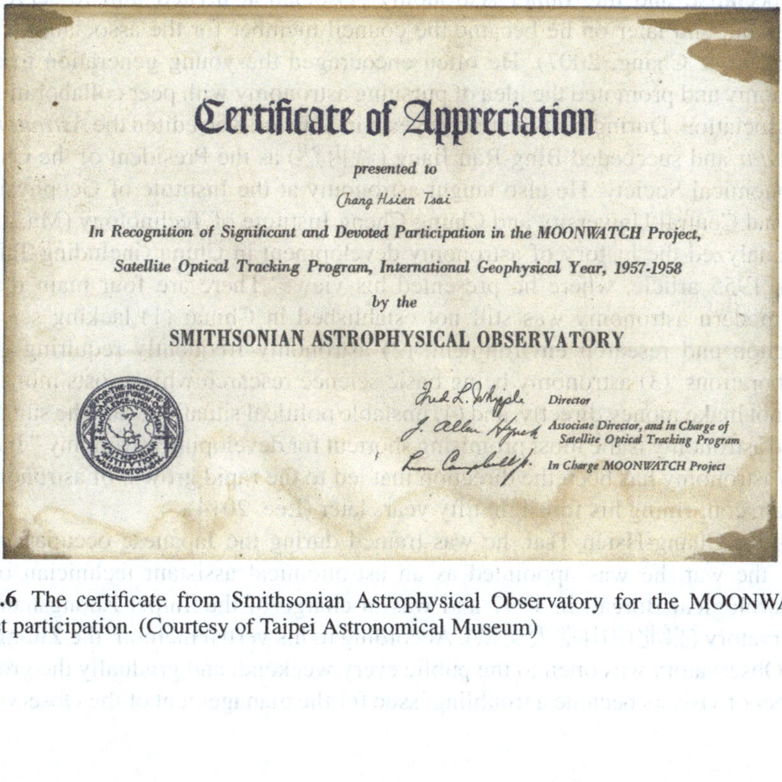

Certificate of Appreciation

presented to

Chang Hsien Tsai

In Recognition of Significant and Devoted Participation in the MOONWATCH Project, Satellite Optical Tracking Program, International Geophysical Year, 1957-1958

by the

SMITHSONIAN ASTROPHYSICAL OBSERVATORY

Director

Associate Director, and in Charge of Satellite Optical Tracking Program

In Charge MOONWATCH Project

Fig. 9.6 The certificate from Smithsonian Astrophysical Observatory for the MOONWATCH project participation. (Courtesy of Taipei Astronomical Museum)

Fig. 9.7 Ping-Tse Kao and his family. (Courtesy of Chun Kao, 高準, grandson of Ping-Tse Kao)

One of the most important persons in promoting astronomy in the 50s and 60s, in addition to Chang-Hsien Tsai, was Ping-Tse Kao (高平子; Fig. 9.7). A former researcher at the Astronomy Institute of the Academia Sinica, Kao came from mainland China in 1948, and was transferred to the Institute of Mathematics. In 1949, the Taiwan Provincial Weather Institution hired him as Senior Technical Specialist, and the Taipei Astronomy Association invited him to serve as a consultant, and later on he became the council member for the association (Tsao, 1971; Liu & Chang, 2007). He often encouraged the young generation to study astronomy and promoted the idea of pursuing astronomy with peer collaborations in the association. During his twenty-five years in Taiwan, Kao edited the *Astronomical Calendar* and succeeded Bing-Ran Jiang (蔣丙然) as the President of the Chinese Astronomical Society. He also taught astronomy at the Institute of Geophysics of National Central University and Chung Cheng Institute of Technology (Ma, 1989). Kao analyzed the history of astronomy development in China (including Taiwan) in his 1955 article, where he presented his view: "There are four main reasons why modern astronomy was still not established in China: (1) lacking scientific education and research environment, (2) astronomy frequently requiring global collaborations, (3) astronomy being basic science research which costs money but does not make money directly, and (4) unstable political situation," and he suggested "radio astronomy is the most promising shortcut for developing astronomy." Indeed, radio astronomy has been the direction that led to the rapid growth of astronomy in Taiwan, confirming his foresight fifty years later (Lee, 2014).

As for Chang-Hsien Tsai, he was trained during the Japanese occupation era. After the war, he was appointed as an astronomical assistant technician of the Meteorological Bureau in 1947 and was in charge of the Taipei Zhongshan Hall Observatory (臺北中山堂天文臺). According to his verbal memoir, the Zhongshan Hall Observatory was open to the public every weekend, and gradually the growing number of visitors became a troubling issue for the management of the Observatory;

Fig. 9.8 A picture of the Yuanshan Observatory (Courtesy of Taipei Astronomical Museum)

moreover, the nearby shopping malls and the increasing light pollution made it unsuitable for astronomical observations. He pushed the Taipei City Government to plan for relocation. Tsai surveyed four sites, and selected Yuanshan (Shilin district; elev. 291 m) because of its higher elevation and convenient transportation. The Yuanshan Observatory (or Taipei City Observatory, 臺北市立天文臺) was completed in 1963 as shown in Fig. 9.8.

Following the principle of equal emphasis on education and observation, besides being open to the public, the Observatory also organized touring educational events and research activities to enlighten many astronomy enthusiasts in Taiwan. The Observatory also purchased a planetarium and many stargazing instruments, including the famous Nishimura 16-inch (41 cm) reflective equatorial mount telescope, and the 25-cm refracting equatorial mount telescope installed in 1973 (Liu & Chang, 2007). Until the early 1980s, it was still the only observatory in Taiwan and was still led by Mr. Tsai. Despite its small scale, this observatory had played an important role in modern astronomy in Taiwan. Most contemporary Taiwanese astronomers had a first taste of astronomy through this small observatory (Ip, 2017). It ought to be noted that Kao and Tsai were very good friends. They met each other soon after Kao moved to Taiwan in an astronomical talk that Kao hosted. They became colleagues in the Meteorological Bureau in 1947 and worked together for astronomy. When Tsai pushed for the Taipei Observatory in 1959, Kao gave him advice and encouragements when Tsai's proposal was rejected by the government (Tsai, 2007). The Yuanshan Observatory moved to Shilin and became the Taipei Astronomical Museum in 1996 as an educational institution.

Another astronomer who contributed to astronomy in Taiwan during this period of time was Mo Tsao (曹謨), who also came to Taiwan from mainland China and devoted himself to astronomy writing. He was an individual member of the IAU and the author of *New Astronomy Foundation* (1967), *A Dictionary of Astronomy*,

History of Chinese Astronomy (1986), and of the biography of Ping-Tse Kao published in "Comprehensive Studies of Chinese Culture - The Contributions of Chinese Scholars to the Study of Chinese Culture in the Past Sixty Years" (Tsao, 1971), etc. He also prepared an ambitious plan "Towards Intensifying Astronomical Research in Free China" in 1961 for Academia Sinica and looked for collaborations in the US; however, this plan was not successful.

9.4 The Chinese Astronomical Society Resumes Its Activities in Taiwan and Joins the International Astronomical Union (IAU)

The Chinese Astronomical Society was established on October 30, 1922. The first president was Lu Gao (高魯),[5] the Director of the Central Observatory. In 1949, the Nationalist Government moved to Taiwan. Some members of the Society came to Taiwan, and many members remained in mainland China. World War II and the subsequent civil war between the Chinese Nationalist Party (Kuomintang) and the Chinese Communist Party (CCP) made scientific research difficult. The war also split the astronomical society into two parts.

After 1949, the Kuomintang (in Taipei) and the CCP (in Beijing) both claimed to be the sole legal government of China, but the two regimes had no control over the areas occupied by the other party. Because both sides did not recognize each other, they were mutually exclusive in the international arena. Some countries in the world recognized the Beijing regime, and some countries recognized the Taipei regime. This antagonism also affected international organizations unrelated to politics, such as the International Olympic Games, various international cultural associations and scientific organizations. Even if scholars on both sides of the Taiwan Strait agreed that participating in international organizations was important and they would not exclude the other party from joining, the governments of both sides had led to a situation in which international organizations had to choose between China and Taiwan. In the 1950s, due to the international siege of the communist regime, scientific societies in the island of Taiwan were able to join relevant international organizations, causing the related societies in mainland China to voluntarily withdraw and lose the opportunity to participate in international activities.

In February 1958, the Department of Higher Education of the Ministry of Education in Taiwan sent a letter to Ping-Tse Kao, Bing-Ran Jiang, and the Director of the Central Meteorological Institute and Taiwan Meteorological Institute Kenneth T. C. Cheng (鄭子政). The Department asked these three scholars to reorganize the Chinese Astronomical Society. The organization was approved by the Ministry of the Interior on April 3 with more than 30 members (Fig. 9.9), and a preparatory

[5] See, e.g., Montmerle et al. (2022)

Fig. 9.9 The memorial picture of the reorganization meeting of the Astronomical Society of the Republic of China in Taiwan in June, 1958. The first president is [8] Bing-Ran Jiang. The important members include [1] Kenneth T. C. Cheng, [5] Ping-Tse Kao, [11] Mo Tsao, [13] Jie-Chien Huang, [23] Shao-Chin Chang, [38] Chang-Hsien Tsai, [41] Li-Chen Wang. (Courtesy of Taipei Astronomical Museum)

meeting was held on April 25, which decided the English title to be the *Astronomical Society of the Republic of China* (ASROC). At the same time, it was decided to send a letter to apply for membership to the International Astronomical Union (IAU). In July of the same year, the certificate of registration of the organization was obtained (Tao, 2008), just before the IAU General Assembly held in Moscow that year. ASROC became an official member of IAU as a result of the decision of the IAU Executive Committee in 1959, and subsequently astronomers in Taiwan participated in IAU meetings regularly. In the meantime, the representative of China announced its withdrawal from IAU.[6]

However, as the international political climate changed over the years, more and more countries switched their recognition to the regime of mainland China and broke their diplomatic relations with Taiwan. In 1971, the General Assembly of the United Nations officially approved the CCP government as the official Chinese

[6] See Chap. 6, and Montmerle and Zhou (2022).

government and Taiwan's membership in the United Nations was relinquished. Soon after that, Taiwan's membership in various international organizations was challenged. This prompted the Taiwan government to concede that if it continued to insist on Taiwan being the only Chinese representative in international organizations, Taiwan would be completely marginalized. In response to the situation, the Taiwanese government no longer insisted on "only one China," but instead took substantive participation as the main goal, seeking to participate in international organizations on an equal and independent basis with mainland China. Under this concept, when Chinese astronomers applied to IAU to resume participation in IAU, Taiwan's attitude towards this issue was to seek a solution so that astronomers on both sides of the strait could participate in the IAU on an equal footing.

Based on this principle, when China began to inquire about the restoration of Chinese membership in IAU in 1978, Taiwan's task was to find membership names acceptable to both parties. The IAU President Professor Blaauw listed the establishment of a joint Chinese Astronomical Society between the two sides of the strait as one of the options, but both sides understood that such an arrangement was not practical based on situation of division and different governments. Finally, during the 17th IAU General Assembly in Montreal, Canada in 1979, the IAU President Professor Blaauw had bilateral consultations with the Chinese delegation headed by Professor Yuzhe Zhang (張鈺哲) and the representative of Taiwan, Professor Chun-Shan Shen (沈君山; Fig. 9.10), and reached a preliminary consensus. IAU obtained the agreement between China and Taiwan by a follow-up letter. The name of Taiwan's membership was to be "China, Taipei," and the name of mainland China's membership "China, Nanjing." Both are now on a fully equal position about all

Fig. 9.10 Picture of Chun-Shan Shen (at the center, cutting the ribbon), taken in 1982. (Courtesy of Taipei Astronomical Museum)

substantive issues (including representation rights, voting rights, and membership dues payment obligations) (Blaauw, 1994; Montmerle & Zhou, 2022).

In the course of his speech at the closing ceremony of the IAU General Assembly, Prof. Shen's gave a very concise and clear explanation of why Taiwan was persevering in maintaining her separate membership: "The Astronomical Society of the Republic of China has never opposed the participation of the astronomical society from mainland China in the IAU. In fact, as Prof. Blaauw just said, we have tried our best to cooperate with the executive officials of the IAU to promote their membership. We have repeatedly stated verbally and in writing that the Chinese Astronomical Society in Taipei only represents astronomical activities in Taiwan. We do not think, and have never said that we represent astronomical activities in mainland China . . . From the perspective of people in Taipei, although the mainland China and Taiwan share common cultural traditions, there is currently a huge gap between the two sides not only in terms of material living standards, but also in terms of political and economic structures. Unification means that both sides are placed under the same system and the same central government. Beijing insists that Taipei belongs to her local government . . . Mainland China has a population of more than 800 million, and Taiwan has a population of less than 20 million. Therefore, any form of reunification at present must place Taiwan under the system promoted by the mainland. Under such circumstances, negotiating reunification will only bring about unilateral sacrifices to Taiwan, and will not lead to a fair compromise between the two sides . . . . Taipei only recognizes Taiwan as part of China, but never recognizes it as part of the People's Republic of China (PRC). From the Olympics to cultural associations to scientific societies, disputes about "nationality or regionality" are everywhere; Beijing always tries to make her delegations national and squeezes delegations from Taipei to be regional, while Taipei uses everything possible to avoid such arrangements."(Shen, 1979).

Prof. Shen's memoir, *The Postscript of the Floating Life* (浮生後記, Shen, 2004), talked about this process of negotiation: "What we were arguing about is quite ridiculous from our current point of view. At the beginning, the positions of the two sides were quite different. But in the end, the main part was agreed. Two equal and independent astronomical societies are listed in the data of IAU. The names of PRC and ROC do not appear, and even the way of listing is negotiated.

Nanjing was used instead of Beijing because the headquarters of the Chinese Astronomical Society was located at the Purple Mountain Observatory in Nanjing. But the final problem lies in the dispute between "at" and "of" (Fig. 9.11). Beijing disagreed with "at Taipei," and only agreed with "of Taipei," or not to join the IAU. Blaauw was very disappointed, but at the conclusion of the meeting, he still asked me to explain to the meeting on behalf of China. I translated the dispute between "at" and "of" as the dispute between "equal to" and "belonging to," and distinguished between "one country, two systems" and "one country, two governances." I gave such a lesson to astronomers from various countries, including representatives from Beijing. However, I think 99% of the audience was confused (Shen, 2004).

Fig. 9.11 Debate between "of" and "at" for the membership name of Taiwan. (Shen, 2004)

In Chinese (Fig. 9.11),

Chinese astronomical society *at* Taipei is written as 中國天文學會"在"台北, while.
Chinese astronomical society *of* Taipei is written as 台北"的"中國天文學會.

It is only one character difference. However, the meaning is quite different between "在" and "的" in this case. Meanwhile, the two characters "在" and "的" might have similar meaning (like "*in,*" instead of "*of*" or "*at*") in other usages.

It is such a subtle difference that is hard to understand but it bothered the two sides to settle on the IAU membership names. It confused non-Chinese scientists to understand the differences which only Chinese-related people care about. Furthermore, it is ridiculous for scientists to waste their time and effort in such language details. As we know now, the IAU membership names for China and Taiwan have no preposition. This removes the dispute and provides the room for each side to interpret the names. Although the resolution is not perfect and even somewhat strange from our current point of view, it did provide an agreed mode for astronomers in mainland China and in Taiwan to join the IAU-related activities seamlessly. Taiwan astronomers sincerely hope that they do not need to discuss this issue again until the political relationship between China and Taiwan is firmly settled.

In any case, after the admission of Taiwan to the IAU and other international scientific unions, and as the economy grew, science policy started to take shape for good in 1967, when the Commission for Science Development (CSD) was organized under the National Security Council. Ta-You Wu (大猷), who was invited by the President Chiang Kai-Shek to be the CSD Director, drafted a long-term plan for scientific development from the viewpoint of "national need" instead of "academic independence." With the operation of CSD, Earth sciences started to be included as a required course for senior high schools in Taiwan in 1972. This immediately enhanced the knowledge of astronomy for all students and provided basic knowledge for students interested in astronomy. It definitely triggered the new generation of astronomers from Taiwan. Wu was a nuclear physicist and had been called the Father of Chinese Physics. He became the President of Academia Sinica in 1984. During his term, he approved the establishment of a preparatory office for the Institute of Astronomy and Astrophysics which has become the largest astronomical institute in Taiwan.

Professor Chun-Shan Shen was also a pioneer in Taiwanese astronomy during the 60s and 70s, besides representing Taiwan during the IAU membership negotiation with China. He was a leading astronomer and published many astronomical science articles and books in the 1960s, introducing modern astronomical knowledge to the next generation (Ip, 2017). In 1986, Shen and Typhoon Lee (李太楓) organized a large campaign for observations of comet 1P/Halley's seventy-six-year return. It was the largest astronomical outreach program in Taiwan and triggered the general public's passion for astronomy.

9.5 The Rapid Development of Astronomy in Taiwan After 1990

With more attention and resources being focused on science and the new generation astronomers returning from overseas with PhD degrees, professional astronomical institutes finally were established in the 90s.

The National Central University (NCU) was the first university in Taiwan to set up an astronomy-related institute. In 1979, the NCU Department of Physics established the Institute of Astronomy and Physics. Later, Professor Ke-Hua Lu (呂克華) was appointed to assist in the construction of its observatory. In 1981, NCU had a 61-cm reflecting telescope in its campus, which was the largest astronomical telescope in Taiwan at that time and the first observatory in Taiwan dedicated for science studies. In 1984, the Department of Earth Sciences of National Taiwan Normal University (NTNU) was established and began to develop astronomy education. In 1992, the NCU Graduate Institute of Astronomy was established. Meanwhile, the institute started to look for an appropriate observatory site on the island for professional astronomical observations. With the efforts led by Prof. Wen-Hsiang Tsai (蔡文祥), a survey through the high mountainous areas of the island was conducted. After considering the accessibility, light pollution, and weather conditions, Lulin Mountain (elev. 2845 m) was selected as the site for the NCU's professional astronomical observatory in Taiwan. In 1999, the "Lulin Observatory" (Fig. 9.12) was established to set up a 1-meter optical telescope (Lulin One-meter Telescope, LOT). The LOT started its operation in 2003 and is still the largest diameter telescope in Taiwan.[7] Since 2001, other universities have successively established astronomical research institutions, such as the Institute of Astronomy, National Tsinghua University (NTHU) in 2001, the Institute of

[7] Asteroid 145,523 Lulin, discovered in 2006 by Taiwanese astronomers Hung-Chin Lin (林宏欽) and Ye Quanzhi (葉泉志), was named by the IAU after the mountain and the Observatory.

Fig. 9.12 Lulin Observatory of NCU. Credit: Institute of Astronomy, NCU

Astrophysics, National Taiwan University (NTU) in 2003, and the Institute of Space Astronomy and Plasma Science, National Cheng Kung University (NCKU) in 2008 (renamed to "Institute of Space and Plasma Sciences" in August 2013). In addition, National Taiwan University Leung Center for Cosmology and Particle Astrophysics (LeCosPA) was established in 2007, which is a first-level research unit in NTU. Currently, more than 10 universities have faculty members with PhD degrees in astronomy or astrophysics to instruct astronomical courses.

Although the Academia Sinica established the Institute of Astronomy in 1928, most of the members and equipment remained in mainland China, so it was unable to relocate in Taiwan. Until the mid-1980s, a group of local and overseas Chinese astronomers gathered to discuss how to develop astronomical research in Taiwan. The Academia Sinica Institute of Astronomy and Astrophysics (ASIAA) preparatory office was established by Typhoon Lee and colleagues in 1993. Meanwhile, the "Ten-Year Plan to Develop Astronomy and Astrophysics in the Republic of China" was also formulated, leading Taiwan's astronomical research to international standards. Under the leadership of the Preparatory Office team, Taiwan and the Smithsonian Astrophysical Observatory collaborated on the development of the Submillimeter Array (SMA), located on the Mauna Kea volcano of Hawai'i. This is the first international astronomical facility collaboration project in Taiwan. The success of the SMA established an excellent model and later triggered more international collaboration projects that provided access to the advanced observation facilities for astronomers in Taiwan.

Since the establishment of the preparatory office, the ASIAA has prioritized investment in manpower and equipment. It was not until 2009 that the ASIAA acquired its own building, sixteen years after its creation. The ASIAA was officially established in 2010 and currently has close collaborations with many world-class academic institutions and observatories to engage in the most advanced astronomical research. With several projects in Hawai'i, the ASIAA also established

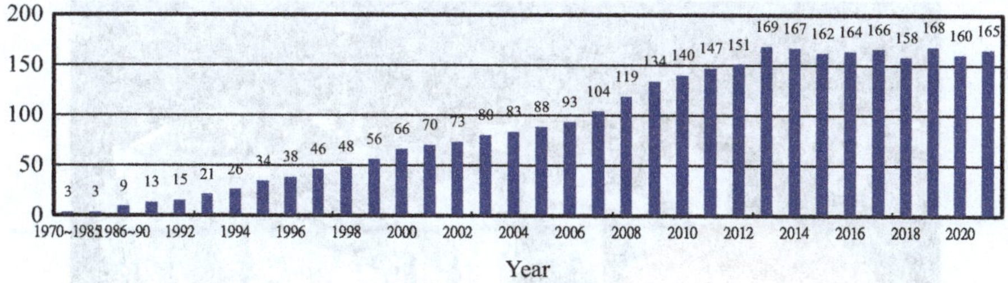

Fig. 9.13 Number of PhDs working in astronomy in Taiwan

a field office in Hilo, on the Big Island. By participating in the construction and development of international astronomical facilities and their instruments, the ASIAA not only gains access to advanced observing facilities for astronomers in Taiwan to enhance their research capability and quality, but also rapidly improves its expertise in telescope-related instrumentation. Research topics carried out at ASIAA range from solar system, star and planet formation, origin of life, to extra-galactic science, cosmology, and supermassive black holes. The goal of the ASIAA is to become an international research institute engaged in frontier projects, competing and collaborating with the leading research groups worldwide. The first decade of ASIAA's development focused on the development of radio astronomical interferometers and related scientific research. The second decade will see the active expansion of back-end instruments of optical and infrared (OIR) telescopes and related scientific research, as well as theoretical astrophysics research. In the third decade, the ASIAA will develop these three fields in a balanced manner. It is worth mentioning that Typhoon Lee, Chi Yuan (袁旂), Fred Lo (魯國鏞), Sun Kwok (郭新), Paul Ho (賀曾樸) and You-Hua Chu (朱有花), who successively took over the directorship of ASIAA, are all members of the Astronomy Planning Committee established in 1991.

With these research institutes, the number of astronomical PhD in Taiwan increased steadily from 1990 to 2013 and reached a plateau of ~160 professional astronomers afterwards (Fig. 9.13). Such manpower and environment enable advanced astronomical studies, education of the new generation scientists, and public education and outreach activities. We hope the community can continue to grow with more astronomy and astrophysics faculty members in physics departments of all universities in Taiwan.

Currently, the Taiwanese community is actively involved in important international collaborations, such as the SMA, the Atacama Large Millimeter/submillimeter Array (ALMA), the James Clerk Maxwell Telescope (JCMT), the Canada France Hawaii Telescope (CFHT), the Subaru Telescope, the Panoramic Survey Telescope & Rapid Response System (Pan-STARRS), the Zwicky Transient Facility (ZTF), the Vera C. Rubin Observatory and the European Extremely Large Telescope (E-ELT). We are also leading the Yuan Tseh Lee Array (YTLA), the Greenland Telescope, and the Transneptunian Automated Occultation Survey (TAOS II) projects. In addition, we have strong connections with the regional

Fig. 9.14 The Taipei Astronomical Museum (Courtesy of Taipei Astronomical Museum)

countries to form the East Asia Core Observatory Association (EACOA) to enhance the regional collaboration and to establish the East Asian Observatory which governs the operation of JCMT.

Meanwhile, the number of amateur astronomers has also increased, and local astronomical associations in the major cities in Taiwan were created. These associations typically arrange regular activities such as stargazing in their local sites. Large yearly events such as star parties and the Messier marathon are also organized with participants from over the whole island. Special events such as solar and lunar eclipses are typically organized jointly by government institutions, universities, schools, and associations. The Taipei Astronomical Museum (Fig. 9.14) and the National Museum of Nature Science in Taichung also provide regular exhibitions and public talks related to astronomy for citizens of different ages. These activities educate the general public with astronomical knowledge and provide essential background to the new astronomical discoveries.

9.6 Conclusions

With the constant endeavor and efforts from dedicated pioneers in the past century, having to deal with complex international situations, astronomy in Taiwan finally flourishes in the twenty-first century. It is a clear example that the development of astronomy requires a stable society that recognizes the importance of science and education, as well as promoting good regional and international relations. We are grateful to have the privilege to explore the universe with the support from the society. We have seen the scientific outcomes can further trigger the passion from the general public for the pursuing of the unknowns and hope this positive feedback can continue to further elevate astronomy in Taiwan.

References

Blaauw, A. (1994). *History of the IAU, The Birth and First Half-Century of the International Astronomical Union*. Kluwer Academic Publishers.

Chang, Y. C. (Zhang, Yuzhe). (1942). Solar eclipse observed in China under the Shadow of Japanese Bombers. *Popular Astronomy, 50*, 198.

Chen, H.-G. (1969). "A Preliminary Study of the History of Astronomy in Taiwan" "臺灣之天文學史事初探". *Taiwan Historica Quarterly,* 臺灣文獻季刊*, 20*, 137.

Chou, H.-J. (2012). "The Science Education of Taiwan Public School during Japanese Colonial Period" "日治時期臺灣公學校理科教育之研究", a master thesis of NTNU. 國立臺灣師範大學碩士論文.

Fan, H.-C. (2007). "An old friend, a period of the early history of the Taipei Observatory – the observation of satellites" "一位老朋友、一段天文臺的早期歷史–人造衛星觀測". *Taipei Skylight,* 臺北星空*, 37*, 33.

Ip, W.-H. (2017). *"The Development of Astronomy and Emergence of Astrophysics in Taiwan"*, The Emergence of Astrophysics in Asia, Historical & Cultural Astronomy Series, Eds. S. Nakamura & W. Orchiston, Springer, p. 257 (ISBN 978-3-319-62080-0).

Lee, T (2014). "Ping-Tse Kao, the antecessor who predicted the development of radio astronomy in Taiwan" "預見我國電波天文學發展的高平子老前輩(代序)", preface of "The name engraved on the moon-astronomer Ping-Tse Kao" "鐫刻在月球上的名字—天文學家高平子", 2014, Beijing: China Light Industry Press, p. 7.

Liu, J.-H. (2021). "Naozou Ichinohe and his unexecuted plans for New Highest Mountain Observatory and Hsiaota Mountain Observatory" "一戶直藏與未竟的新高山天文臺、小塔山天文臺事業", Taiwan Historica epaper, 國史館臺灣文獻館電子報, vol. 204.

Liu, K.-L. (2007). "Taipei Team for Satellite Observation" "臺北人造衛星觀測隊", Taipei Skylight. 臺北星空, *37*, 37.

Liu, K.-L., & Chang, K.-L. (2007). "My days at the Yuanshan Observatory" "我在天文臺的日子". *Taipei Astronomical Museum Magazine, TAIPEI SKYLIGHT,* 臺北星空*, 37*, 19.

Ma, H.-Y. (1989). "Run with the moon: Ping-Tse Kao, the first Chinese astronomer whose name was engraved on the moon" "與月亮一起運行: 第一位留名月球的中國天文學家高平子", China Times Express, 中國時報, (1989 March 13–16).

Montmerle, T., & Zhou, Y. (2022). *China and the International Astronomical Union: Divorce, Separation, and Reconciliation (1958–1982)*, Historical & Cultural Astronomy Series (Cham: Springer Nature Switzerland).

Montmerle, T., Zhou, Y., & Gomas, Y. (2022). *The two lives of Cheng Maolan: From the French "Silk Road" towards Astronomy to the Meanders of Mao's China*. Springer.

Shen, C.-S. (1979). "The dispute between "equal to" and "belonging to"" "「等於」和「屬於」之爭", China Times, 中國時報, (1979 September 13, A2).

Shen, C.-S. (2004) "The Postscript of the Floating Life: One, but not by unification" "浮生後記: 一而不統", Taipei: Global Views – Commonwealth Publishing.

Tao, F.-L. (2008). The Astronomical Society of Republic of China in Taiwan 中國天文學會在臺灣. *Taipei Skylight,* 臺北星空*, 41*, 21.

Tsai, A.-L. (2007). "Taipei City Public Auditorium Observatory: the Predecessor of Taipei Astronomical Museum" "臺北公會堂天文台:「臺北市立天文館」的始祖". *Taipei Skylight,* 臺北星空*, 37*, 9.

Tsao, M. (1971). "Biography of Mr. Ping-Tse Kao" "高平子先生傳略". Retrieved from "The name engraved on the moon—astronomer Ping-Tse Kao" "鐫刻在月球上的名字—天文學家高平子", 2014, Beijing: China Light Industry Press, p. 19.

Wang, P.-K. (2016). "Discussions on the records of fortune and disaster in the Taiwan local historical book in Ching dynasty" "清代台灣方志中的〈祥異(災)〉書寫析論". *Tunghai University Library Journal,* 東海大學圖書館館刊*, 9*(41).

Wang, Y.-H. (2019). "The 80th Anniversary of Taipei City Public Auditorium Observatory" "臺北公會堂天文台80周年", Taipei Skylight, 臺北星空, 89, 24.
Wang, Y.-H. (2020). "The Promotion and Enlightenment of Astronomy in Taiwan" "臺灣天文學推展與啓蒙", The Communication for Taiwanese Studies, 臺灣學通訊, 115, 16.

Shiang-Yu Wang received his Ph.D. degree in electronic engineering from National Chiao-Tung University, HsinChu, Taiwan in 1999. His dissertation focused on quantum well and quantum dot infrared detectors. He is a Research Fellow and Deputy Director of the Academia Sinica, Institute of Astronomy and Astrophysics, Taiwan. He is also the President of the Astronomical Society. His research interests are mainly in the optical and infrared astronomical instrumentation and TransNeptunian Objects.

Yun-Chun Wei received her Bachelor of Science in Physics and Master of Arts in Library Science from the National Taiwan University, Taipei, Taiwan. She is currently the Academic Informationist in Academia Sinica Institute of Astronomy and Astrophysics (ASIAA) in Taiwan.

Mei-Yin Chou received her PhD from the University of Virginia, USA, in 2010. She is now the Education and Public Outreach (EPO) Project Scientist in Academia Sinica Institute of Astronomy and Astrophysics (ASIAA) in Taiwan.

Chapter 10
Development of Astronomy in the Korean Peninsula

Hyung Mok Lee

Abstract Korea has a long tradition in astronomy going back to the pre-historic period. There is a large number of dolmens with star-like cup-marks carved on them. Astronomical events such as solar eclipses, special configurations of the planets and moon, brightening of Venus, sunspots and guest stars including supernovae, novae, and comets are recorded in history books of several dynasties covering about 2000 years. The Royal Observatory that was in charge of publishing accurate calendar existed without interruption for more than 1000 years until the collapse of Joseon dynasty in 1910, following the Japanese colonization of Korea. However, the development of modern astronomy has been significantly delayed because of the turbulent history in the 19th and 20th centuries.

In 1945, Korea became divided in two countries, North Korea and South Korea, on both sides of the 38th parallel. Western astronomy was occasionally introduced to Korea since eighteenth century mostly through China, but it was not widely taught until after the "Korean war" between the two countries (1950-1953). In South Korea, university level astronomy education started in the late 1950s, the Korean Astronomical Society was founded in 1967 and the National Astronomical Observatory was established in 1974 together with the installment of the first research grade telescope. In North Korea, Pyongyang Astronomical Observatory, effectively its National Observatory, was established in 1957. Astronomy and astrophysics in South Korea are now in mature state thanks to the aggressive investments by the government, enthusiastic support by the general public and hard works by the pioneers. South Korea became a member of the IAU in 1973 and now has a membership of more than 180 astronomers. In the meantime, there was very little contact between North and South Korea even in scientific activities including astronomy. North Korea became a member of the IAU in 1961, twelve years earlier than South Korea, but there have not been many occasions for astronomers in both sides to communicate with each other within the IAU or outside of it. South Korea

---

H. M. Lee (✉)
Korea Astronomy and Space Science Institute, Daejeon, South Korea

Department of Physics and Astronomy, Seoul National University, Seoul, South Korea
e-mail: hmlee@snu.ac.kr

© The Author(s), under exclusive license to Springer Nature Switzerland AG 2022
T. Montmerle, D. Fauque (eds.), *Astronomers as Diplomats*, Historical & Cultural Astronomy, https://doi.org/10.1007/978-3-030-98625-4_10

will be the host of the General Assembly of the IAU in 2022 in Busan, and it is looking for visits of the North Korean astronomers. This article summarizes the early development of modern astronomical institutions in both South and North Korea, with very short accounts of the historical overview and current status.

10.1 Astronomy in Ancient Korea

Korea has a long history with notable astronomical heritage. It is difficult to summarize several aspects of the archeoastronomy and history of astronomy in pre-modern era. More comprehensive coverage of this subject can be found in the review articles by Park (2000), Nha (2000, in Korean) and Nha et al. (2017) among others. We also refer to the 40th Anniversary book (History of KASI since 1974, 한국천문연구원 40년사) published by Korea Astronomy and Space Science, 2014), Fifty Years of Korean Astronomical Society (한국천문학회 50년사; Korean Astronomical Society, 2015) for more detailed accounts of the development of modern astronomical institutions with brief summary of pre-modern astronomy. Here we only take short glimpses of astronomy before the twentieth century in Korea as a brief introduction of the traditional science in ancient Korea.

According to the oldest history book, the first kingdom called Joseon was established by Dan-Gun in 2333 BC in the northern part of Korea. Joseon existed for about 2000 years, but there is not much written account on the history of this period. The Three-Kingdom[1] era started around the first century BC and there are abundant records, ruins, and monuments from this period. Astronomical records also appear in the history book called Samguksagi (삼국사기, 三國史紀, History Book of Three-Kingdoms) which covers the period between 57 BC and 935 AD. This book was completed in 1145 during the Goryeo dynasty which existed from 918 to 1392.

Among the oldest remains that contain astronomical information are dolmens. There is a large number of dolmens throughout the world, and about 40% of them are found in Korea (Jensen, 2016; Meyerhoff, 2013). Not all of them have been examined yet, but some of them have mysterious cup-marks on the top. These are thought to be depiction of the constellations in the sky. Figure 10.1 shows one clear example: the constellations of Big Dipper and Draco are seen from a stone that was recovered in the Adeugi Village in the central part of Korea (Park et al., 2001; Yang, 2015).

[1] Three kingdoms are Shilla in South Eastern, Baekje in South Western and Goguryeo in Northern, including Manchurian area, parts of Korea. The Shilla Kingdom united the entire country by conquering Baekje in 660 AD and part of Goguryeo in 668 AD. The period between 668 and the collapse of the Shilla Kingdom in 935 is often called the United Shilla Period, but in this article we will collectively call the entire period covered by the Samguksagi (i.e., 57 BC–935 AD) as the Three-Kingdom era. The Goryeo Kingdom was inaugurated in 918 and became the sole dynasty in Korea by occupying the Shilla Kingdom.

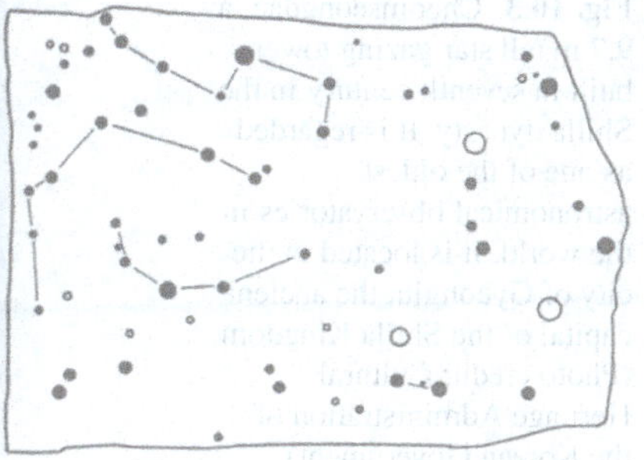

Fig. 10.1 An example of depiction of stars on the stone plate inside a dolmen. This particular piece of the flat stone was recovered from Adeugi Village in the central part of South Korea, just before it was submerged by an artificial lake. Lines connecting some dots are intended to show that these are not random dots but real (northern) sky (photo taken from Park et al., 2001)

Fig. 10.2 Depiction of constellations in the walls of the tomb from the Goguryeo Kingdom. The right circle is the Sun and left circle is the Moon. Also the Big Dipper is depicted in the bottom and the top parts. The dates are not well known but estimated to be middle of fifth century. (photo credit: Korea Creative Content Agency, https://kocca.kr)

While the dolmens are thought to be the remains of the pre-historic periods of Neolithic and Bronze Ages spanning from around fourtieth century BC to around fifth century BC, the tradition of depicting the constellations continued throughout history. Inside the ancient tombs from the Goguryeo Kingdom (BC 37–AD 668) in the northern part of Korea, the walls often contain constellations. An example is shown in Fig. 10.2.

More elaborate astronomical records are found in history books. There are about 240 astronomical events including Solar and Lunar eclipses, motions of the planets, the appearance of comets, meteors, and aurorae are recorded in the Samguksagi. The first solar eclipse that appeared in the Korean history is the one in 45 BC, and 65 additional events can be found from the Samguksagi. Most of the solar eclipses

Fig. 10.3 Cheomseongdae, a 9.7 m tall star gazing tower built in seventh century in the Shilla dynasty. It is regarded as one of the oldest astronomical observatories in the world. It is located in the city of Gyeongju, the ancient capital of the Shilla Kingdom. (Photo credit: Cultural Heritage Administration of the Korean Government)

in the Samguksagi actually took place on or around the recorded date based on the modern calculations (Park, 2008), although some of them could not have been seen inside the Korean peninsula.

The Samguksagi also contains many astronomical events that do not have counterparts in Chinese history books (or perhaps anywhere else). The examples include the sighting of Venus during the daytime and occultation of the planets by the Moon. These events are found to have been real ones by modern calculations (Park & La, 1994).

The most prominent astronomical monument from the Three-kingdom period is the Cheomseongdae (Fig. 10.3), which is a bottle-shaped stone structure that still stands in Gyeongju city, the capital of the Shilla dynasty for almost thousand years until its collapse in 935. The Cheomseongdae means the star gazing tower and was built during the reign of Queen Seondeok (632–647 AD) of Shilla.

The Three-Kingdom era ended by the takeover of Shilla in 935 by the Goryeo dynasty which lasted from 918 to 1392. The Western name of "Korea" is derived from the name of Goryeo. According to the Goryeosa (고려사, 高麗史, History Book of the Goryeo Dynasty), there was an office in the government to carry out astronomical observations and to publish calendars. This office is regarded as the Royal Observatory and was located in the capital (Gaeseong at the time of the Goryeo period, and then moved to Seoul during the Joseon period) and it existed even during the changes of the dynasties until 1910. The number of astronomical events recorded during the Goryeo dynasty is more than five thousands, about a 20-fold increase compared to Three-Kingdom era. The largest number is about the close approaches of planets to the moon, and a planet to another.

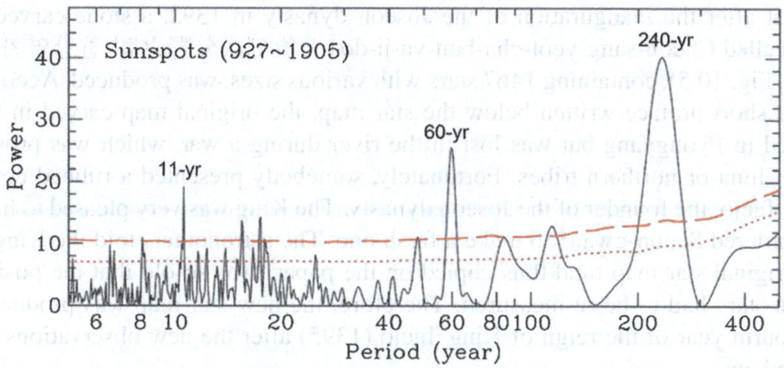

Fig. 10.4 Power spectrum of Korean and Chinese sunspot records in the last millenium (from 927 to 1905). The three most powerful signals correspond to the periods of 240, 60, and 11 years. (Figure taken from Yang et al., 2019)

There are 137 solar eclipse events recorded in the Goryeosa. According to modern calculations, there have been 1130 solar eclipses worldwide during this period. Considering that any solar eclipse can be observed only in the limited areas on Earth, 137 is a substantial number. Such a high rate was possible because astronomers at that time were capable of predicting the eclipses. In fact, there were eclipses that took place where they could not be observed in Korea but they recorded in the Goryeosa. Also, for some eclipse events, there are comments that they were not observed because of bad weather.

There are 42 records of sunspots with their relative magnitudes during the Goryeo period. The time series analysis of the sunspot records showed the well-known eleven-year periodicity (Yang et al., 1998). By combining the sunspot records of Goryeo and Joseon dynasties spanning almost one thousand years, augmented by the Chinese record, 60 and 240-year periodicities are also found (Yang et al., 2019, see Fig. 10.4).

The periodicities much longer than eleven-year can be found only from the observational records spanning a very long time. Previously, Rochard & Rogers (2009) found a 183–243 year periodicity by analyzing the archival data from 1610–2010. Also sudden temperature drops in the early and fifteenth and seventeenth century implied by the frost records in the Chronicles of the Joseon dynasty seem to support a 240-year period of sunspots. This might be related to the 205-year de Veirs-Seuss cycle identified by the carbon-14 record derived from the tree rings (e.g., Wagner et al., 2001; Braun et al., 2005). The sixty-year periodicity in the climate characteristics of Earth's atmosphere is also well known (Veretenenko & Ogurtsov, 2018, and references therein) and it may be closely related to the sixty-year solar cycle found from the Korean historical records covering a long period.

The last Kingdom is the Joseon dynasty (1392–1910) which reinstated the ancient name of Korea. The earlier Joseon dynasty founded in 2333 BC is often called Go-Joseon, meaning the old Joseon.

Just after the inauguration of the Joseon dynasty in 1392, a stone-carved star map called Cheon-sang-yeol-cha-bun-ya-ji-do (天象列次分野之圖 천상열차분야지도, Fig. 10.5), containing 1467 stars with various sizes, was produced. According to the short preface written below the star map, the original map carved in stone existed in Pyongyang but was lost in the river during a war, which was probably with China or northern tribes. Fortunately, somebody presented a rubbed copy to King Taejo, the founder of the Joseon dynasty. The King was very pleased to have it and ordered Seoun-gwan[2] to make a fresh one. The astronomers told the King that the original star map (and thus copied on the paper) was so old that the positions of the stars had to be re-measured. Therefore, the new star map was produced in the fourth year of the reign of King Taejo (1395) after the new observations were carried out.

The changes of stellar positions in the sky are mostly due to the precession of the polar axis. One can therefore determine the epoch of the observations by looking at the distance of the stars from the north celestial pole (NCP) because the star map used the equatorial coordinate system and the center of the star map coincides with the north pole. About 280 stars in the star map have been identified and the study of correlations between the angular distances from the NCP and linear distances from the center of the map for the identified stars showed the following interesting features: the positions of the stars inside the circumpolar circle are consistent with the sky seen at the time of the early Joseon period (late fourteenth century) while those outside the circumpolar circle are closer to the sky in the first century AD (Park, 2008). The interpretation of such a distinction was that the original star map was actually based on the very old measurements (around the first century AD) while the stars within the circumpolar circle had been re-measured by the astronomers of Joseon. Apparently, three years or so were not enough for the astronomers to measure the positions of all the stars with the ancient instruments. Therefore, they may have decided to measure only the stars that could be visible throughout the year.

As more accurate star maps made by Jesuit astronomers in China in the seventeeth century were brought to Joseon, series of revised star maps are locally produced. These maps that incorporate both traditional and European knowledges are summarized in Nha et al. (2017) with their pictures.

The role of astronomy became very important during the Joseon dynasty as the determination of seasons and accurate time keeping were considered to be essential in running the country whose economy largely relied on agriculture. The number of officers at the Royal Observatory increased significantly and the scope of the work

[2] Seoun-gwan was the name of the Royal Observatory from the late part of Goryeo to the early Joseon periods. The Royal Observatories can be traced back to around tenth Century in the early Goryo Dynasty. There might have been similar organizations in the royal courts of earlier dynasties of Three Kingdoms, but no official records remain. Seoun-gwan became Gwansang-gam in 1466 and lasted until the collapse of the Joseon dynasty in 1910. The Royal Observatories belonged to the central government and were located in Seoul, the capital, within the royal palaces or close to them.

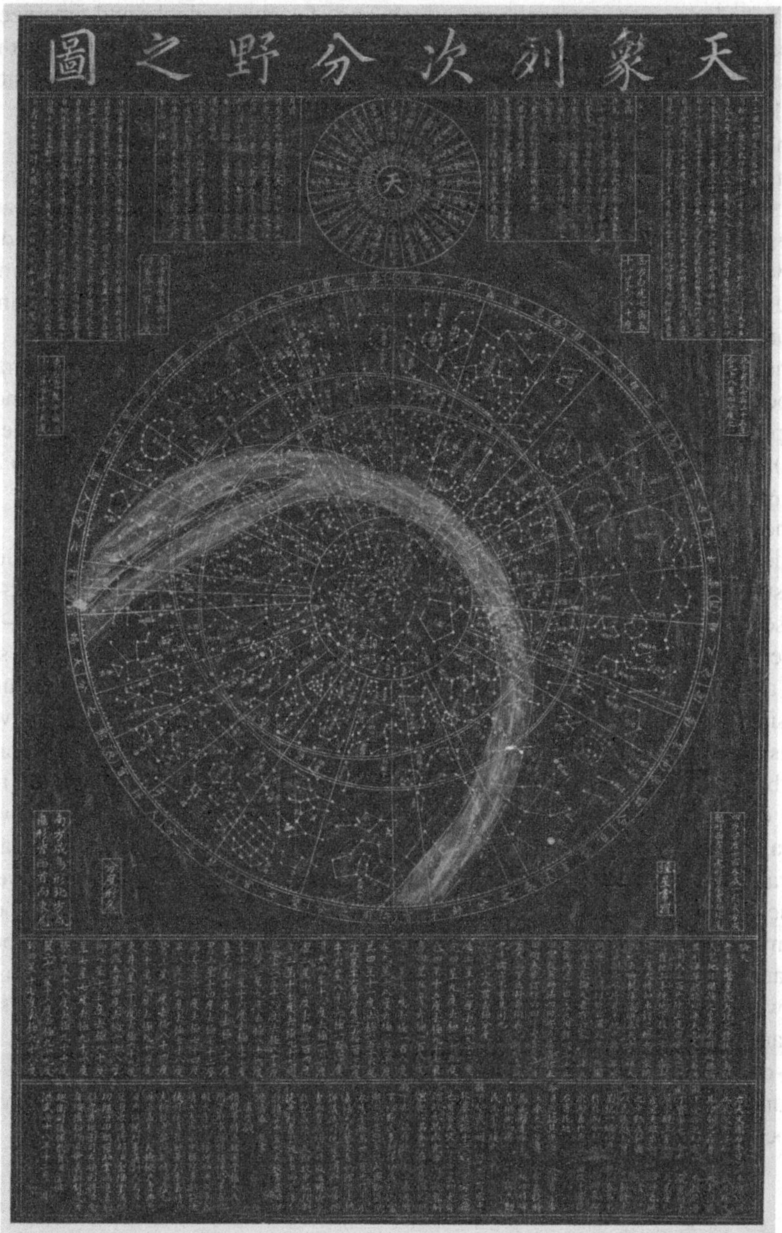

Fig. 10.5 Print of the Chon-sang-yeol-cha-bun-ya-ji-do (天象列次分野之圖, 천상열차분야지도), of the seventeenth century version. The original version was carved in 1395, but it had too many scratched as time goes, and the exact copy was carved during the reign of King Sukjong. The white translucent arc represents the Milky Way. (Photo credit: Seoul Museum of History)

was expanded to encompass the manufacturing of accurate clocks and time keeping. The capability of predicting the apparent motions of the planets, Sun, and Moon was necessary and reached a high level by advancing the observational instruments and calculational methods.

The development of astronomy was particularly rapid during the reign of King Sejong (1418–1450). He led the construction of astronomical instruments, accurate clocks, and calendars that can predict the eclipses and motions of the heavenly bodies. Unfortunately, most of the instruments produced during his period do not survive until today presumably due to the wars with Japan and China in the middle of the Joseon period. The clocks and astronomical instruments produced during the Joseon dynasty are summarized in a monograph by Needham et al. (1986).

The number of astronomical events recorded in the history books has accordingly exploded to more than 20,000 and some of these data have been utilized for modern astrophysical research (a comprehensive summary can be found in Stephenson, 2017), in particular supernovae (e.g., Stephenson, 2004; Stephenson & Green 2002; Koo & Park, 2017; Vogt et al., 2019; Hoffmann & Vogt, 2020) and novae (e.g., Stephenson & Yau, 1987; Shara et al. 2017).

The economic and political situation from the middle period of the Joseon dynasty became very bad because of two wars, first with Japan during 1592–1598 and second with a newly established Qing dynasty (China) during 1636–1637. While western countries transformed into modern society through various stages such as scientific revolution and industrial revolution, Korea, like other countries in Asia except Japan remained as a quiet kingdom isolated from the outside world. Occasionally, the news about Western countries and their scientific achievements was brought into Korea. For example, a telescope has been introduced in the early eighteenth century (at the time of King Yeongjo) by a diplomatic envoy to China, under the name of Cheonli-gyeong, literally meaning a glass that can see several hundred kilometers. Several progressive scholars tried to absorb the Western sciences and ideas from the eighteenth century, but they never became a major stream. Eventually, Korea wanted to open up its borders to the outside world in the late nineteenth century, but it was too late. Japan was able to transform itself to a westernized country and invaded the Korean peninsula in 1910 with a strong military force. Korea was not able to protect itself from this Japanese aggression. Very detailed accounts of the astronomers, astronomical observatories as well as instruments, early encounters with western astronomy during the Joseon dynasty can be found from Nha et al. (2017).

10.2 Shutting Down of the Royal Observatory and the Dark Age (1910–1945)

Gwansang-gam, the Royal Observatory of the Joseon dynasty, was forced to shut down by the Japanese after the annexation of Korea. In 1910, the Japanese imperialism established a meteorological observing station, which essentially replaced the Gwansang-gam, in Incheon, a port city located at about 40 km west of Seoul. The role of this station was to carry out the nation's meteorological service and the annual publication of the almanac under the name of Korean Forklore Almanac (朝鮮民曆). The first director was Yuji Wada. He visited Gyeongju in 1909, explored Cheomseongdae and claimed that it was indeed an astronomical observatory functioning like a modern observatory. In addition, even after retirement, he continued his study of Korean astronomy and meteorology using the Korean ancient records[3] (Wada, 1911) .

The shutting down of the Royal Observatory also meant the end of the traditional astronomy in Korea. It can be said that 1910–1945 was the dark age of Korean astronomy. Hardly any research was done and no observational records remain during this period. We can only catch a glimpse of the astronomical records left by foreigners as described below.

The following is a brief summary of those who played major roles during the dark age. More extensive biographical information of the people who carried out astronomical research and education during this period can be found from Nha et al. (2017).

In 1915, Yeonhee College,[4] a Western-style educational institution, was established in Seoul, and two professors, A. L. Becker (1879–1979) and W. Carl Rufus (1876–1946), were invited to take charge of science education. Becker came to Korea in 1903 as a professor at Yonhee College as well as at Soongsil College[5] in Pyongyang. He taught physics, mathematics, and astronomy. Rufus, who joined Soongsil College in 1907, taught astronomy, mathematics, and physics, later at Yonhee college together with Becker. During his stay in Korea, Rufus introduced the stone-carved star chart Cheon-sang-yeol-cha-bun-ya-ji-do (see Fig. 10.5) in his English articles (Rufus, 1913, 1915).

However, after living for ten years in Korea, Rufus resigned from all positions and returned to the University of Michigan, the USA in 1917 because of his grievance at the policy of education enforced by the Japanese governor-general of Korea.

Won-Chul Lee (1896–1963) was an undergraduate student while Rufus was at Yonhee. He went to the USA in January 1922 and became an undergraduate

[3] Korea has abundant records on natural phenomena such as astronomical events, heavy storms, precipitation, etc., in the history books. During the reign of the King Sejong, accurate rain gauge was invented and used to measure the amount of precipitation in well calibrated scale. As a result, substantial amount of rainfall records remains for various regions.

[4] Yonhee College became Yonsei University, which has been one of the most prestigious private universities in Korea.

[5] Soongsil College was established in 1897 in Pyeongyang, but had to shut down because of depression by Japanese in 1938. It was re-established in Seoul after the Korean war. It is now called Soongsil University.

student at Albion College[6] in Michigan, then entered the University of Michigan as a graduate student of Carl Rufus in June 1922. Apparently, he was asked to take an undergraduate course in order to become acquainted with the American educational system. He obtained his PhD in 1926 with a thesis entitled "Motions of the Atmosphere of Eta Aquilae," where he proved that η Aquilae is a pulsating star. Later this work was published in the Publications of the Observatory of the University of Michigan (Lee, 1932) with small modifications.

Won-Chul Lee was the first Korean to obtain a PhD degree in science from a Western University. He returned to Korea immediately after finishing his PhD and taught mathematics, science, and astronomy in Yonhee College, his alma mater. He installed a 6-inch refractor on the roof of the college building and used it for astronomy education (Fig. 10.6). However, he was forced to resign from his post in 1938 because he was not cooperative with the Japanese government. The telescope he had installed was also confiscated by the Japanese in 1942 and thus disappeared from the campus. While he was teaching at Yonhee College, he was also active in public outreach: he eagerly gave lectures on astronomy to general public in 1930s, mostly at the YMCA[7] building in the central part of Seoul.

Rufus came to Korea again in 1935 to spend his sabbatical year and explored ruins and remnants related to astronomy in Korea. In 1936, he submitted a paper entitled 'Korean Astronomy', which is a summary of his study on the traditional astronomy of Korea during his stay, to the Korean studies committee of the Royal Society of Oriental Studies (Rufus, 1936). It was the first comprehensive report of the Korean astronomy to the outside world.

One of the peculiarities at that time was the work of Japanese scholar Iijima (飯島忠夫) who analyzed the solar eclipse records in the Samguksagi (Three-Kingdom History Book) in 1939. He claimed that these records were blind copies of Chinese material. Based on his prejudice, he raised the suspicion on the genuineness of the Korean history books. Of course, his claim turned out not to be true based on modern calculation of the eclipses as well as other ancient records of astronomical events.[8]

---

6 Won-Chul Lee went to Albion College, probably by the recommendation of Becker and Rufus because both of them did their undergraduate studies at Albion.

7 Young Men's Christian Association. The YMCA in Seoul was founded in 1903 and constructed its three-story building Jong-Ro in 1908 with the fund donated by the royal family of Korea and its members. The original building was burned down during the Korean war and seven-story new building which is still standing today was reconstructed after the war. Many public lectures, including astronomy by Won-Chul Lee for the education of general public, were held in the YMCA building.

8 Indeed, Korean eclipse records during the first millennium are found to be the most accurate among three countries in East Asia (China, Korea, and Japan). Most of the recorded solar eclipse events in the Samguksagi are found to be the actual events based on the modern calculations. Samguksagi also contains many proved astronomical events that do not appear in Chinese books.

Fig. 10.6 Six-inch refractor installed on the roof of the Yonhee college building in 1928. The person on the right with a tie, a fedora hat and glasses is Won-Chul Lee. (Photo courtesy: Yonsei University)

10.3 University Education and Foundation of the Korean Astronomical Society After the Liberation (1945–1967)

After the Japanese defeat in 1945, the almanac-making job was handed to the Gwansang-dae.[9] Won-Chul Lee was appointed as the first head of the Gwansang-dae (which became Korean Meteorological Administration [KMA] in 1990 after a few changes in its name) and he actively promoted the publication of the almanac. During the Japanese colonial period, the traditional lunar calendar was prohibited and therefore the calendar published at that time was only solar calendar. However, after the liberation, both lunar and solar calendars were included in the almanac. The

[9] The observatory (Gwansang-dae) here was derived from Gwansang-gam, the Royal Observatory of the Joseon dynasty. However, the main duty of Gwansang-dae in the modern Korean government was to be a meteorological agency, although there was a small department within Gwansang-dae dedicated to the publication of the calendar and time keeping.

publication of the calendar is now a responsibility of Korea Astronomy and Space Science Institute (KASI) as described in Sect. 10.5.

Even after the liberation, astronomy remained stagnant partly because of the turbulent political situation. Korea was divided into North and South Koreas as a result of the separate elections in 1948, after having a loosely separated period since 1945. Both in the North and South, Korean economy remained in a miserable condition for more than a decade even after the liberation because of the exploitation of the country by the Japanese occupation, especially toward the end of the World War II, and later by the devastations of the Korean war. South Korea began investing significantly on advanced-level education by establishing national university system, and astronomy was already taught at the Physics Education Department at Seoul National University since late 1945 by Rak-Bok Lee, who had graduated from the department of astronomy at the University of Tokyo in 1944.[10] He went to North Korea during the war and became a founding director of Pyeongyang Astronomical Observatory in North Korea (see below).

However, there was no university able to offer an academic degree in astronomy until 1958, when Seoul National University (SNU) formally opened a dedicated Department of Astronomy and Meteorology.[11] Yonsei University also opened the Department of Astronomy and Meteorology in 1968. Yonsei University became the new name of Yonhee College which had a long tradition of education of astronomy, and opening a department dedicated to astronomy was very natural. SNU and Yonsei were the two pillars of astronomy education and research until the early 1980s when other universities began teaching advanced-level astronomy. Sungkyunkwan University in Seoul briefly had a Department of Cosmic Physics in the late 1960s but it lasted only for a few years.

As the number of astronomers grew to more than 20, it became necessary to have an academic society for astronomy in the early 1960s. We can guess the situation at that time through the article entitled “Remembrance of the founding of the Korean Astronomical Society (KAS)” written by the late Professor Kyung Loh Yu (1917–1997) of SNU (Yu, 1976). According to this memoir, preparatory meetings for the establishment of the KAS were held at SNU in 1961 and 1962 while two Professors of SNU (Jong June Hyun [1927–2015]) and Kyung Loh Yu) were studying in the USA.[12] Attendees included Won-Chul Lee, but the founding of the KAS was not achieved in those years because of the absence of Profs. Yu and Hyun of SNU which was the most important institution in astronomy at that time.

---

[10] The author is grateful for Prof. Takashi Onaka at the University of Tokyo for checking of Rak-Bok Lee’s record at the University of Tokyo.

[11] Initially the both astronomy and meterology were at the same department at Seoul National University, but these two fields separated in 1975.

[12] In the early period of modern astronomy education at SNU starting from 1958, there were no faculty members having received formal graduate level education in astronomy. Profs. Hyun and Yu were only two faculty members for astronomy and they were sent to American Universities in order to get training. Dr. Won-Chul Lee had obtained his PhD in astronomy from the University of Michigan in 1926, as mentioned earlier, but he was not directly involved in astronomy education after he became a director of the Korean Meteorological Agency in 1945.

Fig. 10.7 A photo taken after the inaugural meeting of the Korean Astronomical Society on March 21, 1965 at Seoul National University. Although the quality is low, it is possible to identify everybody. For example, Those marked with red circles are Profs. Jung June Hyun (left) and Kyung Loh Yu (right) who were two founding members of Department of Astronomy at Seoul National University. (Photo courtesy: Korean Astronomical Society)

Eventually, the inaugural general meeting was held in 1965 after Profs. Yu and Hyun returned from the USA and several students graduated from the Astronomy Department at SNU. As we can see from the photo taken after the inaugural meeting (Fig. 10.7), many physicists and meteorologists helped in establishing the KAS. They actively participated in the founding work of the society even before the promoter meeting. According to Prof. Kyung Loh Yu, physics professor Jeong-Heum Kim (1927–2005) of the Korea University was involved in drafting the statutes and calling the meeting. The first president elected at the inaugural meeting was Prof. Jung June Hyun and Prof. Kyung Loh Yu became the second president.

KAS launched a journal under the title of "The Journal of the Korean Astronomical Society" (JKAS in short) in December 1968. Initially, JKAS was published once a year until 1979 and the articles could be written either in Korean or English. The first issue carried four research articles, two of them in English and the other two in Korean, and one review article. One of the research articles was on the supernova observational records in ancient Korea, and it was quickly referenced by van den Bergh & Dodd (1970). All the articles published in JKAS are now available via SAO/NASA Astrophysics Data System.[13] JKAS also included news

[13] https://ui.adsabs.harvard.edu

related to the astronomical community in the early days. The news articles appeared in the first issue were about the funding of the National Astronomical Observatory. KAS began publishing the Bulletin of the Korean Astronomical Society (BKAS) separately twice a year from 1976 after which JKAS became a journal dedicated only to research articles. JKAS[14] is now published six times a year and accepts only English articles.

10.4 First Contacts Between North and South Korean Astronomers, and the IAU

Since Korea became a "divided country" in 1948, deeply split by the Korean War (1950–1953), there were neither formal meetings nor academic exchanges between astronomers of the two parts due to the strong political tensions. Sometimes their rivalry made the two Koreas intensely compete in international stages. Both Koreas joined the International Council for Science (ICSU) at the same time in 1961, probably upon recommendation by the ICSU, but joining in the International Astronomical Union (IAU) was more complicated.

North Korea was also admitted as a member of the IAU in 1961, at its XIth General Assembly held in Berkeley, USA, but South Korea was unable to do so. As we have seen in the previous section, there was no representative organization for astronomy in South Korea. In contrast, the Pyongyang Astronomical Observatory (PAO), which is a national observatory of North Korea, was established in 1957. In South Korea, a National Astronomical Observatory was not even under discussion at that time. There was a professionally trained astronomer, Dr. Won-Chul Lee, but he was busy with his job as a head of meteorological administration. As mentioned earlier, there was another astronomer, Rak-Bok Lee, who was initially in South Korea, teaching astronomy at the SNU before the Korean war but he voluntarily went to North Korea during the war and became a founding director of PAO. He retired from his post in 1970. PAO was the North Korean representative of the IAU.

When the news came out that North Korea was planning to attend the IAU General Assembly in Berkeley, South Korean astronomers also wanted to participate. However, without any representing body, it was very difficult to join. This incident precipitated the discussions on the necessity of the Korean Astronomical Society and the National Astronomical Observatory among South Korean astronomers.

Eventually, a few Korean astronomers attended the XVth IAU General Assembly, where South Korea was admitted as a member, in Sydney, Australia in 1973. Since then South Korean astronomers are attending the General Assemblies regularly.

The first IAU-sponsored meeting where both South and North Korean astronomers appeared was the fourth Asian-Pacific Regional IAU Meeting held in Beijing in 1987. At that time, getting a visa to China for a South Korean national was

[14] http://jkas.kas.org/

very difficult since there was no diplomatic relation between South Korea and China. A number of astronomers registered for the meeting, obtained official invitations, and applied for visa in Hong Kong. Unfortunately, not everybody was granted entry to China and those who were turned down have to return to Korea from Hong Kong. The lucky seven people flew to Beijing and successfully attended meeting. They even met North Korean astronomers and exchanged some conversations.

The seventh Asian-Pacific Regional IAU Meeting (APRIM) was held in Busan in August 1996. This was the first IAU sponsored meeting ever held in South Korea and the organizers of this meeting sent an invitation to PAO through IAU headquarter because it is not possible to send mail from South Korea to North Korea. However, there was no answer from North Korea. There were two more IAU meetings (IAU Symposium 197 in the island of Jeju in August 1999, and the twelfth APRIM in Daejon in August 2014), but nobody from North Korea attended these meetings.

Probably the reason was that North Korea was not a member of the IAU anymore at those times. Indeed, following the collapse of the USSR in 1991, North Korea, which depended in part on its support, stopped paying its dues to the IAU, and thus, the IAU had to take the statutory decision to terminate its membership in 1996.[15]

In South Korea, the liberal party had been in power for about ten years from 1998. During this period, the South Korean government took many reconciliatory measures toward North Korea including an establishment of industrial complex in Gaeseong in North Korea, an ancient capital of Goryeo dynasty just across the border, and opening of the Keumgangsan (Mountain) Tourist Area in the east coast just across the border to South Korean citizens. As the relationship between South and North became warmer, in May 2008 North Korea sent, under the auspices of the IAU, two young astronomers from the PAO to the eighth Pacific Astrophysics Conference (PRCSA-2008) in Phuket, Thailand. PAO astronomers also participated the First International Astronomical Summer School in Ulaanbaatar, Mongolia in July 2008, as described in the IAU Report (Guinan & Marschall, 2009).

In August 2012, Sok Jong, Director of the PAO came to XXVIIIth IAU General Assembly in Beijing and presented a paper entitled "Strategic Plan of Development of Astronomy in DPRK" (Jong, 2015), and as a result, North Korea, after twenty years of absence, was re-admitted as an IAU member. This report contained the manpower, facilities, and current research interests of PAO and the strategic plan for astronomy in North Korea. During this General Assembly, many astronomers from both Koreas met in person, but no formal discussions were held on inter-Korean astronomical exchanges. Three years later, two poster papers by North Korean astronomers were presented at the next IAU General Assembly in Hawaii through Prof. George Miley, a former IAU Vice President, at the University of Leiden, but the authors could not come to the meeting.

[15] For details, see the Introduction chapter of this book.

Fig. 10.8 Audience at an astronomy lecture by John Hearnshaw (right, second row) at Kim Il Sung University during his visit to North Korea in September 2012. (Photo courtesy: John Hearnshaw)

Just after the General Assembly in Beijing in 2012, Prof. John Hearnshaw[16] at Canterbury University, New Zealand went to North Korea organized by Korea-New Zealand Friendship Society. He spent the first week of September in 2012 in North Korea where he gave three lectures of two hours each at Kim Il Sung University (Fig. 10.8), had meetings with astronomers from Pyongyang Astronomical Observatory and with officials from the Academy of Sciences, in addition to a numerous cultural visits. He may have been the first astronomer ever to visit North Korea from western countries. He also brought 109 astronomical books, including several volumes of the proceedings of IAU Symposium as a gift for the North Korean astronomers and students.

During the IAU GA in 2015 at Honolulu, South Korea was announced to become a host of the IAU GA in 2021 in the city of Busan. This will be the first IAU GA to be held in Korea, and the astronomical community in South Korea wanted to take advantages of this event to promote exchanges of astronomers. The KAS has been conducting planning and research for the revitalization and collaboration of the inter-Korean astronomical field. In 2019, South Korean astronomers presented topics on the joint projects for astronomy such that South and North Koreas could collaborate with each other as part of their presentation on astronomical

[16] The author is grateful to Prof. John Hearnshaw (former IAU Vice-President, 2018–2021) who has provided his article written for the publication in New Zealand at the University of Canterbury (https://www.scoop.co.nz/stories/SC1211/S00029/north-korea-opens-its-doors-to-uc-scientist.htm). The paragraph about his visit to North Korea is based on his article.

collaboration and science diplomacy between the two Koreas at the meeting of the American Association for the Advancement of Science (AAAS) in Washington DC, USA.

For the purpose of inter-Korean astronomical collaboration, South Korean astronomers plan to promote collaboration in such fields as astrophysics and historical astronomy in ways that do not violate UN sanctions on North Korea, namely, academic conferences and joint research. Eventually, they would like to gradually expand the scope of collaboration by installing observational facilities such as optical and radio telescopes in North Korea, which has many potential high-altitude observing sites.

It is worth noting that the collaboration between the two Koreas is promising as they share more than 2000 years of astronomical interests dating back when the country was united, and many researchers from both sides are studying historical astronomy. The KAS in South Korea is working to build an international collaboration network in order to successfully invite North Korean astronomers to the 2022 IAU General Assembly in Busan.

10.5 The Founding of the National Astronomical Observatory

The Royal Observatory that had ceased to exist since 1910 was not recovered after the 1945 liberation for a long time. It became necessary for the new Republic of Korea to have a national observatory, but it was not an easy task to establish an organization that had been closed for such a long time. The main purpose of the Royal Observatory in earlier periods was calendar making and time keeping, but the such a role is only small part of the modern observatory.

The effort to set out a National Observatory in South Korea started only after the KAS was established. The first important task for the KAS was to reinstate the National Observatory. KAS submitted a petition to the government to establish the Observatory right after its foundation in 1965.

Upon receipt of the petition, the government responded positively. The Ministry of Science and Technology (MOST) was established in April 1967 within the government and MOST enacted a stipulation for a Committee for the Founding of the National Astronomical Observatory (Committee hereafter) in March 1968 and appointed the Vice-Minister as Chair and nine specialists, mostly astronomers and physicists, as its members. The Committee drafted a five-year plan for the establishment of the National Astronomical Observatory (NAO). The initial plan was rather ambitious: the NAO would be composed of 75 employees with five divisions equipped with one 122-cm reflector, one 61-cm reflector, two 41-cm reflectors, and one 20-cm photographic zenith tube. However, this plan was not fully realized because of the limited budget.

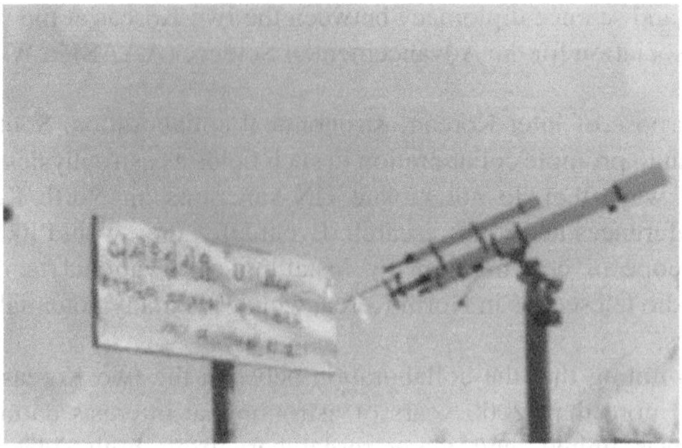

Fig. 10.9 A 4-inch telescope installed for test observations on top of Yangja-san, together with a hand-written welcome message for the visitors. (Photo courtesy: Byung-Ryul Oh, who wrote the message and was one of the observing staff)

Following the five-year plan, the Committee was in charge of selecting the telescope site. In October 1968, the government hired four young astronomers (all of them graduate students from Seoul National University and Yonsei University) as staff members to carry out the various works related to the construction of the observatory. The ranking in the government of these staff members was rather low considering the level of qualifications, but they accepted the offer.

The Committee chose two candidate sites near Seoul and the newly appointed staff members were asked to carry out the test observations. Both sites were mountains located south of Seoul, Mangdeok-san (542 m above sea level) and Yangja-san (704 m). The staff members began test observations on these sites from March 1969 for about ten months using small telescopes (see Fig. 10.9). During this period, the observing team recorded the amount of precipitation, wind speed, temperature and humidity, and took pictures of diurnal motions of the stars in order to obtain the seeing and its variation with time.

Together with these data, a report on the summary of the test observations was submitted to the Committee. The Committee selected Mangdeok-san after carefully studying the report in 1970. However, the observing conditions became degraded in a matter of years because of massive housing projects and construction of a military air strip near the selected site. Seoul was expanding rapidly at that time as a result of the industrialization. The other site was only slightly farther away and could not be an alternative. Eventually, the Committee decided to abandon the earlier choice and to look for other places in the fall of 1971. They no longer insisted that the site had to be close to Seoul.

In the meantime, the government allocated a budget of about $100,000 for the purchase of a telescope for the fiscal year 1972. This was another turning point because in the early 1970s the atmosphere among the government officials was not favorable for establishing an observatory. The Committee had to hurry in selecting the site and in deciding the telescope specification. The meeting for the discussion of the new site was held in autumn 1971 and some members recommended Sobaek-san, a mountain located in the border between the Gyeongsang and Chungcheong Provinces. It is about 250 km South-East of Seoul and the highest point is more than 1300 m above sea level. The Committee agreed to visit the site with weather specialists for inspection.

A journey to Sobaek-san for three days starting from November 12, 1971, was quickly arranged. The envoy included most of the Committee members, one professor in meteorology at Seoul National University, two professors outside of the committee, and the four staff members. Fortunately, there was a road almost to the top of the mountain because there was a military camp in one of several peaks of around 1300 m–1400 m elevation along the Sobaek-san Range which extends over about 45 km. The envoy visited a few peaks on November 13. The preferred site for the observatory was Birobong (1439 m, Fig. 10.10) which is a very gentle peak capable of accommodating several large telescopes. The envoy asked the commander of the military camp to record the weather conditions, and returned to Seoul.

On February 18, 1972, the government called a Committee meeting to make a firm decision for the foundation of the NAO. The meeting was presided by the Vice-Minister who proclaimed that the project to establish the NAO was now at the crossroads and the Committee should make a practical decision to prevent becoming a failure even though the allocated budget was insufficient. This meant that the astronomers had to surmount the scale-down observatory compared to the initial plan. Following the suggestion by the Vice-Minister, the Committee made the following decisions:

- The project for establishing the NAO should proceed.
- In order to purchase the telescope with the allocated budget of $100,000, the committee had to decide the specifications of the telescope by the end of April 1972.
- The site of the telescope had to be decided by the end of June and test observations had to be carried out in order to confirm the decision.
- A detailed budget plan for next several years based on the development roadmap has to be made.
- A foreign specialist should be invited for consultation in 1972.

The minutes of this meeting were formally presented to the Minister for approval. However, the selection of the telescope was slightly delayed and the test observations could not be carried out because of lack of funding.

As for the foreign expert, Dr. Goro Ishida of the Okayama Astrophysical Observatory of Tokyo Astronomical Observatory (which became National Astronomical Observatory of Japan in 1988) was invited in 1972 from April 29 to May 9 for consultation. He visited three possible sites including Sobaek-san together with a few members of the Committee. In May, the Sobaeksan range is covered with azalea under very mild temperature. On the date of the visit, the sky was particularly clear. Ishida was so impressed by the scenery and weather condition of Sobaek-san, that he strongly recommended one of the peaks for the site of the observatory. In particular,

Fig. 10.10 Sobaek-san range seen from Yeonhwa-bong where telescope is currently located. The peak shown by an arrow is the Biro-bong, the first choice by the Committee. (Photo by H. M. Lee)

Fig. 10.11 Dr. Goro Ishida (second from left) from Okayama Astrophysical Observatory at Gyeongju National Museum together with Korean colleagues. He visited Korea to assist the site selection for the optical observatory site in 1972. After the site visits he went to historical cities including Gyeongju, the ancient capital of the Shilla Dynasty for about 1000 years until AD918. (Photo courtesy of Byung-Ryul Oh, at right)

he suggested that the Birobong should be protected because of its beauty, and the observatory could be constructed on Yeonhwabong (1394 m), another peak between the military camp and Birobong. He spent his last few days to visit historical cities such as Gyeongju (Fig. 10.11). Ishida wrote an article about his journey and recommendations in the Astronomical Herald (天文月報) which is the monthly bulletin of the Astronomical Society of Japan (Ishida, 1972). His report was also submitted to the Ministry and the Committee subsequently made a final decision for the telescope site as Yeonhwabong in Sobaek-san on May 25, 1972. The Minister made a formal approval on June 7.

In retrospect, skipping test observations before the construction of the telescope was regrettable because the weather conditions at Sobaek-san were found to be not so good after several years of real observations with the telescope, in spite of the fact that the military officer near the observatory site submitted a very positive weather report to the Committee. According to his report, number of clear nights was very large, but it turned out that the astronomically clear night is only a small fraction of it. On the other hand, there is not much variation within Korean peninsula either. Another telescope site, Bohyun-san (1124 m) located about 100 km south-east of Sobaek-san was selected more carefully in the late 1980s in order to install a 1.8-m telescope. The weather condition in Bohyun-san is better than Sobaek-san, but not drastically. Sobaek-san was designated as a national park in 1988, and the telescope site is well protected.

The specification of the telescope was also decided in the meeting of the site selection on May 25, 1972. It was a rather complicated decision because the opinion was divided between refractor (lenses) and reflector (mirror). After heated discussions, the matter was put on the vote. The result was in favor of reflector by 6 to 2. The committee recommended to purchase a reflector of diameter between 50 to 75 cm, depending on the price. A sub-committee was formed to come up with the detailed specification and vendor of the telescope. The sub-committee recommended to purchase Ritchey-Chretien telescope of 61-cm from Boller & Chivens, a manufacturer of high precision telescopes and spectrographs in Pasadena, California. Several Korean astronomers living in the USA had provided useful information to the sub-committee, including the names of reliable telescope vendors. The recommendation of the sub-committee was promptly approved by the Committee on July 3, 1972. Everything was done rather quickly because of the restriction on the use of the allocated budget. Indeed, the contract for the supply of the telescope between the Office of Supply and Boller & Chivens was made on December 26, 1972.

The 61-cm telescope was delivered to Incheon, a seaport near Seoul on November 25, 1973, while the construction of the building and dome started in summer 1973 and was nearly finished by winter (see Fig. 10.12). The installation of the telescope was completed on November 7, 1974. There was a small problem with the paddle switch for the operation of the telescope just after the installation, but it was quickly fixed by the observatory staff. This was the first research-grade telescope in Korea. It received a lot of attention from the media and general public. The Korea Post issued a commemorative stamp in 1978 (see Fig. 10.13) when the observatory was formally dedicated for research. This telescope is still functioning and used for scientific research with a CCD camera. Its main targets are the variable stars, solar system objects, and transit of exoplanets, comets, and supernovae.

The process for the establishment of the National Astronomical Observatory (NAO) as one of the government organization concurrently proceeded along with the construction of the observing stations in Sobaek-san. The plan for the NAO

Fig. 10.12 The Observatory building at Yeonhwa-bong, Sobaeksan under construction in early 1973. (Photo courtesy Jaehan Lee of KASI)

Fig. 10.13 Commemorative stamp issued by the Korea Post in 1978 to celebrate the official opening of the Sobaek-san Observatory

organization that would be composed of 27 members was drafted during the spring of 1973. However, the government was also seeking a possibility to put the NAO within the Seoul National University and sent a formal letter to SNU asking whether they were interested in it. Maybe the government had the Japanese model in mind: the Tokyo Astronomical Observatory used to belong to the University of Tokyo before it became a National Astronomical Observatory of Japan (NAOJ) in 1988. Fortunately, SNU refused such an invitation, so that NAO was created as an independent organization. The cabinet meeting held on August 31, 1974, approved the development plan of the NAO and proclaimed it on September 13, 1974. The government named Young-Key Min, who was an assistant professor at the University of Alabama, as the first director of the NAO, and he took the position in July 1975. He remained in his office until 1985.

The NAO changed its name a few times: it became the Institute for Space Science and Astronomy (ISSA) in 1986, the Korea Astronomy Observatory in 1991, and finally the Korea Astronomy and Space science Institute (KASI) in 1999. The headquarters was moved from Seoul to its current location in Daejeon in 1986. KASI now operates several telescopes in South Korea and abroad: the Taeduk Radio Astronomical Observatory (TRAO, 14-m millimeter wave telescope completed in 1988 at Daejeon headquarters), a 1.8-m optical telescope completed in 1996 in Bohyun-san, the Korea VLBI Network (KVN) composed of three 21-m telescopes located in Seoul, Ulsan, and on the island of Jeju, the Korea Microlensing Telescope Network (KMTNet) composed of three 1.6-m optical telescopes located in Cerro Tololo (Chile), Siding Spring (Australia) and Karoo Town (South Africa), and a 1.5-m robotic telescope in Mt. Lemmon (Arizona, USA). It also runs a network of small telescopes in Moroco, Israel, Mongolia, and the USA for monitoring potentially hazardous objects like meteors or asteroids. KASI is a founding member of the Giant Magellan Telescope and operating partner of the Gemini Observatory in Chile. KASI also developed small space telescopes and operated them successfully: Far-ultraviolet IMaging Spectrometer (FIMS) in 2006, Multi-purpose Infra-Red Imaging System (MIRIS) in 2013, and Near Infrared Spectroscopic Surveyor (NISS) in 2018. KASI is now developing an ambitious Space Telescope Spectro-Photometer for the "History of the Universe, Epoch of Reionization and Ices" Explorer (SPHEREx) which is a Middle-Class Explorer of NASA led by Caltech/JPL. For more details about KASI please visit its homepage at http://kasi.re.kr/.

NAO started with about 27 members in 1974, and KASI currently is composed of more than 400 employees.

10.6 Current Status of Astronomy in North Korea

As stated earlier, North Korea established the Pyongyang Astronomical Observatory (PAO) in 1957, about seventeen years earlier than in South Korea.[17] The head-

[17] At that time, North Korea's economy was in a much better shape than in South Korea, because the country was heavily supported by the USSR in particular. Eventually, this support stopped with the collapse of the USSR in 1991.

quarters are located in Daeseong-san in the northern part of Pyongyang. We have very limited information regarding the background of such a rapid development in modern astronomy in North Korea. One possible explanation was the influence of Rak-Bok Lee, who was trained in the University of Tokyo and taught at Seoul National University for several years before the breakout of the Korean War. He became a founding director of the PAO and remained in his office until his retirement in 1970. After his retirement, he lived in his hometown at Hamgyung Province.

South Korean astronomers initiated the detailed studies on the astronomy in North Korea in the early 1990s sponsored by the Korean Federation of Societies of Science and Technology and the results are summarized by Min (1993). This article remained the main resource for the list of institutions and facilities for research and education of astronomy, astronomical education system in elementary school to high school, professional as well as introductory level education at universities, public outreach, names of the astronomers, the research articles published between 1975 and 1993, periodicals and monographs in astronomy, as well as astronomical terminologies compared with the South Korean ones.

The latest information on the astronomy in North Korea also can be found from an article by Sok Jong (Jong, 2015) which is based on his report at the 2012 IAU General Assembly in Beijing. According to this article, PAO is the only national astronomical research institution that plays a leading role in professional research work and public enlightenment in North Korea. In addition, PAO is also responsible for the determination of the national standard time and calculating the almanac. Accordingly, PAO annually publishes the Astronomical Almanac and calendar. Recently, Yang et al. (2018) made a comparative study between the Almanacs published in South and North Korea and found that there are substantial differences in them: the North Korean almanac is for astronomers or navigators while the South Korean one is for general public. In that sense, the North Korean Astronomical Almanac contains supplementary information as a Nautical Almanac. In South Korea, the Nautical Almanac is separately published by the Korea Hydrographic and Oceanographic Agency, under the Ministry of Oceans and Fisheries.

As of 2012, PAO has about 40 research and engineering staff, including 6 professors and 11 Ph-D level astronomers. It is composed of departments of theoretical astrophysics, solar-terrestrial physics, radio astronomy, astrometry and astro-geodynamics, ephemeris, astronomical instruments, and Korean ancient astronomy. There is also an astronomical data service center in the PAO. PAO operates two observing stations: Sunan Geomagnetic Observing station and Jaesan Solar Radio Observing Station. The main observing facilities in operation as of 2012 include a 13-cm refractor, a photoelectric transit instrument, a horizontal multi-channel spectrograph and 2 radio telescopes. They seem to be no major improvement from the list of the instruments which appeared in an earlier article by Min (1993) who has collected all available information in the early 1990s.

PAO publishes a Bulletin of Astronomy (천문학 통보, see Fig. 10.14) which is the only journal dedicated for astronomy in North Korea. We have only a partial list of papers published in this journal, and all papers are written in Korean.

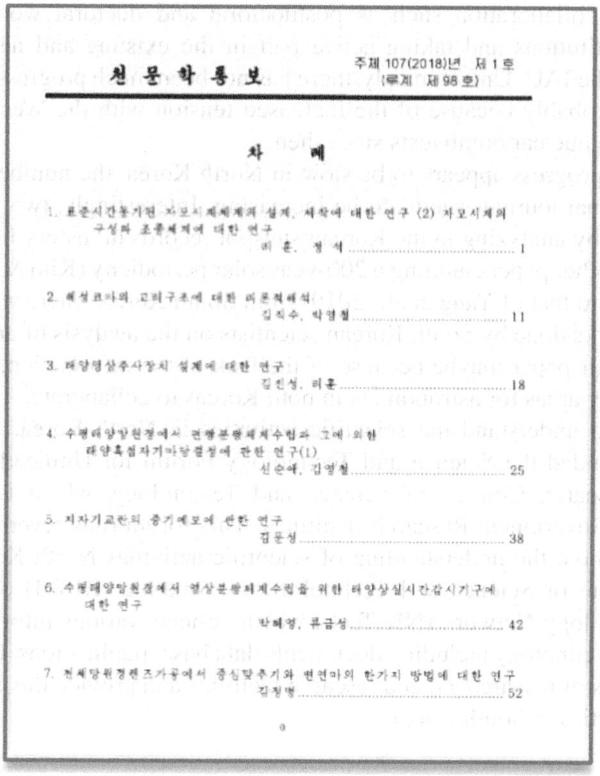

천문학통보

주체 107(2018)년 제 1 호
(루계 제 98 호)

차례

0

Fig. 10.14 An inner cover page of the 2018 issue of Bulletin of Astronomy published by the PAO

Jong (2015) reported that the international collaboration has developed substantially in recent years. The purpose of the collaboration is to perform the cooperative research, to train young researchers by means of modern instruments and observational data. However, most of the collaborations are done with Chinese astronomers and institutes. North Korean astronomers also visited Italy (the International Center for Relativistic Astrophysics at Pescara, and the International Center for Theoretical Physics at Trieste), and the University of Leiden in the Netherlands.

Jong (2015) also revealed the long-term vision and strategic plan for the development of astronomy in North Korea. In recognition of the appreciation for astronomy as a tool for technical advances and intellectual capacity building, PAO aims to conduct cutting-edge professional research in astronomy and to educate all people in North Korea to a sufficient level to secure the benefit of astronomy. The goals of PAO were to raise the research level of several selected key areas of astronomy to an international standard and to build an infrastructure that can sustain a full-scale program for school education and public outreach. The areas of astronomy selected for the special focus are stellar astrophysics, solar physics, cosmology, and applied astronomy. PAO also emphasizes the importance of the international collaboration such as postdoctoral and doctoral works in foreign advanced institutions and taking active part in the existing and newly launched programs of the IAU. Unfortunately, there has not been much progress in the human exchanges, probably because of the increased tension with the Western countries ignited by the nuclear bomb tests since then.

Although progress appears to be slow in North Korea, the number of papers in the international journals seems to be increasing. Interestingly, two North Korean astronomers, by analyzing in the Korean sunspot records in history books, recently published another paper claiming a 200-year solar periodicity (Kim & Kim, 2020), a result similar to that of Yang et al. (2019) from South Korea. There was no mention about the works done by South Korean scientists on the analysis of ancient sunspot records in their paper maybe because of the lack ~~of~~ the information. Clearly there could be many areas for astronomers in both Koreas to collaborate.

In order to understand the scientific activities in North Korea, South Korean scientists founded the Science and Technology Forum for Unification within the National Research Council of Science and Technology, which is a managing body of the Government Research Institutes. This forum runs several programs in order to promote the understanding of scientific activities North Korea. Also the Korea Institute of Science and Technology Information (KISTI) operates North Korea Technology Network (NK Tech) which collects various information on the science and technology including documents data base, publications, list of research papers, names of researchers, and research facilities and provide this information to interested parties in South Korea.

10.7 Concluding Remarks

In spite of the long history of astronomy in Korea, modern astronomy was developed relatively recently. The Japanese occupation, from 1910 until the end of WWII, the Korean war and the subsequent split of the country made the situation very difficult. There have been a number of pioneers in professional astronomy starting from the late 1920s, but high level education became possible only after the Korean War which ended in 1953. In South Korea, universities began opening of astronomy-related departments from 1958. The lack of well-trained professors continued for a few decades, and consequently many students went abroad, mostly to the USA, for graduate studies and the research capability began to grow rapidly from the late 1970s. The Korean Astronomical Society was established by the early generation of astronomers, helped by scientists in other fields such as physics, geology, and meteorology in 1965, giving rise to the mechanism for scientific exchanges. The KAS worked very hard to establish the National Astronomical Observatory, which was inaugurated in 1974 together with the installation of the first research-grade telescope of 61-cm diameter. Since then, the NAO aggressively invested on the observational facilities while universities supplied well-trained astronomers.

The number of founding members of the KAS was about 20 and grew to more than 500 active members today (2020). The Seoul National University started the first astronomy department in 1958 without a graduate program (the master's program started in 1966 and the first PhD students were admitted in 1979), and now eight universities run a PhD program.

Korea has been divided country for more than seventy years without much contact between the two Republics. North Korea founded the PAO, its national observatory, in 1957 while South Korea launched the NAO in 1974. Joining the IAU was thus first done by North Korea in 1961, twelve years earlier than South Korea. However, the development of observational infrastructure and manpower in South Korea became much faster because of the rapid expansion of the South Korean economy. South Korean astronomers are actively seeking opportunities of scientific collaboration with North Korea, but the political situation remains difficult and did not improve for several decades. The upcoming IAU General Assembly to be held in August 2022 in Busan, South Korea, should be a good place for astronomers in two Koreas to gather together. When KASI started the construction of the Korea VLBI Network, it was envisaged that an additional telescope could be placed in North Korea in order to increase the baseline. This plan is still valid, and can be realized when the inter-Korean relationship becomes better.

Acknowledgments The author is indebted to many senior astronomers for their stories about the early development of modern astronomy in Korea. The "Fifty years history of the Korean Astronomical Society" published by the Korean Astronomical Society in 2015, together with "History of KASI since 1974" by KASI were the basic guidelines of this article. Some details regarding the foundation of the National Astronomical Observatory were taken from the 'Memoirs of the forty years of Korean Astronomical Society', published by the KAS in 2005. I was fortunate enough to serve as a President of the KAS at the time of its 50th anniversary in 2015 and a president of KASI at the time of writing this article. Hong-Jin Yang and Insung Yim at KASI have searched for the historical facts about the South-North relationship during last several decades and provided these information for this article.

References

Braun, H., Christl, M., Rahmstorf, S., Ganopolski, A., Mangini, A., Kubatzki, C., Roth, K., & Kromer, B. (2005). Possible solar origin of the 1,470-year glacial climate cycle demonstrated in a coupled model. *Nature, 438*, 208–211.

Guinan, E. F., & Marschall, L. A. (2009). Teaching for astronomical development. In Karel A. van der Hucht (Ed.), *The International Astronomical Union reports on astronomy, transactions of the international astronomical union*, Vol. XXVIIA, p.436.

Hoffmann, S., & Vogt, N. (2020). Counterparts of Far Eastern Guest Stars: Novae, supernovae, or something else? *MNRAS, 496*, 4488.

Ishida, G. (1972). "韓國 國立 天文臺:ソウルの 11日間"("Korean National Observatory: 11 Days in Seoul"). *Astronomy Herald*, August issue, p. 203 (in Japanese).

Jensen Jr., J. (2016). *Earth Epochs: Cataclysms across the Holocene*. John Jensen. p. 276.

Jong, S. (2015). Strategic plan of development of astronomy in DPRK. In T. Montmerle (Ed.), *Highlights of astronomy*, vol. 16, XXVIII IAU General Assembly, August 2012.

Korea Astronomy and Space science Institute (KASI). (2014). *한국천문연구원 40년사(History of KASI Since 1974, in Korean)*. KASI.

Korean Astronomical Society (KAS). (2015). *한국천문학회 50년사(50-year History of the Korean Astronomical Society (in Korean)*. Seoul National University Press.

Kim, C.-J., & Kim, J.-S. (2020). Solar activity cycle of ~200 years from Mediaeval Korean records and reconstructions of cosmogenic radionuclides. *MNRAS, 492*, 384.

Koo, B. C., & Park, C. (2017). Supernova Remnant Cassiopeia A. In A. Alsabti & P. Murdin (Eds.), *Handbook of Supernovae*. Springer. https://doi.org/10.1007/978-3-319-21846-5_50

Lee, D. W. (1932). Motions in the atmospheres of eta aquilae. *Publications of the Observatory of the University of Michigan, 4*, 110.

Meyerhoff, J. (2013). *Mysterious megalithic structures*. Lulu Books, ISBN 978-1-304-65092-4.

Min, Y. K. (1993). 북한의 천문학 교육 및 연구현황 분석 연구(Astronomical Education and Research in North Korea). *Publications of The Korean Astronomical Society, 8*, 207 (in Korean).

Needham, J., Gwei-Djen, L., Combridge, J. H., & Major, J. S. (1986). *The Hall of Heavenly Records: Korean Astronomical Instruments and Clocks, 1380-1780*. Cambridge University Press.

Nha, I.-S. (2000). *한국천문학사(The History of Astronomy in Korea)*. Seoul National University Press (ISBN 9788952101792, in Korean).

Nha, I.-S., Nha, S. L., & Orchiston, W. (2017). The development of astronomy and the emergence of astrophysics in South Korea. In T. Nakamura & W. Orchiston (Eds.), *The emergence of astrophysics in Asia: Opening a new window on the universe*. Springer.

Park, C., Lee, Y. B., & Lee, Y. J. (2001). 청원 아득이 고인돌 유적에서 발굴된 별자리판 연구(A Stone Star Chart from a Delmen at Adeugi in Chongwon). *Journal of the Korean History of Science Society*, 23, 3 (in Korean).

Park, C. (2008). *Astronomy traditional Korean science*. Ewha Woman's University Press (ISBN-10: 8973007793).

Park, C., & La, D. (1994). Confirmation and historical consequence of astronomical records in Samguksagi. *The Korean Jouranl for the History of Science, 16*(2), 167–202.

Park, S. R. (2000). History of astronomy in Korea. In H. Selin & S. Xiaochun (Eds.), *Astronomy across cultures: The history of non-western astronomy* (p. 409). Springer.

Rufus, W. C. (1913). The Celestial Planisphere of King Yi Tae Jo. *Transactions of Korean Branches of the Royal Asiatic Society, 4*, 3.

Rufus, W. C. (1915). Korea's Cherished Astronomical Chart. *Popular Astronomy, 23*(4).

Rufus, W. C. (1936). Korean astronomy. *Transactions of the Korea Branch, The Royal Asiatic Society, 26*, 1–23.

Stephenson, F. R. (2017). Highlights of science of Joseon Dynasty as viewed by a Western scholar. In Il-Seong Nha, Wayne Orchiston and F. Richard Stephenson (Eds.), *The History of World Calendars and Calendar-making, Proceedings of the international conference in commemoration of the birth of Kim Dam* (pp. 13–20). Yonsei University Press.

Stephenson, F. R. (2004). Korean observations of the supernova of A.D. 1604. In W. Orchiston, F. R. Stephenson, S. Débarat, & I.-S. Nha (Eds.), *Astronomical instruments and archives from the Asia-Pacific region* (pp. 15–20). Yonsei University Press.

Stephenson, F. R., & Green, D. A. (2002). *Historical Supernovae and their Remnants*. Oxford University Press.

Stephenson, F. R., & Yau, K. K. C. (1987). Four Korean guest stars observed in A.D. 1592. *Quarterly Journal of the Royal Astronomical Society, 28*, 431–444.

Shara, M. M., Iłkiewicz, K., Mikołajewska, J., Pagnotta, A., Bode, M. F., Crause, L. A., Drozd, K., Faherty, J., Fuentes-Morales, I., Grindlay, J. E., Moffat, A. F. J., Pretorius, M. L., Schmidtobreick, L., Stephenson, F. R., Tappert, C., & Zurek, D. (2017). Proper-motion age dating of the progeny of Nova Scorpii AD 1437. *Nature, 548*, 558.

Veretenenko, S. V., & Ogurtsov, M. G. (2018). 60-Year cycle in the earth's climate and dynamics of correlation links between solar activity and circulation of the lower atmosphere. *Geomagnetism and Aeronomy, 58*, 973

Vogt, N., Hoffmann, S. M., & Tappert, C. (2019). On the possibilities of classical nova identifications among historical Far Eastern guest star observations. *Astronomische Nachrichten, 340*, 752.

Wagner, G., Beer, J., Masarik, J., Muscheler, R., Kubik, P. W., Mende, W., Laj, C., Raisbeck, G. M., & Yiou, F., (2001). Presence of the Solar de Vries Cycle (~205 years) during the Last Ice Age, *Geophysical Research Letters, 28*, 303.

Wada, Y. (1911). Scientific memoirs of the Korean meteorological observatory. *Nature, 85*, 341.

Yang, H.-J. (2015). Astronomical Aspects of the Korean Dolmens. In C. L. N. Ruggles (Ed.), *Handbook of archeoastronomy and ethnoastronomy* (p. 2149). Springer.

Yang, H.-J., Park, C.-G., Kim, R.-S., Cho, K.-S., & Jeon, J. (2019). Solar activity and climate change during the last millennium recorded in Korean chronicles. *Journal of Atmospheric and Solar-Terrestrial Physics, 186*, 139.

Yang, H.-J., Park, C., & Park, M.-G. (1998). 고려시대 흑점기록과 오로라 기록에 보이는 태양활동 주기(Evidence for the Solar Cycle in the Sunspot and Aurora Records of Goryeo Dynasty). *Publications of The Korean Astronomical Society, 13*, 181 (in Korean).

Yang, H.-J., Yim, I.-S., Lee, Y. B., & Choi, H. K. (2018). 남북 천문분야 협력방안 기초연구 II: 남북한 천문 역서 비교(A Study of Cooperation between South and North Korea in the Field of Astronomy II: Astronomical Almanac). *Publications of The Korean Astronomical Society, 33*, 1 (in Korean).

Yu, K.-L. (1976). 천문학회 창립 회고기(Memoir of the Founding of the Korean Astronomical Society). *Bulletin Korean Astronomical Society, 1*, 2 (in Korean).

Hyung Mok Lee was born in Ansan, Korea in 1956, attended Seoul National University and holds PhD in Astrophysics from Princeton University. He worked at Canadian Institute for Theoretical Astrophysics from 1986 to 1989 and returned to Korea to take a junior faculty position at Pusan National University where hosted the fifth Asian Pacific Regional Meeting of the IAU which was the first IAU sponsored meeting held in Korea. He has served as a Chief Editor of the Journal of the Korean Astronomical Society between 2004 and 2009 and President of the Korean Astronomical Society in 2014–2015. He was appointed as the President of the Korea Astronomy and Space Science Institute (KASI) from January 2018 to January 2021. He has pioneered the infrared space astronomy in Korea by promoting a collaboration on AKARI, a Japanese infrared space telescope which was launched in February 2006. He is also a leader of the Korean Gravitational Wave Group, a member of the LIGO Scientific Collaboration.

Hyung Mok Lee has been associated with the Department of Physics and Astronomy at Seoul National University since 1998. He is a member of the International Astronomical Union and Korean Academy of Science and Technology. He received numerous awards including John H. Polanyi Prize from Ontario Government of Canada, and Alexander von Humboldt Award from Humboldt Foundation. He is currently a member of the editorial boards of Publication of the Astronomical Society of Japan and Research in Astronomy and Astrophysics published by the Chinese Academy of Science.

Chapter 11
Germany's Difficulties in Becoming a Member of the IAU

Roland Wielen

Abstract After the foundation of the International Astronomical Union (IAU), it took nearly thirty-two years until Germany became a national member of the IAU. This unfortunate delay was due to both international politics and discussions within Germany (e.g. about the future role of the Astronomische Gesellschaft). The first Statutes of the IAU of 1919 excluded Germany, Austria and other countries from membership for political reasons. When the IAU Statutes were revised in 1931, Germany's membership was possible, but could not be realized because of financial problems. In the mean time, German astronomers (as individuals) were often able to attend General Assemblies of the IAU as invited participants and to work in IAU Commissions as members. After World War II, Germany became a member of the IAU in 1951, Austria in 1955. In 1962, East Germany was accepted as an additional member. The reunification of Germany in 1990 clarified this situation. The member of the IAU is now one unified Germany.

Keywords International Astronomical Union · Astronomische Gesellschaft · Membership · Germany · Austria · Hungary · Turkey · Bulgaria · Max Wolf · August Kopff · Einstein

---

This chapter is republished by permission from the IAU from "Roland Wielen, "Germany's difficulties in becoming a member of the International Astronomical Union". In: Christiaan Sterken, John Hearnshaw, and David Valls-Gabaud (Eds.), Under One Sky: IAU Centenary Symposium. Proceedings of the 349th Symposium of the International Astronomical Union (2018). Cambridge University Press, 2019, p. 205–213)".

R. Wielen (✉)
Astronomisches Rechen-Institut, Zentrum für Astronomie, Universität Heidelberg, Heidelberg, Germany
e-mail: wielen@ari.uni-heidelberg.de

© The Author(s), under exclusive license to Springer Nature Switzerland AG 2022

T. Montmerle, D. Fauque (eds.), *Astronomers as Diplomats*, Historical & Cultural Astronomy, https://doi.org/10.1007/978-3-030-98625-4_11

11.1 Problems Caused by the First Statutes of the IAU of 1919

The IAU was founded in July 1919. The parent organization of the IAU and of other scientific unions was the International Research Council (IRC).

The first Statutes of the IAU from July 26, 1919 (IAU Transactions, Vol. I (1922), p. 219) determine in Section I.2 that the admission of countries to the IAU shall be subject to the Regulations of the IRC. According to these IRC Regulations, there were three categories of countries. These rules had to be followed by the IAU.

(1) The "Allied Powers" of World War I as founding members:
England, France, USA, and other countries.
(2) "Neutral Countries" as possible additional members:
Holland, Sweden, Switzerland, and others.
(3) Excluded from membership were the "Central Powers" of World War I:
Germany, Austria, Hungary, Bulgaria, and Turkey.

The exclusion of the Central Powers was based purely on political motivations and was in obvious conflict with the aim of a really international scientific cooperation. This procedure was strongly criticized by the neutral countries and also by many scientists from the Allied Powers. Some of these discussions are described by Blaauw (1994) in detail in his book on the early history of the IAU.

11.2 Problems Within Germany

Even if the first Statutes of the IAU had allowed the membership of Germany, such an offer would very probably not have been accepted immediately by Germany. There were strong hesitations within Germany against an IAU membership for various reasons:

(a) The "Astronomische Gesellschaft (AG)" existed already as an internationally orientated society (see Sect. 11.3). The future role of the much older AG in the presence of the new IAU was unclear and probably disappointing for Germany.
(b) The German-speaking scientists demanded from the IAU (and other Unions) that the German language should be accepted by the IAU on the same level as English and French, as it had been the case at international scientific meetings before World War I.
(c) The membership of Germany in the IRC (as a condition for the IAU membership) was refused by Germany. The IRC was considered as a mainly political body, with a chauvinistic attitude, disregarding the interests of Germany.
(d) The financial situation in Germany after World War I was so difficult that Germany was essentially unable to pay an IAU membership fee.

11.3 The Astronomische Gesellschaft (AG)

The AG was founded in 1863 as an international astronomical society. A survey of the history of the AG is given by Schmeidler (1988).

While the AG certainly had its center in Germany (Dick 2000), there were many foreign members and officers. We give here a few examples:

Otto Struve ("the Elder") from Pulkovo (Russia) was President ("Vorsitzender") of the AG from 1867 until 1878. He had been a member of the Board ("Vorstandsmitglied") of the AG already since the foundation of the AG in 1863.

Elis Strömgren (Denmark) was President of the AG from 1921 until 1930, and a Board member since 1914.

Especially, remarkable in our context are the various functions of Arthur Stanley Eddington (England) in the IAU as well as in the AG. He was Vice President of the IAU from 1925 until 1928, Board member of the AG from 1928 until 1930, Vice President ("Stellvertretender Vorsitzender") of the AG for three terms from 1930 until 1937, and finally President of the IAU from 1938 until his death in 1944.

Famous members of the AG from non-German speaking countries were for example:

F. Tisserand (France): member 1869–1896†, and sometimes Board member of the AG

S. Newcomb (USA): member 1873–1909†, and sometimes Board member

H. Poincare (France): member 1905–1912†

- ⋆ W. de Sitter (Holland): member 1909–1934†
- ⋆ B. Lindblad (Sweden): member 1922–1945, and sometimes Board member
- ⋆ O. Struve (USA): member 1923–1965†, and sometimes Board member

H. Shapley (USA): member 1925–1945, and sometimes Board member

- ⋆ Bengt Strömgren (Denmark): member 1926–1987†, and sometimes Board member

(⋆ indicates that he has acted also as President of the IAU).

Due to the largely international character of the AG, there was (initially) an obvious conflict between the intended roles of the IAU and the AG.

As a side note: Albert Einstein was a member of the AG since 1921, but he never became a member of the IAU, especially not of IAU Commission No. 1 on Relativity.

11.4 Steps of the IAU Toward a German Participation

In 1926, the political situation improved significantly. Germany joined the League of Nations (a forerunner of the UN) and became a permanent member of its Council. Both the IRC and the IAU were under some pressure to take into account this new development.

The IAU Executive Committee wrote in a letter of July 1926 to Elis Strömgren, President of the AG, that a possible application of Germany to join the IAU "will be very welcome to the Union" (Blaauw 1994, p. 84–86). However, the Statutes of the IAU required still the membership of a country in the IRC. Hence this offer was merely a friendly, but effectless gesture of the IAU toward Germany.

A more realistic way to include German astronomers into the work of the IAU was taken in 1928 by W. de Sitter (Holland), President of the IAU from 1925 to 1928. He invited individual astronomers from non-member countries (especially, from the Central Powers) to attend the IAU General Assembly at Leiden in July 1928 as guests. Fourteen astronomers from Germany, one from Austria, and two from Hungary followed this invitation.

In September 1931, in an Extraordinary Meeting at London, the Statutes of the IAU were finally revised in order to abolish the former restrictions on the membership in the IAU. Section II.3 of the Statutes now determines that a country not adhering to the International Council *(formerly IRC, then ICSU)* may adhere *(nevertheless)* to the IAU (IAU Transactions, Vol. IV (1933), p. 295). This revision of the IAU Statutes in 1931 paved the way for an IAU membership of Germany, at least in principle.

11.5 Steps of Germany Toward an IAU Membership

The revision of the IAU Statutes in 1931 caused a realistic discussion in Germany about the desirability of a membership in the IAU.

Obviously, the probable changes in the Statutes of the IAU were known in Germany in advance. On March 7, 1930, an ad hoc committee was established at a meeting in Babelsberg near Berlin. It was called the "Elferausschuss" because of its eleven members ("elf" in German). Chairman of the Elferausschuss was Max Wolf (see Fig. 11.1). Wolf was at the same time President of the AG (from 1930 to 1932). A detailed biography of Wolf is written by Freiesleben (1962).

The task of the Elferausschuss was to discuss the possibility of Germany's membership in the IAU and to negotiate with the IAU about the details of such a joining. A very good overview on the work of the Elferausschuss is given by Wolf in his application to the German Foreign Office in 1931, reproduced by Freiesleben (1962, p. 165–168). In summary, the Elferausschuss concluded that a German membership in the IAU was very desirable. For financing the IAU membership fee, the German astronomers would have to rely on a support by the government.

In 1932, the Elferausschuss was transformed into a more formal committee, the "Ausschuss der deutschen Astronomen," abbreviated as ADA. The subtitle of the ADA was already "National Committee of Astronomy" for the IAU. Some details on the ADA are given in Figs. 11.2 and 11.3 (based on Lüdtke (1933, 1937)).

Fig. 11.1 Max Wolf (1863–1932). Director of the Heidelberg Observatory. President of the AG 1930–1932. Chairman of the "Elferausschuss" 1931/1932 and of the ADA 1932

Ausschuß der deutschen Astronomen (National Committee of Astronomy, Internat. Astron. Union). Gegr. 1932 in Berlin-Babelsberg. — Mitglieder: Die Direktoren der Staatl. astron. Institute und zugewählte besonders verdiente Fachleute. — Zahl der Mitglieder: 23. — *Vorsitzender:* vacat. *Stellvertr. Vors.:* Prof. Dr. Richard Schorr [Sternwarte Bergedorf].

Fig. 11.2 Information on the "Ausschuss der deutschen Astronomen" (ADA) from 1933. Reprinted from the compendium (Lüdtke 1933, p. 181, Ed. G. Lüdtke).
English translation: Committee of the German astronomers (National Committee of Astronomy, IAU). Founded in 1932 at Berlin-Babelsberg. Members: The directors of the state-owned astronomical institutes and co-opted, especially deserved experts. Number of members: 23. Chairman: vacant. Deputy Chairman: Prof. Dr. Richard Schorr (Observatory Hamburg-Bergedorf)

Ausschuß der deutschen Astronomen (1932). Mitglieder: Die Direktoren der Staatl. astron. Institute und zugewählte besonders verdiente Fachleute. Zahl der Mitglieder: 18. — *Vors.:* Prof. Dr. Richard Schorr [Sternwarte Bergedorf].

Fig. 11.3 Information on the "Ausschuss der deutschen Astronomen" (ADA) from 1937. Reprinted from the compendium (Lüdtke 1937, p. 130, Ed. G. Lüdtke).
English translation: Committee of the German astronomers (1932). Members: The directors of the state-owned astronomical institutes and co-opted, specially deserved experts. Number of members: 18. Chairman: Prof. Dr. Richard Schorr (Observatory Hamburg-Bergedorf)

11.6 Germany's Statement at the IAU General Assembly 1932 at Cambridge (USA)

Since Wolf became seriously ill in 1932 (he died in October 1932), August Kopff (see Fig. 11.4) was entrusted with explaining the German position during the Fourth IAU General Assembly, held at Cambridge, Massachusetts, USA, September 2–9, 1932. Sir Frank Dyson (Astronomer Royal, England) was then President of the IAU. The speech by Kopff at this General Assembly is published in full length

Fig. 11.4 August Kopff (1882–1960). Director of the Astronomisches Rechen-Institut at Berlin and since 1945 at Heidelberg. German delegate at the IAU General Assembly 1932 at Cambridge, Massachusetts, USA. For a detailed biography of Kopff, see Wielen & Wielen (2013)

and in German in the IAU Transactions, Vol. IV (1933), p. 216 (Fig. 11.5). Kopff reported that Germany is very much interested in becoming a member of the IAU in the future, but at present, this is impossible for financial reasons (the required 3600 Gold Francs for Germany's IAU membership fee were not available).

We find it very remarkable that Kopff's speech was delivered in German and that the text was reprinted in German in the IAU Transactions. Was this a signal that German was at least tolerated as a working language within the IAU ? Today, this question is, of course, rather obsolete.

11.7 The Period from 1933 to 1939

During the period from 1933 to 1939, there was no progress in Germany's desired IAU membership. At least, no relevant documents were available to us.

The retardation was most probably caused by a continuing lack of finance. We have no indications that the delay in Germany's IAU membership was connected to the NS government, since other sciences were allowed to apply for membership in their corresponding Unions, e.g., Geodesy and Geophysics in 1937 for the IUGG (Torge 2007). The cooperation of individual astronomers with the IAU was also not hampered by the government or by other organizations.

The IAU was always hopeful regarding to Germany's membership. F. Schlesinger (USA), IAU President from 1932 to 1935, said at the Paris General Assembly in 1935: "I sincerely trust that they *(i.e. Germany)* will have joined us before our next meeting *(in 1938)*" (IAU Transactions, Vol. V (1936), p. 276). And E. Esclangon (France), IAU President from 1935 to 1938, still expressed at the Stockholm General Assembly in 1938 "the hope that Germany would soon join the Union" (IAU Transactions, Vol. VI (1939), p. 318).

During World War II (1939–1945), the question of Germany's IAU membership was not acute for obvious reasons.

I am going to invite Dr Kopff, whom we are delighted to welcome as a representative of German astronomy, to address the Assembly.

DR KOPFF:

HERR PRÄSIDENT, MEINE DAMEN UND HERREN,

Wie Sie soeben gehört haben, ist vor einiger Zeit ein deutsches nationales Komitee gebildet worden, der *Ausschuss der deutschen Astronomen*, der sich die Angabe gestellt hat, den Eintritt der deutschen Astronomie in die Internationale Astronomische Union durchzuführen. Der Eintritt ist beschlossen worden; aus finanziellen Gründen allein hat er bisher nicht erfolgen können. Doch werden sich auch diese Schwierigkeiten hoffentlich bald überwinden lassen.

Ich habe den Auftrag, Ihnen die Grüsse des Ausschusses der deutschen Astronomen und aller deutschen Astronomen zu überbringen, insbesondere auch die Grüsse unseres verehrten Vorsitzenden Prof. Max Wolf. Er hat bisher die Verhandlungen über unseren Eintritt in die Union geführt, und zu unserem grossen Schmerz hat eine schwere Krankheit ihn verhindert, selbst hierher zu kommen.

Sie haben bereits zahlreiche deutsche Astronomen zu Mitgliedern in die verschiedenen Kommissionen gewählt, und Sie haben es uns dadurch ermöglicht, noch enger als bisher mit Ihnen zusammenzuarbeiten. Für diese Aufnahme in die Kommissionen möchte ich Ihnen besonders danken. Herzlich danken auch möchte ich Ihnen, Herr Präsident, für die freundlichen Worte, die Sie uns und vor allem unserem verehrten Prof. Max Wolf gewidmet haben.

Zu unserem lebhaften Bedauern ist es dieses Mal für uns nicht möglich gewesen, in grösserer Zahl in den verschiedenen Kommissionen anwesend zu sein. Aber als der einzige Vertreter aus Deutschland darf ich Ihnen wenigstens im Namen der deutschen Astronomen einen guten Erfolg für diesen Kongress wünschen.

Fig. 11.5 Text of the speech delivered by August Kopff (Berlin) as the German delegate at the IAU General Assembly 1932 at Cambridge, Massachusetts, USA. Reprinted from IAU Transactions, Vol. IV (1933), p. 216. English translation: Mister President, Ladies and Gentlemen, As you have just heard, some time ago a German national committee has been established, the *Ausschuss der deutschen Astronomen (Council of the German astronomers)*, which aims at carrying out the entrance of the German astronomy into the International Astronomical Union. The entrance has been resolved; for financial reasons alone, the entrance could not take place so far. But hopefully, it will be possible to overcome those difficulties in the near future. I have been entrusted with bringing you the greetings of the Council of the German astronomers and of all German astronomers and especially the greetings from our distinguished chairman Professor Max Wolf. He has conducted the negotiations on our entrance into the Union up to now, but, to our great regret, a grave sickness has prevented him from coming here himself. You have already elected many German astronomers to members of the various commissions. Thereby, you have made it possible for us to cooperate with you even stronger than before. For this admittance, I would particularly like to thank you. I would also like to cordially thank you, Mister President, for your friendly words which you have addressed to us and to our venerable Professor Max Wolf. To our great regret, it was this time impossible for us to attend the various commission meetings in a larger number. As the only representative from Germany, I would like to wish you at least in the name of the German astronomers a good success for this congress

11.8 Solution After World War II

After the end of World War II in 1945, it took some years to establish a German administration. The Federal Republic of Germany was founded in 1949. This was a basic precondition for discussing and especially for financing a membership of Germany in the IAU. The monetary reform of 1948 (from Reichsmark to Deutsche Mark) had already made the payment of an IAU membership fee in hard currency at least possible.

The IAU General Assembly at Zürich in August 1948 was attended by 8 astronomers from Germany as invited guests. Their travel expenses were generously covered by Switzerland. In his speech at the General Assembly, Harold Spencer Jones (Astronomer Royal, England), IAU President from 1944 to 1948, said: "We look forward to the renewal of collaboration with German and Austrian colleagues" (IAU Transactions, Vol. VII (1950), p. 16).

In 1951, Germany became finally a national member of the IAU. Only a few relevant documents on this procedure are available. In the "Mitteilungen der Astronomischen Gesellschaft," 1950, p. 3, it is reported from the AG meeting in September 1950 that the entrance of Germany into the IAU is urgently desired ("dringend erwünscht" in German). It was expected that the corresponding negotiations between the President of the AG (F. Becker) and the President of the IAU (B. Lindblad) will achieve the desired effect rather soon. According to Blaauw (1994, p. 162), the IAU General Secretary, Bengt Strömgren, informed the AG in a letter of 17 March 1951 that the IAU recognized the AG as the adhering body for Germany. This letter was acknowledged by the President of the AG in his letter of 24 April 1951. A voting of an IAU General Assembly on Germany's membership was not necessary.

Although the negotiations were formally carried out by the President of the AG, Friedrich Becker (Bonn), we think that August Kopff (Heidelberg) was the real driving force for the entrance of Germany into the IAU. Kopff was Vice President of the AG and internationally very renowned.

For the IAU, it was certainly an important signal from the large American astronomical community that Kopff was elected as an "Honorary Member" of the American Astronomical Society (AAS) in June 1949 during the the eighty-first meeting of the AAS (see Gingrich 1949). This honor was especially remarkable, since it was given to a German astronomer only four years after the end of World War II. It certainly indicated that German astronomers should now be fully accepted internationally.

Kopff had the important advantage that he had very good personal relations to three leading persons of the IAU, namely to Harold Spencer Jones (Astronomer Royal, England), IAU President from 1944 to 1948, to Bertil Lindblad (Sweden), IAU President from 1948 to 1952, and to Bengt Strömgren (Denmark), and then IAU General Secretary from 1948 to 1952. The reasons for Kopff's good relations to Harold Spencer Jones and Bertil Lindblad are described in detail in some of our books (Wielen & Wielen 2012, 2013, 2016). Bengt Strömgren was a Board member of the AG from 1939 to 1945. During this period, Kopff was President of the AG.

It is an irony of history that the AG became in 1951 the adhering body for Germany in the IAU. About thirty years ago, the IAU and the AG had been competitors as international astronomical societies (see our Sect. 11.3). The IAU and the AG seem to have something of close relatives. Taking up a comparison made by de Sitter at the IAU General Assembly at Leiden in 1928, who considered the diminishing relation between parents (IRC, ICSU) and children (IAU), Elis Strömgren (President of the AG) called the AG an "old aunt" (alte Tante) of the IAU. The old aunt (the AG) wished the strong child (the IAU) good luck on its way (Vierteljahrsschrift der Astronomischen Gesellschaft, 63. Jahrgang (1928), p.254–255).

The formerly persistent problem of financing the IAU membership fee for Germany could also be solved satisfactorily. The annual costs were (and are) covered by the ministry of the German Federal Government responsible for such international relations.

The other countries from the former Central Powers became also members of the IAU after World War II: Hungary in 1947, Austria in 1955, Bulgaria in 1957, and Turkey in 1967. The slight delay in the IAU membership of Austria (relative to Germany's entrance) was probably caused by financial problems.

11.9 Problem During the Cold War Period

Since 1951, when Germany had become a member of the IAU, the astronomers from West Germany and East Germany were listed for many years commonly under the heading "Germany," "Allemagne," or "Deutschland." This was, for example, the case at the IAU General Assemblies at Rome 1952, at Dublin 1955, at Moscow 1958, and at Berkeley 1961.

However, the Cold War brought a new problem for Germany's IAU membership. In 1962, East Germany (GDR) was accepted as a new member by the IAU. Suddenly, the IAU had now "two Germanies" among its members, instead of initially none or later one ! East Germany created its own adhering organization. For this reason, the AG, which had still members from East Germany, had to be replaced by the "Rat Westdeutscher Sternwarten" (RWS), Council of West German Observatories, as the adhering organization of the Federal Republic of Germany for the IAU.

11.10 Final Solution Since 1990

In October 1990, the reunification of Germany (Deutsche Wiedervereinigung) took place. East Germany became a part of the Federal Republic of Germany, and thereby

East Germany disappeared as a member of the IAU. Since 1990, there is only one unified Germany member of the IAU. We expect that this is now the final solution for the long-standing problem of Germany's membership in the IAU !

Due to the disappearance of East Germany, the IAU lost the corresponding membership fee. Although the loss was rather small, many countries claimed that the future fee of the unified Germany should be the sum of the previous fees of West Germany and East Germany. However, the unified Germany remained in the old membership category (VII, 20 units) of West Germany, since this category was still adequate, even if based on the new, somewhat increased number of adhering astronomers (only the USA was in the then highest category VIII with 30 units). In any case, the German ministry financing the fee had made it very clear that it could not accept a significantly higher contribution of Germany than that which West Germany had paid before, because of all the other expenses caused by the reunification. The present author was the National Representative of the reunified Germany at the IAU General Assembly 1991 in Buenos Aires. When he communicated the point of view of Germany on its fee, this German position was criticized by some of his colleagues. They regretted that the basic IAU unit of contribution had to be increased (albeit very slightly, by about 4/1000 only) because of the loss of East Germany's fee (see IAU Transactions, Vol. XXI B (1991), p. 113). Due to such a rise, all the other countries had then to pay (at least indirectly) for Germany's reunification !

The adhering organization of the reunified Germany for the IAU is the "Rat Deutscher Sternwarten" (RDS), Council of German Observatories. In 2012, the RDS became an organ of the AG. Usually, the President of the AG is now simultaneously also the Chairman of the RDS.

11.11 Conclusion

We conclude this chapter with a remarkable statement from Otto Struve (USA). He was President of the IAU from 1952 until 1955. In his speech (IAU Transactions, Vol. VIII (1954), p. 25), as President Elect at the end of the IAU General Assembly 1952 in Rome, he said:

"... there was the admission of Germany. This action represented the healing of a wound which had threatened the very life of the Union from its beginning."

Addendum

Joachim Wambsganss, President of the Astronomische Gesellschaft (AG) and Chairman of the Rat Deutscher Sternwarten (RDS), gave at the IAU Centenary Day in Paris, October 4, 2019, a talk on *Germany joining the IAU: A complicated*

Story. He proposed to publish here a reprint of the paper by Roland Wielen on *Germany's difficulties in becoming a member of the International Astronomical Union*, because Wambsganss' talk was largely based on Wielen's presentation at IAU Symposium No. 349. The paper by Roland Wielen was published in *Under One Sky: IAU Centenary Symposium. Proceedings of the 349th Symposium of the International Astronomical Union, held in Vienna, Austria, August 28–31, 2018. Edited by Christiaan Sterken, John Hearnshaw, and David Valls-Gabaud. Cambridge University Press, 2019, p. 205–213*. The paper presented above is essentially a reprint of this chapter, but with a few changes, mainly in the style of the references to the IAU Transactions.

References

Blaauw, A. (1994). *History of the IAU: The Birth and First Half-Century of the International Astronomical Union* (296 p). Kluwer Academic Publisher: London.

Dick, W. R. (2000). *Acta Historica Astronomiae, 10*, 148.

Freiesleben, H. C. (1962). *Max Wolf: Der Bahnbrecher der Himmelsphotographie. 1863–1932* (241 p). Wissenschaftliche Verlagsgesellschaft: Stuttgart.

Gingrich, C. H. (1949). *Popular Astronomy, 57*, 307.

Lüdtke, G. (Ed.). (1933). *Minerva. Jahrbuch der gelehrten Welt*, 31. Jahrgang, 1933, 1. Abteilung, 1. Band (A-L), Forschungsinstitute, ... Walter de Gruyter: Berlin/Leipzig, p. 181.

Lüdtke, G. (Ed.). (1937). *Minerva. Jahrbuch der gelehrten Welt, 32*. Jahrgang, 1937, Abteilung Forschungsinstitute, ... Walter de Gruyter: Berlin/Leipzig, p. 130.

Schmeidler, F. (1988). *Die Geschichte der Astronomischen Gesellschaft*. Jubiläumsband 125 Jahre Astronomische Gesellschaft, Hamburg, 69 p.

Torge, W. (2007). *Die Geschichte der Geodäsie in Deutschland* (p. 295). Walter de Gruyter: Berlin / New York.

Wielen, R., & Wielen, U. (2012). *Von Berlin über Sermuth nach Heidelberg. Das Schicksal des Astronomischen Rechen-Instituts in der Zeit von 1924 bis 1954 anhand von Schriftstücken aus dem Archiv des Instituts*. Published in electronic form on the Open Access Platform HeiDOK of the University of Heidelberg (Universitätsbibliothek), 400 p. (with an abstract in English). http://www.ub.uni-heidelberg.de/archiv/14604. http://nbn-resolving.de/urn:nbn:de:bsz:16-heidok-146043. https://doi.org/10.11588/heidok.00014604.

Wielen, R., & Wielen, U. (2013). *August Kopff, die Relativitätstheorie, und zwei Briefe Albert Einsteins an Kopff im Archiv des Astronomischen Rechen-Instituts*. Published in electronic form on the Open Access Platform HeiDOK of the University of Heidelberg (Universitätsbibliothek), 253 p. (with an abstract in English). http://www.ub.uni-heidelberg.de/archiv/15653. http://nbn-resolving.de/urn:nbn:de:bsz:16-heidok-156534. https://doi.org/10.11588/heidok.00015653.

Wielen, R., & Wielen, U. (2016). *Astronomische Ephemeriden, Navigation und Krieg: Die erstaunliche Zusammenarbeit der Ephemeriden-Institute von Deutschland, England, Frankreich und den USA im Zweiten Weltkrieg nach Dokumenten im Archiv des Astronomischen Rechen-Instituts*. Published in electronic form on the Open Access Platform HeiDOK of the University of Heidelberg (Universitätsbibliothek), 253 p. (with an abstract in English). http://www.ub.uni-heidelberg.de/archiv/22021. http://nbn-resolving.de/urn:nbn:de:bsz:16-heidok-220213. https://doi.org/10.11588/heidok.00022021.

Roland Wielen was born in Berlin in 1938. He received his diploma in physics in 1962 at the Freie Universität Berlin. In 1963, he became a staff member of the Astronomisches Rechen-Institut at Heidelberg. 1966 PhD in Astronomy, 1969 Lecturer and 1974 Adjunct Professor at the Universität Heidelberg. He worked in 1972 also at the Observatoire de Nice and in 1974/75 at the Hamburger Sternwarte. In 1978, he was appointed full professor of astrophysics at the Technische Universität Berlin, and in 1985 full professor of theoretical astronomy at the Universität Heidelberg and director of the Astronomisches Rechen-Institut. His main fields of research are stellar dynamics and astrometry. From 1982 to 1985, he was president of IAU Commission 33 on Galactic Structure and Dynamics. He acted in 1991 as National Representative of Germany at the XXIst IAU General Assembly in Buenos Aires, the first IAU General Assembly after the reunification of Germany in 1990. Since 2004 he is a professor emeritus. Together with his wife Ute Wielen, he has written a series of books on various aspects of the history of astronomy.

Part III
During and After the Cold War

Chapter 12
Building International Bridges with Astronomy

George Miley

Abstract During the past fifty years, astronomy and the cosmic perspective have been important in building bridges between countries and peoples. This article tells some "science diplomacy" stories with which the author was involved, or witnessed from close by. These include:

- The development of long-baseline radio interferometry between the West and the USSR in the mid-sixties at the height of the Cold War,
- A visit to astronomical institutions in the People's Republic of China in 1973 during the final stages of the Cultural Revolution,
- Negotiating with the Democratic People's Republic of Korea to re-join the International Astronomical Union in 2012,
- Setting up the IAU Office of Astronomy for Development in Cape Town, South Africa, in 2013 and the "Universe Awareness" program for very young children in 2004.

12.1 Introduction

It is a pleasure to contribute to the proceedings of "Astronomers as Diplomats: When the IAU builds bridges between nations." I am writing from two very different vantage points. First, during the last half century, I have participated in and/or witnessed several informal activities in which astronomers have helped build bridges between countries that were not political friends and were relatively isolated by the international community. Secondly, as the IAU Vice President responsible for Education and Development between 2006 and 2012, I was involved in efforts to bring astronomers from the Democratic People's Republic of Korea, a previously neglected part of our "pale blue dot," back into the IAU.

G. Miley (✉)
Sterrewacht, Leiden University, Leiden, The Netherlands
e-mail: miley@strw.leidenuniv.nl

© The Author(s), under exclusive license to Springer Nature Switzerland AG 2022
T. Montmerle, D. Fauque (eds.), *Astronomers as Diplomats*, Historical & Cultural Astronomy, https://doi.org/10.1007/978-3-030-98625-4_12

12.2 Radio Astronomers Bridging the "Iron Curtain"

In 1963, when I began as a young Irish postgraduate student at Jodrell Bank, astronomy was beginning to break out of its optical "prison." Two distinct astronomical cultures were emerging. On the one hand, there were optical astronomers, who worked with small telescopes and photographic plates and usually had little involvement in or interest with advanced technologies and engineering. On the other hand, there was radio astronomy that was being pioneered by a breed of creative ex-radar engineers. Many were back from the second world war, used to working in groups and pushing electronics to the limit. These engineers turned their radar antennas to the sky in a version of "swords to ploughshares." There was little contact between the "died in the wool" often arrogant optical astronomers and the new breed of "upstart" radio astronomers. Although the IAU set up Commission 40 (Radio astronomy) in 1948, it was 1982 before a radio astronomer (Robert Hanbury Brown) became IAU President. There were exceptions. In Europe, Jan Oort realized the importance of radio astronomy, developed it as part of Dutch astronomy as a whole and thereby built a much-needed bridge between the two astronomical cultures. In the USA, there were also far-sighted optical astronomers who saw radio astronomy as a new window on the Universe, including Jesse Greenstein, Otto Struve, and Bart Bok and in the USSR brilliant theoretical astrophysicists, such as Iosif Shklovsky who became fascinated by the spectacular new objects that radio astronomers were finding.

Bernard Lovell was one of the new breeds of radio astronomers. Building on his experience of second world war radio and radar engineering, he set up the Jodrell Bank Experimental Station within the University of Manchester and during the 1950s, he realized his vision of a huge radio telescope, with the completion of a 250-foot diameter parabolic dish in the middle of the UK Cheshire countryside. The "Mark I" antenna, subsequently christened the "Lovell" Telescope was finished just in time to observe the first Sputniks and the Soviet moon probes.

Lovell used his large telescope (Fig. 12.1) to make observations of the thrilling Soviet space probes in a series of high-profile PR campaigns during the 1950s and 1960s. For the British, Jodrell Bank was an inexpensive chauvinistic symbol of UK participation in the space race. This was important because an enormous cost overrun almost forced the closure of the facility. It was saved partly because of Lovell's high visibility Sputnik "outreach" activities and partly because of the ability of the Mark 1 radio telescope to spy on the USSR.

Lovell turned to an "intelligence" offshoot of the UK government to help solve his financial problems—the Government Communications Headquarters (GCHQ) at Cheltenham. The staff of GCHQ were regular unpublicized visitors to Jodrell Bank. We postgraduate students noticed that these shadowy unannounced visitors frequently used the 250 ft telescope to observe the moon when it was visible simultaneously from both the UK and the USSR! Although it was obvious what they were doing, as the 1960s "peaceniks" we avoided thinking about possible

Fig. 12.1 The Lovell Telescope at Jodrell Bank Observatory, here in 2016. It was completed in 1957, in time to help Soviet scientists in following their space probes during the "Moon race" (Credit: Benjamin Shaw; Jodrell Bank Centre For Astrophysics, Creative CommonsAttribution-Share Alike 4.0 International license. Original in color.)

connections between these GCHQ visits and the funding of our postgraduate student salaries and grants!

But simultaneously with Jodrell Bank's involvement in the UK's defense program, Lovell was talking to Soviet scientists about astronomy and long baseline interferometry. These talks led to a unique effort in political bridge building that took place during the cold war. Lovell describes the saga in his book "Out of the Zenith" (Lovell, 1974). Supported and cajoled by Sir Bernard, Jodrell staff members including Henry Palmer, Barrie Rowson, Bryan Anderson, and a battalion of postgraduate students pioneered the development of long baseline radio interferometry (Palmer et al., 1967).

The goal was to measure the sizes and structures of the spectacular newly discovered radio sources. Most distant radio-loud quasars and radio galaxies were unresolved with short-baseline interferometers that linked radio telescopes together using cables. However, measuring the sizes of these extragalactic radio sources was crucial for answering one of the most intriguing astronomical questions of the time. The dependence of radio source sizes on redshift was a test that could help distinguish between relativistic and steady state cosmologies and their radically different philosophical ideas about the evolution of the Universe. The need to measure the sizes of compact radio sources led Jodrell radio astronomers, with their expertise in radio link interferometry to push to longer and longer baselines (e.g., Allen et al., 1962). This work culminated in 1967 with a baseline of 127 km between the Mark 1 Telescope at Jodrell and the radar dishes of the Royal Radar Establishment, Malvern, RRE (Palmer et al., 1967). However, even this baseline was not long enough to spatially resolve several small powerful radio sources. The solution involved breaching the iron curtain.

It is difficult to realize now how solid the iron curtain was in the 1960s and how rare it was to travel across it. For most westerners the Soviet Union was a land shrouded in mystery and we knew more about Mars than about China. However, Lovell went to the USSR in 1963 as a guest of the Soviet Academy and again in 1966. He visited the Soviet deep tracking station and had discussions with a group that included one of the world's foremost theoretical astrophysicists, Iosef Shklovsky, at the Lebedev Institute in Moscow. Shklovsky was extremely excited by the discovery of quasars and predicted that their sizes could be smaller than 0.8 seconds of arc. During discussions between Lovell and Soviet scientists and engineers, including Shklovsky, Matveyenko, Kardashev, and Sholomitsky, it was clear that to resolve many radio sources, baselines of thousands of kilometers would be needed. Radio link interferometers would need to be replaced by interferometers with recording systems and accurate clocks at each participating telescope (Very Long Baseline Interferometers—VLBI). Reduction of the interferometric data could then take place by replaying the telescope signals through a correlator at a later time. This was first proposed by Matveyenko et al. (1965).

Lovell and the Soviets also realized that a way to achieve the necessary long baselines would be to combine telescopes at Jodrell Bank with appropriate antennas in the east, including radio telescopes in the USSR. In 1966, Bryan Anderson started developing a tape-recording interferometer system. However, the Jodrell team was beaten to it by astronomers in the USA and Canada.

Although the Canadians (Broten et al., 1967) were first to obtain fringes from a tape-recording interferometer, it was the US groups from NRAO, MIT and Cornell Ken Kellermann, Barry Clark, Marshall Cohen, and Jim Moran who were most successful in pushing the VLBI technique to longer and longer baselines in an attempt to spatially resolve the enigmatic quasi-stellar radio sources. In the quest to do so, they braved cold war politics, developed collaborations with the Soviets, and succeeded in establishing VLBI baselines from the USA to the Crimea.

An important player in this bridge-building saga was Ken Kellermann (Fig. 12.2a), then a young American radio astronomer. In 1965, he had just finished a postdoctoral position in Australia and managed to travel home by way of the USSR. Bart Bok, Director of Mount Stromlo Observatory helped him obtain an invitation to visit from the USSR Academy of Sciences and Marcel Minnaert, then President of the IAU Commission 38 on the Exchange of Astronomers sent him a cheque from the IAU to help finance the trip. (No proposal needed in those days!) During Ken's visit he met some of the most prominent Soviet astronomers, including Vitkevitch, Shklovsky (Fig. 12.2b), Kardashev, Matveyenko, and Sholomitsky. In the Autumn of 1965, Marshall Cohen, Ken Kellermann and colleagues started developing the US tape-recording interferometer system, tested it internally within the USA and then between the USA and Sweden. In the Spring of 1968, Cohen and Kellermann wrote to Vitkevitch suggesting a VLBI experiment between the USA and the USSR. This begun the long-lasting collaboration in which radio astronomy truly bridged the "iron curtain" (e.g., Kellermann & Cohen, 1988) and which warm relations were established between radio astronomers in the USA and the Soviet Union,

Fig. 12.2 Ken Kellermann and Iosif Shklovsky, two of the most prominent bridge builders in the 1960s between the USA and the USSR. *Left.* K. Kellermann, here at IAU Symposium #58, Canberra, Australia, Aug.12-15, 1958, on "The Formation and Dynamics of Galaxies" (Credit: American Institute of Physics, John Irwin Slide Collection). *Right.* I. Shklovsky (date unknown; Credit: Sternberg Astronomical Institute, Moscow University)

including Shklovsky's appointment as 1968 Jansky Lecturer at the US National Radio Astronomy Observatory.

In 1968, after finishing my PhD in the long baseline group at Jodrell Bank I went to NRAO as a postdoc and witnessed Ken and the US VLBI group setting up the first interferometer experiment between the USA and the USSR. I remember traveling on the daily shuttle from the Green Bank telescopes at the heart of rural West Virginia to NRAO headquarters in Charlottesville and chatting to the local shuttle driver on the way. He was flabbergasted and extremely worried, because he was transporting a rubidium frequency standard in his shuttle and had just learned that this "atomic clock" was to be shipped to the communist Soviet Union. He was concerned that his shuttle had been implicated in this highly dangerous act! In this way the quest to understand spectacular quasars and their role in cosmology led astronomers to establish a physical bridge across the iron curtain and begin a warm West-East scientific collaboration at the height of the Cold War.

The role of astronomers and radio interferometry in stimulating globalism culminated in 2019 with the Event Horizon Telescope that spanned our planet to make the first image of a massive black hole—in the nucleus of the galaxy M87

(Virgo A). Also, the US–USSR collaboration of the sixties eventually led to the first space VLBI mission (RadioAstron).

I am convinced that in the future the quest to explore the Universe will lead to radio interferometers that span the Earth–Moon system. During the next century, driven by the need to image smaller and smaller structure in the Universe, baselines will eventually bridge the solar system. It is to be hoped that that such adventures of human exploration be carried out in a global context to provide an inspirational collaborative bridge across nations and cultures rather than in a spirit of competition.

12.3 Radio Astronomers Bridging the "Bamboo Curtain"

Radio astronomers also played a major role in breaking China's isolation from world astronomy during the 1960s and the 1970s. The story of China's withdrawal and subsequent re-joining of the IAU was well told in Chapter 9 of Adriaan Blaauw's history of the IAU's first half century (Blaauw, 1994). The withdrawal of the People's Republic of China followed a decision made by the IAU Executive Committee and General Assembly to recognize Taiwan's claim to independent membership, despite the persistent arguments by the People's Republic of China (PRC) that this was unlawful because Taiwan was an integral part of China and the adherence of Taiwan to the IAU was implicit in the membership of the PRC. Taiwan formally applied for IAU membership in 1958, although the level of astronomy in Taiwan was minimal.

It is my view that Taiwan's application was the result of a political policy led by the US State Department under its hard-line Secretary of State John Foster Dulles, to remove the PRC from all international organizations and to isolate this Communist country from the world. The US policy of isolation coupled with the application from Taiwan put the IAU in an extremely difficult position and forced it to take sides in one of the most contentious international political issues of the day—the definition of China. Unlike that of the West, the PRC's view was that Taiwan is an integral part of China. There was therefore no possibility that the PRC with its "one China" ideology could have accepted the admission of Taiwan to the IAU. In addition, the level of astronomy on the island was small. Taiwan's application to join the IAU was accepted in the IAU Executive Committee in September 1959 by 5 votes to 2. This unfortunate decision resulted in the withdrawal of China and its relative astronomical isolation for almost twenty years.

The isolation was at its peak 13 years later in the Autumn of 1973, when I visited China as a guest of their Academy after the IAU General Assembly in Sydney. I was not the first astronomer to visit China since their enforced isolation. The radio astronomer W. N. Christiansen of Australia ("Chris") made several trips to China during the 1960s and helped them build a 16-element radio array at the Miyun reservoir close to Beijing (Fig. 12.3). Chris also played an important role in arranging my invitation by the Academy under my host, the distinguished Chinese

Fig. 12.3 The radio array beside Miyun reservoir close to Beijing in 1973 during the late stages of the Cultural Revolution. (Photo George Miley)

Fig. 12.4 The Peking radio astronomy group, with its leader Professor Wang Shouguan in the center, in front of a 6-meter dish being built for the radio array. (Courtesy Sky & Telescope, Photo George Miley)

radio astronomer Professor Wang Shouguan[1] (Fig. 12.4). As a thirty-one-year-old radio astronomer, the three-week visit behind the mysterious bamboo curtain just as the Cultural Revolution was drawing to a conclusion was an amazing experience

[1] See Wang Shouguan, this volume, Chap. 8.

Fig. 12.5 In 1973 during the Cultural Revolution at Hsin-Lun (now written Xinglong), 150 km North of Beijing, a 60-cm Cassegrain reflector was being constructed, see as the prototype for a future 2-m telescope. (Photo George Miley)

that I described in an article for "Sky and Telescope" and several articles published by the NRAO Observer and the Irish Independent (e.g., Miley, 1974).

I visited astronomical facilities in Shanghai, Nanking, and Beijing. There I saw and related how the astronomical institutions were left relatively unscathed and were even expanded during the Cultural Revolution (Fig. 12.5). When I asked what was the basic motivation was for China in doing astronomical research, I received a threefold answer that could readily be given today by astronomers all over the world. First there was the direct importance of applications of astronomy such as timekeeping and solar work. Secondly, there was the awareness that advances in applied science must rest on a solid foundation of pure research. Thirdly, there was China's belief that the knowledge of man's place in the Universe is likely to weaken superstitious beliefs and encourage a scientific attitude. I returned from my three-week visit to China determined to do all I could to help break the isolation of Chinese astronomers and bring them into the world astronomical community.

The problems and the efforts to overcome them are well described by Xiaowei Liu in his chapter[2]. The IAU President, Adriaan Blaauw worked hard to bring the PRC back into the IAU. In his history of the first fifty years of the IAU, Adriaan describes a document submitted to him by "a colleague from Leiden" when he took up the IAU Presidency in September 1976. I was that colleague and my document was entitled "On the Legality of the adherence of 'The Republic of China (Taiwan)' to the IAU."

[2] This volume, Chap. 5. See also Liu (2020).

I argued that "the 1959 IAU decision to allow 'The Republic of China' (Taiwan) to adhere to the organisation contravened the statutes of the IAU and was therefore invalid" and that "even if the original decision to admit 'The Republic of China' had been legally permissible, the continued adherence of 'The Republic of China' as representing Taiwan violated the IAU statutes." Although I learned that Adriaan Blaauw sent my document to ESO lawyers for comment, I still believe that the 1959 decision of the IAU Executive Committee to admit Taiwan, despite the absence of appreciable astronomical activity on the island was unwise and of questionable legality. The application of Taiwan forced the IAU to take sides on a political issue (the definition and representation of China) and mingle in one of the most contentious political issues of the day—still a sensitive political matter. The Chinese view of the "China Crisis" is well described in Xiaowei Liu's chapter. It is unrealistic to suppose that an organization such as the IAU can make decisions about this type of issue completely independent of international and national politics, given that the organization and the research of the astronomers that it represents are dependent for funding by political decisions taken by national governments.

Despite considerable unrest about it among IAU members, the question of Chinese IAU membership lingered in the IAU for several more years. The impasse was finally settled when the Peoples Republic of China re-joined the IAU at the Patras General Assembly in 1982. I contend that this is an issue where diplomacy by astronomers outside the IAU did more to bring astronomers together than the official IAU organization.

12.4 Astronomers Bridging the 38th Parallel: Returning North Korea to the IAU Family

Although North Korea (the Democratic People's Republic of Korea) joined the IAU in 1961, its membership lapsed in 1995 because of non-payment of their dues for five years. In 2006, after becoming IAU Vice President with the portfolio for Education and Development, I regarded it as important to encourage North Korea to re-join the IAU. I had invaluable help in achieving this goal from several IAU officials and members, including Ed Guinan, who was then Chair of the IAU Teaching for Astronomy Development (TAD) Programme Group, Michèle Gerbaldi, then Chair of the International School of Young Astronomers (ISYA), and Karel van der Hucht, the IAU General Secretary.

In March 2007, at the ISYA held in Malaysia, preliminary discussions were held between Dr. Ri Jin Yong, the Science and Technology Counsellor at the DPRK Embassy in Kuala Lumpur and Michèle Gerbaldi and Ed Guinan. Dr. Yong arranged for two young DPRK astronomers to attend the ISYA meeting. Afterwards, Karel van der Hucht added the PAO to the list of libraries that receive complementary IAU publications.

More than a year after the Malaysian ISYA several DPRK astronomers from the Pyongyang Astronomical Observatory (PAO) attended the first IAU astronomical summer school to be held in the People's Republic of Mongolia from 21 to 26 July 2008. On the following day I held formal talks with Professors Jong Sok (Director, PAO), Kim Mun-Song, and Nam Sok-Chon at the Puma Hotel, Ulanbaatar. Besides the possibility of the DPRK re-joining the IAU, our discussions covered the situation of astronomy in the DPRK and future needs and plans of the PAO.

I learned that the PAO was founded on 7 March 1957; it had about 40 academic staff, 14 MSc students, and 11 PhD students. Its main activity was solar physics, with some interest in radio astronomy and theoretical cosmology. Kim il Sung University, the largest university in the DPRK also has an astronomy group as a sub-section of the physics department, with about 8–10 astronomers.[3]

Dr. Jong Sok told me that the astronomical priorities of the DPRK included:

- The need for extended visits by several DPRK astronomers to foreign institutes, where they could acquire expertise in modern high-technology astronomy and use the acquired knowledge to improve the standard of astronomical research in the DPRK.
- Acquiring experience in frontline astronomical data analysis software such as IRAF.
- Broadening the interests of the PAO into new areas, such as stellar astrophysics.
- Acquiring a small solar telescope.
- Acquiring an adequate library of textbooks and access to modern journals.
- The organization of a national astronomy school for interested physics students and schoolteachers.
- Hosting lectures in the DPRK by outstanding astronomers from abroad.

Jong Sok also said that DPRK astronomers wished to re-join the IAU as soon as possible and that he already had talks with the DPRK Academy about the matter. However, the annual dues would be a substantial burden, even the fees for interim membership. He raised the possibility is that application from the DPRK could be postponed for a few years, during which the infrastructure of the PAO could be brought up to a sufficient state to warrant IAU membership.

In reply, I expressed my hope that the DPRK would re-join the IAU at the earliest possible opportunity. As far as the IAU is concerned, improvement of the infrastructure would not need to take place before the DPRK re-joins. I also stressed that the IAU now regards the world development of astronomy as part of its core mission and hoped to raise the budget for such activities substantially in the future. Furthermore, developing countries usually receive more in direct benefits from IAU membership than the cost of their relatively modest dues.

On 29 July 2011, I sent an official letter from me as IAU Vice President to Professor Jang Chol, President of the Academy of Science of the DPRK, inviting the DPRK to apply to re-join the IAU. In this letter I referred to the fruitful discussions

[3] See this volume, Chap. 10 by Hyung Mok Lee.

that I had with Professor Jong Sok in Mongolia three years previously. I was delighted that the DPRK was eventually re-admitted to official membership of the IAU at the 2012 IAU General Assembly in Beijing. We were particularly pleased at Leiden Observatory to host fruitful several-month visits by two North Korean astronomers in 2013, with the help of a grant from the IAU Office of Astronomy for Development. I hope that many astronomers from the DPRK will attend the 2021 General Assembly of the IAU at Pusan, South Korea[4]. It would be an excellent opportunity for them to develop lasting collaborations with astronomers throughout the world, particularly with their South Korean colleagues.

12.5 Concluding Remarks

I have always believed that our professional vantage point gives astronomers a special duty to combat nationalism and help bring the peoples of our world together. Here I have described three very different examples in which astronomy has contributed to building such much-needed bridges. The view of astronomy as an essential resource in framing human ideology was also an important aspect in setting up the IAU Office of Astronomy for Development in Cape Town, South Africa, in 2013 and the "Universe Awareness" program for very young children in 2004, that has now morphed into the "Pale Blue Dot" project. The perspective of our planet as a tiny speck within our solar system, our galaxy, and our Universe, can play a special role in helping to solve the many problems that threaten the survival of our species and to advance the UN Sustainable Development Goals-SDG16 (Peace), SDG4 (Education), and SDG13 (Climate Action). Given the current vulnerability of our planet to nuclear war and human vandalism, such efforts have never been more essential.

My crazy dream for the distant future is that we astronomers can assist these efforts by exposing world leaders, their children and grandchildren to a simple appreciation of the enormity, beauty, diversity, and history of the cosmos. Our goal should be to stimulate the sense of reverence and awe for the Universe and the relative unimportance of individual countries, current disputes, and the temporary nature of human lives. Maybe the United Nations, the IAU, and some of its member states could take the lead in creating such a program, in which world leaders and their families would be invited to spend a night at one of our most inspiring observing sites and have individual discussions about the cosmos with some of our most brilliant enthusiastic and loquacious populizers of astronomy?

Acknowledgements I am grateful to Thierry Montmerle, Pedro Russo, and Richard Schilizzi for important comments on the manuscript and particularly to Ken Kellermann for his account of how the US–Soviet collaboration in VLBI took place at the height of the cold war.

[4] Postponed to the summer 2022 due to the SARS CoV-2 pandemic.

References

Allen, L. R., et al. (1962). Observations of 384 radio sources at a frequency of 158 Mc/s with a long baseline interferometer. *Monthly Notices of the Royal Astronomical Society, 124*, 477.

Blaauw, A. (1994). *History of the IAU: The Birth and first half-century of the International Astronomical Union* (p. 296). Kluwer Academic Publ.

Broten, N. W., et al. (1967). Diameters of some quasars at a wavelength of 66.9 cm. *Nature, 216*, 44–45.

Kellermann, K. I., & Cohen, M. H. (1988). The origin and evolution of the NRAO-Cornell VLBI system. *Journal of the Royal Astronomical Society of Canada, 82*, 248–265.

Liu, X. (2020). The China Crisis. In C. Sterken, J. Hearnshaw, & D. Valls-Gabaud (Eds.), *Proc. IAU Symp. 349, Under One Sky: The IAU Centenary Symposium* (pp. 222–227). Cambridge University Press.

Lovell, S. B. (1974). *Out of the Zenith: Jodrell Bank 1957-1970*. Harper & Row.

Matveyenko, L. I., Kardashev, N. S., & Sholomitski, G. V. (1965). *Radiophysica, 8*, 651.

Miley, G. K. (1974). Astronomy in China Today. *Sky and Telescope, 47*, 148–152.

Palmer, H. P., et al. (1967). Radio Diameter Measurements with Interferometer Baselines of One Million and Two Million Wavelengths. *Nature, 213*, 789–790.

George Miley is an Irish-Dutch Emeritus Professor of Astronomy at Leiden University and previous Director of Leiden Observatory. His multi-wavelength research focuses on radio galaxies, some of which are the most distant known objects in the Universe. He initiated LOFAR, a new-generation Dutch-European radio telescope and has participated in several space missions, including the Hubble Space Telescope.

After graduating in physics from University College Dublin in 1963, he obtained a PhD in radio astronomy from the University of Manchester, Jodrell Bank and has occupied positions at the US National Radio Astronomy Observatory and the NASA-ESA Space Telescope Science Institute, the latter as Head of Academic Affairs. Among the awards that he has received are an Honorary Doctorate from Trinity College Dublin, an Honorary Fellowship by the Royal Astronomical Society, a Knighthood in the Order of the Netherlands Lion and the Royal Shell Oeuvre Prize.

In 2004 with the support of a Royal Netherlands Academy of Sciences professorship, Miley initiated the Universe Awareness programme to further globalism and the UN Sustainable Development goals by exposing young children to the enormity and wonder of the Universe. As vice-president of the International Astronomical Union, Miley designed the IAU 2010–2020 strategic plan "Astronomy for Development" and helped establish the IAU Office of Astronomy for Development (OAD) at the South African Astronomical Observatory in Cape Town in 2011.

Chapter 13
From *Luna-3* to *Apollo*: The IAU and the "New Moon"

Thierry Montmerle

Abstract In October 1959, the Soviet *Luna-3* space probe made a successful flyby of the Moon, and photographed its far side. A "new Moon" was revealed, that had never been seen from the ground. Feeling like the ancient explorers, the Soviet Academicians decided to give names to a number of features appearing on the *Luna-3* photographs. This created a tension with the IAU, which had fixed a centuries-old tradition for naming lunar features in 1935, and thus expected to be consulted. Nevertheless, it was difficult not to recognize the importance of the *Luna-3* achievement, and the first names on the far side of the Moon were approved at the IAU General Assembly in Berkeley in 1961.

The Kennedy 1961 declaration for "landing a man on the Moon and returning him safely to the Earth" before 1970 started a peaceful, yet fierce competition between the two "superpowers", soon to be called "the Moon race". Increasingly fine details of the lunar surface were expected to be revealed, calling for a greater need for names, and the IAU created in 1964 an ad hoc "Working Group on Lunar Nomenclature", which included Soviet and American astronomers. The problem was the deadline set by Kennedy, which made some key decisions urgent, even overriding the established IAU procedures.

Seen from the outside, the Working Group was successful: in 1970, a list of over 500 crater names of the far side was approved at the Brighton General Assembly; in 1973, a Resolution re-defining the criteria for lunar nomenclature for the whole Moon was approved at the Sydney General Assembly. However, behind the scenes, and as told in this chapter on the basis of IAU archives, bitter conflicts arose, which brought the Working Group on the verge of breakdown. It appears that NASA played a decisive role for its recovery. With the Moon race over, the Working Group was then disbanded, to give way to a new, global Working Group on "Planetary System Nomenclature" (WGPSN), including the Moon, which is still very active today.

T. Montmerle (✉)
Institut d'Astrophysique de Paris, Paris, France
e-mail: montmerle@iap.fr

© The Author(s), under exclusive license to Springer Nature Switzerland AG 2022
T. Montmerle, D. Fauque (eds.), *Astronomers as Diplomats*, Historical & Cultural Astronomy, https://doi.org/10.1007/978-3-030-98625-4_13

13.1 Introduction: The *Luna-3* Flyby and the Soviet Challenge to the IAU

Aiming at the Moon, after a near-miss with *Luna-1* and the crash of *Luna-2*, the Soviet *Luna-3* spacecraft made a successful flyby on Oct. 7, 1959, at an almost grazing distance of 9500 km, sending back two dozen black-and-white pictures of its "far side." The technology (shooting and processing on a 35-mm film, followed by scanning of the images and transmission back to Earth with a TV camera) was extremely innovative, but suffered from the same defects as the TV broadcasts of the time: the (analogic) images were of rather poor quality, and details were difficult to identify. Nonetheless, the news quickly spread around the world: for the first time in the history of mankind, Earthlings could see the "hidden" side of the Moon, confirm that it was spherical rather than flat (after all, it always presented the same "face"), and start a new, challenging relation with our natural satellite.

Thus, besides the known ("near") side of the Moon, there was a new world that terrestrial telescopes would never been able to see, as it was visible only from space. Although traveling to the Moon had been imagined by writers many times since the seventeenth century–following the invention of the telescope and the first maps of the Moon–it would take only a decade after *Luna-3* that an *Apollo-11* astronaut would send back a historic message while setting foot on the lunar surface: "*One small step for man, and a giant leap for mankind.*"

But in 1959, the Soviets could relish their leadership over the USA in space technology: exactly two years before, they had celebrated the 1917 October Revolution by launching the Earth's first artificial satellite, *Sputnik-1*, weighing 86 kg, on Oct. 4; just a month afterwards, they had put in orbit *Sputnik-2*, weighing 508 kg; both preceded the launch, on Feb. 1, 1958, of the first American satellite *Explorer-1*, having a miserable, suitcase-like weight of 14 kg ... And now, with the 280-kg *Luna-3* lunar probe, they were offering a "new Moon" to the world.

The *Luna-3* images (Fig. 13.1) were then examined by the Soviet Academy of Sciences (the "Soviet Academy" for short in what follows), which had set up a special Committee, under the leadership of Academician Aleksandr Mikhailov, one of the top scientists behind the *Sputnik* program. This Committee identified 252 lunar features, of which 152 belonged to the far side. Continuing (they believed) the historical tradition of the terrestrial explorers, they were the first to give *names* to 18 of their most conspicuous topological features. All were of scientists of undisputable international standing (with many non-Soviets, like Pasteur or Edison, or even the fifth-century mathematician Tsu Ching-Chi, the first Chinese name on the Moon), or famous promoters of lunar travel (Jules Verne, Konstantin Tsiolkovsky). This list was released to the public[1] and published in Moscow the following year, in the "*Atlas of the Far Side of the Moon*" (Barabashov, Mikhailov and Lipsky 1960).

[1] Burchett and Purdy (1961), p. 60.

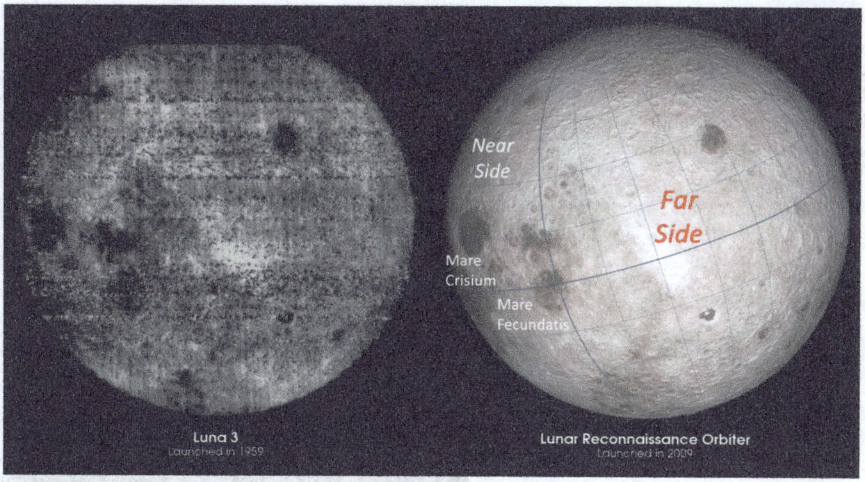

Fig. 13.1 *Left*: *Luna-3* flyby image of 1959 (frame 29), compared with a modern image (*right*) taken fifty years later by the *Lunar Reconnaissance Orbiter* (LRO). Note that the left 1/3 of the image is of the Moon's nearside ("Mare Crisium" and "Mare Fecunditatis"), but seen from a new perspective, invisible from the Earth. (Source: https://svs.gsfc.nasa.gov/4109, Credit: NASA's Goddard Space Flight Center Scientific Visualization Studio; labels added by the author)

This release came as a bad surprise for the IAU, for many reasons of fundamental importance. First and foremost, since its creation in 1919 the IAU had appointed a Commission (No.17) on "*Lunar Nomenclature.*" This Commission had endorsed, at the time of the fourth IAU General Assembly in Paris (1935), a two-decade-long, meticulous effort led by the British selenographer (and former school teacher) Mary Adela Blagg (1858–1944), which had started in 1913 with the publication of a *Collated List of Lunar Formations*[2] and ended with the publication of the official catalog and map of the Moon (*Named Lunar Formations*, Blagg & Müller, 1935).[3] By doing so, the IAU had established itself as *the* authority on lunar names, so, at least implicitly, it expected to be consulted on the matter.

At the time of the *Luna-3* flight, however, Commission 17 was no longer in charge of lunar nomenclature (which had been essentially a dormant activity since this nomenclature had been fixed in 1935), but was responsible for "*The*

[2] The full title was: "*Collated List of Lunar Formations Named or Lettered in the Maps of Neison, Schmidt, and Mädler Compiled and Annotated for the Committee,*" by Mary A. Blagg, Samuel Arthur Saunder (1913) (Blagg & Saunder, 1913).

[3] Somewhat surprisingly in retrospect, the Blagg & Müller publication of 1935 was not subject to a Resolution in Paris recommending its use: the only Resolution alluding to it was a Resolution submitted by Commission 17 and voted at the previous General Assembly (Cambridge, USA, 1932), recommending "*that the list of lunar formations and the list of revisions of the Collated List of Miss Blagg be printed, and that a sum of 5000 gold francs be allocated for this work.*" (Transactions IAU vol. IV, https://www.iau.org/static/resolutions/IAU1932_French.pdf).

Fig. 13.2 The Soviet Academician Aleksandr Aleksandrovich Mikhailov (1888–1983). (Source: https://alchetron.com/Aleksandr-Aleksandrovich-Mikhailov; License CC-BY-SA)

Motion and Figure of the Moon," i.e., lunar geodesy, so that the "new Moon" was placed under the purview of Commission 16 ("*The Physical Study of Planets and Satellites*").[4] In 1958, this Commission was presided by G.P. Kuiper, a Dutch-born American astronomer with a long-time interest in the Moon (who would found the Lunar and Planetary Laboratory in Tucson, Arizona, in 1960), and had 40 members (including Kuiper), five being Soviet astronomers, hence having connections with their Academy of Sciences (which ruled all of Soviet science). Furthermore, Mikhailov himself (Fig. 13.2), one of the co-authors of the *Atlas*, was a former member of the IAU Executive Committee as Vice-President (1944–1948), and was now a member of its "Special Nominating Committee"[5] (1958–1961); and Barbashov, also a co-author, was a member of Commission 16. So the Soviet astronomers had no shortage of communication channels between their Academy and the IAU, but obviously, they did not (or were instructed not to) use them.

Second, the Soviets did not follow the well-established tradition, fixed in 1935, that lunar feature names should be of astronomers or scientists having made important contributions to lunar studies (e.g., mathematicians). In fact, *none* of the new names belonged to that category, some of them even being on the verge of celebrating military activity: while Edison, Maxwell, or Sklodowska-Curie ("Madame Curie") had dedicated their lives to fundamental research, Kurchatov

[4] Returning to 1919 basics, in 1964 Commission 17 was reorganized to gather all of lunar science under the name "*The Moon,*" while Commission 16 continued to be in charge of "*Planets and Satellites,*" except the Moon.

[5] This Committee is a "search committee" looking for possible names to be proposed as members of future Executive Committees.

was the father of the Soviet atom bomb, tested only ten years before . . . And to top all these names, a clear, elongated, feature was named "Montes Sovietici," which could easily be taken as a political allusion to the nationality of the discoverers, a property that had always been strictly avoided.[6]

This caused quite a stir within Commission 16. Such a *fait accompli* was nothing less than a challenge to the IAU authority by the Soviet Academy, but in view of the importance of the *Luna-3* results, and of the publicity around them, an open conflict was out of the question. Nonetheless, the IAU had to react: since IAU astronomers meet every three years in their General Assemblies, the previous one having taken place in Moscow in 1958 (after the *Sputnik-1* success, but before *Luna-3*), the next opportunity to show leadership was for Commission 16 to elaborate a Resolution to be voted at the next General Assembly. This was due to take place in 1961 in Berkeley, California, offering the opportunity to update the 1935 catalog and maps by Blagg & Müller.

This two-page Resolution (No 2 of Commission 16)[7] gives many recommendations about the designations of lunar features. Worthy of note here are: (1. *a*) *craters must bear names of* <u>*deceased*</u>[8] *astronomers or famous scientists; . . . ;* (1. *f*) . . . *Features can still be named after the nearest crater, followed by an uppercase letter of the latin alphabet*[9] . . . ; (2.) . . . *No new designations should be added to the existing ones for lunar features observable from the Earth . . . ;* (3) *For lunar features not observable from the Earth, it is recommended to adopt the designations introduced in the Atlas of the Far Side of the Moon, in the way presented in the following table and map* [see Fig. 13.3].

So, while the IAU appeared to have ceded to the Soviet Academy by accepting the *fait accompli*, it had at least adapted the original work by publishing its own simplified map (which included lunar coordinates; see additional information in the figure caption), and established revised rules of nomenclature, valid in particular for new results from space–with in mind, conceivably, to force the Academy or other institutions to follow them in the future.

At any rate, this early sticking point demonstrated that lunar nomenclature, after centuries of tradition, essentially of academic and cultural nature, was now running the risk of being political, which, especially in the ambient "Cold War" climate, was extremely dangerous in view of its impact on the public. To prevent this from happening, in the subsequent years up to and including the post-*Apollo* era, the IAU

---

[6] By an irony of nature, this feature, seen lighter than its surroundings on *Luna-3* photographs, was later found to correspond only to a flat terrain which appeared to be *darker*, so this "political" name was quietly dropped. See below, Sect. 13.5.

[7] Transactions IAU, vol. XI B (https://www.iau.org/static/resolutions/IAU1961_French.pdf).

[8] In italics in the original (roman) text.

[9] This is called the "Mädler" nomenclature system (Beer & Mädler, 1837), to which we shall return at length below. It was the one used by Blagg & Müller.

Désignation	n° du catalogue	Coordonnées sur la carte longitude °	latitude °
Tsu Chung-Chi	1	141	+18
Kurchatov	2	144	+32
Jules Verne	5	151	−37
Mendeleev	10	167	− 2
Popov	71	99	+14
Hertz	74	101	+11
Edison	83	100	+24
Lobachevsky	88	112	+ 9
Pasteur	95	111	−10
Sklodowska Curie	112	102	−23
Tsiolkovsky	151	131	−22
Lomonosov	177	99	+28
Joliot Curie	183	93	+25
Maxwell	185	99	+30
Giordano Bruno	208	103	+36
Montes Sovietici	103	(de 111	+19
		(à 124	− 5
Mare Ingenii	150		
Mare Moscoviense	152	149	+27

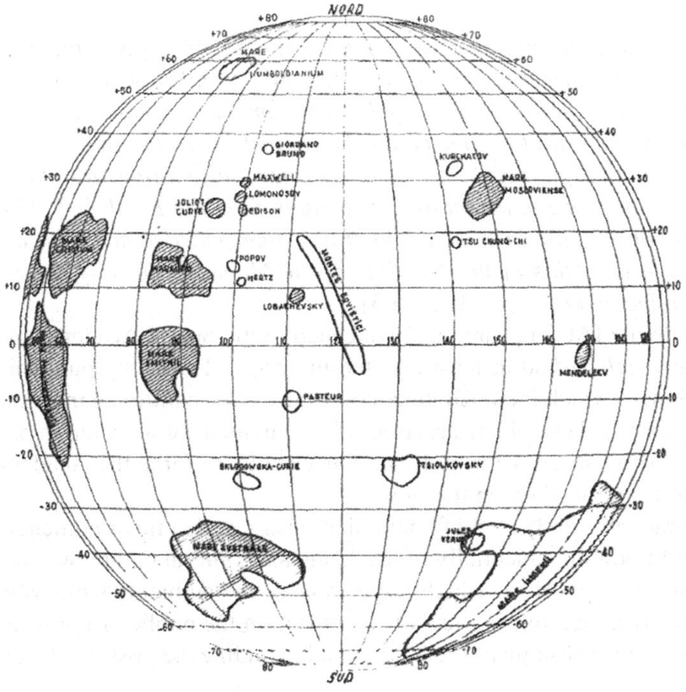

Commission 16, Résolution no. 2: Carte indiquant les désignations de la surface lunaire non observable de la Terre.

Commission 16, Resolution no. 2: Chart showing the recommended nomenclature for the surface

Fig. 13.3 Map published in the Resolution No. 2 proposed by Commission 16 at the Berkeley General Assembly (1961). Note that the title of this map [here incompletely scanned in English from the IAU website, see https://www.iau.org/static/resolutions/IAU1961_French.pdf], should read in full: "*Chart showing the recommended nomenclature for the surface of the Moon not observable from Earth.*" This is somewhat misleading, since (to the left), Mare Crisium and Mare Fecunditatis can be seen from Earth (see Fig. 13.1) near the Eastern limb of the Moon, albeit under a grazing viewing incidence would play a major role, if at times hectic, to cool down several conflicts, both external[10] and internal.

13.2 The Moon Race and the Need for Lunar "Travel Guides"

Indeed, going to the Moon soon became a major strategic–even patriotic–goal for the two "superpowers." Shortly after his inauguration, on May 25, 1961, President Kennedy pressed the US Senators to engage the country into the so-called "*Apollo*" lunar *manned landing* program, a far cry from the "simple," crewless lunar flyby of *Luna-3*. In his address,[11] Kennedy referred to the recent Soviet successes, the need for the USA to be "first," and declared: "*I believe that this nation should commit itself to achieving the goal, before this decade is out, of landing a man on the Moon and returning him safely to the Earth.*" The italics are mine, here to emphasize the most important point: "before this decade is out" meant "before 1970"! Once the corresponding budget had been approved, Kennedy could make the announcement public, in his electrifying "Address on the Nation's Space Effort" of Sep.12, 1962, at the Rice University Stadium in Houston, Texas, a year after the IAU General Assembly in Berkeley: "*We choose to go to the Moon!*"[12]

So this is how what would be called "the Moon race" started. For the IAU, the constraint was that it would set a *mandatory timescale*: not only follow the American and Soviet lunar space programs and results, but also synchronize its recommendations, in the form of Resolutions when appropriate, at its General Assemblies, i.e., on a three-year cycle, a short timescale indeed, under a lot of pressure (from the public, from space agencies, etc.) to issue decisions quickly. This meant that there would be strict milestones from then on: the summers of 1964 (Hamburg), 1967 (Prague), 1970 (Brighton), and 1973 (Sydney). The strain would be increasingly severe, and as we shall see, proved almost fatal to the IAU as the *Apollo* missions were coming to an end.

For the taxpayers, it turned out to be a very costly race (just under 300 billion of today's dollars were spent by each of the two countries over ten years[13]), and there were also tragic accidents on both sides (although never on the way to the Moon).[14]

---

[10] On the conflict with the United Nations about "Extraterrestrial Names" in the period 1971-1982, see Chap. 16 by TM in this volume.

[11] https://www.jfklibrary.org/learn/about-jfk/historic-speeches/address-to-joint-session-of-congress-may-25-1961.

[12] https://www.youtube.com/watch?v=g25G1M4EXrQ.

[13] https://www.history.nasa.gov/SP-4029/Apollo_18-16_Apollo_Program_Budget_Appropriations.htm; for the USSR, see the CIA document, released on Sep. 29, 2006: https://www.cia.gov/readingroom/docs/CIA-RDP66B00403R000400080002-6.pdf.

[14] https://en.wikipedia.org/wiki/List_of_spaceflight-related_accidents_and_incidents.

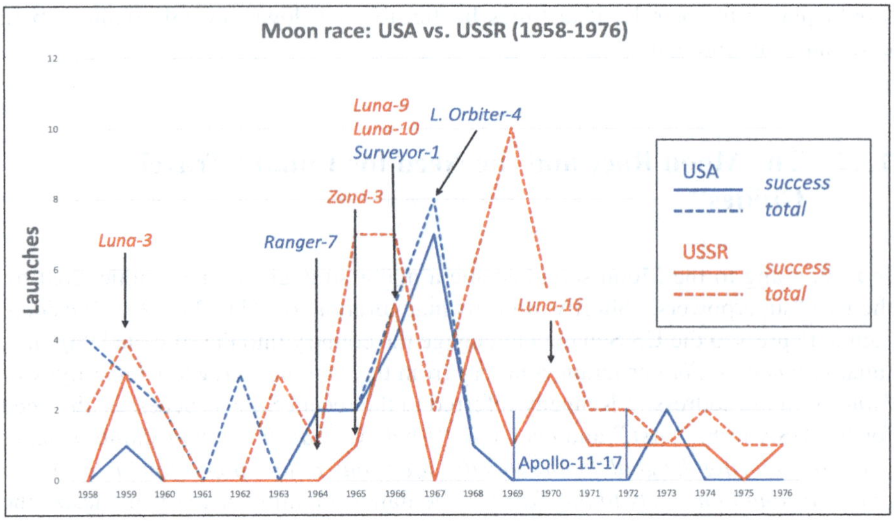

Fig. 13.4 The "Moon race" between the USA and the USSR: number of launches (in one-year steps) from 1958 (*Luna-1*: first Soviet attempt to go to Moon) to 1976 (last *Luna* mission, soil sample return: *Luna-24*). American missions are in blue, Soviet ones in red; successful missions are visualized with continuous lines, the total number with dashed lines. (Data from https://en.wikipedia.org/wiki/List_of_lunar_probes, last accessed Jan. 8, 2022). Graph by the author

In his Rice address, JFK had said: "*We choose to go to the Moon in this decade, not because it is easy, but because it is hard . . .*" Hard it was indeed: there were many mission failures, although the proportion would decline with time, from about 75% initially, to 0% at the end of the decade (for the end phase of the *Apollo* program; but 90% for the Soviet program in 1969, in a desperate attempt to attract attention[15]).

Figures 13.4 and 13.5 illustrate the ups and downs of the lunar space programs of both the USA and the USSR, and how the IAU General Assemblies happened to be interleaved with the successive missions.

The Moon was indeed a new world to explore, and before visiting it, it was necessary to create sets of *maps*, as accurate as possible, at different scales, depending on the quality of the data (the photographic film "resolution," improving with time), down to a (large) roadmap scale (1:250,000). The next step was to identify the nature of topological features (craters, mountains, rifts, etc., but also "seas"–not real seas of course, but extended, flat, dark areas, etc., a historical tradition in lunar maps, see below), in other words, to create Moon "travel guides." "Editions" would be updated from one step to the next: first, flybys as for *Luna-3*, later impactors and landers, then crewless orbiters; and finally manned orbiters and

[15] *Luna-15* was a robotic mission to bring back lunar material, exactly at the same time as *Apollo-11*, but it crashed on the lunar surface (see below, Sect. 13.5). The Soviet space program became disorganized after the death of its head, Serguei Korolev, in 1966 (see Chap. 14 by M. Marov).

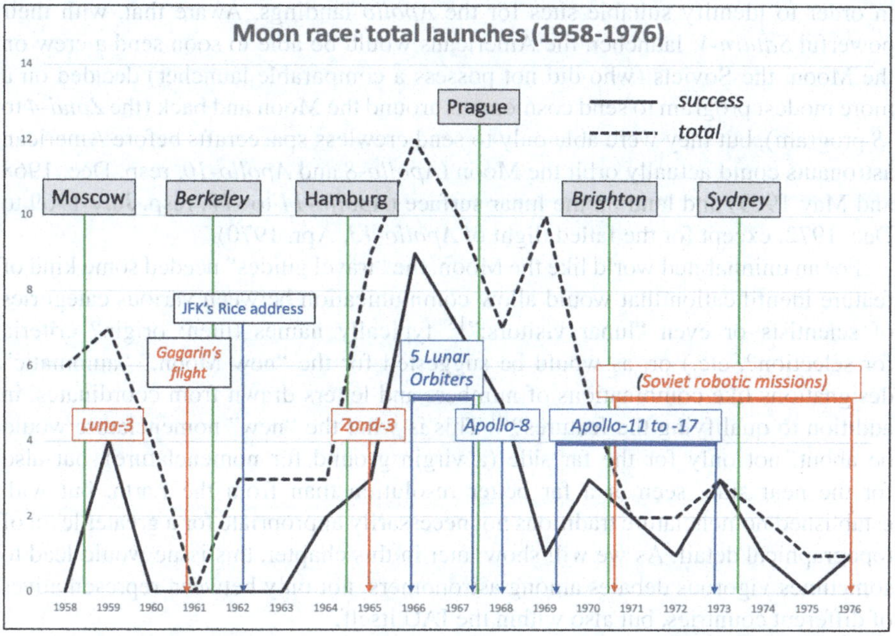

Fig. 13.5 Same as Fig. 13.4, but summing up all lunar missions. *Top labels:* date and location of IAU General Assemblies; italicized names indicate the ones in which important nomenclature Resolutions were voted: Berkeley (first feature names for the far side of the Moon), Brighton (first catalog of 513 crater names for the far side), and Berkeley (acceptance of non-scientist, commemorative names for the whole lunar surface). *Other labels:* most important events and missions during this period, as discussed in the text. Graph by the author

landers, even "jeep"-like rovers (the USA), or robotic ones (the post-*Apollo* Soviet lunar program).[16]

More precisely, as the Moon race intensified, not only was the "far side" of the Moon studied again by the Soviets (with the *Zond-3* flyby mission of 1965; more in the next section), but for the first time, the Moon could be observed globally, i.e., with spacecrafts in orbit around it, allowing a systematic mapping, including of the lunar poles. The first to succeed were the Soviets again, with *Luna-10* (Apr. 3, 1966), followed by *Luna-11* later in 1966, which carried scientific instruments but unfortunately no camera.[17] The five American *Lunar Orbiters,* launched during the same period (1966–1967), were all successful and sent back 3000 video photographs, taken from a wide variety of orbits (with altitudes from 6100 km down to 50 km), allowing to go down to details from 300 m to 1 m in size! The main mission of these Orbiters was to draw a complete map of the Moon,

[16] See Chap. 14 for highlights. For technical details, see, e.g., Lund (2018).

[17] *Luna-12*, also launched in 1966, had a TV camera, but only 4 photographs were published.

in order to identify suitable sites for the *Apollo* landings. Aware that, with their powerful *Saturn-V* launcher, the Americans would be able to soon send a crew on the Moon, the Soviets (who did not possess a comparable launcher) decided on a more modest program to send cosmonauts around the Moon and back (the *Zond-4* to *-8* program), but they were able only to send crewless spacecrafts before American astronauts could actually orbit the Moon (*Apollo-8* and *Apollo-10,* resp. Dec. 1968 and May 1969) and land on the lunar surface (*Apollo-11* to *-17*, resp. July 1969 to Dec. 1972, except for the failed flight of *Apollo-13,* Apr. 1970).

For an uninhabited world like the Moon, the "travel guides" needed some kind of feature identification that would allow communication between various categories of scientists or even "lunar visitors:"[18] typically names (then: origin? criteria for selection?, etc.) or, as would be suggested for the "new Moon," "automatic" designations like combinations of numbers and letters drawn from coordinates, in addition to qualifying the features?[19] This is what the "new" nomenclature would be about, not only for the far side (a virgin ground for nomenclature), but also for the near side, seen at a far better resolution than from the Earth, but with established nomenclature traditions not necessarily appropriate for a greater level of topographical detail. As we will show later in this chapter, this issue would lead to sometimes vigorous debates among astronomers, not only between representatives of different countries, but also within the IAU itself.

13.3 The *Zond-3* Naming Conflict with the Soviet Academy

The next IAU General Assembly took place in Hamburg three years after Berkeley, in the summer of 1964. In the meantime, as summarized in Fig. 13.5, nothing much had happened on the front of launches towards the Moon, it was a time for preparations. There was a total of 9 launches by the USA and the USSR, but 7 failed, and the two which were successful (*Ranger-7* and *Ranger-8*) happened just before the General Assembly, too late to be taken into consideration.

During the Hamburg General Assembly, no Resolution was proposed by either Commissions 16 or 17. However, an internal reorganization was worthy of note, because it considerably influenced the subsequent years of the IAU work on lunar nomenclature: sub-Commission 16a ("*Nomenclature and cartography of the lunar surface*"), headed by G. Kuiper, which was under the purview of Commission 16 ("Physical study of planets and satellites"), was, logically, transferred as a "Working Group" (WG) on "*Lunar nomenclature*," under the purview of Commission 17, simply re-named "*The Moon.*" Donald Menzel (1901–1976), the newly elected

[18] An expression used by C. de Jager, who was IAU General Secretary in the years 1970–1973: letter of Sep. 15, 1972 to D. Menzel, IAU archives; see Chap. 16.

[19] See Chap. 16 for discussion on alternate nomenclature systems proposed by the UN Working Group on Extraterrestrial names.

Fig. 13.6 Banquet at an astronomy Symposium in Utrecht (1963). Two members of the "Working Group on Lunar nomenclature" were present: Donald Menzel (second from the left), and Marcel Minnaert (second from the right). The other guests were Jan Hendrik Oort (first on the left), past President of the IAU (1958–1961), Charlotte Moore Sitterly, an American solar astronomer, and Albrecht Unsöld, a German stellar spectroscopist (hidden behind the candelabrum). (Credit: Hans Nieuwenhuijzen, Creative Commons Attribution-Share Alike 4.0 International)

President of this Commission (see Fig. 13.6), was at the time Director of the Harvard College Observatory in Cambridge, near Boston. He was a theoretical astrophysicist, mainly involved in the study of the Sun. He had also worked for the Department of Defense during WWII and was a science-fiction author (which would show up in nomenclature issues, see below).[20]

Menzel took also the chairmanship of the newly created WG, which comprised two other members: Mikhailov, whom we met above (Fig. 13.2) as a co-author of the first *Atlas of the Far Side of the Moon* at the time of *Luna-3*, and the Belgian-born spectroscopist Marcel Minnaert (1893–1970), established in Utrecht, the Netherlands (Fig. 13.6). The fourth WG member, *ex-officio*, was the younger Audouin Dollfus (1924–2010), his French successor as President of Commission 17 in 1967 for six years (see below), who had already been also President of Commission 16 from 1961 to 1964 (Fig. 13.7).[21] Together, Menzel and Dollfus would have a decisive influence on lunar nomenclature, and on solving (or creating, in the case of Menzel) the difficult crises that occurred during their mandates.

The "triennium" following the Hamburg General Assembly would be very different for Commission 17 and its newly created WG, since, going from 1964 to 1967 (General Assembly in Prague), it would encompass the "Golden Year" 1966, in which no less than 12 lunar missions took place (nine were successful). Commission 17, by way of Menzel (always a prolific writer) delivered an unusually

[20] For more details, see Menzel's Obituary by Goldberg and Aller (1991).

[21] Id. for Dollfus (in English): Cabrol et al. (2010), Rosenfeld (2010), McKim (2011).

Fig. 13.7 The fourth member of the "Working Group on Lunar nomenclature:" Audouin Dollfus. He was always wearing a "*béret basque*;" he is here seen in his balloon capsule after a flight at 14,000 m altitude (a world record at the time) to make UV observations of Venus, on Apr. 22, 1959. (Credit: Personal archives of Ariane Dollfus, daughter of A. Dollfus, with her kind permission) This capsule can be seen today on display in the *Musée de l'Air et de l'Espace*, at Le Bourget airport, North of Paris

long Report, published in the IAU Transactions.[22] Its Section B ("*Lunar Probes*"), compiles the "*Hard Landings*," the "*Soft Landings*," the "*Far Side of the Moon*," and the "*Orbiters*" (*Luna-10* and *Orbiter-1* to *-4*), while Section C ("*Cartography*") mentions the *Ranger* maps (they were impactors taking high-cadence photographs at close range shortly before crashing on the lunar ground), and photographic maps from ground-based telescopes (assembled in the *System of Lunar Craters*, initiated by W.G. Arthur and G. Kuiper of the Lunar and Planetary Laboratory in Tucson, which was completed in 1966).[23]

As for the far side of the Moon (of no direct relevance for the *Apollo* program), in July 1965 the Soviet probe *Zond-3* repeated the *Luna-3* flyby and obtained 25 high-quality complementary images of its unknown as yet "back side" (Fig. 13.8). All the other results concerned the near side of the Moon: nomenclature issues for the smaller topological features would soon be raised by the *Lunar Orbiters* images, and big problems clearly laid ahead. Menzel lucidly wrote in his report (p. 345):

[22] Transactions IAU Vol. XIII B, pp. 341–374.

[23] https://the-moon.us/wiki/System_of_Lunar_Craters.

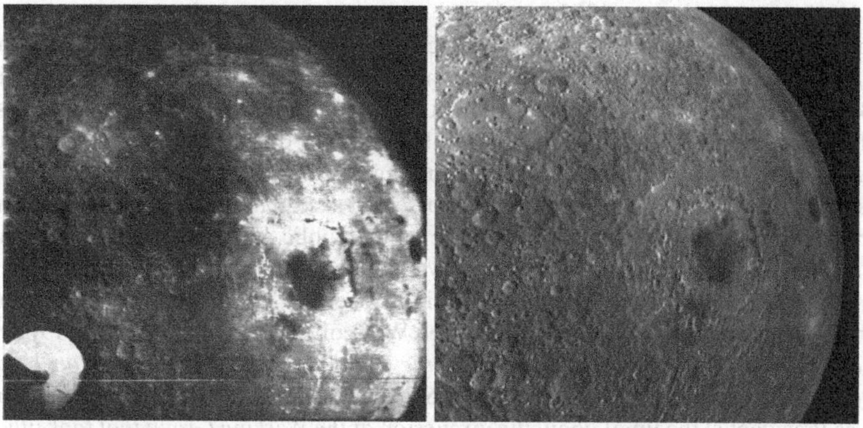

Fig. 13.8 *Left: Zond-3* photograph of the far side (1965) taken from a distance of 9200 km (source: http://mentallandscape.com/C_CatalogMoon.htm, site maintained by D.P. Mitchell). *Right:* corresponding LRO image, which has been adjusted by the author for comparison using NASA's 3D model of the Moon (https://solarsystem.nasa.gov/resources/2366/earths-moon-3d-model/). The conspicuous, big impact crater (900 km in diameter) is filled with a dark "sea" (flat terrain), called "Mare Orientale." The area first observed by *Luna-3* (Fig. 13.1) is to the left of the pictures

The enormous amount of new data related to the mapping of lunar craters, obtained from the space programs of the U.S.A. and the U.S.S.R. necessitates a complete review of the problem of nomenclature of various lunar features. Recommendations made by special committees of the National Academies of the U.S.A. and U.S.S.R. furnish a starting point for consideration of the problem by the Sub-committee on Lunar Nomenclature of Commission 17 of the IAU under the chairmanship of Z. Kopal. As the result of correspondence undertaken prior to and meetings held during the General Assembly of the IAU in Prague, recommendations will be made to the General Assembly concerning nomenclature for features of the reverse side of the Moon and also of special features on the visible surface.

A poll of the membership of Commission 17, undertaken by its president, indicates that the majority favors extension of the present system of nomenclature to the reverse side of the Moon, with craters named for deceased scientists, with special attention to astronomers.

The first charts of the far side of the Moon ("LFC:" Lunar Farside Chart) were produced by NASA in Aug. 1967 (i.e., when the IAU General Assembly was taking place in Prague). LFC-2 mixed the "*Orbiter-3* to *-5*" and "*Zond-3*" results, at a scale of 1:10,000,000 (1 cm = 100 km). Already at that scale, the number of craters is very large (several thousand), thus the perspective of having to find names for so many of them was daunting. But the next triennium, until the 1970 General Assembly in Brighton, would also include the two next *Apollo* missions, this time "the real ones" (i.e., with a crew)–actually three: not only *Apollo-8*, due for Christmas 1968, but another orbital flight as a rehearsal for a landing (*Apollo-10*, in May), and two *Apollo* landings in 1969 (*Apollo-11* in July, of course, but also *Apollo-12* in November). The tremendous worldwide impact they made on the public put an unexpected pressure on Commission 17 and its WG. Now they had only three years to deliver.

But for the time being, an embarrassing issue was raised by Mikhailov about the *Zond-3* images. In a letter to his three WG colleagues dated Nov. 28, 1968,[24] he writes: "*Allow me to inform you of the opinion of the Commission on lunar nomenclature formed by the Academy of Sciences of the USSR* [of which Mikhailov was President at the time] *about the discussion of the Working Group that took place recently in New York. The Commission thinks that it would be proper to maintain the tradition according to which the privilege of naming objects belongs to the author or group of authors who first discovered these objects or published the results obtained. This tradition has been maintained since the time of the early explorers of the Moon as well as after the organization of the IAU* (as for instance by Blagg & Müller, 1935)."

Mikhailov continued:

> It also seems justified to accept the choice of objects proposed by the Commission of the Academy of Sciences of the USSR. There is no doubt in their reality as they all are without any exception clearly depicted on the photographs obtained by the automatic station "Zond-3" (1965) and published prior to the General Assembly of the IAU in Prague (August 1967). *The nomenclature proposed by the Commission and the data obtained from the investigation of the photographs of "Zond-3" have been widely distributed* in the publication "Atlas of the far side of the Moon" (part 2), Map of the entire Moon and the lunar globe.

[The italics are mine.] This sounded very much like a reminiscence of the *Luna-3* story of 1959, as if nothing had happened since then, with names of features discovered on the far side of the Moon being hastily made public without consulting the IAU . . . (see above, Sect. 13.1).

On Dec. 3, Menzel, who had not yet received Mikhailov's letter, wrote to his WG colleagues and to Dollfus: "*I am very pleased, on the whole, with what our meetings accomplished in Cambridge [Harvard] and in New York. I only wish that*

---

[24] Unless otherwise mentioned, in what follows all the letters or excerpts between quotes are from the IAU archives. For this particular topic (the Moon), these archives were carefully maintained by Dollfus himself at the Paris Observatory and later transferred to the nearby IAU Secretariat at the Institut d'Astrophysique de Paris ("*the Dollfus Files*"). As far as one can judge, they are essentially complete for the periods corresponding to his mandates as President of Commission 16, then Vice-President and President of Commission 17, i.e., covering without interruption the period 1958-1973 !. They thus provide a unique testimony of the backstage of the work of these Commissions during that critical period.

all such meetings could be so effective and conducted in an atmosphere of friendly cooperation."

Then, on Dec. 9, Menzel finally received the letter, and thunder stroke. He wrote to the WG an angry, 4-page reaction:

> I am astonished and somewhat dismayed by Professor Mikhailov's letter of 28 November, 1968. There was also a Lunar Nomenclature Commission appointed by the National Academy of Sciences of the U.S.A. Had we taken the viewpoint proposed by Professor Mikhailov in his letter, we could have come up with a much larger list than that proposed by the Academy of the U.S.S.R., in view of the great discoveries that are made by the Lunar Orbiters. I resisted all attempts of our National Academy to propose such a list, feeling that it was usurping the authority of the Working Group on Lunar Nomenclature of the I.A.U." [. . .] "The fact that the Russian maps and globes have had relatively wide distribution in no sense constitutes an authorization for their adoption. [. . .] I recall discussing this question with you and you concurred with me that formal adoption of any list would have been premature, at the time." [. . .] "Thus far I have not felt that it was necessary to consult or inform the Nomenclature Committee of our Academy of Sciences concerning our decisions. I wanted to keep all discussion and decisions on an international plane, without reference to the kind of priorities that your Academy is insisting on.

Before going on about other matters, he told Mikhailov: "*I have been extremely careful, throughout my Presidency of Commission 17 and continuing as Chairman of our Working Group, to avoid any sense of nationalism. I shall, therefore, defer asking your Academy of Sciences for advice, until I hear from you.*" [The italics are mine.]

Seeing a conflict brewing, Minnaert tried to cool down the situation. In a long letter dated Dec. 12, 1968, he replied to Menzel and the WG: "*Our working group did not wish a Soviet sector and an American sector on the Moon, but a broad solution. It seems to me that our list* [agreed in New York] *is to be preferred, simply because we have been able to make use of the Soviet list and to combine it with our own knowledge of the history of astronomy. Nevertheless our list is not perfect, discussion is possible.*"

But, fortunately, the conflict completely lost its substance with the *Lunar Orbiter* maps and the subsequent arrival of the American astronauts around and on the Moon. They were the first humans to "live" (if only for a short time) under the (gravitational) influence of another celestial body. The impact of this event on the Apollo-8 astronauts turned out to be much more important on the public. This created another crisis, of a totally different nature: how was the IAU to cope with name suggestions made by the public at large, and conveyed by the press?

13.4 The *Apollo-8* Public Naming Crisis

Unbeknownst to them, given the wide publicity about their flight, (and notably the iconic "live" Earthrise picture taken by W. Anders on Dec. 24, 1968, dubbed "one of the most significant photographs of all times;" Fig. 13.9), the *Apollo-8* astronauts would trigger a public debate about "lunar names," amplified by the reaction of the American press to the flight. The starting point was an article of the *New York Times*

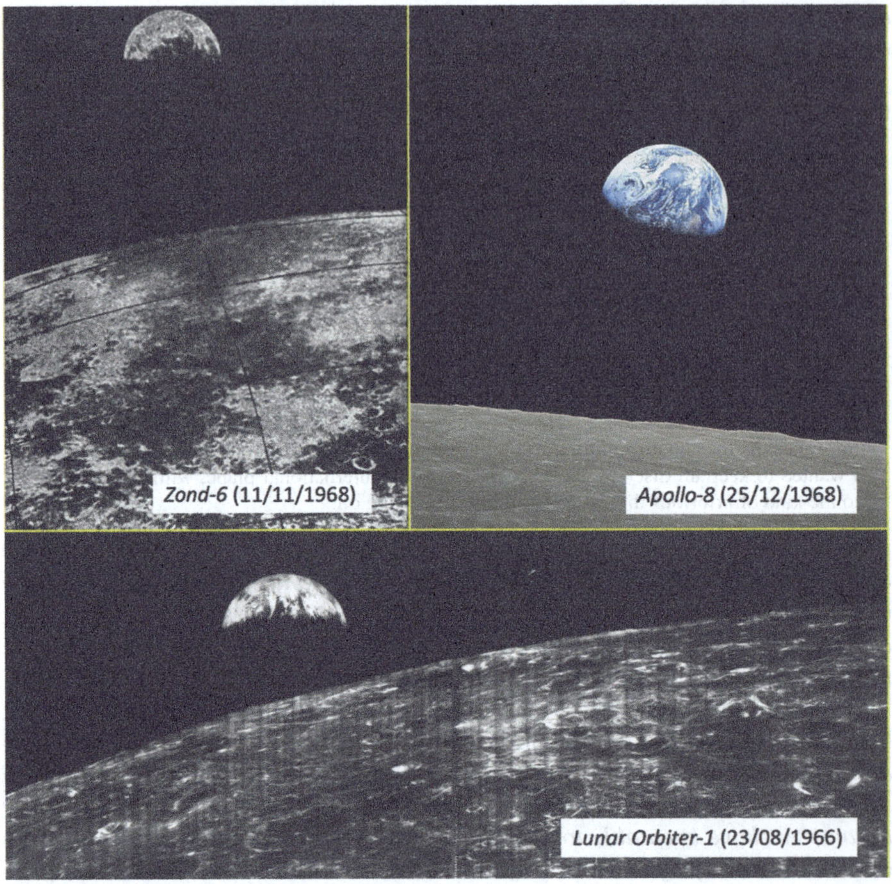

Fig. 13.9 A history of "Earthrises." The iconic, hand-held Hasselblad *Apollo-8* image of Christmas 1968 (*top right;*https://www.hq.nasa.gov/office/pao/History/alsj/a410/AS8-14-2383HR.jpg) was not the first photograph of a lunar "Earthrise:" the first one had been obtained with *Lunar Orbiter-1* in 1966 (*bottom;*https://www.nasa.gov/multimedia/imagegallery/image_feature_623.html). *Zond-6* obtained another six weeks before *Apollo-8* (*top left:*http://mentallandscape.com/C_Zond06_2.jpg). But the first two images were in black-and-white, taken automatically and not in person by astronauts or cosmonauts

on "Orbiting Astronauts," dated Dec. 25, 1968 (p. 37), which Menzel circulated along with a letter dated Dec. 30, with the following warning:

> It seems that the Astronauts, as a means of amusing themselves during their long orbiting of the moon, attached unofficial names to a number of colleagues, and even for Houston (Texas) and Washington (D.C.). Since that time, newspapers and news magazines have hounded me for my opinion as to whether or not the I.A.U. would accept their designations."
> "I have ascertained, first of all, that NASA Headquarters has been extremely ethical in their handling of the naming. They have insisted it was only an amusement of the Astronauts and had no official status. I pointed that, to accept their names would be an enormous departure from standard procedure.

The drawing in the *New York Times* article, displaying an elongated map of the area overflown by the spacecraft, shows a dotted-line perimeter at the border between the Moon's near side and far side, near Tsiolkovsky crater: indeed, "Houston," the *Apollo-8* crewmember names (F. Borman, J. Lovell, and W. Anders), and also the names of two of their colleagues in the Astronaut team, E. M. See and C.A. Bassett, who had died in a plane crash while on a training flight in 1966. But that was not all: on p.36 of the same issue, a shorter article is entitled: "*Moon Landmarks Named in Dozens – Astronauts Honor Officials, Friends and Dead Fliers.*" The article starts with mentioning the three astronauts who had died in the *Apollo-1* pad fire in 1967, but also Lovell's wife, Marylin, whose name he assigned to a small mountain on the near side of the Moon. The list continues with NASA officials, etc. (No wonder NASA could not accept those names!)

The list, an "amusement" as it may have been, raised two problems for the IAU. On the one hand, deceased persons were eligible for a crater name, and astronauts who died on duty were thus worth considering; and on the other hand, naming oneself can be a joke, but as Menzel remarks in the same letter:

> I can see, however, that we might make some commemorative naming of this epic flight. The newspapers here, of course, are full of it. And I have seen many editorials urging that the IAU accept these names, suggested by the astronauts themselves." [...] "One final point that I made to all of my callers was that I hoped to avoid having a Russian sector and an American sector on the moon. I felt that any naming should be truly international and that all nations be represented. I also mentioned my own preference for selecting an area that we might call the Basin of Astronauts. I did not elucidate as to what we meant by Basin nor did I attempt to give the Latin equivalent.

On Jan. 6, 1969, Menzel went one step further:

> We might address ourselves the one possibility that this memorable flight brings up, namely the possibility of departing from our precedent of naming craters only for deceased people and assign three craters to the names of the three astronauts that participated on this flight, with the understanding that we are not creating a precedent.

But pressure from the press was unrelenting. On the same day, *Newsweek* ran an article entitled: "*Moon Names: a New Lunar Who's Who?,*" explaining fairly accurately the nomenclature rules of the IAU, but concluding:

> And the agency [NASA] says that if it ever does put any of the names before the IAU for approval they will be those of dead persons." [...] "Perhaps so. But it seems possible that the designations made by Borman and his crew may stick –after all, they got there first– unless the IAU acts soon. After that, the matter is in the hands of cartographers.

On Jan. 9, Minnaert replied to Menzel:

> The wish of the American public to see the names of Borman, Lovell, Anders on the Moon is of course understandable, but we should be very cautious before accepting this suggestion. As soon as we transgress the official rules lots of proposals will be made. Let us remember

> that the Soviet colleagues have never proposed the name of Gagarin (before he died),[25] though his heroism is at least comparable to that of Borman and his companions.

On the other hand, there was no news yet from Mikhailov and the Soviet Academy, so the WG could not take any decision.

But the public pressure was still being felt. On Apr. 30, two weeks before the scheduled flight of *Apollo-10*, the Boston daily *Christian Science Monitor*, published an article[26] entitled: "*Names for the moon really run the gamut.*" It was preceded by a brief foreword: "*The following article tells how naming lunar landmarks is posing a bit of a problem.*" The core of the article comments on the *Apollo-8* astronauts' naming suggestions, with again a warning: "*If the IAU delays action and new rules on moon names too long, the custom and practice will make what are only temporary names permanent.*"

Unfortunately, the Soviet Academy still kept silent, so in spite of the contacts Menzel had with the press, nothing could be done, except, probably, to repeat that the IAU was the only naming authority, a position firmly supported by NASA, and that without the agreement of the Soviet representative no decision could be taken. But the clock was ticking, and the successful landing of *Apollo-11*, only two months after the circumlunar orbiting flight of *Apollo-10,* gave a new dimension to the lunar nomenclature problem.

13.5 The Problem of Honoring Astronauts and Cosmonauts

The crux of the issue for the IAU was thus how to deal with the pressure of the American public to honor the *Apollo* astronauts having reached the Moon. While having been so far neutral and strictly followed the IAU rules, NASA decided to come into the game and apply its own pressure. Or rather, that of its new Administrator, Thomas O. Paine, who had been appointed on March 5, 1969, and confirmed by the Senate on March 10. His official NASA biography states[27]:

> During his leadership the first seven Apollo manned missions were flown, in which 20 astronauts orbited the earth, 14 traveled to the Moon and four walked upon its surface. Many automated scientific and applications spacecraft were also flown in U.S. and cooperative international programs.

So Paine, a former research engineer in materials, had been personally deeply involved in the organization of the *Apollo* program. In an official letter sent to the President of the (US) National Academy of Sciences (NAS) during the summer of 1969,[28] Paine argued:

---

[25] Gagarin was killed in a jet fighter accident on March 17, 1968, i.e., only a few months before the *Apollo-8* flight.

[26] Found in the IAU archives.

[27] http://astronautix.com/p/paine.html.

[28] For some reason, the official letter kept in the IAU archives does not mention the exact date.

> Nine astronauts are recognized throughout the civilized world for their outstanding pioneering accomplishments in direct observation, description and photography at close range of many lunar farside features, followed by landing and exploration on the lunar surface. (. . .)[29] I would like (. . .) to recommend that they be appropriately honored by having their names ascribed to significant lunar features they were the first to see. In the light of their special contribution to lunar exploration, it would seem uniquely appropriate *to take this action as an exception to the sound general practice of so honoring only the deceased.*

[The italics are mine]. Then he asks the NAS President his endorsement to forward this recommendation for consideration by the "US National Committee for the IAU",[30] and "*if suitable to them,*" by the "Lunar Nomenclature Committee of the IAU."

More importantly, perhaps, is Paine's recommendation "*for the names of several astronauts, both American and Russian, who lost their lives in active service in the program to explore the Moon,*" quoting the three astronauts killed in the *Apollo-1* pad fire (see above), and cosmonaut V. Komarov, killed during reentry aboard *Soyuz-1* (all in 1967). While, not surprisingly, Paine follows in essence the main suggestions of the *Apollo-8* astronauts, he also introduces an idea that they had failed to have: to honor also Soviet cosmonauts, at least a deceased one, Komarov.[31]

In fact, Paine's letter was indeed forwarded to the IAU, and the case was discussed between Commission 17 and its Working Group. From a letter sent by A. Dollfus (President of Commission 17) to the WG members on Nov. 21, 1969, we discover that, at last, the Soviets were back in the WG, with the appointment of Academician Boris Yulievich Levin in replacement of Mikhailov (who had resigned after the conflict with Menzel).[32] Some correspondence may be missing from the IAU archives, but Dollfus's letter suggests that the answer to Paine's request should be a cautious one: to accept only to give to (undefined) craters the names of the *Apollo-11* astronauts, and only provided the WG agreed to make an exception to the "deceased persons" rule (the point specifically made by Paine). And Dollfus's letter answers the question: Levin having sent a negative answer by "cablegram" (presumably on behalf of the Soviet Academy), the WG has no choice but to oppose the exception. Again a deadlock caused by the Soviets . . .

Clearly, in spite of some form of "neutrality" being constantly advocated by scientists, especially in this Cold War period, the issue had now become fully political: it was unthinkable that the Academy of Sciences of the USSR would have given their answer without discussing it first with the Soviet government. On the other hand, the USA had won the Moon race, and the American public wanted the ultimate recognition: that the names of "their" astronauts, in addition to their

---

[29] Here Paine quotes the names of the *Apollo-8*, *-10*, and *-11* astronauts.

[30] This is here the committee of American astronomers in charge of the relations with the IAU. Each country has a similar committee, which is renewed every triennium.

[31] Perhaps surprisingly, Paine does not mention Gagarin, but it may be because Gagarin's death did not happen while he was a cosmonaut on duty (see Note 25), contrary to Komarov.

[32] Letter of D. Menzel to Commission 17 in December 1969, commented in Chap. 16; see also below. It may also be simply because Mikhailov had reached an old age (he was 81 at the time).

footsteps, should be engraved forever on the lunar surface for the whole world to see. Even the last-minute attempts by the Soviets in the Moon race had failed: they had launched their (automatic) *Luna-15* spacecraft three days before *Apollo-11* in the hope of being the first to bring lunar samples back to Earth, but, after 52 orbits, it unfortunately crashed *during the time* Neil Armstrong and Buzz Aldrin were walking on the Moon![33],[34]

Fortunately, and perhaps inspired by the end of Paine's letter, Menzel made a diplomatic breakthrough. In two separate 2-page letters (dated Dec. 1, 1969), the first to the WG and the other to all members of Commission 17, he explains his plans. In fact, he *opposes* the advices given by *both* the WG and by the US National Committee! And he comes with a spectacular proposal.

First, Menzel describes the context: he acknowledges (i) the negative answer given by the Soviet Academy of Sciences; (ii) Dollfus's acceptance of this answer, consistent with WG discussions previously held in Paris in September; (iii) the US National Committee, which, on the other hand, "urges to accept" to recognize the *Apollo-11* astronauts. He concludes: "*We are all very disturbed by the attitude of the Soviet Academy, which is here interpreted as an attempt to withhold due recognition from the successful U.S. lunar landing.*" But something must be done. Further down, he adds:

> I can tell you that I have had heavy mail urging me to accept the unofficial names assigned by various astronauts, and significantly pointing out the acceptance of the Sea of Moscow, etc. *Various other groups, notably the International Academy of Astronautics (. . .) are questioning the authority of the I.A.U. to assign names.* In particular, they feel we have not done enough to recognize the astronauts themselves.

[The italics are mine.]

So, in view of these officials, but contradictory recommendations, Menzel deemed them as advisory only, and he "*does not feel bound*" to accept them. Then comes the outline of his proposal:

> In the interest of international harmony, I offer the following proposal: that we include the names of more –not fewer– living astronauts: 12 in all, six U.S., and six USSR. These twelve names should all be associated with outstanding "firsts." In addition to the astronauts of Apollo 11, I propose those of Apollo 8, the first three to orbit the moon.

[Of course, the number 12 was chosen to include those six!] (. . .)

> The I.A.U. convention [of 1935, revised after *Luna-3* at Berkeley in 1961] did not contemplate lunar landings and probably did not take into account the entire space program." "So let's be generous. I point out that Mare Astronautarum and its environment contain an abundance of unnamed craters. Let us move forward into cooperative space

[33] *Luna-15* crashed in Mare Crisium (the Sea of Crises!), about 500 km NE of the *Apollo-11* landing site (http://solarsystem.nasa.gov/missions/profile.cfm?MCode=Luna_15).

[34] Actually, 1969 was a very bad year for the Soviet lunar program: out of 10 spacecrafts planned, eight were lost due to launch or Earth orbit failures, and *Luna-15* crashed on the Moon. (https://en.wikipedia.org/wiki/List_of_lunar_probes#1968-1970). Only the *Zond-7* circumlunar flyby of Aug. 11 that year was successful.

research in the spirit of the editorial that appeared in this morning's New York Times, a copy of which I enclose.[35]

[The conclusion of this editorial reads: "*For the future the United States will gain most by taking the lead to make the moon a planet of international cooperation, on which men working together for peaceful objectives will set a useful example for a badly divided earth.*"][36]

The second letter, this time addressed to Commission 17, was presented as an "interim report" of the WG. Having in mind the pledge to produce a catalog of crater names on the far side of the Moon at the next IAU General Assembly in Brighton the following summer, Menzel first explains the methodology, starting with a call, dated Apr. 25, 1968, for suggestions from the Commission members themselves. Further recommendations were requested from 68 National Academies of Science "*and other relevant bodies of 55 countries from every continent (except Antarctica)*". And "*special advisory committees on lunar nomenclature were established in both the U.S.S.R. and the U.S.A. These bodies made many new, interesting and useful suggestions to the Working Group.*" This information shows that, in spite of the fact that the WG comprised only four members (including the President of Commission 17, A. Dollfus, *ex officio*), there was behind it a large network representing essentially all nations in the world (well beyond the IAU, which had only 48 "National Members" in 1967: the UN comprised 118 member states in 1965).[37]

So large was the pool of names, stored "*on a single card catalog*" (!) that the WG reduced the list to about 600, compiling short biographies, with the help of Barbara Bell (1922–2017) from Harvard College Observatory,[38] Menzel's home institution. Special maps of the far side of the Moon were provided by the Astronautical Chart and Information Center (ACIC) on behalf of NASA, and the adopted names were attributed in alphabetical order from West to East (i.e., as seen from the Earth) in latitude zones 20° wide (10° towards the poles), with the exception of prominent craters, attributed to "individuals of special distinction." (Note that, at this stage, only scientists, or famous individuals having a link with the Moon, such as writers,

[35] Issue dated Monday, December 1, 1969. Article by Harry Schwartz, entitled: "*President Nixon's Opportunity to Bridge the 'Lunar Gap'*".

[36] A perhaps optimistic statement, even though Nixon was starting to withdraw 150,000 troops from the maximum of 450,000 that were present in Vietnam that year.

[37] The important exception was China, which until 1971 when the "People's Republic of China" was admitted to the UN, was officially the "Republic of China" of 1912, reduced to the island of Taiwan after 1949. But four Chinese names, presumably suggested by Taiwanese astronomers, were added to the one given after the 1959 *Luna-3* flyby, Tsu Chung-Chi; see above, Sect. 13.1 and Fig. 13.2.

[38] Her biography can be found at https://baas.aas.org/pub/barbara-bell-1922-2017/release/1?collectionSlug=6ixhl5lq.

were considered, following the IAU Resolution adopted at the Prague General Assembly in 1967.)[39]

The most notable exception was the introduction of "Mare Astronautarum," the "Sea of Astronauts," a "dark, broad area" of the far side, which is not specified in the report. As a start, Menzel recommended naming three craters in this area after the *Apollo-11* astronauts. But as the date of the Brighton General Assembly was approaching (it was due to begin on Aug. 28, 1970), there were more internal discussions within the WG and with Commission 17, which resulted in significant modifications to the original plans. One was to change the spatial distribution of crater names: after consultation with cartographers, who did not like the idea of an alphabetical distribution, the names were attributed to craters randomly, but taking their size into account, the most prominent craters being attributed to the most distinguished names, for instance, Nobel laureates.

As for the astronauts, the "Mare Astronautarum" was dropped in favor of a very extended (diameter 537 km) crater named "Apollo," in the southern hemisphere of the *far side*. This carter has many smaller craters, six of which were then attributed both to the three living astronauts of *Apollo-8,* and to the deceased astronauts of *Apollo-1*. Then the names of the three astronauts of *Apollo-11* were attributed to small, but very visible craters on the *near side* of the Moon, next to their Mare Tranquillitatis landing site, the site itself being named "Statio Tranquillitatis" (Fig. 13.10). Then, concerning the cosmonauts, there was a similar group of craters surrounding the "Mare Moscoviense" discovered by *Luna-3* (276 km in length), and it was agreed to honor six living, prominent cosmonauts (of course less well known to the public): Konstantin Feoktistov, Alexei Leonov, Andriyan Nokolaev, Vladimri Shatalov, Valentina Terechkova,[40] and German Titov (for details, see Menzel et al., 1971). Komarov and Gagarin, both deceased, were attributed, respectively, a small crater near Mare Moscoviense, and a large crater (diameter 265 km) in the far side southern hemisphere (Korolev was honored by a very large crater near the equator, 437 km in diameter) (Fig. 13.11a, b). Also, the site of the first Soviet hard landing (*Luna-2,* 1959) was named "Sinus Lunicus," and that of their first soft landing (*Luna-9*, 1966), "Planitia Descensus."

Ironically, as it turned out, Menzel was contradicting himself, by eventually promoting a diplomatic, "two-sector" solution which he had explicitly opposed in his letter of Dec. 30, 1968, to the WG (see above, Sect. 13.4)!

Also worth mentioning was some "cleaning up" by the Working Group: the Curie family having had the extraordinary privilege to boast four Nobel laureates (for *five* prizes: Pierre and Marie, *née* Skłodowska, 1903, Physics; Marie again, 1911, chemistry; Irène, their daughter, jointly with her husband Frédéric Joliot, Physics,

[39] The final list of 513 names was published separately after the Brighton GA in 1970 (Menzel et al., 1971). It includes names of science-fiction writers such as Cyrano de Bergerac (1619–1655), next to the Gagarin crater, and H.G. Wells. Jules Verne had already been assigned a crater after *Luna-3* (Fig. 13.3). Recall that Menzel was a science-fiction writer himself...

[40] The first woman in space. Her flight had started on June 16, 1963, and lasted 70 hours and 43 minutes. (https://en.wikipedia.org/wiki/Valentina_Tereshkova).

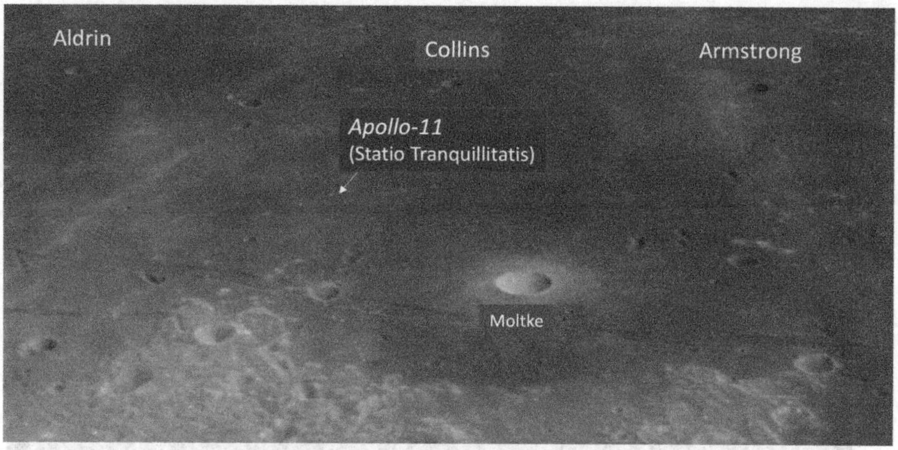

Fig. 13.10 Craters near the *Apollo-11* landing site, bearing the name of its astronauts, side view from *Apollo-16*. (Blow-up of "Apollo Image AS16-M-1389;" credit NASA/JSC/Arizona State University: https://wms.lroc.asu.edu/apollo/view?image_id=AS16-M-1389.) Labels added by the author

1935) posed some sort of a problem . . . So the names attributed in 1961 by the Soviets were adjusted to avoid confusion and allow them to keep "their" craters: "Joliot-Curie" became "Joliot," "Sklodowska-Curie" [dropping the Polish ł] became "Sklodowska." The crater "Curie" is at the limit between the near side and the far side, near the North pole, and stands for Pierre (as a result, perhaps for lack of an appropriate alternate name, Irène Joliot-Curie does not have a crater of her own on the Moon, but she does have one on Venus!).

Another clean-up took place for some topological features, and this is how the "Montes Sovietici" of 1961, a name which had been more or less controversial since then, simply disappeared: as already mentioned, while visible in the *Luna-3* photographs, these features corresponded, under a good solar illumination, to a large, elongated area darker than its surroundings. And by a twist of history, this reality was discovered by *Zond-3* in 1965, so this other potential "diplomatic problem" vanished by itself with the help of Nature . . .

All in all, in spite of both internal and external difficulties, Commission 17 and its WG on Lunar Nomenclature had done their job and fulfilled its task, as promised in Prague three years earlier. The outcome at the Brighton General Assembly was the vote of a surprisingly brief Resolution (No.8): "*The International Astronomical Union accepts the approximately 500 names proposed by IAU Commission No. 17 on the Moon for designating the craters on the far side of the Moon.*" In particular, the Resolution makes no reference to the exception of naming of 12 craters after living persons –six astronauts and six cosmonauts. This will have to wait for the next year, with the publication of the complete list of names, explanations, and short

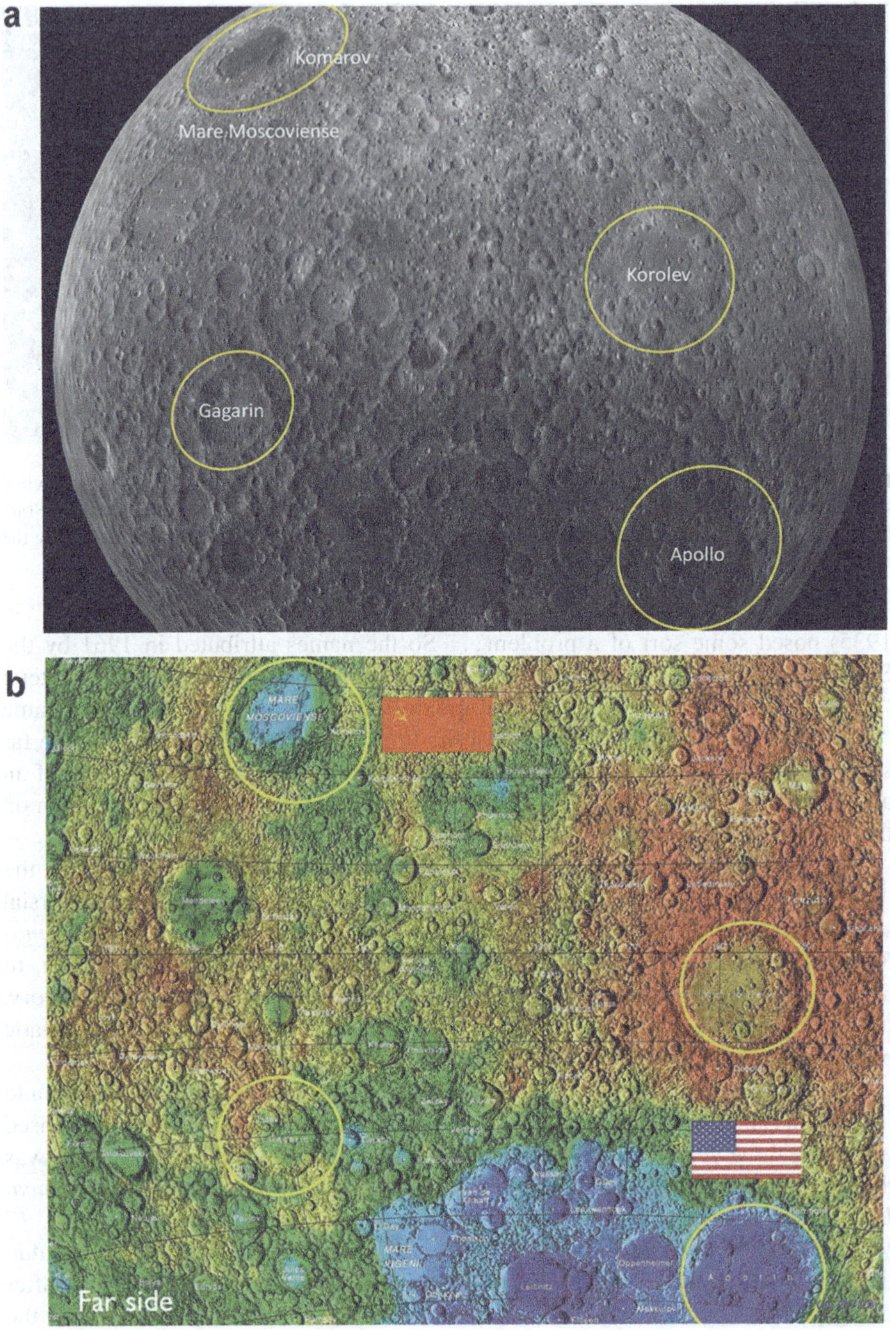

Fig. 13.11 (**a**) Far side of the Moon: the astronaut and cosmonaut separate "graveyards" (LRO). (**b**) Same as (**a**), superimposed on the topographical map of the Moon. (Source: https://pubs.usgs.gov/imap/i2769/ . Credit USGS). The labels on Figs. 13.11a, b have been added by the author

biographies, in *Space Science Reviews*.[41] On the other hand, the sad news was that one of the WG members, Marcel Minnaert, passed away on Oct. 26, only a few months after the Brighton GA, which he could not attend due to illness.

13.6 The Perilous Road to Sydney

During the General Assembly in Brighton in 1970, elections took place as usual, and the Officers of the Commissions changed, while the organization of the Working Groups remained at the discretion of the Commissions. As for Commission 17 ("The Moon"), however, the President remained the same, Audouin Dollfus. The Vice-Presidents were Boris Levin (USSR), and Gerald Kuiper (USA) (Fig. 13.12). Levin had been Vice-President of Commission 16 during the previous triennium, and Kuiper had been a member of both Organizing Committees of Commissions 16 and 17, Dollfus having been also an Organizing Committee member of Commission 16. Commission 17 still had three Working Groups (WG): WG1 (Nomenclature) was still chaired by Donald Menzel, with Levin and Dollfus (ex-officio); Minnaert was not replaced immediately, but only in April 1972, by Harold Masursky (Fig. 13.13), an astrogeologist from Flagstaff (Arizona).[42] All these astronomers had had high IAU responsibilities and knew each other very well, and were good colleagues, until strong disagreements would arise "on the road to Sydney."

In addition, Commission 17 co-opted 10 "consulting members" from different fields now of interest for lunar studies (as advisors to the three WGs), and two ex-

Fig. 13.12 Gerald Kuiper (1905–1973), here in 1964. After having been at Yerkes Observatory, he became Director of MacDonald Observatory. In 1963, he founded the Lunar and Planetary Laboratory in Tucson, Arizona. (Photo H. van Gelderen, Dutch National Archives; Creative Commons Attribution-Share Alike 3.0, Netherlands)

[41] *Op. cit.* (Note 39).

[42] The two other WGs were: WG2 "*Shape and Motion of the Moon*" and WG3 "*Geology and geophysics of the Moon.*"

Fig. 13.13 Harold Masursky (1922–1990) here in 1980. He was one of the first astrogeologists of the USGS Astrogeology Science Center in Flagstaff, Arizona, founded in 1963. (Photo: C. Springmann, Nat. Geo. 157, 18 [1980], Wikimedia Commons; original in color)

officio members, coming from COSPAR.[43] The Commission as a whole comprised 59 members, including the Officers and two women (only), unfortunately seen in the role of "secretaries" in spite of their degrees.[44]

On the other hand, at the top of the IAU leadership, the Executive Committee had a new President, Bengt Strömgren from Denmark (1908–1987), and a new General Secretary, Cornelis (Kees) de Jager, from the Netherlands (1921–2021), who would follow closely the work of Commission 17 and its WG on lunar nomenclature during his term. C. de Jager would be also President of COSPAR in 1972, strengthening the ties between both organizations (which already had in common an IAU-COSPAR Working Group on "Laser study of the Moon," its Chairman being one of the ex-officio member of Commission 17).

Also created by ICSU, in 1970, at the initiative of D. Menzel at early stages (ca. 1965), in relation with the development of the *Apollo* program, and pushed by Dollfus who would be its first Chairman, was an interdisciplinary "*Inter-Union Commission for the Moon–Commission Inter-Union de la Lune*" (IUCM-CUIL, better known as "*Inter-Union Commission for Lunar Studies*"). IUCM held its first session in Paris in May 1971, and its first General Assembly met in Nov. 1972.[45] It played a role in the background of the Commission 17 activities, through the action of Dollfus, who chaired both: indeed, reference to the IUCM is found many times in his correspondence.

---

[43] COSPAR (the "Committee on Space Research") had been established by ICSU in 1958 in London, following the launch of *Sputnik-1*.

[44] Barbara Bell, already mentioned, who worked with D. Menzel at Harvard College Observatory, had obtained a PhD (Harvard, 1951); Welsh-born Barbara Middlehurst had earned a masters degree (Cambridge, UK, 1947), and at the time worked with G. Kuiper in Tuscon. Both appear to have never married.

[45] Transactions Vol. XV B (1974), p.203. IUCM, which was active until 1976, comprised scientists from geophysics (IUGG), geology (IUGS), radio science (URSI), and COSPAR.

Last but not least, what shaped the 1970–1973 triennium for Commission 17 in particular in terms of context, was the evolution of the lunar space programs. On Sep. 2, 1970, budget cuts from Congress forced NASA to cancel the *Apollo* program, the *Apollo-17* flight (1972) then being the last (instead of the planned *Apollo-20*; T. Paine, the NASA administrator, resigned as a result two weeks afterwards).[46] With the exception of the *Explorer-49* radioastronomy orbiter launched in 1973, this would be the last American mission to the Moon until 1994 (with the orbiter *Clementine*). NASA instead put a priority on solar system exploration, using the Moon as gravity-assist (*Mariner-10* flew to Venus and Mercury in 1973).

The *Apollo* era was coming to an end. In contrast, the USSR, having "lost" the Moon race, opted notwithstanding to continue to do science on the Moon, notably with rovers (they invented the concept, with their *Lunokhod* robotic vehicles).

Undoubtedly, the WG work on "traditional" nomenclature of the far side of the Moon, based mainly on maps from *Lunar Orbiter* data drawn up for NASA by the professional cartographers of the Aeronautical Chart and Information Center (ACIC) in St. Louis, Missouri (a service of the US "Department of the Air Force"), was a success. Later scrutiny by Commission 17 cartographers and selenologists from the Lunar and Planetary Laboratory, mainly British-born Ewen A. Whitaker,[47] and David W.G. Arthur, did find a few errors or inconsistencies, like names attributed to faint craters, or unnamed large craters, but overall the group was satisfied. Commenting on a meeting of the WG that took place in Houston on March 9, 1973 (to which we shall return below), Kuiper wrote to Dollfus:

> During the IAU meeting at Prague in 1967 the decision was made that the [WG] be composed of three senior astronomers, Drs. Minnaert, Mikhailov, and Menzel, not directly involved with lunar space missions, mapping, or tectonic surface studies. This dissociated the [WG] from the immediate personal impacts during the exciting days when the details of the lunar Far Side were revealed for the first time. This was a good decision, and the roughly 500 names for the largest Far-Side craters so collected and assigned, were approved by the IAU in 1970.

Thus, an atmosphere of consensus seemed to prevail.

However, a year after Brighton, two challenges would seriously compromise this consensus.

First, a totally unexpected event took place. A letter (dated July 21, 1971), addressed to Dollfus (with copy to Menzel) arrived, boasting the letterhead of the United Nations. It was signed by A.M. Komkov "Vice-President, UN Group of Experts on Geographical Names," and started with the following: "*I have the honour to inform you that the UN Group of Experts on Geographical Names, following the directions of the Secretary-General* [of the UN] *has included the problems of nomination of extraterrestrial topographic features in its scope. In this connection, a special Working Group has been set up and I have been named its coordinator." (...) "I believe that the progress in studying the problem of*

[46] *Op. cit.* (Note 27).

[47] 1922–2016. Author of the reference book "*Mapping and Naming the Moon*" (Whitaker, 1999).

nomination of extraterrestrial topographic features (. . .) will be a success provided close collaboration of our working groups is organised. (. . .)." In other words, *the UN was frontally challenging the IAU in its long-standing tradition of giving names to celestial objects and their features.* On the other hand, the letter contained also an implicit warning to the IAU: astronomers were not considered as "owners" of the Moon anymore, they had to open themselves to other sciences (which, to a certain extent, was already the goal of the IUCM), and to the public at large.[48]

Second, in a letter dated Oct.4, 1971, Mr. R.W. Carder, who was one of the Commission 17 consultants, but through whom exchanges about NASA mapping issues took place, informed Dollfus that the *Apollo-15* astronauts had given "informal names" to surface features in the vicinity of the *Apollo-15* landing site in advance of their flight, located between Hadley Rille and Mount Hadley (already named in Blagg & Müller, 1935). As Fig. 13.14 shows, a large number of very small features (craters, hills, etc.) can be seen, down to sub-km scale, as opposed to scales down to 10 km at best for the craters named on the near side during telescopic times. Although obviously not concerning the whole lunar surface, Commission 17 and its WG1 now faced the mind-boggling issue of naming the so-called "smaller topographic features at the scale of direct exploration." Not only that: the problem was more generally of "who would have the privilege (the astronauts, like the ancient explorers?), or the authority (the IAU, as the international institution where the scientific expertise was supposed to lie since 1935? Or else the UN, representing all the countries in the world?) to give names to the lunar landscape?"

Conceptually, the problem was not new. In the mid-nineteenth century, the German amateur astronomer Johann Heinrich von Mädler (1794–1874), together with a wealthy banker, Wilhelm Beer (1797–1850), had built his own private observatory in Berlin, and set out to obtain the best maps of the Moon available, which took nine years to complete (1828–1836). The result was published as *Allgemeine Selenographie* (1837), the Moon being represented in four sheets of one quadrant each, about 1.20 m on a side (!), with coordinates (and North down). The nomenclature was based on names of famous astronomers (including in antiquity), following a tradition established by a Jesuit priest, Father Giovanni Riccioli, in 1651, that had gained wide acceptance, and based on maps drawn by his fellow S.J., Father Francisco Grimaldi. However, Mädler's precise observations revealed a multitude of smaller details, mainly craters, near features already named.

In order to take into account the coexistence of large and small features in a given area of the lunar surface, Mädler invented a system of so-called "satellite features" (in general craters), identified by *letters*, associated with a nearby more conspicuous one, named "parent feature." Originally, the letters were isolated; they could be Roman or Greek, uppercase or lower case, meant to give a clue to the nature of the features. However, this system proved to be rather intractable (and often needless),

[48] A detailed account of this episode, which lasted ten years and led to a major crisis between the IAU and the UN, is given in Chap. 16. However, it will play an important role here and be mentioned several times.

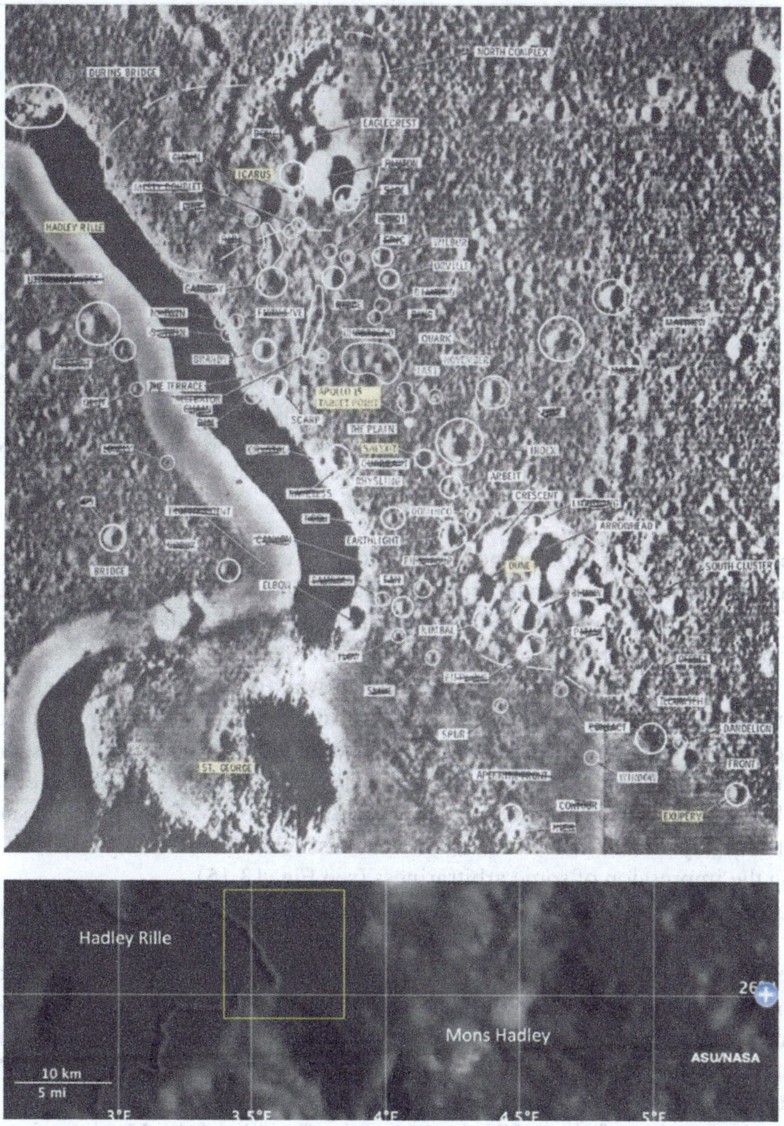

Fig. 13.14 ***Top:*** *Apollo-15* astronaut "tour guide" names, assigned in preparation for their landing and map location (background: LRO images; label "APOLLO-15 TARGET POINT" towards the center of the image. IAU archives). LRO is now flying at an altitude as low as 50 km, hence the tremendous amount of details it can reveal. ***Bottom:*** Blow-up by the author of image centered on the nearby Béla crater (Credit: USGS, https://planetarynames.wr.usgs.gov/Feature/947)

so in practice, only the idea of capital letters remained: satellite features were thus simply ordered by increasing distance from their parent feature. But at some point, the letters associated with one parent feature would occupy the same "territory" as

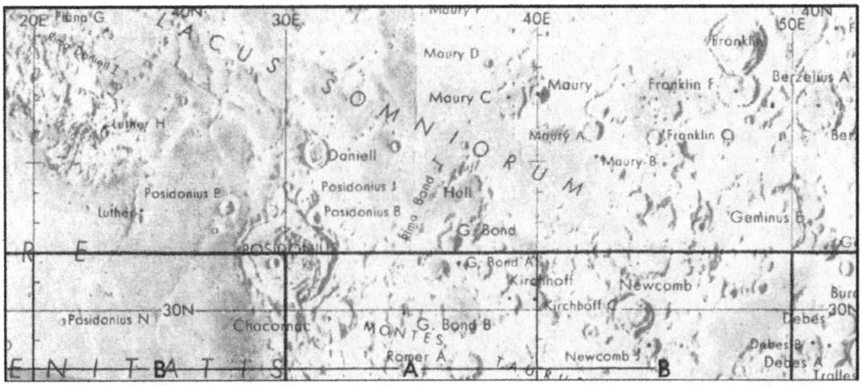

Fig. 13.15 *Lacus somniorum* region and the (simplified) Mädler system: parent craters (here POSIDONIUS, Maury, etc.) and their lettered "satellite" craters (from a map published by NASA, printed Dec.11, 1972; found in the IAU archives). See also Fig. 13.17 below

those associated with another, not to mention the cases of "craters within craters" or located themselves on crater rims, generating inevitable confusion –recall that, even in 1935, at the time of the Blagg & Müller catalog and atlas, everything (maps, lettering, coordinates, etc.) were drawn by hand![49]

A simplified solution to the Mädler system was found, by keeping only roman uppercase letters, always written along with the explicit name of their parent feature, and identified with a dot. But in the process, some letters, for which the association was found to be ambiguous, were simply dropped, so that in the end even the simplified Mädler system had many inconsistencies (e.g., missing letters, satellite craters equally distant from named craters but associated to one only, etc.), all in all giving the impression of some arbitrariness (see Fig. 13.15).

To these misgivings of the Mädler system, everybody agreed. But after all, so long as only the most prominent near side features were properly identified, and named according to certain rules, giving other names to smaller features, lettered or not, was not deemed so important, except locally, for instance, due to specific scientific motivations. By 1967, over 700 proper names had been assigned to various features (mostly craters), and about 8000 features were defined as "satellites" and eventually designated by adding letters.

But the naming of about 500 craters on the far side of the Moon having been successfully accomplished, arose a consistency issue: even though the two sides had major differences in their topography (the far side being almost everywhere covered with craters, the near side being dominated by large "flat" areas, classified as "Oceanus," "Mare," etc.), the Moon was a single celestial body and it was logical that it should have a unified description, not only of the topographic features ("mons," lacus," etc.; that unification had been done early, as soon as the orbiters

[49] For a detailed history of lunar mapping and nomenclature, see Whitaker (1999).

had mapped the whole sphere), but also of the nomenclature process itself. As for the far side, and given the huge number of craters (and the limited time available until the Brighton General Assembly when the results were to be published), the WG not even dared to think about some adaptation of the Mädler system, they just drew a list of proper names and their assignments.

So, somewhat paradoxically, the simple nomenclature adopted for the far side (no Mädler system) raised new problems *for the near side*: in view of the necessity of addressing the nomenclature of small-scale features discovered from space, and invisible with Earth-based telescopes (smaller than 10 km, say), what should be the direction to take: (*i*) revise in-depth the well-established Mädler (lettering) system, e.g., using some form of code associating double letters, numbers in the form of short-hand coordinates, etc.? or (*ii*) drop the Mädler system altogether (progressively if necessary) in favor of a much-enlarged list of proper names, i.e., an extension of the existing large-scale nomenclature adopted for the far side, but down to some predefined small-size limit?

13.7 Civil War

Unfortunately, a deep fault appeared within the WG, and within the Commission, about these questions. On May 17, 1972, the WG held a meeting in Madrid, gathering Menzel (Chairman), Dollfus, Masursky (his first WG meeting as a replacement for Minnaert), Carder (ACIC), and W. Brunk from NASA, as observer. Levin was absent, but had had prior correspondence with Menzel and Dollfus.

Summarizing the situation, Menzel strongly argued in favor of dropping the Mädler lettering system altogether and replace it with names, even if that meant finding up to 20,000 candidates. He also mentioned the *Apollo-15* astronauts request as an illustration for the need of feature names as a help for lunar navigation and soil sample collecting (see above, Fig. 13.14). Of course, not all of these "informal" names were acceptable, on various grounds (names already existing, not following the standard IAU rules, inappropriate or ambiguous, etc.), but provided they were submitted somehow to the IAU for approval, possibly after screening by NASA, they could be considered valid. Alternate suggestions, like codes or coordinates (as proposed by the UN Working Group on Extraterrestrial Names, in particular), were found to be impractical and cumbersome for users, and rejected offhand.

The minutes of this meeting were circulated, and on Aug. 28 Levin replied to Menzel, expressing general agreement, with two reservations. A procedural one: "*My feeling is that at the present moment the most urgent problem is to define the whole IAU policy in respect of all aspects of astronomical literature, as the policy inside the IAU, and that in respect of the UN and other organizations. Of course, it is for the IAU Executive Committee to decide, not Commission 17, or our WG itself.*" And a practical one: "*I am too not a proponent of the conventional method of naming craters by additional roman letters. (. . .) One must have some statistics of craters and possible names before to propose some decision.*"

In an 8-page "Memorandum" dated Oct. 21, written in preparation for the next WG meeting due to take place in Paris on the following November (27–28), and circulated for comments "to a large number of interested individuals" (unspecified, but which suggests it was sent even beyond the WG and Commission 17), Menzel summarizes his arguments, suggesting to extend the name list to non-scientists: writers, artists, singers, painters, etc.[50] (including women, very poorly represented in the existing far side list because there were so few female scientists to choose from).

The flak was immediate and came from the Lunar and Planetary Laboratory, where Kuiper and his colleagues had, in particular, prepared the *Photographic Lunar Atlas* a decade before (1960). Not being invited to the Paris meeting because they did not belong to the WG (and perhaps somewhat acrimonious for not having been part of the far side nomenclature, in spite of Kuiper's later reassuring words of 1973 mentioned above), their reaction to Menzel's Memorandum was extremely violent.

In a letter dated Nov. 22, Whitaker wrote to the WG a long letter vigorously opposing Menzel's proposal to drop Mädler's system and questioning the competence of the WG:

> The problem of the assignation of names to small craters in certain very restricted areas is a cartographic and selenological one, to be undertaken by those who are competent in this field. While the IAU should offer guidance in the choice of names, it should have no input in the actual assignment of those names. The Working Group (...) clearly has no mandate for this task of name assignments.

Kuiper, in a "Personal" letter to Dollfus (Nov. 29) went even further:

> A dichotomy has developed between the professionals in this field, concerned about the real problems of the Moon, and a person like Menzel who could go on forever naming things as a separate art, without much relevance or concern to the realities of lunar science and space missions. I feel that the I.A.U. should withdraw from this latter type of activity and that the Working Group which appears to have been dominated by Menzel be dissolved.

Arthur concurs, writing to Kuiper the day before (Nov. 28):

> We have a serious problem in relation to the distribution of names on the far side. (...) This distribution of names was performed on the basis of a hastily compiled map by men with no experience in these matters, no acquaintance with the Orbiter materials, and no sense in the limitations of the data. (...) I am not clear as to how this mess should be straightened up, or who does the systematic work leading to this." (...) "The working group has far exceeded both its mandate and its capabilities. I think it should be brought to an end.

Thus, a kind of civil war had erupted between Americans: Menzel in the North (New England), and LPL in the South (Arizona). They had allies, however, as we'll soon see: the French Dollfus on Menzel's side, the Soviet Levin on LPL side. Then, believe it or not, the Cavalry went to the rescue of Menzel, in the form of a newcomer, Col. Strickland of Washington, D.C., on behalf of NASA: "*The situation on lunar nomenclature is changing very rapidly* [writes Menzel to the WG

[50] Actually, the far side list did already include the *Luna-3* names of non-scientists, approved in 1961 (see Note 39).

on Dec.8]. *I had a phone call from Colonel Strickland, of NASA Headquarters, concerning the problems of lunar mapping. Colonel Strickland is head of all projects for lunar mapping, based primarily on NASA photography obtained with the Lunar Orbiters and Apollo missions.*" "*He tells me that, starting in February or March, they will start preparing a new series of maps on the scale of 1:250,000.*[51] *Complete coverage will require about 200 sheets. I further understand that first attention will be given to the equatorial regions, accurately photographed by the Apollo missions.*" [Note that, at the time of writing this letter, the astronauts of *Apollo-17*, the last of the program, were actually touring the Moon on their lunar "jeep."] Further down, Menzel writes:

> This puts a great deal of pressure on our Working Group to come up with a list of 2000 names between now and March 1. (...) I feel it is the responsibility of our Working Group to provide the names requested." (...) "I repeat my previous statement that the list (...) should include the names of great writers, playwrights, musicians, biologists, medical, explorers, engineers, inventors, philanthropists, and possibly others.

Then it seems that Menzel somehow got carried away, and he crossed the red line: "*Because of the urgency of the call, I cannot even wait for the approval of the Executive Committee of the IAU.*" (The italics are mine.) Not only that: on Dec. 13, he circulated another, brief "memorandum," that went even farther beyond the limits of IAU procedure, claiming: "*NASA authorities have requested the I.A.U. Working Group on Lunar Nomenclature to provide a list of names of distinguished persons, for assignment to lunar features of hitherto unnamed or designated merely by literal suffixes as a satellite crater to a major feature, such as Hercules A or Littrow BC." (...) "The Working Group now proposes to extend recognition to individuals in fields such as the following:* [list: writers, poets, etc.]," ending with: "*To ensure that the representation be as international as possible, this request is being disseminated, as far as possible, to all countries in the world. (..) Women as well as men of distinction should be included." (...) "Please give this matter your most URGENT attention.*" (The italics are mine; the uppercase letters are in the original.)

The memorandum was signed "A. Dollfus, B.J. Levin, H. Masursky, Donald H. Menzel, Chaiman," implying that the WG as a whole, and the IAU authorities, had approved Menzel's initiative. But this was not the case, and it caused quite a drama. On Dec. 22, The General Secretary, Kees de Jager, likely after consultation with the President, Bengt Strömgren, and other Executive Committee members, replied in a polite, but irritated tone:

> I have received your memorandum of December 13 regarding the naming of lunar craters. (...) I feel it is my duty to inform you that normally all external activities under the authority of the IAU have to be guided by the General Secretary. This is normal practice, which has recently been stressed and confirmed again by the Executive Committee.

[51] This scale corresponds to regional road maps (1 cm for 2.5 km), like the British OS maps, the French Michelin Road Atlas, etc. These maps are usually identified by some geographical name, in addition to the cartographer's reference number.

> Particularly with regard to an item as important as the naming of lunar craters I think it might have been advantageous to the case you serve when this memorandum would have been distributed through the official IAU channels.

(Later Menzel disclosed that the channel he had used was the US State Dept., by way of personal relations, which aggravated the case.)

Then the naming war turned "international." While the IAU archives have no record of possible immediate reactions from Dollfus and Masursky, they do include two identical telegrams sent by Levin from Moscow, respectively dated Jan. 15 and Jan. 18, 1973, to Dollfus and de Jager. Figure 13.16 shows the telegram sent to de Jager, where one can read Levin's violent response: "PROFESSOR MENZEL WITHOUT AUTHORIZATION OF THE WORKING GROUP . . . STOP THIS MOVE INTEND FORCE IAU ACCEPT WITHOUT APPROVAL COMMISSION 17 AND GENERAL ASSEMBLY . . . NAMING LUNAR CRATERS OPPOSED BY KUIPER MYSELF AND OTHERS STOP STRONGLY PROTEST STOP . . . REPORT DATED DECEMBER ELEVEN EXPOSES MAINLY MENZEL'S OWN VIEWS NOT WORKING GROUP DECISIONS STOP." At the same time, in a letter dated Jan. 15, Levin develops his reply: "*I was surprised by the content and style of the last numerous letters and memoranda by Prof. Menzel. Previously he acted as Chairman of the Working Group of Commission 17 of the International Astronomical Union, but now seems to act as a representative of NASA.*" (. . .) "[The Dec.13 memorandum] *is an*

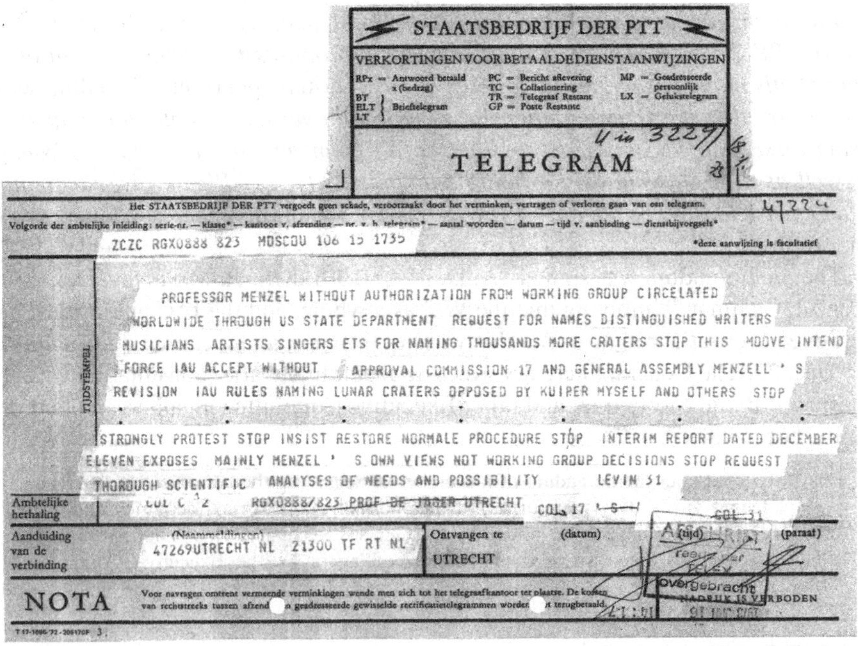

STAATSBEDRIJF DER PTT

VERKORTINGEN VOOR BETAALDE DIENSTAANWIJZINGEN

RPx = Antwoord betaald x (bedrag) PC = Bericht aflevering MP = Geadresseerde persoonlijk
TC = Collationering
BT, ELT, LT = Brieftelegram TR = Telegraaf Restant LX = Gelukstelegram
GP = Poste Restante

4 in 3229

TELEGRAM

Het STAATSBEDRIJF DER PTT vergoedt geen schade, veroorzaakt door het verminken, vertragen of verloren gaan van een telegram. 47724

Volgorde der ambtelijke inleiding: serie-nr. — klasse* — kantoor v. afzending — nr. v. h. telegram* — aantal woorden — datum — tijd v. aanbieding — dienstbijvoegsels*

ZCZC RGX0888 823 MOSCOU 106 15 1735

*deze aanwijzing is facultatief

TIJDSTEMPEL

PROFESSOR MENZEL WITHOUT AUTHORIZATION FROM WORKING GROUP CIRCELATED WORLDWIDE THROUGH US STATE DEPARTMENT REQUEST FOR NAMES DISTINGUISHED WRITERS MUSICIANS ARTISTS SINGERS ETS FOR NAMING THOUSANDS MORE CRATERS STOP THIS MOOVE INTEND FORCE IAU ACCEPT WITHOUT APPROVAL COMMISSION 17 AND GENERAL ASSEMBLY MENZELL ' S REVISION IAU RULES NAMING LUNAR CRATERS OPPOSED BY KUIPER MYSELF AND OTHERS STOP

STRONGLY PROTEST STOP INSIST RESTORE NORMALE PROCEDURE STOP INTERIM REPORT DATED DECEMBER ELEVEN EXPOSES MAINLY MENZEL ' S OWN VIEWS NOT WORKING GROUP DECISIONS STOP REQUEST THOROUGH SCIENTIFIC ANALYSES OF NEEDS AND POSSIBILITY LEVIN 31

Ambtelijke herhaling: COL C 2 RGX0888/823 ~~PROF DE JAGER~~ UTRECHT COL 17 ~~COL 31~~

Aanduiding van de verbinding: (Naammeldingen) 47269UTRECHT NL 21300 TF RT NL

Ontvangen te UTRECHT (datum) (tijd) (paraaf)

overgebracht

NOTA Voor navragen omtrent vermeende verminkingen wende men zich tot het telegraafkantoor ter plaatse. De kosten van rechtstreeks tussen afzender en geadresseerde gewisselde rectificatietelegrammen worden niet terugbetaald.

NADRUK IS VERBODEN

Fig. 13.16 The telegram sent from Moscow by Levin to Kees de Jager in Utrecht on 18/1/73. (IAU archives) *attempt to force us [underlined] to assign thousands of new names. I regard as necessary to withdraw this memorandum." (...) "No one has yet convinced me that it is necessary to increase 3-5 times the number of named craters." (...) "I continue to be against the use of names of writers, painters...*" [etc.] (The italics are mine.)

13.8 NASA Comes to the Rescue

Spectacular as it seems from these reactions, the Menzel initiative and his obvious disrespect of IAU procedures somewhat surprisingly seem not to have had serious consequences in the end. In fact, given the short deadline he had given to receive name suggestions (a two-month delay!), and the rather cryptic way of formulating his memo as seen from outside, it seems that the response was weak, if any–at least there is no trace of it in the archives. Another reason may be that, beyond the flare-up, the Sydney General Assembly was approaching, and NASA was really eager to start the ball rolling about the identification of their 1:250,000 scale map sheets, even though the *Apollo* flight program was now terminated. Much remained to be done, and the only way forward was to renew support for the WG and Commission 17 work, in spite of the difficulties. Menzel had been the mastermind of the diplomatic settlement of the USA-USSR conflict about the nomenclature of the far side of the Moon, but obviously he had been not so diplomatic with his colleagues of the IAU... As it turned out, it would be up to the Commission 17 President, Dollfus, to engineer a solution such that his Commission and the WG would deliver in Sydney (in a way covering up the crisis), as successfully as they had done in Brighton.

On Feb. 23, 1973, an official letter on behalf of the National Academy of Sciences (NAS) was sent to Menzel. It looked very much like a circular indicating a definite roadmap on lunar nomenclature. It was signed by the NAS "Staff Officer for IAU Affairs," E.R. Dyer, Jr., and copied to a variety of people: on top, C. de Jager for the IAU, then D.S. Heeschen, C.R. O'Dell, and C.R. Tolbert for the US National Committee for the IAU (they were respectively the USNC Chairman, Vice-Chairman, and Acting Secretary), then Dollfus, Kuiper, Levin, and Masursky, and finally to N.W. Hinners, "Deputy Director & Chief Scientist, Lunar Exploration (Apollo)." Thus, it had all the appearance of an "act of law" to stop the civil war. Menzel was designated not as a "representative of NASA," as Levin had formulated, but as having an official mandate from NAS, to be followed by NASA. Of the seven numbered items in Dyer's long letter, each had some instructions for the recipients. The most important statement regarding the mission of the WG was clear: "*I am hereby requesting on behalf of the USNC for the IAU to go ahead and act on (...) the assignment of names to features in* [20-30] *quadrangles of the series of lunar maps NASA plans to publish this spring." (...) "subject only to pro-forma approval, first by IAU Commission 17, and then by the General Assembly of the IAU.*" Dyer, while recommending to use names, did not take sides on the issue of their origin, as long as they had IAU approval: on this point, it was the WG task to decide first.

Based on these developments, the WG was on the safe side: it would not to be "brought to an end," as the LPL group had wished (albeit admittedly not officially requested), but rather put back on the right track. With the ACIC cartographers providing the maps on behalf of NASA (this was the main reply to the LPL criticism of "incompetence" of the WG), finding new names for smaller craters was a well-defined task of a reasonable scale, even if the estimate of "20-30" eventually turned out to be insufficient (it would be more like 250, many sheets covering no crater already named), quite different from Menzel's initial "all-out" suggestion to find 2000 names with a March 1 deadline: the number was far smaller, and the deadline was now extended by several weeks to the Sydney General Assembly in August.

The WG had been provided with a preview of the maps, a fraction being shown here (Fig. 13.17, version of Dec. 11, 1972). The NASA cartographers had superimposed grids following the paths of *Apollo-15* and *Apollo-16,* and indicated in the corner their identification system: LAC (for "Lunar Chart") "quadrangles" of 1:1,000,000 scale, identified by numbers, here "58," (later also by names)[52] divided in four quadrant identified the letters "A" to "D," running clockwise, themselves divided in 4 sheets each of 1:250,000 scale, unambiguously identified with the numbers "1" to "4," also running clockwise, so each sheet had a "license plate number," here "58C3." If one or more named craters were present, the sheet would bear the name of the most conspicuous one, and that would be it. If no names crater were present, one had to find a name. It may have happened, for instance in Mare regions, that no crater was apparent, but that was rare, then some other feature, like a "rille" (crevasse: see Fig. 13.14, bottom), could be named instead. Whatever arguments the LPL had developed, it was hard not to be convinced (*i*) that the NASA straightforward sheet identification system by "license plates" was very convenient, and (*ii*) that most sheets already having a complementary, "operational" name identification common for road maps (here based on crater names), the naming system could easily be extended to the other sheets, elegantly replacing Mädler's awkward lettering system in the relevant areas.

All these issues, and others, were thus on the agenda on the next WG meeting, which took place in Houston on March 9. Present were Menzel, Dollfus, and Masursky; Levin was again absent but had sent documents in advance. In addition, two guests joined the group, Kuiper and Col. Strickland; the ACIC representative, R.W. Carder, was absent.

As a result, Menzel produced minutes in the form of a thick, 19-page document, too long to be presented in detail here. The main conclusions, however, were: (*i*) Kuiper admitted that the Mädler system needed revision, and that the far side and the near side of the Moon required the same feature naming system; Menzel replied by stating that 3/4 of the Mädler letters were not used; (*ii*) Menzel argued again in favor of adding new names, citing once more the UN pressure to broaden their selection

[52] See the Lunar Chart Series: https://www.lpi.usra.edu/resources/mapcatalog/LAC/.

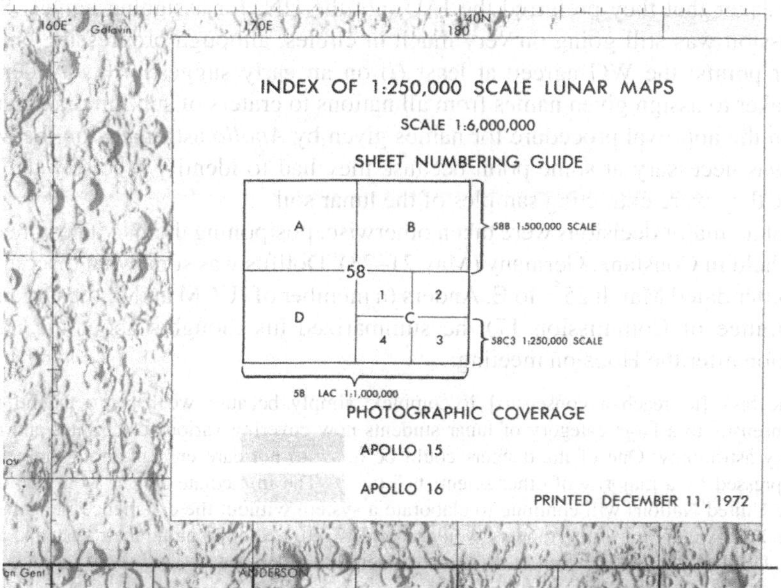

Fig. 13.17 Part of the NASA map of Dec. 11, 1972 (the same as in Fig. 13.15) showing the 1:1,000,000 and 1:250,000 grids, subsequently named *Provinces* and *Regions*, respectively

critera; he also argued that names of non-scientists would add a "human" side to the names, and enhance the "position" of the IAU within the public; (*iii*) Strickland insisted again on the necessity to find new names for the "nameless" 1:250,000 scale maps (about 250), and to assign about 2000 (mostly existing) names to the 1:1,000,000 map; he said the IAU "*should be the cognizant body*;" (*iv*) referring to the UN challenge to the IAU, Dollfus mentioned that the USSR had indicated to de Jager that they preferred the IAU, not the UN, for assigning names. So the discussion was still going on very much in circles, although progressing on some minor points: the WG agreed at least (*i*) on an early suggestion by Anders and Whitaker to assign given names from all nations to craters of sub-km diameter, and (*ii*) on the approval procedure for names given by *Apollo* astronauts on the Moon; this was necessary at some point because they had to identify precisely the place where they were extracting samples of the lunar soil.

But no major decisions were taken otherwise, postponing them to a new meeting, to be held in Constanz, Germany (May 21–23). Dollfus was somewhat discouraged. In a letter dated March 25[53] to E. Anders (a member of IUCM and of the Organizing Committee of Commission 17), he summarized his thoughts about the naming situation after the Houston meeting:

> The task [to reach a consensus] is complex, simply because we have to extend the consensus to a large category of lunar students now covering various disciplines and not only astronomy. One of the dangers could be (...) to not care enough about the needs expressed by a majority of other scientists." (...) "The immediate danger would be that the United Nations will continue to elaborate a system without the experience we gained, the NASA mapping cartographers would be obliged to elaborate alone their systems, and the USSR cartographers could follow on different lines. (...).
>
> I have just received a memorandum by Prof. Kuiper, dated 17 March 1973, written after the IAU Working Group Conference in which he was invited as an expert. His view is still that the lunar nomenclature elaborated at the ground-based telescopic observation scale, and for the use of the astronomers of the pre-Apollo period, should be left intact and all extension to smaller features still postponed for three years more (next IAU meeting). This view will obviously be fully discussed once again at our next IAU Conference; but it disregarded all the steps ahead already agreed upon. *I am afraid it might again cast a doubt about the real ability of IAU astronomers to decide alone on the needs expressed for a smaller than telescopic-scale lunar features nomenclature.*

(The italics are mine.)

Time was running out, something had to be done to unlock the situation. Given the proximity of the Constanz meeting (six weeks later) and, above all, of the Sydney General Assembly, where the Commissions had to formally produce their triennial reports, Dollfus came up with an idea: instead of trying desperately to build a large list of names, with no method in sight, try to pin down some recommendations (to be agreed on), then formalize them in the form of a Resolution submitted to the General Assembly. After all, as already mentioned, Dollfus was a veteran of IAU procedures. The archives keep an annotated copy of a "working document for the draft of a possible resolution" to be studied at the next WG meeting, dated Apr. 6, 1973. After the usual "*noting that*...," "*remembering that*..." etc., come decisions and recommendations: "*decide to refuse any further development of the Mädler system*... ," "*assign new names for each sheet of NASA maps at*

[53] The copy kept in the IAU archives mention March "5," but this must be a typo, since that date preceded the Houston meeting. It could not be on the 15th either, since Dollfus mentions a letter from Kuiper received on March 17. Since no month has 35 days, the true date must be March 25.

1/250,000 scale . . . ," "recommends to study the increase of the number of craters named after commemorative names (. . .) which could include such individuals as writers, composers, etc . . . .," "recommends for the case of the front side for which a capital letter was already used, to study ways of preventing errors in simultaneous use of capital letters and newly assigned commemorative names," etc. This was a diplomatic way to recommend the adoption of new proper names without reference to science, for the lunar map sheets that needed them, and consider them to progressively replace the lettered craters inherited from the Mädler system.

At the Constanz meeting, after much discussion,[54] the spirit of Dollfus's working document was adopted. To help in drafting the Resolution for Sydney, NASA introduced a new, somewhat poetic, vocabulary: the large areas covered by the 1:1,000,000 quadrangles would be called "*Regions*," the smaller ones, at 1:250,000 scale, "*Provinces*" (see Fig. 13.17).

Eventually, Commission 17 and the WG had found a way to present themselves in an orderly fashion, at least seen from the outside. The Sydney "Resolution on Lunar Nomenclature" presented by Commission 17 is quite long and more detailed than Dollfus's first draft, and some points are rather technical, reflecting the huge amount of work necessary to be ready in time. It includes a map of the NASA regions coverage and a sketch of the *"Region"*-*"Province"* relation. It can be found in full in the IAU website.[55] For our purpose here, the three main sections (out of 10) are as follows:

> (7) Whereas previously assigned lunar names have been, with few exceptions, those of distinguished deceased scientists, future assignments may also contain the names of distinguished, deceased contributors to human culture and knowledge, such as writers, painters, musicians, etc., chosen on an international basis.[56] Excluded, however, are political, military, and religious figures, as well as modern philosophers.
>
> (8) Very small features requiring identification for some special reason may be assigned, with IAU approval, first names, male or female, from an international list. Excluded, however, are names having more than three syllables.
>
> (9) For operational purposes, such as designating areas specially explored or sampled, Apollo missions have assigned names to certain features in the vicinity of the respective landing sites. The list was corrected and revised by the Working Group in consultation with the astronauts and Principal Investigators of the Apollo missions. The IAU requests that future lists of this type be submitted to the IAU in advance of the mission.

---

[54] Unfortunately, Menzel did not record the attendees, but the minutes were sent to Levin, Masursky, Dollfus, Anders (WG), de Jager, Strömgren (IAU), Dyer, Goldberg (NAS), Heeschen, Hinners (USNC), Col. Strickland and G. Kuiper.

[55] All the Resolutions of the 1973 General Assembly in Sydney can be found in: https://www.iau.org/static/resolutions/IAU1973_French.pdf (actually in both French and English). The Commission 17 Nomenclature Resolution, as well as other Resolutions on Moon research, can also be found, with more details, in its report (Transactions Vol. XV B, 1974, pp. 109–115).

[56] Ironically, with the "unauthorized" first naming of features on the far side of the Moon following the successful *Luna-3* flyby, the Soviets were the ones who had invented the first steps in that direction.

In the meantime, the nomenclature issue (and the role assigned to the IAU) had grown to include planets, following the successful, albeit still limited, explorations of Mars, Venus and Mercury by Soviet and American space probes. As a result, at the Sydney General Assembly a major restructuring took place: a large "*Working Group for Planetary System Nomenclature*" (WGPSN), with "Task Groups" each assigned to the different planets and to the Moon, was established by the General Assembly, and fell under the purview of the Executive Committee. It was thus an "independent" structure, yet directly supervised by the Executive Committee, which may have had in mind that, sometimes, Working Groups in charge of sensitive issues could have a tendency for free-rolling, and that they had to be kept under control... And it was now one of the tasks of the WGPSN to cope with the crisis with the UN that would last until 1982, when, finally, a UN Resolution recognized the authority of the IAU on the so-called "extraterrestrial names."[57]

The WGPSN still exists today, and remains very active: its naming decisions, after approval, are recorded in the "Gazetter of Planetary Nomenclature" managed for the IAU and NASA by the *US Geological Survey*.[58]

13.9 Epilog

Yet, it seemed that a strange curse had been placed on the defunct WG1 of Commission 17 on the occasion of its new incarnation as the WGPSN: a few months after the Sydney GA, on Dec. 23, 1973, G. Kuiper passed away, and similarly D. Menzel died on Dec. 14, 1976, i.e., right after the next (Grenoble) General Assembly. As already mentioned, Minnaert himself had passed away in October 1970, also following a General Assembly (Brighton)... Of course, this simply meant that the generations pass, and new generations take over. But the context of lunar nomenclature was changing, too: the Cold War was warming up. The SALT treaty, signed on May 26, 1972, in Moscow by Richard Nixon and Leonid Brezhnev stopped the arms race, and American and Soviet scientists started to collaborate, in particular in space research (first *Apollo-Soyuz* docking in July 1975).

And then the Moon race was really over–so much so that the Moon surface was "abandoned" for forty years, until new actors appeared: India, Israel, and of course China. The rover *Yutu* (玉兔, the Jade Rabbitt), a payload of the *Chang'e-3* mission (*Chang'e*, 嫦娥, is the goddess of the Moon) landed on the Moon on Dec. 14, 2013. It was the first rover to operate on the Moon since *Lunokhod-2* in 1973,[59] although its travel was limited to 114 m due to operational difficulties. *Yutu-2* (*Chang'e-4*

[57] See Chap. 16. For an early, public report by its Chairman on the activities of the WGSPN and its relations with the UN Working Group on Extraterrestrial Names, see Millman (1976).

[58] https://planetarynames.wr.usgs.gov/.

[59] Strictly speaking they are not quite comparable: the car-sized *Lunokhod-2* weighed 840 kg, whereas the *Yutu* series have a length of 1.5 m and weigh 140 kg, a bit less than the Martian rovers.

mission) landed on the far side (the first to do so), inside the Von Kármán crater, near the South pole, on Jan. 3, 2019. It traveled 120 m but lost power prematurely. Features photographed around the landing site were given Chinese mythological names, which were approved by the WGPSN and the IAU a month later.[60]

Before that, other "commemorative names" were approved by the WGPSN on the far side of the Moon, in particular those of the three *Soyuz-11* cosmonauts who lost their lives after leaving the Soviet space station *Salyut-1* in 1971, in the Mare Moscoviense area, and those of the 14 astronauts who were the victims of the *Space Shuttle* accidents (*Challenger*, 1986, and *Columbia*, 2003), within the *Apollo* crater (see above, Figs. 13.11a, b).

As of today, one finds from the *Gazetter of Planetary Nomenclature* that no less than 9163 lunar feature names are recorded, including 7113 "satellite features," i.e., lettered names. The number of "commemorative names" has grown to 2050, from about 850 at the time of the Sydney General Assembly. The Mädler system is thus quite resilient, new names progressively replacing older lettered names, but the rate is slow. On the other hand, lettered craters must be kept available for naming by future generations... IAU's WGPSN will take care of it.

Acknowledgments It is a pleasure to acknowledge the assistance of Mme Ginette Rude, Archivist at the IAU Secretariat in Paris until 2016, who directed my attention to the *Dollfus Files* and triggered my interest in this episode of the history of the IAU. I am also deeply indebted to Ms. Ariane Dollfus-Babinet for interesting correspondence about her father Audouin Dollfus, and for permission to show one of the pictures she sent me of his balloon flights (see Fig. 13.7).

References

Barabashov, N. P., Mikhailov, A. A., & Lipsky, Yu. N. (1960). *Atlas of the far side of the moon (Acad.Sci.USSR), translated as "An Atlas of the Moon's Far Side, The Lunik III Reconnaissance*" (p. 143). Interscience Publishers.

Beer, W., & Mädler, J. H. (1837). *Allgemeine Selenographie* [Full title: *Der Mond nach seinen kosmischen und individuellen Verhältnissen, oder allgemeine vergleichende Selenographie*; or: *The moon according to its cosmic and individual relationships, or general comparative selenography]*. Simon Schropp.

Blagg, M., & Saunder, S. A. (1913). *Collated List of Lunar Formations Named or Lettered in the Maps of Neison, Schmidt, and Mädler, Compiled and Annotated for the Committee*, by Mary A. Blagg & Samuel Arthur Saunder (Edinburgh, 1913; reprinted by the International Association of Academies, Lunar Nomenclature Committee, University of Michigan Library, 2015).

Blagg, M. A., & Müller, K. (1935). *Named Lunar Formations* (with W.H. Wesley, S.A. Saunder, and J.H.G. Franz). P. Lund, Humphries.

Burchett, W., & Purdy, A. (1961). *Cosmonaut Yuri Gagarin, first man in space*. Panther Books (188 p.).

Cabrol, N., et al. (2010). *AAS Division for Planetary Sciences*. https://dps.aas.org/content/obituary-audouin-dollfus-1924-2010

[60] See the IAU Press Release: https://www.iau.org/news/pressreleases/detail/iau1901/.

Goldberg, L., & Aller, A. H. (1991). Donald Howard Menzel. *NAS Biographical Memoirs, 60*, 149–168.

Lund, T. (2018). *Early exploration of the moon, ranger to apollo, Luna to Lunniy Korabl* (391 p.) Springer Nature Switzerland.

McKim, R. (2011). Audouin Charles Dollfus 1924–2010. *Astronomy & Geophysics, 52*, 1.44.

Menzel, D., Minnaert, M., Levin, B., Dollfus, A., & Bell, B. (1971). Report on Lunar Nomenclature by the Working Group of Commission 17 of the IAU. *Space Science Reviews, 12*, 136–186.

Millman, P. M. (1976). Topographic nomenclature on planetary bodies. *Icarus, 29*, 155–157.

Rosenfeld, R. A. (2010). Obituary: Audouin Dollfus (1924–2010) *J.R.A.S. Canada, 104*, 251–252.

Whitaker, E. A. (1999). *Mapping and naming the moon* (p. 242). Cambridge University Press.

Thierry Montmerle has been General Secretary of the IAU (2012–2015). This made him familiar with IAU history and procedures, as well as with "backstage" information about the work of successive Executive Committees and Commissions. He participated in the IAU Centenary celebrations in 2019 by co-organizing the "Astronomers as Diplomats" conference in Paris, on which this book is founded. His scientific domain is space research in high-energy astrophysics, specializing in the role of energetic particles and photons (X-rays, gamma-rays) on the interstellar medium, and the resulting star and planet formation and evolution processes. After doctoral and post-doctoral studies in Montreal, Canada, he held a position at the Saclay Center for Nuclear Studies near Paris, France (operated by CEA, the French Atomic Energy Commission), where he spent most of his career.

Chapter 14
After the Moon Race: Russian Robotic Missions and Lunar Surface Features

Mikhail Marov

Abstract The Soviet-American competition in space exploration culminated in the Moon race at the threshold of 1960–1970s. It was driven by political motivations and serves as a good lesson for the next generations. After being the leader, the former USSR eventually let the USA win the ambitious program to land a man on the lunar surface. Instead, the Soviets successfully implemented a program of robotic Moon exploration, including the world's first automatic lunar sample returns and rover operations on the lunar surface. Both *Apollo* expeditions and robotic missions greatly advanced our knowledge of the Moon and planetary sciences in general.

The International Astronomical Union (IAU) contributed to scientific exploration of our closest neighborhood, astronomers playing an especially important role in the mapping/naming the Moon surface features, and getting insight into the lunar/planetary astronomy, geology and geochemistry. This activity was mostly conducted in the framework of the IAU "Working Group for Planetary System Nomenclature" (WGPSN). The productive work of the WGPSN allowed to identify and select the sites for safe landings on the Moon, and at the same time, to accommodate requirements of scientific interest. It paved the road for the future Moon missions, as well as for missions to other planets. The experience gained at the WGPSN is an invaluable legacy of the international cooperation and in particular, in resolving complicated scientific-cultural and political problems, a key of diplomacy. The author was a member of the WGPSN for several decades, and describes here his own experience and that of colleagues in this group as a good example of astronomers' actions as diplomats. The author also gives some other illustrations drawn from his personal career, combining professional skills as space physicist/astronomer with diplomatic activities.

M. Marov (✉)
Vernadsky Institute, Russian Academy of Sciences, Moscow, Russia

© The Author(s), under exclusive license to Springer Nature Switzerland AG 2022
T. Montmerle, D. Fauque (eds.), *Astronomers as Diplomats*, Historical & Cultural Astronomy, https://doi.org/10.1007/978-3-030-98625-4_14

14.1 Introduction

Historically, the Moon race in the 1960–1970s has been one of the most impressive milestones and exciting events of space exploration. While driven by political motivations, both American Apollo expeditions and Soviet robotic missions greatly advanced our knowledge of the Moon and planetary sciences in general. The International Astronomical Union (IAU) contributed to the scientific exploration of our closest neighbor, mostly through mapping and naming the Moon surface features, and getting new insights into lunar and planetary astronomy, geology and geochemistry.

The fundamental politico-historical question is why the Soviets let the Americans win the great competition to be the first nation to land a man on the Moon. The Soviet Union's space program was years ahead of the USA's in the 1950–1960s: the first Earth's artificial satellite (*Sputnik-1*, 1957); the first flights to the Moon, including taking the first images of its far side (*Luna-3*, 1959); the first man in space (Y. Gagarin,1961); the first woman in space (V. Terechkova, 1963); the first extravehicular activity (EVA: A. Leonov, 1965); the first photographic flyby of the Moon (*Zond-3*, 1965); the first landing on the lunar surface (*Luna-9*, 1966); the first precursors of orbital stations; etc., to mention a few. However, Soviets lost the Moon race and space leadership for the subsequent years.

Among several reasons to explain the Moon race failure by the USSR we distinguish the following items:

- First, the unexpected sudden death in 1966 January of the Chief Designer of the USSR rocketry and space program, head of the famous OKB-1 Design Bureau (now Energia Rocketry-Space Corporation, RSC) Sergey Korolev. He was the founder and actual leader of the Soviet space program, with enormous engineering and organizing capabilities.
- Second, the lack of political will and a rather poor management of the manned flights program to the Moon, resulting in wasting money, and the inefficient governance of the project. The situation was badly enforced by unjustified competition of chief designers in the implementation of N1-L1 (orbiting, LOK) and N1-L3 (landing, LK)[1] initially headed by Sergey Korolev. The project of an orbiting flyby[2] of the Moon (LOK module with two cosmonauts was proposed by Vladimir Chelomey using his Proton launcher. The system was in a good shape and successfully tested by a few Zond unmanned missions.
- Third, the imperfect design of the Korolev's N-1 heavy rocket with the first stage equipped of 30 engines of 150-ton thrust each (Nikolay Kuznetsov design), instead of the 600-ton thrust engines (Valentin Glushko design). The choice of

[1] LOK and LK are Russian acronyms Lunnyi Orbitalnyi Korabl' and Lunnyi Korabl' meaning Lunar Orbiting Vehicle and Lunar Lander, respectively.

[2] Basically, the LOK flyby trajectory represented very high elliptical Earth's orbit with no residence around the Moon.

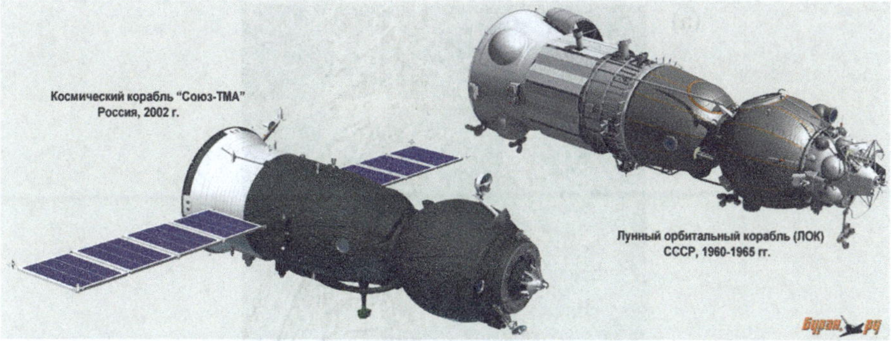

Fig. 14.1 Transport Modified Antropometric Vehicle Soyouz TMA) for Earth's orbiting flights (design better fitted for tall cosmonauts and more comfortable soft landing) and its modified version [Лунный Орбитальный Корабль ЛОК : Lunnyi Orbitalnyi Korabl' LOK] (*Lunar Orbital Spaceship* LOK). (Credit: RSC Energia)

Kuznetsov's engines in 1961 was hardly justified, and resulted from bad personal relationships between the former friends Korolev and Glushko .

– Fourth, the inefficient leadership of Korolev's successor Vasiliy Mishin, who was incomparable with Korolev regarding organizing capability.

In the second half of the 1960s some serious problems appeared in the lunar manned flight preparation. Four test launches of the N-1 rocket with LOK and LK mock ups at the top turned out unsuccessful though at the forth test in June 1971 the rocket flew nearly 107 s and lifted to 40 km. Yet, while the great efforts have been undertaken to succeed in the program of man landing, and thus to chase Americans with Apollo missions, an understanding of Soviets grew that there was no chance to compete with Americans. Political motivations were lost and soon after US President Nixon in 1973 canceled Apollo program, the Soviet Government decided in 1974 to close the lunar manned project.[3] Nonetheless, despite obvious failure of its implementation, nowadays, we admit the project itself as one of the most remarkable examples of great engineering skill of the former USSR space generation and historical space legacy.

It is worth noticing that the Soviet lunar program was quite successful with the Moon orbiting missions using Zond spacecraft as LOK (L1) precursor. The LOK concept was based on a slightly modified TMA version of Soyouz spacecraft, widely used for manned flights in the Earth's orbit, Fig. 14.1. Indeed, the LOK orbiter was basically adapted to a lunar expedition and successfully tested in a few unmanned flights as Zond missions. It was particularly important that, as was the case for Zond vehicles, a Proton, rather than an N-1 rocket, could be used for manned expeditions with LOK (Fig. 14.2). However, it was affected as a competitor for the Moon landing because of personal relations between chief designers, so the expedition

[3] The official decision was taken in 1976.

(a)

(b)

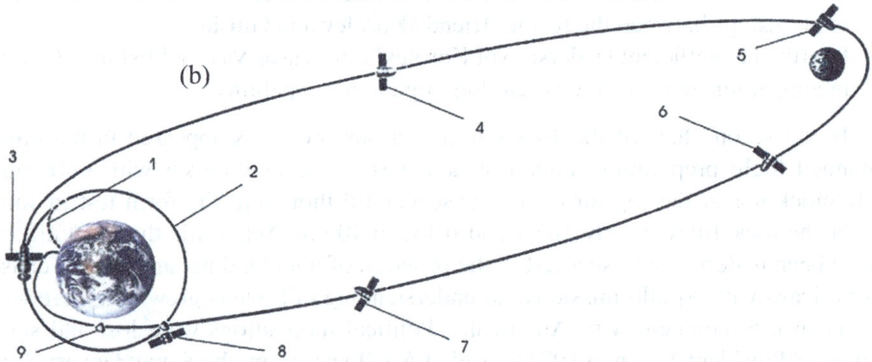

Fig. 14.2 (**a**) Proton rocket with Zond vehicle (piloted LOK equivalent) at the launch pad. (Credit: RSC Energia); (**b**) Scheme of the Moon flyby with Zond vehicle and/or the LOK crew. 1—launch; 2—Earth's orbit; 3—vehicle; 4—Earth-Moon trajectory; 5—the Moon flyby; 6,7,8—corrections of return trajectory; 9—entry Earth's atmosphere and atmospheric drag. (Credit: TSNIIMASH)

did not occur. The LOK program abandonment, as well as LK, is regarded as an unforgivable and short-sighted policy of the former Soviet Union leadership.

Figure 14.3 shows the impressive configuration of the N-1 (11A52) rocket of 2500 t of start mass and of 110 m height, resembling the American Saturn V used for Apollo manned flights. In my mind, despite a few failures, the N-1 could have reached an exploitation phase, had some additional improvements been introduced timely and testing continued. As far as the lunar lander LK is concerned, it was successfully tested in the Earth's orbit and performed maneuvers simulating landing on the Moon, lift-off and rendezvous with orbital module (Fig. 14.4). Cosmonauts commissioned to orbit and land on the Moon were selected and were experienced

Fig. 14.3 Heavy rocket N-1 at the assembly (*left*), and in the launch pad (*right*). (Credit: RSC Energia)

Fig. 14.4 Lunar lander LK (Лунный посадочный модуль). (*Left*). Engineering mock up. (*Right*) Under assembly in the montage facility. (Credit: RSC Energia)

enough to undertake these missions. A more detailed description of the overall situation concerning the manned lunar project and in particular, technical details of the N-1 L3 project can be found elsewhere (see, e.g., RSC Energia, 1996; SSC Tsniimash-70, 2016).

An important segment of the American Apollo and planned Soviet manned flights was the landing sites selection on the Moon surface, to accommodate the requirements of both safety and scientific interest. Four sites were selected for the

Fig. 14.5 Sites on the Moon surface selected for landing of the Soviet expeditions based on the returned images and IAU assigned features. (Credit: RSC Energia and US Geological Survey)

Soviet expeditions to the Moon (Fig. 14.5). Lunar surface images taken by Soviet Zonds and American Lunar Orbiters were of principal importance for landing sites selection and also for the detailed study of the Moon surface properties by the astronomical community. The study ran in parallel with mapping and refinement of the Moon coordinate systems, and naming the most prominent features to complement ancient names assignments. The main IAU body responsible for this work was the "Working Group of Planetary System Nomenclature" (WGPSN), which significantly contributed to the Moon scientific space exploration. Its role is discussed in more detail in the last sections.

14.2 Soviet Robotic Missions

Korolev's death dramatically affected the overall space research in the former Soviet Union. The situation deteriorated further after the death of another leader of the Soviet space program and former President of the USSR Academy of Sciences, Mstislav Keldysh, in 1978. In fact, space science collapsed after the so-called "perestroika" initiated by Mikhail Gorbachev, followed by the disintegration of the former Soviet Union in 1991, and political and social-economic turmoil in the country, the consequences of which had not yet been overcome. Now, looking back to the period between mid-1960 and 1970s, we must pay tribute to the Head and General Designer of the Lavochkin Association, Georgy Babakin, who initiated an excellent lunar-planetary robotic program based on the Korolev's heritage and Keldysh's support. This innovative program allowed the Soviet Union to undertake outstanding missions to the Moon, Venus, and Mars, and to take a pioneering role in their studies. We regard Babakin leadership in the innovative robotic missions' development as the glorious pages of the Soviet space exploration.

As was said above, by the end of 1960s, it became clear enough to people dealing with the manned Moon flight program, and eventually understood by the Soviet leaders, that there was no chance to chase Americans in this challenging endeavor. An alternative approach to compete with the USA was suggested by Georgy Babakin: to develop urgently and launch to the Moon a new generation of lunar robotic missions, in order to take automatically and return back to Earth samples of lunar soil, and to land on the lunar surface mobile vehicles—rovers. The latter would be able to travel up to several tens of kilometers and study the Moon surface under the Earth's ground guidance and remote control. This excellent idea was supported by Mstislav Keldysh, and finally approved by both Communist Party leaders and the Government.

This enterprise opened a new era in planetary exploration. The vehicles were successfully developed and launched concurrently with Apollo missions. The first attempt with *Luna-15* was undertaken on July 13, 1969, one week before the first successful Apollo 11 landing, so as to be ahead the American manned mission. Unfortunately, *Luna-15* crashed when approaching the Moon surface before landing in the planned site of Mare Crisium. *Luna-16* launched in 1970 was the first that returned back to Earth automatically samples of lunar soil from Mare Fecunditatis. It was followed by launches of *Luna-20* in 1971 and *Luna-24* in 1976, the latter taking samples of lunar regolith from 2.4 m deep with specially developed boring device, conserving the scientifically important soil column stratification. In parallel, the first lunar satellites were launched to study the lunar relief and other natural features of our closest space neighborhood.

Figure 14.6 shows the robotic vehicle's profile for sample return: general configuration, lift-off from the Moon (artist's view), and return capsule with lunar soil delivered on the Earth's surface, while Fig. 14.7 shows sites of *Luna-16, -20,* and *-24* landing on the Moon surface. *Luna-24* clearly demonstrated the importance of landing site selection: it was fortunate to land only a few meters from a steep crater rim, as it is evident in the images taken decades later by NASA's orbiter Lunar Reconnaissance Orbiter (LRO, launched in 2009).

The returned lunar samples were placed in the vacuum chamber of the Vernadsky Institute laboratory for follow-up in-depth investigations (see Fig. 14.8). The multiyear study tremendously advanced our knowledge of the Moon origin and evolution, which was made possible by the integrated efforts of the scientific

Fig. 14.6 Robotic vehicle for automatic return of lunar samples (*Luna-16, -20, -24*). (*Left*) General view with slave manipulator (robotic arm) for taking samples with return capsule at the top. (*Center*) Lift-off of the return rocket from the Moon surface (artist's view). (*Right*) Return capsule with lunar soil on the Earth's surface. (Credit: Lavochkin Association)

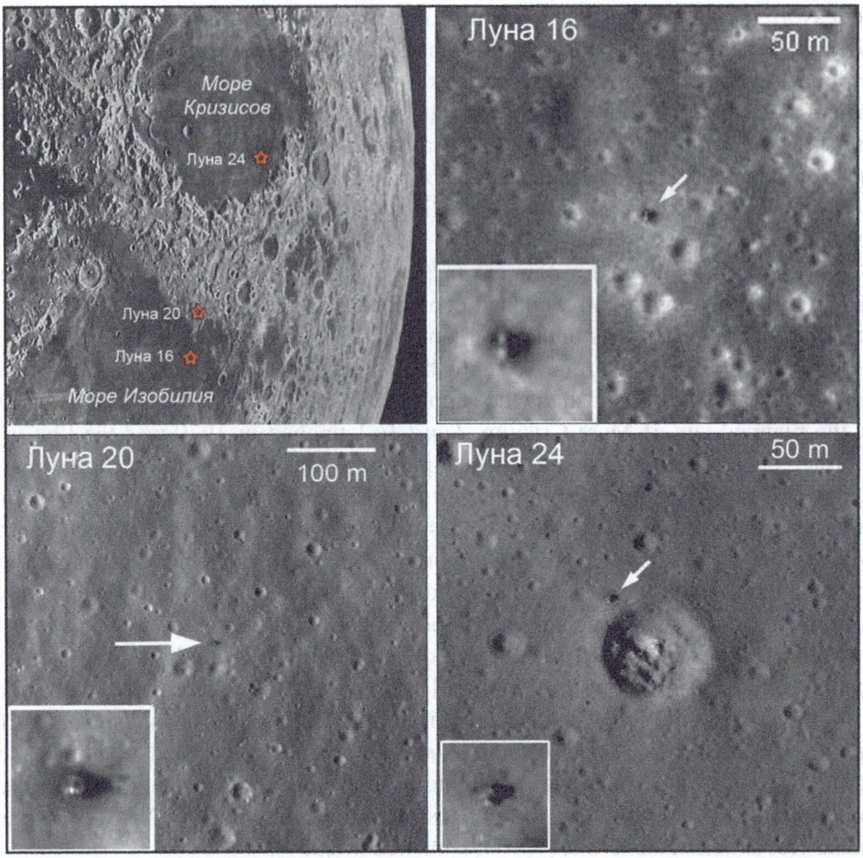

Fig. 14.7 Regions of the *Luna-16*, *-20,* and *-24* landing on the Moon surface. *Luna-24* (lower right) was in a real jeopardy because landing a few meters from a steep crater rim. This confirms how important an accurate knowledge of surface topography at landing site selection is for the mission success. (Images taken by *Lunar Reconnaissance Orbiter*, credit: NASA)

community, and in particular, by an Agreement about the Apollo and Luna samples to be exchanged between Soviet and American scientists.

The first lunar rover *Lunokhod-1* (*Luna-17*, Fig. 14.9, left) was launched in 1970 and *Lunokhod-2* (*Luna-21*, Fig. 14.9, right) in 1973. *Lunokhod-1* operated on the Moon surface in Mare Imbrium 11 lunar days (10.5 Earth's months), travelling slightly more than10.5 km. *Lunokhod-2* operated nearly four lunar days inside crater Le Monnier, East of Mare Serenetitatis, and covered a distance of about 42 km. More detailed description of the robotic lunar vehicles for lunar samples return and travelling over lunar surface, with many technical details, can be found elsewhere (see ASV Lavochkin Association, 2010; Huntress & Marov, 2011; Keldysh & Marov, 1981; Tsniimash-70, 2016). Results of the lunar samples in-depth study are summarized in (Vinogradov, 1975).

Fig. 14.8 (*Left*) Lunar samples of *Luna-16* in the vacuum chamber of the Vernadsky Institute laboratory. Second from the left—George Babakin, second from the right—Mikhail Marov. (*Right*) Mikhail Marov witnessing process of the lunar soil investigation, 1970. (From the Author's archive)

Fig. 14.9 Lunar rovers *Lunokhod-1* (*left*) and *Lunokhod-2* (*right*). (Credit: Lavochkin Association)

Lunokhod operations required especially accurate knowledge of lunar surface relief and soil properties. Many astronomers and planetary geologists dealt with detailed morphological analysis of images taken by the lunar orbiters and contributed to the study. They compiled diagrams of crater size-frequency distribution and variations of the spatial crater's density, analyzed variations of spatial rock fragments density, etc. Astronomers progressed in Moon science, with application to space research in general through detailed mapping of lunar surface morphology, associated with the origin and evolution of different topological features and terrains. The results allowed us to place important constraints on landing sites selection in terms of scientific importance and operational safety, keeping in mind numerous obstacles like small craters, rocky outcrops, heaps of stones distribution, etc. However, the practical implementation of complicated tasks is always risky and can be in jeopardy. This is what happened to *Lunokhod-2*. On April 20, 1973, the rover entered and maneuvered in the crater eastward of Fossa Recta with its solar battery lid in the open position. This was a great mistake of the operators, and turned out to be fatal for the vehicle's further activity: when getting out of the crater, the lid touched upon the crater's rim and some soil slipped down on the lid's surface. At the end of the fourth lunar day before entering the lunar night, the lid responding technical requirements was closed, and the soil powdered the radiator. Its efficiency to radiate heat in space dropped dramatically, and at the beginning of fifth lunar day, *Lunokhod-2* got overheated and broke down.

Both *Lunokhods* allowed us to make important science on the Moon surface and provided close up views of the lunar landscape. Wheel tracks and footprints left behind by the rovers and astronauts on the Moon will preserve for millions of years because of weak erosion of the lunar surface for millenia. Examples of some images transmitted by *Lunokhod-1* are shown in Fig. 14.10a, including spectacular images of tracks left behind by *Lunokhod-1 maneuvering* on the Moon surface (Fig. 14.10b). Traces of the route of *Lunokhod-2* were transmitted by the US LRO spacecraft (see Fig. 14.10a, right).

Robotic missions to the Moon were the climax of the lunar program in the former Soviet Union. It was an ambitious enterprise with important science benefits and a demonstration of great engineering skill. The full description of the overall history of the Soviet space program, highlighting the lunar-planetary missions, is given in the book by Mstislav Keldysh and Mikhail Marov "Kosmicheskie Issledovaniya" (Space Research) published by Nauka in 1981. An in-depth analysis of this endeavor and in particular, many technical details of Luna missions is given in the book of Wesley Huntress and Mikhail Marov "Soviet Robots in the Solar System. Mission Technologies and Discoveries" published in 2011 by Springer-Praxis. Both books covers are shown in Fig. 14.11. Let us note that automatic sample return operations conducted fifty years ago have not been repeated elsewhere, until the Chinese lunar sample return mission Chang'e 5 in 2020.

The Russian planetary space program experienced several crises in the following decades but certainly survived. Russian scientists contributed to the Moon study by installing instruments for searching water using neutron monitoring technique on board of NASA's lunar vehicles Lunar Prospector and LRO. It allowed tracing hydrogen contents in the lunar surface layer of about 1m deep associated with water deposits in soil of the subpolar regions. Areas nearby the South pole are of primary interest because of the presence of water ice mostly in deep, shadowed craters, and deserve more in-depth study by astronomers and geologists (Fig. 14.12). There was announced in 2019 rather ambitious national plan of the Moon study with a new generation of landers in the polar regions of primary interest for further exploration (Fig. 14.13). The Russian Space Agency (RSA) announced a new start of manned missions to the Moon around the early 2030s with an ultimate goal of the Lunar Base deployment; its tentative infrastructure could be as it is shown in Fig. 14.14. It might provisionally concur with the USA, ESA, and Chinese plans including installation of underground shelters to protect astronauts from the harmful cosmic radiation. These plans imply a growing responsibility of astronomers, space

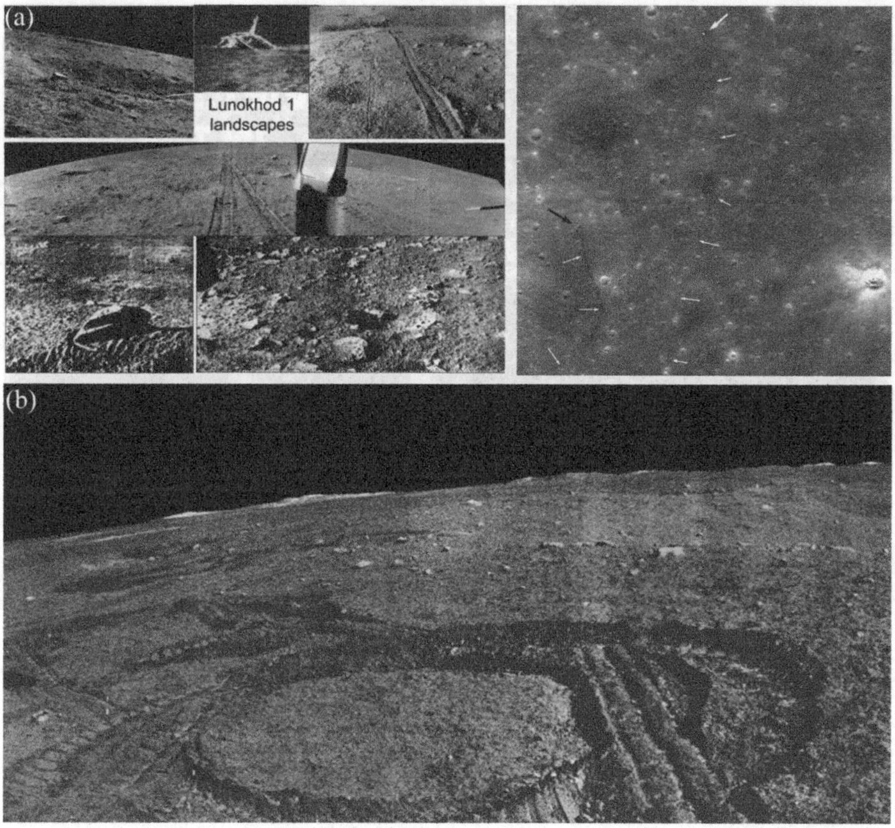

Fig. 14.10 (**a**) Examples of images of the lunar surface transmitted by *Lunokhods*. (*Left*) *Lunokhod-1* landscapes, pitted terrain, and rover tracks (Credit: Lavochkin Association). (*Right*) Route of *Lunokhod-2* on the Moon surface (shown with white arrows; black arrow: the final parking place). (Image taken by LRO, Credit NASA); (**b**) Tracks of *Lunokhod-1* maneuvering on the lunar surface. (Credit: Lavochkin Association)

scientists, and planetary geologists for an accurate Lunar Base site selection, in order to accommodate both safety and utilization requirements including water ice/volatiles and other local resources abundance for the practical needs of the base inhabitants. Radio astronomical facilities deployed on far side of the Moon constitute a challenging milestone for basic astronomical science.

14.3 Astronomers as Diplomats: A Personal View

We address diplomacy as an important segment in every human endeavor, which generally benefits human society for centuries, putting aside some exceptions. Astronomy and space science integrate very large communities, which are inter-

Fig. 14.11 (*Left*). Book of Mstislav Keldysh and Mikhail Marov «Космические исследования» ["*Space Research*"] describing Soviet space program (Nauka PH, 1981). (Right). Book of Wesley Huntress and Mikhail Marov "*Soviet Robots in the Solar System. Mission Technologies and Discoveries*" (Springer-Praxis, 2011), containing detailed description of the Soviet lunar and planetary robotic missions. Credit: NAUKA PH and SPRINGER-PRAXIS

national by definition. Astronomers address many problems to be resolved based on cooperation between various teams that require negotiations and hence, need diplomatic skills to overcome controversies and join efforts to achieve results of the mutual interest and importance.

However, initially this was not the case as far as space is concerned. Space exploration began as product of the progress of the rocketry and missiles industry, in the background of a growing Cold War tension. Paradoxically, amazing technological progress open for humans unfolded perspectives of discoveries of the mysteries of nature, in parallel with using space as a new sphere of confrontation and thus promoting new risks and vulnerabilities to the very existence of our civilization.

The Cold War paradigm motivated space exploration during its first decades and was perceived as an arena of competition. It was driven by ambitious concepts to achieve superiority in this technologically advanced field of human space activity. Americans experienced moral and political damage brought by the launch of Soviet first satellite and Gagarin flight, decided to compensate it with an ambitious plan to land a man on the Moon and returning him safely back, a challenge in which the Soviet Union could not stay uninvolved for political reasons. This marked the beginning of the Moon race. Under these circumstances, any diplomatic effort to reduce the confrontation seemed simply useless. The first steps toward finding

Fig. 14.12 Contoured region nearby the lunar South Pole of special scientific interest and applications. Astronomers currently regard it as priority for the Moon future exploration and exploitation, because of the presence of water ice, which could be used for lunar bases. (Credit: Vernadsky Institute)

cooperation at the state level were undertaken only in 1971 when US NASA delegation visited Soviet Academy of Sciences in Moscow in order to sign the first in the history document on cooperation in space, which established the new relationship between nations. It is amazing that the event happened when successful Apollo program was yet continued and Soviets culminated with their recognized Lunar, Venus and Mars robotic missions. Basically, the step was a demonstration of ending confrontation of superpowers in space area (called by some media "tenant in space") and touched upon related issues such as Lunar race, which became politically insensible. In other words, it promoted the basic reason for allowing a non-competitive approach.

As far as the scientific community is concerned, it cooperated traditionally in the study and exploration of space throughout all those years, without need to be diplomats in scientific discussions, often at person-to-person level at the scientific meetings and by correspondence. Astronomers played a leading role in the exchange of research results, and in discussions of new ideas combining personal interests with the problems of universal importance. In my case, I had unique opportunities to have friendly relationship with many outstanding scientists over the world and at the same time to meet leading people in the field of space research, discussing with them and their collaborators hot topics and cooperative items (Fig. 14.15).

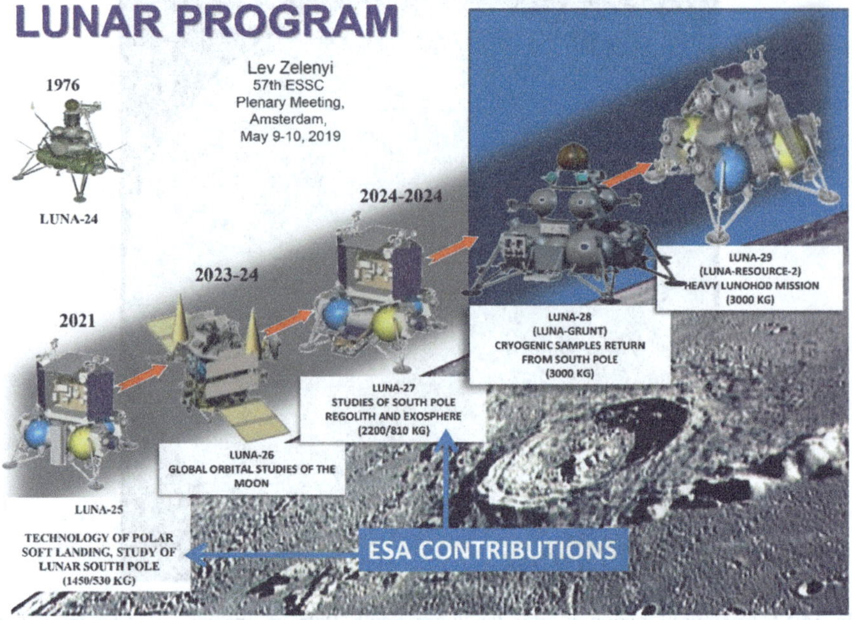

Fig. 14.13 Russian National Lunar space program with automatic missions, including possible contributions from ESA as of 2019. LUNA28 and LUNA 29 are blue print missions. (Credit: ROSCOSMOS and Space Research Institute IKI)

Fig. 14.14 Provisional Lunar Base infrastructure: Russian concept. (Credit: ROSCOSMOS)

Unfortunately, after the end of the Cold War in 1989, the world faced new challenges and motivations. It opened wide perspectives for diplomacy. Along with the increasingly conceptual approach of commercialization, including space

Fig. 14.15 Mikhail Marov with the Director of NASA Jet Propulsion Laboratory (JPL) W.Pikering at the time of Mariner 9 flight in the Mars' orbit. JPL, Pasadena, California, 1971. (From the Author's archive)

projects, the problems of competition and striving for superiority in the world have not disappeared, while running in parallel with a limited cooperation between nations, especially what USA and Russia is concerned. Simultaneously, with the globalization of exchanges in the world, the awareness of the vulnerability of the home planet from space hazards has been gradually growing, which is best understood by astronomers. That is why their role in the diplomatic field and in solving these and other global political problems is of key importance, which is intrinsically related to preservation of safe environment and life on the home planet.

Needless to say, I do not intend to discuss these problems here in any detail. Nevertheless, I may say that during my scientific career, I pursued such a goal with an underlying paradigm that cooperation is better than confrontation. I participated jointly with my colleagues in various international bodies, meetings, and negotiations, on which I acquired a diplomatic skill while being an astronomer. Here are some views in the context of this book, based on the personal experience, which seems worth to share.

Indeed, my personal career in space physics and astronomy has been tightly related with diplomatic activities. In the domestic environment, I gained a great experience of finding compromise between scientists and engineers, being in a multiyear position of the Project Scientist for our Moon, Venus, and Mars missions. In the international arena, this included participation in the different delegations negotiating space programs and projects. The most remarkable experience I gained was through my positions in the IAU, as Vice-president and President of Commission 16 (Physical Study of Planets and Satellites) throughout 1994-2000, and President of Division III (Planetary Systems Sciences), the third largest of twelve

IAU Divisions with seven Commissions, two Working Groups and important science highlights. Specifically, during my term (2000–2003) quite hot discussions began between the Division members and interested astronomers concerning "a planet" term definition, the new category of extrasolar planets and the new status of Pluto as a member of Kuiper belt (dwarf planet) rather than ordinary Solar system planet and introducing "plutoids" term. Serious efforts were undertaken to reach a consensus years ahead. I kept my activity as the Division OC member during 1997–2009. Since 2012, I have been nominated a Chair of the UNESCO Tentative Working Group on technology heritage related to space exploration and simultaneously, a Chair of the IAU Commission C.4 Working Group 3 "Heritage of space exploration" as a segment of the UNESCO Astronomy World Heritage Initiative. This experience was complemented by chairing and membership in the several COSPAR and IAGA (International Association of Geomagnetism and Aeronomy) Commissions/Working groups. In particular, since 2003 I have been serving as IAU representative in the COSPAR Scientific Commission B "Space Studies of the Earth-Moon System, Planets and Small Bodies of the Solar System." extended for a new term in 2021.

Another experience in the diplomatic area I gained from the co-chairing jointly with Jacques Blamont (then top adviser for CNES, the French Space Agency), the Soviet-French project of launching high-altitude balloon in 1984 to study in detail the clouds of Venus. The project was successfully developed and partially manufactured (Fig. 14.16) but its launch was abandoned by a voluntarist decision of the former Director of the Soviet Space Research Institute Roald Sagdeev, to which I strongly objected while Blamont yielded, probably under CNES and Sagdeev's pressure (Blamont, 1987).

It is worth emphasizing the important role played by IAU astronomers in the implementation of lunar and planetary projects. As said above, the main IAU body responsible for mapping, studying and naming of relief and morphological features on planetary bodies was the Working Group of Planetary System Nomenclature (WGPSN), of which I have been a member for a few decades, since the early 1970s until 2011 when I stepped down. The WGPSN was a new body set up at the Sydney General Assembly in 1973, which replaced the former WG of IAU Commission 17 (The Moon) dealing with lunar nomenclature. This decision raised the status of the newly created group, since the WGPSN reported directly to the IAU Executive Committee, and consisted of several Task Groups (TG) dealing with the different solar system planets, their satellites, and minor bodies, one of them being the Moon TG. The WGPSN consisted of distinguished astronomers coming from all over the world with recognized reputation and excellent scientific and historical-cultural background, including ancient mythology and cultural roots of many nations. It led to truly international meetings in different countries, specifically during IAU General Assemblies. Outstanding scientists have guided the WGPSN: Peter Millman (Canada), Harold Masursky (USA), Kaare Aksnes (Norway); currently Rita Schulz (Germany) chairs it. Throughout the years, this body has been a brilliant example of integration of scientific knowledge, diplomacy, responsibility, and very friendly relationships (Fig. 14.17).

Fig. 14.16 Soviet-French project of the Venus atmosphere and clouds study with a large balloon. Venus balloon prototype testing at the Toulouse Space Center, 1983. *Left to right:* French Project manager M. Gérard (?), R. Kremnev (Babakin's Deputy, Lavochkin association), and the project scientific co-chair M. Marov. (From the Author's archive)

Of specific importance was the diplomatic skill of participants when resolving some complicated problems involving selection of personalities, national interests, historical-chronological events, cultural and linguistic issues, etc. This objective approach and respectful attitude was the key in cooperation of the members involved. The WGPSN experience serves as an excellent example of diplomacy to pursue all sorts of endeavors, even beyond astronomy. When I stepped down the body I received warm letters from the WGPSN colleagues, one of them from the distinguished USA astronomers and good friend of mine Bradford Smith is shown below:

Fig. 14.17 (**a**) Historic photo of the 1985 meeting of WGPSN at Abastumani Observatory, Republic of Georgia, Soviet Union. At the **top row:***André Brahic* (in between of the central

02.12.2011
"My dear friend Misha,
I am in Hawaii and just read your message about stepping down from the WGPSN. I received this news with a certain amount of sadness, but with the realization that everything must eventually come to an end. It brings back memories of the early days of the WG in which you, Hal and I disregarded the pressures of national and international politics and worked together as colleagues and friends. I can well remember the many good times we shared, such as running the Olympic track together in Greece. Diane and I pleasantly remember visiting you and Natasha at your home in Moscow, and your visit with us here in Hawaii. It is with tears in my eyes that I write this.
I wish you only the very best in the future, and I hope that we will still correspond occasionally.
С любовью,
Brad"

In my memory, the most impressive actions of astronomers as diplomats in official international relations are those of the Joint Working Group on Outer Space, Moon, and Planets, established in 1971 in the framework of the abovementioned Agreement between Soviet Academy of Sciences and NASA, signed by the USSR Academy President Mstislav Keldysh and NASA Deputy Administrator George Low (Fig. 14.18). I had been nominated co-Chair of the WG for the Soviet side, and Assistant NASA Administrator Noel Hinners (later Vice-President of Martin Marietta, Co.) for the American side. Delegations of the WG met twice a year, alternatively in Moscow and Washington, D.C., to discuss mutual programs of space activity and future projects. Negotiations ran in a very friendly atmosphere. A good relationship between WG members was established for years, which maintained

Fig. 14.17 (continued) person, a *local astronomer* wearing a Georgian traditional wool hat, and *Slava Shevchenko* of Shternberg Astronomical Inst., Moscow, who wears sunglasses); to his left *Junber Lominadze*, Abastumani Observatory, hosting the meeting. In the **first row** (*left to right*): *Victor Tejfel* (Fesenkov Astrophysical Inst., Alma-Ata, Kazakhstan), *Merton Davies* (Rand Corp., California, USA), *Kaare Aksnes* (Oslo Univ., Norway), *Harold Masursky* (USGS, Flagstaff, Arizona), *Toby Owen* (at that time at Stony Brook Univ., NY, USA), *Brad Smith* (at that time at New Mexico, USA) and *Rita Beebe* (Univ. of California, USA) left of him. Behind him in the **second row** in left are: *Sasha Basilevsky* (Vernadsky Inst., Moscow, USSR) and *Nikolay Erpylev* (Inst. for Astronomy, Moscow) and to his right is *Peter Millman* (Natl. Research Council, Canada) and *Yury Tjuflin* (Central Inst. of Cartography, Moscow, USSR). To the right in the **first row** is *Mikhail Marov* (at that time at Keldydh Inst. of Applied Mathematics, Moscow, USSR), the WGPSN meeting organizer; (**b**) WGPSN meeting in Norway, 2005. *Left to right:* Rita Schulz (Germany), George Burba (Russia), Brad Smith (USA), Guy Consolmagno (Vatican Observatory), and Mikhail Marov. (Figure 14.17 (a and b) from the Author's archive)

Fig. 14.18 Meeting of the USA NASA and USSR Academy of Sciences delegations, Moscow, 1971. (*Left*) NASA Deputy Administrator G. Low, Director of Space Research Inst. Acad. G. Petrov, and M. Marov. (*Right*) After signing Agreement on cooperation in outer space. *From the left:* G. Low, M. Marov, Acad. E. Fedorov, Acad. A. Vinogradov. (From the Author's archive)

Fig. 14.19 (**a**) Negotiations between Soviet and American members of the WG on *Outer space, Moon, and Planets*. *Left*: Dr. Ichtiaque Rasool (Goddard Space Inst., NY) and NASA interpreter; *Right*: Director of Space Research Inst. Acad. George Petrov and Prof. Mikhail Marov, 1973. (**b**). M. Marov and NASA Assistant Administrator N. Nogle, 1971. (From the Author's archive)

by the distinguished WG leaders despite quite hot discussions (Fig. 14.19). The climax of the activity was a real-time exchange of the data obtained by Mariner 9 and Mars 5 missions through a tele-bridge between Deep Space Network antennas in Evpatoriya, Crymea, and Pasadena, California, with Dr. Gerry Soffen (later on Viking missions Project scientist) and myself at both ends.

It was an exciting period of adventure in space that marked cooperation between superpowers, instead of years of chasing and confrontation. Unfortunately, it was terminated abruptly in 1980 when the Reagan Administration came to power and new tensions on space research rose with its "Strategic Defense Initiative" (1983), sometimes also called "Star Wars." In the new situation, a special Committee named "Scientists against nuclear threat" was set up in USSR in the 1980s and I took part in discussions with American delegations headed by Susanne Eisenhower the problems of the tensions alleviation.

I view the current world situation through the prism of those historical facts. Space age manifested many innovative breakthroughs in the development of our civilization. Human society benefits and at the same time becomes increasingly dependent on the global space systems of communication, navigation, meteorology, geodesy, and monitoring of natural environment, also closely related to military interests and mutual inspection of the existing international threats performance. At the same time, humankind is highly vulnerable to adverse space factors. Let us stress again the point that we are becoming more aware of potential threats coming from the spatial environment of the Earth, such as cosmic radiation, solar flares, asteroid-comet threat. Security and protection from potential natural hazards are issues comparable to international tensions. The tensions are caused by the continuously increasing dependence of our lives on the political and socio-economic situation and confrontation in the world, which could be influenced by military space networks. There is no other opportunity to attenuate tensions and avoid catastrophes but reasonable and clever diplomacy, meaning negotiations instead of hot war and threats to life. Based on these realities, the responsibility of scientists for the fate of our minute planet in the endless space and in particular, the role of astronomers in the modern world cannot be overestimated.

14.4 Conclusions

The Moon race of the 1970s was one of the most exciting milestones in space exploration. The former USSR yielded to the USA in an ambitious endeavor to set foot on the Moon surface, which the Americans successfully did six times with the Apollo lunar program. Instead, the Soviets undertook robotic missions to the Moon, with three successful Luna sample return missions and two *Lunokhod* rover missions. Both Apollo human and Luna robotic missions greatly advanced the lunar and planetary sciences and were the most ambitious projects of Moon exploration for years ahead.

Astronomers contributed to this process through the mapping planetary surface morphology and naming relief features, in particular in framework of the activities of the IAU Working Group of Planetary Science Nomenclature (WGPSN). These activities allowed us to identify and select the sites for safe landings on the Moon prior, during and after Apollo and Luna missions, and at the same time, to accommodate requirements of scientific interest. It also paved the road to the future Moon missions as well as missions to other planets. The experience gained at the WGPSN is an invaluable legacy of the international cooperation and in particular, in resolving delicate scientific-cultural and political problems, a key of diplomacy.

Author brought some examples of his participation in the different international bodies and in the various national and international conferences and/or delegation meetings negotiating space programs and projects. He hopes that his diplomatic activity through this participation promoted astronomical collaboration and benefited progress in space exploration.

This historical outlook testifies an important role that astronomers played in society unification and advancing space sciences despite political tensions with ultimate goal to ensure progress of the human civilization. No less important is their position in the modern world, facing many controversial problems including potential threats coming from space, as well as forthcoming Moon-Mars utilization, search for extraterrestrial intelligence (SETI), etc. One may hope that the role of astronomy in prosperity of our society will gradually increase.

Acknowledgments The author acknowledges fruitful discussions with his colleague Dr. Alexander T. Basilevsky and Prof. Lev Zeleny of some historical events related to Lunar Race and Soviet robotic missions, as well as plans of the future Moon exploration with spacecraft. This work was performed in the framework of Vernadsky Institute scientific research program supported by the governmental funds. The author also acknowledges the invitation of the organizers of the IAU@100 Symposium in Paris to write this Chapter.

References

Автоматические космические аппараты для фундаментальных научеых и прикладных исследований, Редакторы Г.М. Полищук и К.М. Пичхадзе. НПО им. Лавочкина, МАИ-Принт, Москва 2010. [Avtomaticheskie Kosmicheskie Apparati dlya Fundamental'nikh Nauchnikh i Prikladnikh Issledovaniy. Redaktori G.M. Polytshuk i K.M. Pichhadze. NPO im. Lavochkina, MAI Print, Moskva 2010: *Automatic Space Vehicles for the Fundamental Scientific and Applied Research,* 2010, Lavochkin Association, Eds. G.M. Polytshuk and K.M. Pichhadze. MAI-PRINT, Moscow 2010].

Blamont, J. (1987). *Venus dévoilée, Voyage autour d'une planète*. Editions Odile Jacob.

Виноградов А.П. (Редактор). [Vinogradov, A.P., redactor]. Космохимия Луны и планет. Труды советско-американской конференции по космохимии Луны и планет. НАУКА, Москва, 1975. [Kosmokhimiya Luni i planet. Trudi Sovetsko-Amerikanskoy konferehtsii po kosmokhimii Luni i planet, NAUKA, Moskva, 1975 : *Proceedings of the Soviet-American Conference on Cosmochemistry of the Moon and Planets.* Vinogradov, A.P., Ed., NAUKA, Moscow, 1975].

Huntress, W. T., & Marov, M. Y. (2011). *Soviet Robots in the Solar System. Mission technologies and discoveries*. Springer-Praxis.

Келдыш М.В., & Маров М.Я. (1981). Космические исследования. НАУКА, Москва. [Keldysh M.V. and Marov M.Ya.,. Kosmicheskie issledovaniya. NAUKA, Moskva, 1981: Keldysh M.V., and Marov M.Ya. *Space Research.* NAUKA, Moscow, 1981.]

Космический научный центр ЦНИИМАШ-70. Редакторы Горшков О.А., Анфимов Н.А. и Панишкин Н.Г.. Т.1. Королев-РИНФО, 2016. [Kosmicheskiy Nauchniy Tsentr TSNIIMASH-70/ Redaaktori Gorshkov O.A., Anfimov N.A. i Panichkin N.G. Korolev-RINFO, 2016 : *Space Scientific Center TSNIIMASH-70,* Eds. Gorshkov, O.A., Anfimov, N.A., and Panichkin N.G. v.1. Korolev- RINFO, 2016].

Ракетно-космическая корпорация ЭНЕРГИЯ (1946–1996), т. 1. Редактор Ю.П. Семенов. Королев, 1996. [Raketno-Kosmicheskaya Korporatsiya ENERGIYA (1946-1996). T. 1. Redaktor Yu. P. Semenov. Korolev, 1996 : *Rocketry-Space Corporation ENERGIA (1946-1996),* Ed. Yu. P. Semenov, v.1, Korolev, 1996.]

Mikhail Marov Born in 1933, Moscow. Graduated from Moscow State Technical University, 1958; PhD, 1964; Full Doctorate degree, 1970; Full Professor, 1977; Corresponding Member of USSR Acad. Sci., 1990; Full Member of Russian Acad. Sci., 2008. Full Member of Internat. Acad. Astronautics, 1982. Affiliations: RSA Energia (1958–1962), Keldysh Inst. of Applied Math. (1962–2008), Vernadsky Inst. of Geochemistry (2008-now). Great contribution in the planning and implementation of the Soviet space research program (1960s–1980s). Project scientist of the Soviet "Luna", "Venus", and "Mars" missions. Professional interests: hydromechanics, astrophysics, space research, planetology, cosmogony. Pioneered experimental Venus and Mars research taking first *in situ* measurements in their atmospheres. Developed theoretical problems of the mechanics of multicomponent reacting gases, turbulence of inhomogeneous media, and planetary aeronomy, alongside with methods for their mathematical modeling. Published over 280 scientific papers and 18 monographs. Editor-in-chief of Solar System Research journal, Member of the Chairman of the Commission for K. E. Tsiolkovsky scientific heritage. Actively participated in the international scientific bodies IAU and COSPAR. Elected Vice President and President of IAU Com. 16 (1994–1998) and President Division III (1998–2001). Awarded with national Lenin, State, and Demidov prizes and numerous orders and medals, as well as international Galabert prize, A. Siff prize, COSPAR W. Nordberg medal. For outstanding scientific achievements, most recently awarded with one of the highest awards of the Russian Academy of Sciences—Keldysh Gold medal.

Chapter 15
The *International Virtual Observatory*: The Result of Joint Efforts, Coordination, and Diplomacy

Olga B. Dluzhnevskaya

Abstract Astronomy is one of the first sciences, which since ancient times began to accumulate observational data. They were stored in catalogs at astronomical observatories and for a long time, due to the specifics of astronomical research, did not depreciate over time. However, the volume of catalogs in the second half of the twentieth century, using the capabilities of photography, began to increase significantly, and to facilitate access to these data, it was decided to place them all in one astronomical institution. So in 1973, the "Centre de Données Stellaires" (Stellar Data Center; CDS) was created in Strasbourg, France, combining stellar catalogs accumulated in the collections of European observatories. CDS founded two branches: in NASA, the Astronomical Data Center, Maryland, USA (ADC) for the countries of the American continent, and the second—the Soviet Center for Astronomical Data (CAD)—in Moscow, in the Institute of Astronomy of the Russian Academy of Sciences, for astronomical institutions of USSR and Socialist countries.

In subsequent years, there has been a rapid increase in the number of resources, both terrestrial and spatial, containing large amounts of astronomical data of various types: images, spectra, time series, numerical characteristics of astronomical objects, etc., requiring the creation of a structure on a global scale, which became the International Virtual Observatory (IVO). The IVO Project was prepared by IAU Commission 5 and approved at the XXIVth General Assembly in Manchester (2000). Currently the IVO unites 20 national virtual observatories, one of which is the Russian Virtual Observatory (RVO), which is presented in this paper. In the process of organizing work within the IVO, it was decided in 2002 to create an "IVO Alliance" (IVOA) - an administrative organization which is responsible today for the development and implementation of standards for the exchange of various types of information between project participants throughout the world.

O. B. Dluzhnevskaya (✉)
Institute of Astronomy of the Russian Academy of Sciences, Moscow, Russia
e-mail: olgad@inasan.ru

© The Author(s), under exclusive license to Springer Nature Switzerland AG 2022

T. Montmerle, D. Fauque (eds.), *Astronomers as Diplomats*, Historical & Cultural Astronomy, https://doi.org/10.1007/978-3-030-98625-4_15

15.1 Astronomical Data: Catalogs, Databases

At first, astronomical data were the results of observations with the naked eye of the Sun, the Moon, planets, and bright stars, obtained with primitive instruments. Then, after the invention of the telescope by Galileo Galilei and with the subsequent construction of astronomical observatories equipped with optical telescopes of various systems, the opportunities for studying celestial bodies significantly improved. The possibility of observing ever fainter stars, the number of which began to increase dramatically, opened up. The results of observations began to be recorded in the form of star catalogs (starting with the catalog of Hipparchus, in the second century BC!) containing their names (for bright stars) and coordinates, but in modern times also their photometric characteristics (magnitudes) in different wavelength ranges. To date, most observatories have maintained their observational programs for many decades, since photometry is very useful in the study of unsteady objects, such as variable stars, novae, or supernovae; also, a comparison of the positions of stars in old catalogs with modern data allows to measure their proper motions (i.e., their apparent displacement over the sky)—in that case, the difference in the epochs of observation can be crucial.

However, it is possible to save the results of observations not only in the form of catalogs. After the introduction of photography into the practice of astronomical observations (a daguerreotype of the Moon obtained by J.W. Draper in 1840), it became possible to save captured photographic (negative) glass plates for many years for repeated studies. Especially important are the celestial surveys performed by a large number of observatories based on international projects, in a way bringing to completion the unfinished *Carte du Ciel* project that had been initiated by Admiral Mouchez at the Paris Observatory in 1887 (Chinnici, 2022).

To save photographic glass plates, libraries (commonly called "glass libraries") were created, and star catalogs derived from them in astronomical observatories began to be transferred to astronomical data centers.

Since the beginning of the 70s and for more than twenty-five years, the centers for astronomical data provided the community with a fundamental service. However, although these centers did not lose their significance, the rapid progress in astronomy during this time identified new problems.

An avalanche-like growth of astronomical data resulted from an enormous increase in the observational capabilities of both ground-based and space observatories. The total area of mirrors of the world's telescopes (three-meter and larger) increased tenfold during the 1970–1980 period, reaching over 1000 m^2.[1] If in 1970, when it came to creating the first astronomical data center, a collection of 20 catalogs from four observatories could be prepared, then in 2010 the total number of catalogs already exceeded 9000. The largest source of astronomical data nowadays is coming from large digitized sky surveys, yielding Terabytes of data for the whole sky (in one

[1] By 2020, with the availability of mirrors over 6m in diameter, the collecting area is about 12,000 m^2, the equivalent of a football field !

observing session), in a single spectral range. Astronomy is now at a pronounced stage of exponential growth in the volume of observations. In 2000, the Sloan Digital Sky Survey (SDSS), obtained with a single 2.5 m telescope in New Mexico, USA, collected in the first two months more digital information (20 Tb) than all the telescopes in the world before it.[2] More generally, Terabyte (and then Petabyte) surveys of the sky, including from space, now cover a wide range of wavelengths, sometimes repeated over short timescales to look for time variability (like surveys of the gamma-ray sky obtained every couple of hours by the *Fermi* satellite); billions of recorded objects (e.g., by the *Gaia* satellite astrometric surveys), hundreds of measured parameters for each object, etc., with, as a result, a total volume of astronomical archives doubling annually.

Simultaneous progress was made in the field of information technology: a thousandfold increase in the speed of computers, a significant reduction in the cost of computing and data transfer, as well as the emergence of high-speed networks, made it possible to consider the possibility of creating virtual observatories, i.e., being able to combine data from all observatories, from the ground, and from space, and make them available at your fingertips.

But before using the capabilities of the virtual observatory, it was necessary to prepare and convert astronomical data with the goal of building a successful prototype first.

15.2 The *Centre de Données Stellaires* and Its Branches for Collecting Astronomical Catalogs

At the end of the twentieth century, due to significant advances in observational, recording, and computational technologies, astronomy faced an avalanche-like increase in the amount of data received. These datasets cover the whole sky in various wavelength ranges, from gamma-rays and X-rays, through ultraviolet, optical, and infrared, all the way up to the radio range. A large number of catalogs have already accumulated in astronomical institutions, yet it was very inconvenient to use them, since they were dispersed at different locations and in different formats. In 1973, six European astronomical institutions decided to create a single center for the storage and distribution of star catalogs, where the star catalogs created at all observatories in the world would be merged together. The Observatory of the University of Strasbourg (France) volunteered to be the place for founding and organizing such a Center.

I learned about the creation of this center of astronomical data by participating in the "2nd European Astronomical Conference" (Trieste, 1974). After I gave a talk on the study of open clusters, Prof. Bernard Hauck, from Geneva, came up to me. He was very famous because he held high posts in international organizations and

[2] https://www.sdss.org/.

in his country, heading two astronomical institutions as Director of the Institute of Astronomy of Lausanne University and of the Geneva Observatory, both located in Sauverny, making up together the largest astronomical center in Switzerland. He told me that in Lausanne, a large number of catalogs had been stored in a computer, with observational data on open clusters, and he suggested sending me an invitation for joint research. He also said that similar data were available at the Strasbourg *Centre de Données Stellaires* (CDS). So this was the first time I heard about the Strasbourg Center. Since this center was international, one associate position was allocated for one-month visits by foreign scientists. I applied successfully and received an official invitation to work in Strasbourg.

The capabilities of the center for working with star catalogs made a great impression on me and, in principle, I agreed with the Director of CDS, Prof. Carlos Jaschek (born in Argentina but having acquired a French nationality), to create a branch of CDS in Moscow. A corresponding agreement was signed between CDS and Astronomical Council of the USSR Academy of Sciences ("Astrosoviet") in 1980. In the structure in which the Soviet Center for Astronomical Data (CAD) was created, I was appointed as its first Director.

Our center had become the second branch of CDS. The first branch, created because of the difficulty of contacting all the observatories spread in the western hemisphere, was the NASA branch of CDS, the Astronomical Data Center (ADC) (located at the Goddard Space Flight Center, Greenbelt, Maryland), which received all new catalogs from Strasbourg and then distributed them to the observatories of both Americas. The director of the American branch of CDS was a very erudite woman, Dr. Jaylee Mead, and Dr. Wayne H. Warren was involved in creating the catalog database. We closely cooperated in our joint affairs and exchanged visits.

Similarly, the CAD in Moscow received catalogs from CDS and distributed them to astronomical institutions of the Soviet Union and Socialist countries. It was also responsible for collecting, analyzing, preparing in accordance with the requirements, and transferring astronomical catalogs prepared in those countries back to Strasbourg (descriptions received by the CDS catalogs and studies carried out using catalogs from the CDS database were published in the Bulletins of this center). A few years later, we were given the opportunity to install in the CAD the most important database of stellar data, SIMBAD,[3] created by CDS.

To manage the activities of the CDS, an International Council of 12 members was created: six representatives of European institutions that contributed to the creation of the Center's catalogs, and six astronomers from other countries of the world. After the creation of CAD within Astrosoviet, I was appointed *ex officio* a member of the CDS International Science Council (ISC CDS), upon the recommendation of C. Jaschek.

Prof. Gart Westerhout, a radio astronomer, at the time Scientific Director of the US Naval Observatory in Washington, was the Chair of this International Scientific Council and its composition consisted of famous scientists from Germany, France,

[3] http://simbad.u-strasbg.fr/simbad/.

Switzerland, and Spain—all men and of ages fifty to sixty. All of them held positions at the Director level or were quite influential scientists in their own countries. Thus, within the Council, of which I was a member for six years (two terms, 1980–1986), I turned out to be the only, not so old, woman, and coming from a "communist" country, as well as, more generally, a representative of the Socialist countries.

I must admit that my appointment did not stir a wave of enthusiasm. I did not have any experience of working in such international organizations, and no one made any concessions either for this reason or because of insufficient language proficiency, rather on the contrary: I had a feeling of "let her suffer, so that she knows where she got into"...

At each meeting of the council (held twice a year), I had to make reports (which were later published in the CDS Bulletins (Dluzhnevskaya, 1980, 1984)) and the questions asked to me were by no means friendly, but with each next meeting I felt more confident and this, apparently, changed the attitude of my colleagues towards me. In between meetings, the French began to approach me first (perhaps feeling more confident as I was fluent in French), then the rest in the course of business conversations, just like a friend and, in general, simply a nice person. We all went to a restaurant to dine together, and for most members, I no longer seemed to be a "communist monster" abandoned as an agent to find out all the secrets of the organization!

However, the Chair of the Council was not completely happy with my presence and was very cautious, but after two or three years, Gart and I developed not only respectful, but, one might say, friendly relations. I am still very grateful to him for his invitation to visit the US Naval Observatory, and for the solemn reception that was given to me there, as well as for the invitation to his birthday party in the circle of relatives and friends that took place in The Hague, his native town, in 1994 during the IAU General Assembly.

Based on the signed agreement, our center began to receive star catalogs available from Strasbourg. As a member of the CDS International Scientific Council, I usually came to Strasbourg twice a year to attend the meetings of this Council (Fig. 15.1), at which reports from all members of the Council were heard, and plans and other issues of the activities were discussed. Using this opportunity to fill the catalog inventory of our center, I figured out in advance which catalogs most needed by CAD users should be included first, and I would send colleagues of CDS a request to prepare copies of these catalogs on magnetic tapes. In those years, transporting these tapes on large and heavy coils, in plastic boxes, was the only way to access this information. Therefore, every time I visited Strasbourg, I took with myself a special large suitcase, in which, after necessary clearance, I usually transported 16 such magnetic tapes across the border, weighing more than 20 kilograms of "scientific information" with recorded catalogs from Strasbourg, and I returned them back in the same suitcase (they were then valuable and were only loaned to us). Little by little, thanks to such physical labor, the inventory of catalogs of our CAD was building up—a "storehouse of knowledge" accumulated in the world by hardworking and strong astronomers (men and women !) every six months.

Fig. 15.1 CDS Council meeting of December 1982: (from left to right) J.-C. Mermilliod (Lausanne), G. Westerhout (USNO), D. Macchetto (STScl), M. Crézé (Besançon), F. Ochsenbein (CDS staff; half hidden), T. Lederle (ARI), O. Dluzhnevskaya (Moscow), Ch. Bruneau (CDS Secretary), G. Paturel (Lyons), C. Turon (Paris), L. Houziaux (Mons), C. Jaschek (CDS Director), J. Delhaye (Paris) and D. Carnochan (invited Starlink representative). On the steps of the Strasbourg Observatory

The main difficulty of our entry into the world of "computerized" scientific community still was the incompatibility of the computers we used at that time with Western ones, and even the tapes that I brought from Strasbourg had to be rewritten first in some kind of closed center that had machines of both types. This is the level at which our cooperation with the Strasbourg Center began.

CAD was the first organization of this kind, and of much significance for science in the USSR Academy of Sciences, since then, all the astronomical institutions of our country and Socialist countries received unlimited and free access to all the astronomical data accumulated from around the world. When I presented my report at a meeting of the Presidium of the USSR Academy of Sciences, it was approved, and the Academy allocated additional funds to our institute for supporting employees and for the purchase of the necessary equipment.

In 1984, our center held an international meeting in Tbilissi, Georgia: "Data collection, analysis, scientific results," which was very representative and very successful (Fig. 15.2) (Proceedings of the colloquium, 1984).

The very creation and first years of the establishment of the Strasbourg Center as a world center of astronomical data occurred mainly due to the energy and purposeful work of two remarkable scientists: C. Jaschek, already mentioned, and the Director of the Observatory of the University of Strasbourg, Dr. Alphonse Florsch. These two people, in addition to common scientific interests and affairs regarding the organization of the Center, worked closely together. Their families were friends.

Fig. 15.2 International meeting: "Data collection, analysis, scientific results", 1984, Tbilissi, Georgia. In the front row, in a white dress, O.B.Dluzhnevskaya; to her right, J. Mead, and in the second row to her left a white shirt, W. H. Warren (Tbilissi, 1984). (Credit: Local Organizing Committee of the 1984 Georgia meeting)

The cooperation with the Strasbourg Center in those days when I was a member of the ISC CDS was not only formal business but also very "humanly warm." You can say I was "patronized" by both Carlos and Alphonse, helping in all matters. I made friends with Daniel Egret, Marc Wenger, François Ochsenbein, and many other CDS staff, and when in Strasbourg, I felt like among close people who tried to do everything for us that was within their power.

In the following years, I corresponded with both Carlos and Alphonse, and when I came to Strasbourg, I always visited Carlos and his wife Mercedes while they lived in Strasbourg, and the Florsches in their country house, until the very last years of their lives. To meet Carlos, who had retired in Spain, I specially went to Salamanca, where he was already seriously ill, widowed, living with his daughter. I was glad that I was able to personally express my feelings of deep gratitude to both Carlos and Alphonse for what they did for us...

15.3 From Glass Libraries to Digital Databases

From the end of the nineteenth century and throughout most of the twentieth, the observations in all astronomical observatories of the world were recorded on delicate photographic glass plate negatives, of sizes typically $8' \times 10'$ (20 cm × 25 cm), up to 35 cm × 42 cm, covering different fields-of-view on the sky depending on the optical characteristics of the telescopes. These plates were assigned an identification number and recorded in observation logs, which indicated the date, coordinates of the center of the photographic plate, and other data. This classification made it possible, if necessary, to find the corresponding records taken many years before. However, it was also necessary to know in which observatory, and when, the photographic plates had been obtained. All this information was stored in observatory logs, and then kept in archives: the "glass libraries" of observatories.

At meetings of the International Scientific Council of CDS, besides reports, various innovative projects were discussed. In 1993, I proposed to create a database of photographic plates stored in all astronomical observatories, which seemed impossible to the Council members. Then, as an example, I asked to receive a copy of such archives from the leadership of the Crimean Observatory of the USSR Academy of Sciences in Nauchny, near Simferopol, and gave a presentation at the next CDS meeting. The plate collections (archives), together with the archives of glass libraries of other observatories in Russia and the Former Soviet Union (FSU) countries, were published in the CDS Bulletin (Dluzhnevskaya, 1994a, 1994b). They laid the foundation for the glass libraries of world astronomical observatories, created at the turn of the new millennium, already within the framework of the IAU Commission 5 (see below).

Of great value for research are the results obtained by observatory teams with identical equipment and covering practically the whole celestial sphere. A classic example of such a project is the abovementioned *Carte du Ciel,(*Chinnici, 2022*)* conducted by astronomers of 20 observatories located in different regions of the globe. They made use of identical "normal astrographs" with a lens diameter of about 30 cm, and photographic plates 16 cm × 16 cm in size and a useful sky field of $130' \times 130'$,[4] for determining the coordinates of stars and compiling a catalog, from 1887 to 1970. In the last years of the twentieth century, astronomers from the Astronomical Institute of Moscow University carried out the digitization of the *Carte du Ciel* catalog (4 million stars) (Gulyaev & Nesterov, 1992), which allows to use of these data in modern research. It was one of the first works of this type in the world.

The photographic plates on which all the work was carried out in previous years degraded with time, and the threat of losing valuable information for modern astronomical research was realized.[5] Only a few of the traditional ground-based observatories had managed to create an archive of observations in a ready-to-use form that is preferred by modern research methods.

At the beginning of the 1990s, many observatories began to build or use automatic machines to digitize photographic plates. An example is the MAMA[6] microdensitometer built at the Paris Observatory. Other observatories used other

[4] For comparison, the Moon has an angular diameter of 30', so the sky field covers about 20 Moons.

[5] This threat is real: on Jan.18, 2016, the Harvard College Observatory collection was flooded by a leaky burst pipe. Fortunately the plates were not damaged. (https://skyandtelescope.org/astronomy-news/flood-threatens-century-old-photographic-plates-0703201623/).

[6] MAMA = *Machine Automatique à Mesurer pour l'Astronomie*. See (Berger et al., 1991).

methods, like large scanners, for instance, the Harvard College Observatory collection of 500,000 plates (DASCH project, started in 2003).[7] This allowed to begin creating a digital database of photographic plates that computers could directly use and correlate with other data.

In those years, our institute led a Multilateral Cooperation with the Socialist countries on the problem of "Physics and the Evolution of Stars." Bulgarian astronomers from the Sofia Astronomical Observatory, led by Dr. Milcho Tsvetkov, undertook the development of a project eventually leading to the creation of an international astronomical database of glass libraries: the "Wide-Field Plate Database: European Plate Archives" (WFPDB), developed over twenty years (from the beginning of the 1980s till 2005–2007).

If the main result of the *Carte du Ciel* project was a star catalog created on the basis of measurements of photographic plates, then the goal of the WFPDB project was the preservation for future scientific research of astronomical data accumulated in the glass libraries of observatories around the world. Any user could receive an image of the sky section of interest (on the desired date, if available) on the Internet and, if necessary, for further study being able to contact the observatory that stores this particular photographic plate. However, photographic plates, as information keepers, turn out to be short-lived (a few decades at most) and require special storage conditions, which may not be ensured at all observatories. In this regard, it was decided to scan these collections and have the data processed accordingly, including observation logs, to provide users with direct access. The scope of the work turned out to be very broad, so it was decided to limit the effort to European observatories and, for starters, photographic plates obtained with telescopes with large fields of view. Plates exposed on wide-angle telescopes (known as "Schmidt" telescopes), as a rule have a size of 30 cm × 30 cm (but larger plates have a size up to 42 cm × 42 cm) and a field-of-view of 8 × 8 degrees on the sky.

Work on the WFPDB took place in collaboration with other European international organizations and projects: CDS, ESO (European Southern Observatory), IVOA (International Virtual Observatory Alliance; see below), GAVO (German Astrophysical Virtual Observatory), UDAPAC (Uccle Direct Astronomical Plate Archive Center), RVO (Russian Virtual Observatory), also HYPERLEDA (database of galaxies), BG_GRID (collection of the plates, AO, Sofiya), etc.

The WFPDB catalog now includes about 2.2 million scanned photographic plates (resolution 1 arcsecond/pxl), which corresponds to approximately 1/3 of all photographic plates stored in world observatories for the period 1872-2005.

By 2007, a large number of European astronomical observatories and institutes took part in the creation of the WFPDB database. Their list was given by Dr. M. Tsvetkov in the report "Wide - Field Plate Database: European Plate Archives" at the conference "DCA_EURO-VO," SOFIA, 24-25.01.2008 DCA(Data Centre Alliance)."

[7] http://dasch.rc.fas.harvard.edu/project.php.

List by Countries Vs. Institutes—Observatories

Armenia: Byurakan
Austria: Vienna
Belgium: Brussels
Bulgaria: Sofia, (WFPDB), Rozhen
Czech Republic: Ondrejov, Ceske Budejovice, Klet
Denmark: Brorfelde
Estonia: Tartu
Finland: Helsinki, Turku
France: Paris, Strasbourg, Nice, Bordeaux, Observatoire de Haute-Provence, Toulouse
Georgia: Abastumani
Germany: Potsdam, Bamberg, Heidelberg, Sonneberg, Bonn, Hamburg, Jena, Tautenburg, Münster, Berlin-DLR, Garching-ESO
Hungary: Budapest-Konkoly, Szombathely
Italy: Padova, Asiago, Roma, Turin, Bologna, Catania, Firenze
Latvia: Riga, Baldone
Poland: Torun, Krakov, Poznan
Romania: Bucharest, Cluj
Russia: Moscow, Zvenigoirod, St. Petersburg, Kazan
Switzerland: Basel
Serbia: Belgrade
Slovakia: Tatranska Lomnitsa - Skalnate Pleso
Spain: Madrid, Valencia
Sweden: Uppsala
The Netherlands: Leiden
United Kingdom: Armagh, Edinburgh, Cambridge, Oxford
Ukraine: Kyiv, Odessa, Nikolaiev, Crimea, Lviv
Vatican: Castel Gandolfo

All the data are available in direct access and can be visualized[8] (Tsvetkov & Tsvetkova, 2012). The distribution of the centers (in celestial coordinates) of the photographic plates contained in the Catalog is shown in Fig. 15.3.

The distribution of the centers of 500,000 plates (out of a total of about 2 million) is shown in Fig. 15.3. (This gives an idea of how the plates included in the database are distributed, without overloading the drawing with a large number of points.) The northern hemisphere (left half of the map) has a better coverage than the southern one (right half), because there are more telescopes. The best coverage is obtained along the equator (dark, "S-shaped" lane), and in selected areas (for instance, the dark "spots" near the South celestial pole correspond to the Magellanic Clouds).

[8] Archive Address: http://www.skyarchive.org.

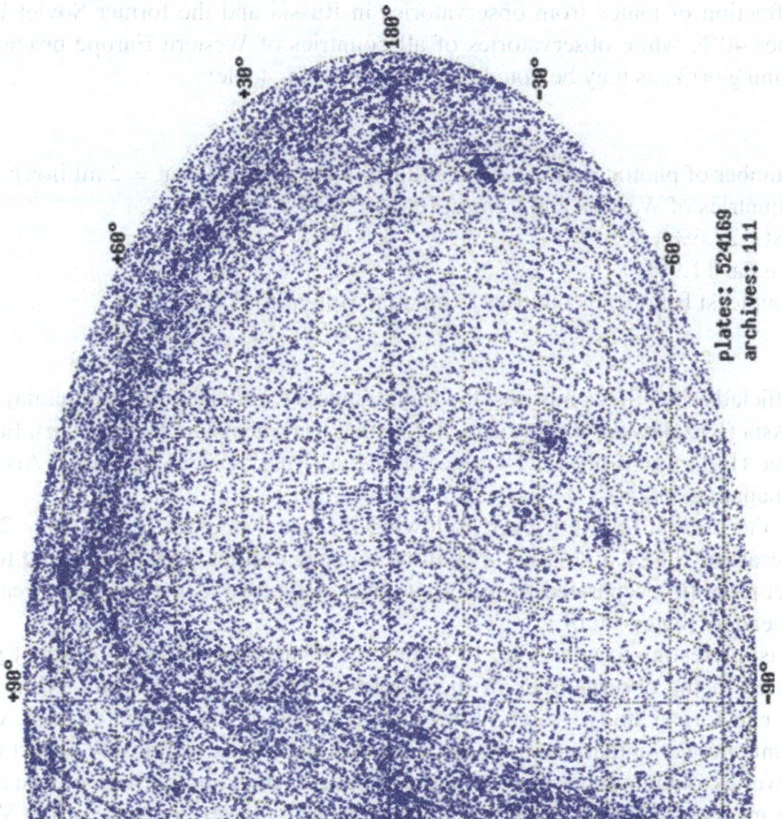

Fig. 15.3 The distribution of 500,000 photographic plates used for the WFPDB catalog on the celestial sphere, in galactic coordinates (Tsvetkov & Tsvetkova, 2012)

The fraction of plates from observatories in Russia and the former Soviet Union reaches 40%, while observatories of all countries of Western Europe provide the remaining 60%, as may be noted from the following table:

Number of photographic plates included in the project (total = 2 million):
Countries of Western and Central Europe: 57.7%
FSU: 27.6%
Russia: 11.9%
Southeast European countries (Rumania, Bulgaria): 2.8%

Officially the FSU comprises: the Baltic States (Estonia, Latvia, Lithuania), Central Asia (Kazakhstan, Kyrgyzstan, Tajikistan, Turkmenistan, Uzbekistan), Eastern Europe (Belarus, Moldova, Ukraine), Eurasia (Russia), Transcaucasia (Armenia, Azerbaijan, Georgia)[9]

In the present text, however, following the usual practice in the early 2000s, the acronym "FSU" and "Russia" will be separate entities. (And now these former Soviet republics, separated from Russia, became known as the Commonwealth of Independent States—CIS.)

It is interesting to note (Fig. 15.4) that the distribution of the number of plates captured in the observatories of the whole world has three maxima. Although a deeper historical study, which is beyond the scope of the present chapter, would certainly be warranted, one may speculate that these maxima closely reflect major discoveries and modern access to new wavelengths domains, including from space, allowing cross-correlations between observations—the very purpose of the Virtual Observatory, as follows.

The first peak, just at the end of WWI, is modest and sits on an overall gradual increase that reflects the rising technology in photographic astronomy, in particular the introduction of storing photographic data closely spaced in time to study variable objects. Then, as of the mid-20s, the sixfold rise, from about 1000 in 1927 to 6000 in 1933, is spectacular: this likely coincides with the birth of extragalactic astronomy (Lemaître was the first to explain, in 1927, the redshift-distance relation just found by Hubble by an expanding universe[10]), following an improvement in the sensitivity of the plates.[11] After a relative decrease in plate taking from the pre- to the post-WWII years until 1960, another rise was even more spectacular, a fourfold increase from about 5000 to 20,000 plates in one single year, 1965: this is the year of publication of the discovery of quasars—the birth of high-redshift cosmology. This discovery could have served as an incentive for different observatories to get

[9] E.g.: https://en.wikipedia.org/wiki/Republics_of_the_Soviet_Union

[10] See the IAU 2012 Press Release: https://www.iau.org/news/pressreleases/detail/iau1812/.

[11] The sensitivity of the plates used for the *Carte du Ciel* project, for instance, was too low to reveal galaxies. See (Chinnici, 2022).

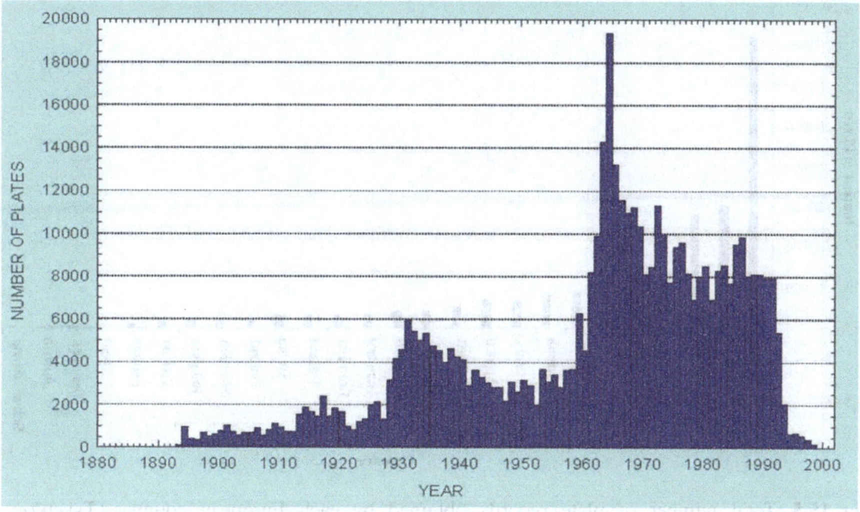

Fig. 15.4 Total number of plates recorded in the world since the first photographic observations (Tsvetkov & Tsvetkova, 2012)

more insight into this and other relevant exciting astronomical events. The relative stability of plate taking in the following twenty years (1970–1990) coincides with the development of space astronomy (the first X-ray satellite, *Uhuru*, was launched in December 1970), and of large Schmidt all-sky surveys (like the northern, second "Palomar Observatory Sky Survey," or the southern "UK Schmidt survey"). The abrupt end of this period coincides with the rapid deployment of digital detectors (CCD's, IR cameras, etc.) in world observatories and with the launch of the *Hubble Space Telescope* in the spring 1990. After the 90s, observatories practically stopped using photographic plates for mapping the sky.

More detailed information on the number of records obtained and scanned at European observatories is provided by histograms showing the total number of records for each country (Fig. 15.5). In this figure, Germany demonstrates a very significant first place; Ukraine has the second place with about twice less. Russia takes an honorable third place. In subsequent years, most observatories also digitized their glass libraries and provided these data to the WFPDB, like the astronomical plate collection of our institute, in Zvenigorod not far from Moscow (Dluzhnevskaya & Tsvetkov, 2012).

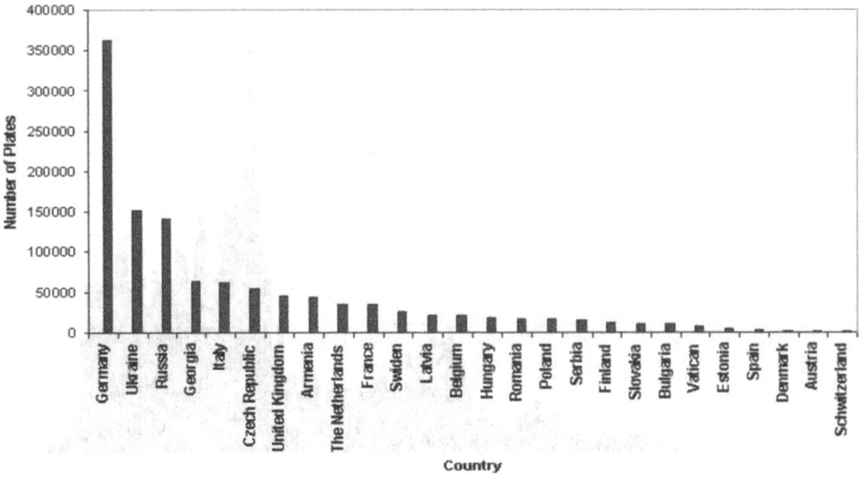

Fig. 15.5 Total number of plate records obtained by each European country (Tsvetkov & Tsvetkova, 2012)

Similar difficulties arose with photographic spectroscopy. The IAU Working Group for Spectroscopic Data Archives of the IAU Commission 5 attempted to set up a World Plate Store of Photographic Spectra in order to safeguard the accessibility of these valuable data for the future, and technological progress in recent years led us to think about the possibility to eventually digitize them also in the future. If so, this would end fulfilling the Resolution (B3) adopted by the IAU General Assembly in Manchester (2000) on "Safeguarding the information in photographic plates" (see also below).[12]

15.4 Archives of Glass Plate Libraries: Part of the UNESCO World Astronomical Heritage?

In 2004, the so-called Thematic Initiative "Astronomy and World Heritage" was created within the framework of UNESCO, which carried out work to preserve various objects related to astronomy: ancient sanctuaries, medieval and modern observatories, astronomical instruments, manuscripts, etc.[13] Over the years that followed, a large number of symposia, business meetings, and meetings of UNESCO national representatives (Focal Points) of the participating countries took place.

[12] https://www.iau.org/static/resolutions/IAU2000_French.pdf.

[13] See today the IAU-UNESCO "Portal to the Heritage of Astronomy:" https://www3.astronomicalheritage.net/.

In September 2011, in Paris, the Astrophysical Institute (IAP) hosted a Colloquium entitled "Preservation of the Astronomical Heritage," which was held under the patronage of UNESCO and was organized by the French Astronomical Society and the Ministries of Foreign Affairs of France and other European countries.[14] The Focal Points of the participating countries of the Initiative, directors of astronomical institutions, and famous scientists from many European countries took part in the work of the Colloquium. Various issues of preservation of buildings of astronomical observatories, telescopes, archives, and other historically significant objects of the astronomical heritage of Europe, as well as the legal side of these issues, were discussed. I was the Focal Point for Russia in this organization since 2004 and made two reports at the colloquium on the astronomical observatories of Russia and Ukraine and on their glass plate libraries.

The first meetings of the colloquium were devoted to the problems of preserving astronomical observatories, but whatever the status of an observatory, even if the most magnificent architectural masterpiece with well-functioning telescopes with perfect optics, the most important criterion in astronomy, as in other sciences, remains the results of their research. Until the middle of the twentieth century, these were mainly photographic plates depicting portions of the starry sky, planets, the Moon, and the Sun, as well as spectra of celestial bodies. Although at present most of the photographic plates accumulated in the archives of astronomical observatories are not used very actively, nevertheless, their value is not decreasing, but rather increasing, especially those obtained as part of international programs.

The problem with preserving these unique observational data accumulated over many decades was the subject of discussions at meetings in, the second half of the colloquium. The role of these archives for scientific research was discussed, as well as the means of preserving astronomical photographic plates and manuscripts. The discussion was very active, as the oldest records are already starting to deteriorate, and urgent measures are needed to save this unique material. As a result of the discussion, it was noted that the archives of astronomical negatives are of lasting value for world astronomy, and over time the value of these collections can only increase. In this regard, the meeting proposed that archives of glass plate libraries should be included in the UNESCO World Astronomical Heritage category.

Unfortunately, the project has not made much progress so far, but in recent years, the astronomical observatories of the world and the IAU have independently done a lot of work to scan plates, digitize them and create glass libraries and databases.

[14] The seat of UNESCO is located in Paris.

15.5 The Activities of IAU Commission 5

After the branch of CDS was created in Moscow, as the Director of this center I became a member of the Organizing Committee of Commission No. 5 of the IAU (in 1982-1991, and again in 2003-2006). This Commission ("Documentation & Astronomical data") is one of the oldest Commissions of the IAU: it was created at its first General Assembly in Rome in 1922.

In 1991, the XXIst IAU General Assembly took place in Buenos Aires, Argentina. Upon the recommendation of Bernard Hauck (who had introduced me to CDS, see above), when, after his duties as Vice-President of the Commission 5 , he was elected President of this Commission, I was elected Vice-President to succeed him. And I began hard work to participate in the leadership of the Commission, which oversees one of the fastest-growing areas of science. We worked with Bernard and other members of the Organizing Committee in amicable terms, and for six years when he was President, and myself Vice-President, a very significant task could be accomplished.

The first three years for all of us and for me, especially, were characterized by the fact that it became much easier and more efficient to discuss all issues together: the electronic mail went into operation. It was a time of rapid development of information technology. The leadership of the Commission contributed in every way to this process in astronomy, and our three centers were the core that determined the main directions of the Commission's activities. In subsequent years, in different countries, regional centers and centers that collect data of a certain type were created and maintained contacts with IAU Commission 5:

- Astronomical Data Analysis Center, Tokyo, Japan (ADAC)
- Beijing Astronomical Data Center, China (BDC)
- InterUniversity Center for Astronomy & Astrophysics, Pune, India (IUCAA)
- Centro de Datos Astronomicos, La Plata, Argentina (CDA)
- European Southern Observatory, Garching, Germany (ESO)
- Infrared Processing and Analysis Center, Pasadena, CA, USA (IPAC)
- Canadian Astronomy Data Center, Victoria, B.C. (CADC)

The responsibilities of these centers included storage (authors usually do not store data themselves for a long time), duplication and transfer of astronomical catalogs to users, as well as processing of astronomical data (samples, cross-identification, visualization), examination (collection of information about resources, reviews), creation of new catalogs and providing advice to catalog creators. CDS itself expanded the scope of stored data and became known in 1983 as the

Strasbourg Center for Astronomical Data (however, retaining the "CDS" acronym, now standing for "*Centre de Données astronomiques de Strasbourg,*" or "Strasbourg Data Center" for short, instead of "*Centre de Données Stellaires*") and its director was Dr. Françoise Genova.

We had close business relations with the leaders of all the centers, but we quickly became friends with Jaylee Mead, Director of the ADC at NASA, and this relationship lasted until her death (2012).

At the meetings of Commission 5 on XXIIrd General Assembly (The Hague 1994) much attention was paid to the activities of data centers created in recent years and establishing contacts between them. Jaylee herself never participated in General Assemblies and ADC was represented at the 1994 GA (The Hague) and the 1997 GA (Kyoto) by the Head of the GSFC Planetary Magnetospheres Laboratory Cynthia Y. Cheung (see Fig. 15.6)—specialist in Space Science mission operations and data management, including multi-mission space science data archive interoperability, applied information technology, robotics, and intelligent control systems.

The meetings of the CDS Council, as I already mentioned, were held twice a year, one in Strasbourg and the other off-site—one of the members invited the others

Fig. 15.6 Three women directors: C. Y. Cheung (representing ADC, USA), O. Dluzhnevskaya (CAD, Russia), F. Genova (CDS, France). The photo was taken during the XXIIrd General Assembly, The Hague 1994. (From the personal archive of O.B.D.)

Fig. 15.7 Participants of the Colloquium "International cooperation in the dissemination of astronomical data," Pulkovo Observatory of the Russian Academy of Sciences, 1996. Among others, in the front row, center, wearing sunglasses: Bernard Hauck (Pres. IAU Com.5, Switzerland). To his right are Françoise Genova (Dir. CDS, France), Elizabeth Griffin (UK), Koichi Nakajima (Dir. ADAC, Japan); to his left are Olga Dluzhnevskaya (Dir. CAD, Russia), Nina Polosukhina (Russia), Jaylee Mead (Dir. ADC, USA). (Credit: Local Organizing Committee of the meeting)

to his or her place. In our country, at my invitation, the first meeting devoted to the problems associated with the activities of our center was held at the Pulkovo Observatory, South of St Petersburg, in 1996 - a colloquium entitled: "International cooperation in the dissemination of astronomical data" (Figs. 15.7 and 15.8), already within the framework of Commission 5 (Proceedings of the colloquium, 1996).

In 1997, at the 23rd General Assembly in Kyoto, I was elected President of IAU Commission 5 (Fig. 15.9) and, at my suggestion, Françoise Genova, Director of CDS (Fig. 15.6), became Vice-President. Thus, our commission began to represent "something completely exceptional"—two women in the commission's leadership, which was at that time very rare. In addition, the Director of ADC, the American branch of CDS, was, as I mentioned, also a woman—Jaylee Mead.

The next key event would happen during the following General Assembly in Manchester, three years later.

Fig. 15.8 Coffee break: Jaylee Mead (USA), Bernard Hauck (Switzerland), Olga Dluzhnevskaya (Russia). Behind them: Francois Ochsenbein (France) and Françoise Genova (France), at the IAU Commission 5 Colloquium "International cooperation in the dissemination of astronomical data" (Pulkovo Observatory, 1996). (Photo from the personal archive of the author)

Fig. 15.9 Presidents of IAU Commission 5: Jean-Claude Pecker, France, 1979–1985 (far right; IAU General Secretary, 1964–1967), George Wilkins, UK, 1985–1991 (far left), Bernard Hauck, Switzerland, 1991–1997 (center) and O.B. Dluzhnevskaya, Russia (1997–2003). (Picture taken during the XXIIrd IAU General Assembly, The Hague 1994, after the election of the President of IAU Commission 5 Bernard Hauck, and of its Vice-President O.B. Dluzhnevskaya. (From the personal archive of the author)

15.6 The *International Virtual Observatory*

IAU Commission 5, with its four Working Groups and three Task Groups, approached the XXIVth IAU General Assembly (Manchester, UK, Aug. 7-19, 2000), as they say, "fully armed" and, as it seems to me now, my greatest credit to this Commission is what I could manage to do at this Assembly.

The event took place during "the year of the millennium." Everyone prepared differently for this amazing date. It was clear that the Assembly would definitely discuss the establishment of a Virtual Observatory, and we had prepared various proposals for discussion. Both European astronomers and American astronomers had prepared their projects.

As explained above, the project to create an international database of glass plate libraries, the "Wide-Field Plate Database: European Plate Archives" (WFPDB) was carried out over the last years of the millennium under the purview of IAU Commission 5. It played a large role in preparation for the Manchester General Assembly.

Of course, all the members and Task Groups of Commission 5 did a great job in preparing for the Global Data Network, but I do not discuss this here, since the present chapter is devoted to the creation of an International Virtual Observatory. Nonetheless there was at least one very important initiative carried out by this time: a "Working Group on Libraries" was created and functioned, uniting the librarians of astronomical institutions of the world as a single network. This network represented a large amount of information and resources, long-term projects for digitization and document preservation, as well as developing an indexing and distribution system for electronic preprints. This initiative allowed any astronomer to almost immediately find an article in any of the networked libraries and get a copy of it. This Working Group was created on the initiative of the heads of libraries of large observatories of the world and was called LISA (Library and Information Services in Astronomy), and organized a number of very important and interesting international meetings within the framework of IAU Commission 5 (Dluzhnevskaya, 1989; Grothkopf et al., 1998; Special Issue, 1995).

The FITS (Flexible Image Transport System) Working Group of our commission developed the FITS format for data exchange in astronomy, i.e., the transfer and storage of images and tabular data, that had already been widely used in the USA, Japan, and other countries.

Over the preceding three years, four ADASS (Astronomical Data Analysis Software and Systems) conferences were held on issues related to the creation of the Virtual Observatory, mainly prepared by our American colleagues. Colleagues from the CDS organized their own conference—"The Virtual Observatory," European Science Foundation (ESF) Exploratory Workshop (20-21 September 1999, Strasbourg, France)

Two projects were the most advanced (USA, Europe), so heated discussions were expected at the Assembly about which VO project was the best. And then I had to show "diplomatic abilities." Unexpectedly for everyone, I quickly assembled the meeting of the Organizing Committee and suggested holding a Business Session at the very beginning of the General Assembly (see the agenda below; according to the schedule, it was supposed as usual to happen at the end of the Assembly, on Aug.19). The development of the Virtual Observatory project was the responsibility of our commission, but there was no corresponding Working Group. At this meeting, I proposed to change the structure of the Commission, by introducing a new Working Group, called "Creation of an International Virtual Observatory." The goals and composition of this Working Group were subsequently discussed and approved by all the participants.

Françoise Genova, then Vice-President of the Commission, was elected Chair of this group.

In this way, open discussions at the General Assembly itself were avoided, and from then on, all projects for the development of the Virtual Observatory fell under the auspices of this Working Group. Thus began a new stage in the activities of Commission 5, the main purpose of which was to create a grandiose structure for the entire world astronomical community—the *International Virtual Observatory* (IVO).

The goal of this project was to combine all world observational databases and other astronomical resources into a single system that would allow to download and analyze any data from any observatory via the Internet from any computer.

Astronomical data sets and other resources should work as a whole. Many projects and data centers around the world began to work towards this goal.

The IVO was thus designed as a powerful software environment in which giant astronomical archives and databases freely distributed around the world would be integrated, along with a user interface including analysis tools and a computing helpdesk. The IVO integrated all significant national and international projects for the creation of virtual observatories, combining the existing archives from ground and space observations, and provide astronomers and possibly other users, as well as the public, with convenient access to them. Indeed, the IVO would also be a powerful tool for the democratization of astronomy, since even small scientific and educational astronomical institutes would share the same opportunities for conducting scientific research based on modern observational material as the leading astronomical organizations in the world.

COMMISSION 5: DOCUMENTATION AND ASTRONOMICAL DATA

Business Session program.

Aug.12

9.00–10.30

President report (10 min.)

Reports from Chairpersons of WG&TG (7×10 min.)

Report from Donna Coletti on the project:

(continued)

"Preservation of Observatory Publications" (10 min.)
11.00–12.30
New structure of Com. 5
New members and new consultants
New SOC
Resolutions
Activity for the next triennium
…… … The principal field of science of IAU Commission 5 are astronomical
data in their development, and one and maybe the most fundamental
of purposes will be the participation in the creation of a Virtual
Astronomical Observatory …

15.7 Today: The IVO Alliance

Each of the existing national projects has its own scientific and technological tasks. While this diversity seems beneficial enough for the success of the IVO, there are also some common elements that need to be harmonized for the effective design of the IVO. "Virtual Observatory France" is an example(Genova, 2019). Most of these common elements are related to data standards and the use of software packages.

The different VO communities, therefore, required a mechanism to facilitate international cooperation in the pursuit of common goals. Such a mechanism was the International Virtual Observatory Alliance (IVOA),[15] which was created in June 2002 with the aim of "facilitate the international coordination and collaboration necessary for the development and deployment of the tools, systems and organizational structures necessary to enable the international utilization of astronomical archives as an integrated and interoperating virtual observatory." (https://www.ivoa.net/)

The IVOA (Fig. 15.10) is an organization that discusses and agrees on the technical standards necessary to make VO possible. It also acts as the basis for the discussion and exchange of ideas and technologies between virtual observatories, as well as a body for the popularization of VO.

The IVOA Executive Committee includes representatives of all collaborating Virtual Observatories (for instance, O. Malkov is the representative of the RVO in the Alliance, see below). The Chairperson is elected from among the representatives and performs his duties for one year, which is preceded by one year as Vice-Chairperson. The Executive Committee meets three–four times a year to discuss goals, priorities and strategies.

[15] http://ivoa.net.

Fig. 15.10 Logo of the IVOA (2005) with the logos of the member countries, clockwise from 12h: Australia, Italy, India, Canada, Spain, Armenia, China, EuroVO, France, Germany, Hungary, Japan, Korea, USA (National Virtual Observatory), Brazil, Russia, Great Britain. (http://www.ivoa.net)

The IVOA's work focuses on standards development. Nine working groups were created in the IVOA structure, each of which was to develop standards for describing astronomical data, objects, processes, and software architecture.

Working groups are created with the participation of representatives of various projects in those areas where it is necessary to define and agree on key standards, the language used by applications to access distributed data in virtual space and interaction technologies. Working groups develop standards using a process modeled after the model of the World Wide Web Consortium, in which working projects proceed to the draft recommendations and, finally, to the adopted recommendations. These recommendations are ultimately approved by the Working Group on the Virtual Observatory of IAU Commission 5. The IVOA also has "interest groups" that discuss experience using VO technologies and provide feedback to working groups.

The IVOA holds two seminars each year: a weekly meeting in the spring, usually in May, and a shorter meeting in the fall, which is either coordinated with the annual ADASS conference or with a regional meeting on the VO project. These meetings provide an opportunity for working groups and interest groups to hold face-to-face discussions and solve more complex technical issues.

Fig. 15.11 Participants in the Alliance meeting, Moscow 2006. Françoise Genova(center). To her right: A. Cherepaschuk (Dir. SAI MSU), O. Dluzhnevskaya (Dir. CAD), O. Malkov (RVO Representative in the Alliance). To her left B. Shustov (Dir. INASAN) on the steps of the Sternberg Astronomical Institute of Moscow State University. (Credit: Local Organizing Committee of the meeting)

Since this is an international organization with high scientific authority and well-deserved recognition around the world, these events are held at a high level with the participation of the leaders of the Academies of Sciences and high-level statesmen. To host an Alliance meeting is considered a great honor for the inviting organization, and each time the issue is resolved through competitive invitations.

In September 2006, a meeting of the Alliance was held in Moscow, at the Institute of Astronomy of the Russian Academy of Sciences, the Sternberg Astronomical Institute of Moscow State University, and the building of the Presidium of the Russian Academy of Sciences (RAS) (Fig. 15.11) (Proceedings of Section 11 of the Joint European and National Astronomical meeting (JENAM), 2000).

Each year, the number of participants in these prestigious international meetings has grown, which can be seen in the picture showing the participants in the Alliance meeting in Beijing, China, in 2007 (Fig. 15.12).

In 2011, at the IUCAA (InterUniversity Center for Astronomy and Astrophysics) Alliance meeting held in Pune, India, the Virtual Observatory of Ukraine was accepted into the IVO upon the recommendation of Russia. From the CIS countries (Community of Independent States), the Virtual Observatory of Armenia has also been included in the IVO.

Every year, in addition to two meetings, the IVO Alliance holds a whole series of events also in different countries and often teleconferences. Sometimes the

Fig. 15.12 Participants in the Alliance meeting in front of the hotel where the meeting was held (Beijing 2007). In the front row in a white shirt and light trousers, F. Genova (VO France), to her right is Dave De Young (NVO, USA), then O. Dluzhnevskaya (RVO, Russia). Behind her is Masatoshi Ohishi (JVO, Japan). Far left, in the far row, in a beard and glasses, Bob Hanisch (IVOA, USA). (Credit: Local Organizing Committee of the meeting)

meetings of the Alliance coincide with conferences of ADASS[16]—which are held annually, and, like the meetings of the Alliance, each year at a different location of the astronomical institutions. The conference has great authority and attracts a large number of participants (200–350 people); it is a forum for scientists and programmers involved in the analysis and dissemination of astronomical data.

In September 2013, an Alliance meeting was held in conjunction with the XXIIIth ADASS Conference, in Waikoloa, Hawai'i (Fig. 15.13) (Dluzhnevskaya et al., 2002). In November 2020, one of the two meetings of the Alliance is scheduled to take place in Granada (Spain) and will be timed to coincide with the ADASS XXXth anniversary planned there.[17]

One of the important elements of the program is the development of communication between developers and users with a wide range of expertise in the production and use of software and systems. The program consists of invited negotiators, oral and demonstration presentations, study guides, user group meetings, and meeting special interest groups.

[16] http://adass.org.

[17] https://adass2020.es.

Fig. 15.13 Participants in the ADASS XXIII conference and Alliance meetings, Waikoloa, Hawai'i, September 2013. (Credit: Local Organizing Committee of the ADASS XXIII conference and Aliance meeting.)

Any national or joint virtual observatory can become a member of this IVO alliance—their implementation is not limited to national borders. The projects use observational data in a wide range of wavelengths obtained from space and ground-based instruments used by the international astronomical community. The goal of each project is to equip astronomers with powerful mechanisms of modern research of observational data, mechanisms that can also be used by our colleagues from related sciences.

Currently, the IVOA includes 20 national and international Virtual Observatory projects created around the world and one intergovernmental organization (ESA) (Fig. 15.14). Membership is open to other national and international projects in accordance with the IVOA guidelines for participation.

Since its inception in 2002, the IVOA has boasted a number of achievements:

- Achieving genuine Global Cohesion in discussing and agreeing on key astronomical standards. The IVOA is endorsed by the IAU as a generally accepted method for developing standards related to astronomical data. These standards are submitted for ratification to Commission 5.
- Creation of a forum for discussion and discussion of astronomical data technology in general, as well as VO standards in particular. Bio-annual "interoperability meetings" are a key element in resolving international disagreements, as well as exchanging ideas and technologies.

Fig. 15.14 IVOA members 2020: 1. Argentina Virtual Observatory (NOVA); 2. Armenian Virtual Observatory (ArVO); 3. UK Virtual Observatory (Astrogrid); 4. Australian Virtual Observatory (ASVO); 5. Brazilian Virtual Observatory (BRAVO); 6. Chinese Virtual Observatory (China-VO); 7. Chilean Virtual Observatory (ChiVO); 8. Canadian Virtual Observatory (CVO); 9. European Space Agency (ESA); 10. European Virtual Observatory (Euro-VO); 11. German Astrophysical Virtual Observatory (GAVO); 12. Hungarian Virtual Observatory (HVO); 13. Japanese Virtual Observatory (JVO); 14. National Virtual Observatory USA (USVOA); 15. Observatoire Virtuel France (VO-France); 16. Russian Virtual Observatory (RVO); 17. Spanish Virtual Observatory (SVO); 18. Ukrainian Virtual Observatory (UKR-VO); 19. Italian Virtual Observatory (V.Obs.it); 20. Virtual Observatory India (vo-i); South African Astroinformatics Alliance: (NSF/SA[3]) (http://www.ivoa.net/about/member-organizations.html)

- Quickly harmonize the initial set of basic standards—a format for exchanging tables, specifying simple directory services and image queries, defining metadata describing resources, a dictionary of standardized column names, and a set of standards for creating VO registries. Services that meet these initial standards are already deployed worldwide. In addition, it has been very successful to define a simple standardized way for applications to transfer messages between themselves, so that they can work seamlessly on VO tasks. At the time of preparation of this document, the IVO agreed on a fundamentally new standard table access protocol, which allows software to send flexible queries to a wide variety of databases.

IVOA is working hard on further standards, including those needed for virtual storage addressing, single sign-on, semantic reasoning, meshing, and the modular implementation of web services.

The creation of an active IVOA made it possible to create a virtual environment that combines the results of astronomical research carried out in astronomical institutions around the world and using access tools developed by the Alliance to use this data to any user.

However, all efforts of the Alliance would be useless if the national parts of the International Virtual Observatory were not updated with new archives, databases, catalogs, and other astronomical data. Information on the activities of the French virtual observatory, as already mentioned, can be found in (Genova, 2019), and the next chapter zooms on the history of creation and the main stages of development of the Russian Virtual Observatory.

15.8 The *Russian Virtual Observatory* (RVO): The First Decade

Virtual observatories began to be created in different countries since 2000, and one of the first was the Russian Virtual Observatory (RVO). The initiators of the creation of the RVO were the CAD in Moscow, and the Special Astrophysical Observatory of the RAS in Zelenchuk, Greater Caucasus. Currently, the RVO project is carried out on the basis of the Agreement signed by the directors of the Institute of Astronomy of the RAS[18] (INASAN), the Institute of Informatics Problems of the RAS, the Sternberg State Astronomical Institute of Moscow State University, and the Special Astrophysical Observatory of the RAS. The base center of the RVO is the INASAN Center for Astronomical Data (CAD)—the Russian branch of CDS.(Dluzhnevskaya et al., 2002; Витковский & Длужневская, 2001)

In December 2001, the Scientific Council on Astronomy of the Russian Academy of Sciences approved the initiative to create a Russian Virtual Observatory, the ultimate goal of which was to provide Russian astronomers with access to extensive sources of data and metadata generated as a result of observational projects. The work on the creation of the RVO is an essential part of the research plan of Section 13 "Databases and Information Support" of Scientific Council on Astronomy of the Russian Academy of Sciences (Chair O.B.D.). The creation of the RVO was also discussed at the sixth General Meeting of the Euroasian Astronomical Society: "International Collaboration in Astronomy," Moscow May 25–June 2, 2002.(Dluzhnevskaya et al., 2003)

The contribution of the RVO to the task of combining world astronomical resources is very significant for Russia. There are about 30 astronomical institutes and organizations in the country, many of which have extensive data archives, and they collaborated with INASAN in the creation of the RVO. One of the advantages of Russian astronomical data is the ability to create long series of observations combined with data from observatories in other countries. This is due to the fact that many observatories in Russia, span nine time zones in longitude (and extend

[18] Founded in 1936.

from about 40° to 80° in latitude), so they are located on the opposite sides of the globe relative to most major world observatories.

The main objectives of the RVO Project are to provide the Russian astronomical community with a convenient and effective mechanism for accessing foreign data sources and to combine Russian astronomical information resources as an important component for integration into the International Virtual Observatory.[19]

Historically, the relevance of the creation of the RVO is determined, first of all, by the fact that with the collapse of the USSR in 1991, Russian astronomy had almost lost its instrumental base, i.e., its observatories close to the southern border of the country. Scientific work at the Abastumani Astrophysical Observatory in Georgia, the Shemakha Observatory in Azerbaijan, the Maydanak Observatory in Uzbekistan was very difficult or even impossible. The observation station of St. Petersburg University in Byurakan, Armenia, and the high-altitude Shorbulak station of the Main Astronomical Observatory (Pulkovo) in the Pamirs (Tajikistan) was closed. The best way out of this situation was the integration of the Russian astronomical community into the world information network of astronomical data, then primarily accessible in western observational archives.

By 2010, there were already 755 catalogs in the CAD database. In order to provide Russian astronomers with convenient access to foreign data sources at the CAD in the very first years after the creation of the RVO, a series of mirror sites of known foreign databases were placed in the CAD, primarily the most important stellar database SIMBAD, as well as ADS, VizieR, INES, VALD, WFPDB and others, and their operation supported. In CAD also, a BDB—database of double stars[20] was developed and installed, containing the most complete information about all categories of binary stars, namely, visual, interferometric, spectroscopic, photometric, astrometric, and positional parameters of 260 000 components of systems of multiplicity from 2 to more than 20 (Kaygorodov et al., 2012; Dluzhnevskaya et al., 2013; Kovaleva et al., 2015). The functioning of the General Catalog of Variable Stars is supported—52,000 reliably identified variables (GCVS)[21] (Samus et al., 2017).

The problem of integration and providing the foreign community with a comprehensive description of the form and content of Russian astronomical information resources, made according to the standards of the IVO, is also being addressed within the framework of the information center system. In particular, work has been completed to create a structured list of Russian astronomical Internet resources, as well as coming from FSU countries, containing more than 50 original astronomical data resources and constantly updated. The list was praised by foreign colleagues. (Dluzhnevskaya et al., 2008; Malkov et al., 2009).

One of the unique CAD obligations to the international community is the processing, verification, and transfer to the CDS of data tables from articles published in Russian astronomical journals, and star catalogs prepared by Russian

[19] http://www.inasan.ru/eng/rvo.

[20] http://www.bdb.inasan.ru.

[21] www.sai.msu.ru/groups/cluster/gcvs.

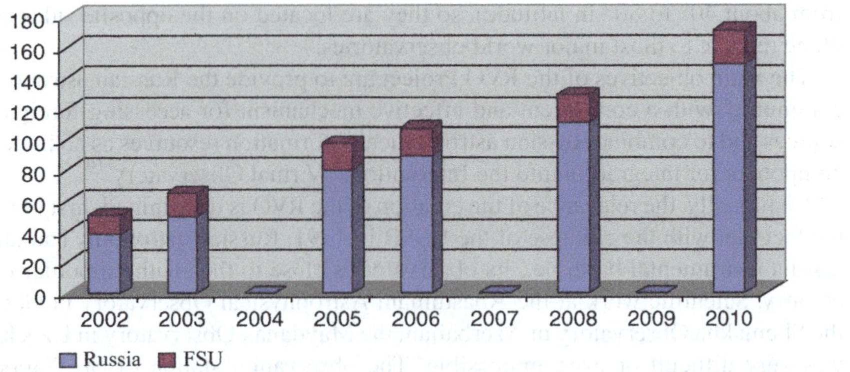

Fig. 15.15 The histogram gives an idea of the growth of the total number of Internet resources created in institutions in Russia and the FSU during the first decade of the RVO (Malkov et al., 2008)

astronomical organizations, including catalogs created by the CAD employees. An exhaustive bilingual list of Russian astronomical institutions has been created. The list also includes some astronomical organizations of FSU countries.

As part of the RVO project, the photographic plates accumulated in Russian observatories (and some FSU observatories) were digitized. Scans are entered into the WFPDB database.

The present organizational structure of the RVO, approved in 2010, includes six Working Groups, the activities of which cover all areas of the RVO. The results of the RVO project are reflected on the constantly updated RVO website.

An analysis of astronomical data resources that are directly accessible on the Internet on sites of astronomical or astronomical related institutions in Russia and the FSU, and created with the participation of the host institution, was prepared at the CAD in 2010 (on the 10th anniversary of the creation of the RVO). Some other types of resources are mentioned separately.[22]

The RVO Astronomical Data Resources Register was first created in 2002 and updated in 2003, 2005, 2006, 2008, 2010 (Figs. 15.15 and 15.16).

Analyzing the available information, the following trends of the astronomical resources of the countries of the former USSR on the Internet can be noted. The bulk of direct access to astronomical resources belongs (from the FSU countries) to the astronomical institutions of Ukraine (a number of observatories present their resources), Armenia, Estonia, Latvia.

- In Armenia (Byurakan Astrophysical Observatory), ArVO was created—a member of the IVO Alliance, as mentioned above.
- In Ukraine, many astronomical institutions are actively opening direct access via the Internet to the created astronomical resources (Main Astronomical Obser-

[22] http://www.inasan.rssi.ru/eng/rvo/rus_res.html.

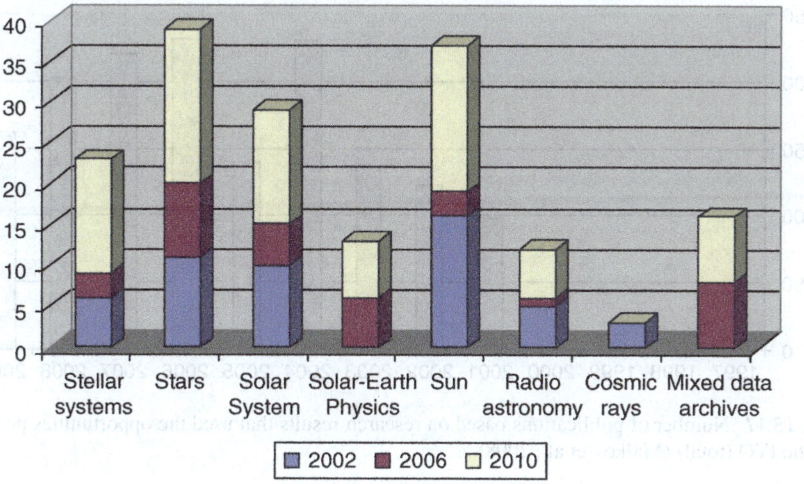

Fig. 15.16 Thematic structure and dynamics of professional Russian astronomical Internet data resources, 2002–2010 (Malkov et al., 2008)

vatory of NASU—Golosiiv, JSC Odessa State University, JSC Kharkov State University, Institute of Radio Astronomy NASU). The observational archives are being digitized and catalogs of photographic plates of glass libraries of GAO NASU, AO OSU (Astronomical Observatory of Odessa State Univ.), are available to one degree or another. The success in the implementation of the goals could be judged as early as 2003, when a demonstration of the capabilities of the first VO prototypes took place, and how much the authority of the IVO project has grown in the world is evidenced by the growth in the number of publications based on research results that used the opportunities offered by the IVO (Figs. 15.17 and 15.18).

In 2010, the latest analysis of astronomical data resources that are directly available on the Internet on sites of astronomical or related astronomical institutions of Russia and the FSU was prepared as part of the RVO. Ukrainian and Armenian astronomers created their own virtual observatories, which, as we have seen, became equal members of the IVO Alliance, astronomers of the Baltic states apparently cooperated with European countries.

In recent years, astronomers from Georgia, Azerbaijan, and the republics of Central Asia have extremely rarely participated in joint events with us and, in practice, have not sent articles to magazines published in Russia (which was the rule). However, the Institute of Astronomy of the Russian Academy of Sciences conducts joint research on the problems of Meteor astronomy and near-Earth space research with Tajik astronomers. Colleagues from the Astrophysical Institute of the Kazakh Academy of Sciences published in 2015 the book "Astrophysics in Kazakhstan," which gives an overview of the work performed in recent years, in virtually all areas of astronomy. A separate monograph is devoted to the study of

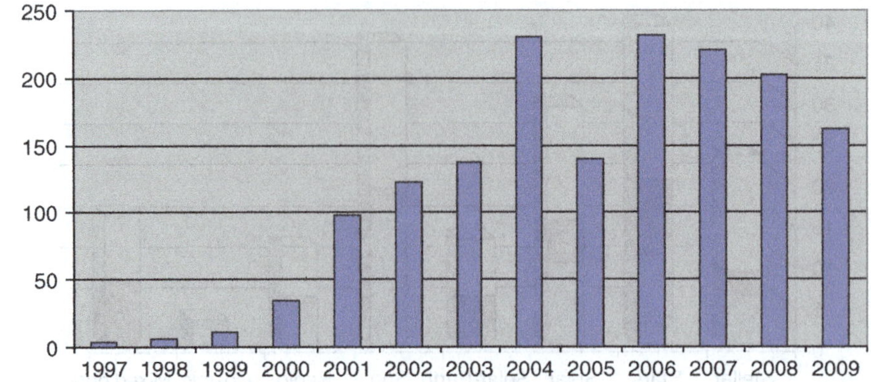

Fig. 15.17 Number of publications based on research results that used the opportunities provided by the IVO (total) (Malkov et al., 2008)

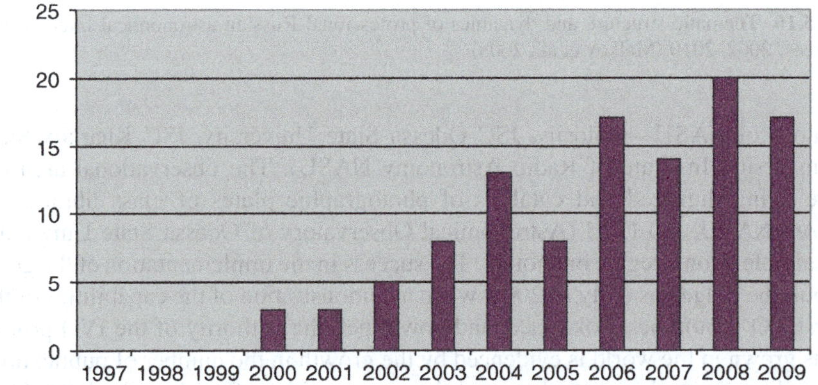

Fig. 15.18 Number of publications based on research results that used the opportunities offered by the IVO in refereed journals (Malkov et al., 2008)

the moon and planets of the solar system. Some Russian observatories still use telescopes installed in Soviet times in observatories of the former Soviet republics. SAO RAS, for example, has instruments in the observatories of Uzbekistan, Georgia, Armenia, Ukraine, Latvia, and Estonia, so that joint work is still in progress.

15.9 Conclusion

The creation of the grandiose project of the International Virtual Observatory, has unfolded over the past twenty years by astronomers of 20 countries. It began back in 1972 with the unification of catalogs of European observatories into a single collection in the Strasbourg *Centre de Données Stellaires* (CDS) (later the *Centre de Données astronomiques de Strasbourg*), along with the creation of two branches: the Astronomical Data Center (ADC), in Maryland, the USA for countries of the American continent, and the Soviet Center for Astronomical Data (CAD), in Moscow, for astronomical institutions of the USSR and of the Socialist countries. These centers provided astronomical data to the astronomers of their geographical regions and transferred the catalogs they prepared to the CDS. The avalanche-like increase in information required the creation of a virtual observatory by the efforts of thousands of highly qualified specialists working in institutions of various levels, the most authoritative of which is, of course, the International Astronomical Union.

The creation of the IVO in 2000 allowed astronomers around the world to receive for their research all the information currently available for objects of interest to them, regardless of when, using what technique, in what spectral ranges they were received and on what media they are stored.

The observations can be the result of space missions, images, and spectra obtained by ground-based optical or radio telescopes, star catalog data, photometric data, tables, etc., as well as digital scans of photographic glass plates from the collections of world astronomical observatories accumulated over a century. A similar approach is applicable, of course, to any science where data is freely available to the public. This model is increasingly being followed by physicists, geologists, and representatives of other sciences; however, astronomers, as it often happens, with their experience of global collaborations over centuries, understood the advantages of such openness before others.

Over recent years a huge amount of observational data has accumulated worldwide. These can be found in article (Malkov et al., 2016). The main goal of that paper is to give a historical sketch and an overview of the main modern trends in construction and functioning of large data sources in astronomy. In particular, a list of major collections of astronomical data, both existing and expected to appear in the near future is presented there.

After several years of intensive technological development, the resources of the International Virtual Observatory have reached a level of maturity sufficient for their use in basic scientific research. We can mention some of the first few studies in the different fields of astronomical science, the results of which were obtained using the capabilities of the IVO. This was the discovery of a new type of stars, the so-called brown dwarfs, a large number of new star clusters, the study of the interstellar medium, the structure of our galaxy, and others (Berriman et al., 2003; Caballero & Dinis, 2008; Caballero & Solano, 2008; Chilingarian et al., 2009; Dluzhnevskaya et al., 2000; Koposov et al., 2008; Malkov, 2003; Malkov et al., 2020; Padovani et al., 2004; Piskunov et al., 2006; Richards et al., 2007; Zolotukhin et al., 2010).

The IVO continued to replenish with data, improve software and structure, and the IVO Alliance (IVOA) became a powerful tool for processing, analyzing, and synthesizing nowadays astronomical databases.

The IVO is also a powerful tool for the democratization of astronomy, since even small scientific and educational astronomical institutions, and even individual users, can get from their computers at no cost essentially the same opportunities for conducting scientific research on modern observational material as leading astronomical organizations of the world.

The current status and capabilities of the IVO can be found in (Allen et al., 2019).

In conclusion, I would like to note that, being enthusiastic about this project, I feel a sense of satisfaction from having contributed to his birth and participated in its development.

It also seems to me that the IVO project, which is the result of joint and friendly work, can serve as an excellent example of how astronomers from more than 20 countries in difficult political and economic situations came together, and worked together fruitfully towards a magnificent creation that helps scientists from all over the world to conduct their research.

This is a great example of how astronomers can simultaneously be diplomats.

Acknowledgments I gratefully recall the founders of the *Centre de Données Stellaires* in Strasbourg, Carlos Jaschek and Alphonse Florsch, and the Deputy Chairman of the Soviet Astronomical Council (Astrosoviet) A.G. Masevich. I have to emphasize and acknowledge an important role of IAU Commission 5 in the promotion of the World Virtual Observatory Initiative underlined by very friendly cooperative efforts of the Commission President Bernard Hauck and the Organizing Committee members. I am thankful to my colleagues from the Russian Virtual Observatory project O.Yu. Malkov, D.A. Kovaleva and P.V. Kaygorodov for permanent assistance and support.

My special thanks to Thierry Montmerle for his invitation to attend the meeting "Astronomers as Diplomats" in Paris, followed by his kind suggestion to present this paper, as well as for his careful editing and refinement of the original version.

References

Allen, M. G., Dowler, P., Evans, J. D., Cui, C., Jenness, T., Merin, B., Berriman, G. B., & Kavelaars, J. J. (for the IVOA Executive Committee and Technical Coordination Group), *The International Virtual Observatory Alliance in 2019*, arXiv:1912.0148IvI [astro-ph.IM], 3 Dec 2019.

About the four million star catalog. Collection of scientific papers edited by Gulyaev A. and Nesterov V., Publishing house of Moscow State University, 1992

Berger, J., Cordoni, J. P., Fringant, A. M., Guibert, J., Moreau, O., Reboul, H., & Vanderriest, C. (1991). Astrometric study of 1221 UV-excess objects at the North Galactic Pole. *Astronomy & Astrophysics Supplements, 87*, 389.

Berriman, B., Kirkpatrick, D., Hanisch, R., Szalay, A., & Williams, R. (2003). Discover of Brown Dwarfs with Virtua Observatories. *IAU Joint Discussion, 25*, 60.

Caballero, J., & Dinis, L. (2008). A revisit to agglomerates of early-type Hipparcos stars. *AN, 329*, 801.

Caballero, J., & Solano, E. (2008). Young stars and brown dwarfs surrounding Alnilam in Orionis and Mintaka (delta Orionis). *Astronomy & Astrophysics, 485*, 931.

Chilingarian, I., Cayatte, V., Revaz, Y., Dodonov, S., Durand, D., Durret, F., Micol, A., & Slezak, E. (2009). A population of compact elliptical galaxies detected with the virtual observatory. *Science, 326*, 1379.

Chinnici, I. (2022). *Precursors to IAU: Paris Observatory and the* Carte du Ciel *project*, this volume (Chap. 1), and refs. therein.

Dluzhnevskaya, O. (1989). *Astronomical Literature Acquisitions in the USSR*. In Proceedings of Colloquium 110 of the International Astronomical Union "Library and Information Services in

Astronomy", held in Washington, DC, USA, 26 July-1 August,1988, US Naval Observatory, Washington, DC, USA, p. 48.

Dluzhnevskaya, O., Kovaleva, D., & Malkov, O. (2008), *Russian Astronomical Data Internet Resources: Present Status*. IVOA Interoperability Meeting, 19-23 May 2008, Trieste, Italy, (invited talk).

Dluzhnevskaya, O., Malkov, O., & Pasian, F. (2000). *Project of a Multi-wavelength All-sky Catalogue*, JENAM-2000. In European Astronomy at the Turn of the Millennium, May 29 – June 3, Moscow, Russia, Abstracts, p. 177.

Dluzhnevskaya, O., et al. (2002). *Russian Virtual Observatory: Information System of the Centre for Astronomical Data*. In Proceedings of the 4th All-Russian Scientific Conference "Digital Libraries: Advanced Methods and Technologies, Digital Collections", volume 1, pp. 202–220, Dubna (Russia), October 15-17, 2002.

Dluzhnevskaya, O. et al. (2003). *Russian Virtual Observatory: Information hub*. Proceedings of the 6th General Meeting of the Euroasian Astronomical Society: International Collaboration in Astronomy: "Astronomical and Astrophysical Transactions", Moscow 25 May- 2 June 2002, volume 22, issues 4–5, pp. 375–382.

Dluzhnevskaya, O. B. (1980). News from the Soviet Astronomical Data Center. *BICDS, 19*, 60.

Dluzhnevskaya, O. B. (1984). Report of the Soviet Astronomical Data Center. *BICDS, 26*, 89.

Dluzhnevskaya, O. B. (1994a). Plate collections and data archived at the Observatories of the former USSR. *BICDS, 44*, 37–51.

Dluzhnevskaya, O. B. (1994b). *Plate collections and data archived at the Observatories of the former USSR*, (Erratum Crimea). *BICDS, 45*, 47–55.

Dluzhnevskaya, O. B., Kaygorodov, P. V., Kovaleva, D. A., & Malkov, O. Y. (2013). "*BDB: The Binary Star Database*" 2014, in ADASS XXIII Conference, eds. N. Manset, P. Forshay. Waikoloa, Hawai'i, Sep 2013, ASP Conf. Ser. Vol. 485, 247–250.

Dluzhnevskaya, O. B., & Tsvetkov, M. K. (2012). The Zvenigorod astronomical plate collection presented in the Wide-Field Plate Database. *Proceedings of the VIIth Bulgarian-Serbian Astronomical Conference (VII BSAC), 1-4 June 2010*, Chepelare, Bulgaria, Publ. Astron. Soc. "Rudjer Boskovic", No 11, pp. 209–218.

Genova, F. (2019). The *Action Spécifique Observatoires Virtuels France* (Virtual Observatory France Specific Action) in the Open Science Context. In SF2A-2019: Proceedings of the Annual meeting of the French Society of Astronomy and Astrophysics, pp. 27–31

Grothkopf, H., Andernach, S., Stevens-Rayburn, M., & Gomez M (1998). Proceedings of a conference held at Puerto de la Cruz, Tenerife, Spain, 1998, ASP Conf. Series 153.

Kaygorodov, P., Debray, B., Kolesnikov, N., Kovaleva, D., & Malkov, O. (2012). The New Version of the Binary Star Database (BDB). *Baltik Astronomy, 21*, 309–318.

Koposov, S., Glushkova, E., & Zolotukhin, I. (2008). Automated search for Galactic star clusters in large multiband surveys. I. Discovery of 15 new open clusters in the Galactic anticenter region. *Astronomy & Astrophysics, 486*, 771.

Kovaleva, D., Kaygorodov, P., Malkov, O., Debray, B., & Oblak, E. (2015). Binary star DataBase BDB development: Structure, algorithms, and VO standards implementation. *Astronomy and Computing, 11*, 119–125.

Malkov, O. (2003). Interstellar Extinction from Large Surveys. *Baltic Astronomy., 12*, 514.

Malkov, O., Dluzhnevskaya, O., & Kovaleva, D. (2008). *Russian Virtual Observatory: Russian and FSU resources to be integrated in the IVO 2009*. In 21st International CODATA Conference, Kyiv [Kiev], Oct 2008, 357–362.

Malkov, O., Dluzhnevskaya, O., & Kovaleva, D. (2009). *Russian Virtual Observatory: Russian and FSU resources to be integrated in the IVO*. In 21st International CODATA Conference, Kyiv, Oct 2008, pp. 357–362.

Malkov, O. Y., Murphy, J., Saha, S., & Zhao, G. (2020) *Parameterization of stars and determination of interstellar extinction from multicolor photometry 2020*. In Proceedings of the IEEE MMLA 1st International Conference on Modeling, Machine Learning and Astronomy(MMLA) conference (Nov 22-23, Bangalore, India), Springer, in press.

Malkov, O. Y., Dluzhnevskaya, O. B., Kaygorodov, P. V., Kovaleva, D. A., Samus, N. N., Sachkov, M. Y., Pakhomov, Y. V., & Kilpio, Y. Y. (2016). Big data in astronomy: Past, present, future (in Russian). *Istoriya nauki i tekhniki (История науки и техники), 12*, 60–74.

Padovani, P., Allen, M. G., Rosati, P., & Walton, N. A. (2004). Discovery of optically faint obscured quasars with Virtual Observatory tools. *Astronomy & Astrophysics, 424*, 545–559.

Piskunov, A. E., Kharchenko, N. V., Roser, S., Schilbach, E., & Scholz, R.-D. (2006). Revisiting the population of Galactic open clusters. *Astronomy & Astrophysics, 445*, 545.

Proceedings of Section 11 of the Joint European and National Astronomical meeting (JENAM), "*Astronomical Data Acquisition, Processing and Storage Modern Facilities*", Moscow, Russia, May 29 – June 3, 2000, Baltic Astronomy, An International Journal, 9(4), Vilnius, Lithuania, 2000.

Proceedings of the colloquium "*Stellar catalogues: data, compilation, analysis, scientific results*", Tbilissi, 10-15 September, 1984, Abastumani Astrophysical Observatory Mount Kanobili BULLETIN 59, Tbilissi, Ac. of Sci. Georgian Republic, SU, 1985

Proceedings of the colloquium "*International Cooperation in Dissemination of the Astronomical Data*", St. Petersburg (Russia), July 2-9,1996, Baltic Astronomy. An International Journal, volume 6, number 2, Vilnius, Lithuania, 1997.

Richards, A. M. S., Muxlow, T. W. B., Beswick, R., Allen, M. G., Benson, K., Dickson, R. C., Garrett, M. A., Garrington, S. T., Gonzalez-Solarez, E., Harrison, P. A., Holloway, A. J., Kettenis, M. M., Laing, R. A., Richards, E. A., Thrall, H., van Langevelde, H. J., Walton, N. A., Wilkinson, P. N., & Winstanley, N. (2007). Using VO tools to investigate distant radio starbursts hosting obscured AGN in the HDF(N) region. *Astronomy & Astrophysics, 472*, 805.

Samus, N., Kazarovets, E., Durlevch, O., Kireeva, E., & Pastukhova, E. (2017). General catalogue of variable stars: Version GCVS 5.1. *Astronomy Reports, 61*(1), 80–88.

Special Issue "*Library and Information Services in Astronomical Data*", ESO, 10-12 May 1995, Vistas in Astronomy, An International Review Journal, 39, part 2, 1995.

Tsvetkov, M., & Tsvetkova, K. (2012). *Wide-Field Plate Database: Latest Results. New Horizons in Time-Domain Astronomy*. Proceedings of the International Astronomical Union, IAU Symposium, 285, 417

Zolotukhin, I., Revnivtsev, M., & Shakura, N. (2010). Infrared identification of 4U1323-619 revisited. *MNRAS., 401*, L1.

Витковский, В.В., Длужневская О.Б., « Российская виртуальная обсерватория. Принцыпы и методы реализации », Всероссийская астрономическая конференция, Тезисы документов, стр. 35, 6-12 августа 2001 г., Санкт-Петербург Vitkovsky, [V.V., & Dluzhnevskaya, O. B. (2001). *Russian Virtual Observatory. Principles and approaches to implementation*. All-Russian Astronomical Conference, Abstracts, p. 35, August 6–12, 2001, St. Petersburg].

Olga B. Dluzhnevskaya born on Jan. 27.01.1936 in Moscow (Russia). Graduated from Moscow State University (1959), Ph.D (1966). Since 1959, scientist of the Astronomical Council of the USSR Acad. of Sci. (currently, Institute of Astronomy of the Rus. Acad. of Sci. (INASAN RAS), since 1980 Director of Astronomical Data Centre (a branch of the Strasbourg Stellar Data Centre) and a member of the International Scientific Council of the Centre (1980–1986).

Professional scientific interests—characteristics of young open clusters based of the theory and calculations of the stellar evolution and different characteristics of stars and interstellar mater, later on with the use of Virtual observatory data. She published about 150 scientific papers in the referred journals.

O.B. Dluzhnevskaya served as Scientific secretary of the Council for bilateral scientific cooperation with France, India, and Finland and Vice-President of the Commission "Physics and evolution of stars" for multilateral cooperation of the USSR Acad. of Sci. with the East European socialist countries (1975–1985).

In 1991, she was elected Vice-President (1991–1994, 1994–1997), and then President (1997–2000, 2000–2003) of the IAU Commission 5 "Documentation and Astronomical Data". She was a Chairman of Section 13 "Databases and information support" of the RAS Scientific Council on Astronomy (2000–2018). "Virtual observatory" was set up which became the key project for both IAU Com.5 and Sec.13 SC RAS.

Since 2004, she serves as the Russian Focal point for UNESCO World Heritage Thematic Initiative "Astronomy and World Heritage" (2004 to present).

Minor planet 2009SH215 was named after "DLUZHNEVSKAYA" in 2011.

Part IV
The IAU, the United Nations, and UNESCO

Chapter 16
What's in a name? When the UN Challenged the IAU (1971–1982), a Hidden Story

Thierry Montmerle

Abstract In 1971, the United Nations "Group of Experts on Geographical Names" (created in 1967) set up a "*Working Group on the Names of Extraterrestrial Topographic Features*", as part of a global initiative on the "Standardization of Geographical Names". Its Chairman, Dr. A.M. Komkov, from Moscow, wrote to A. Dollfus, then President of Commission 17 of the IAU ("The Moon") a letter explaining that his Working Group had been tasked by the United Nations to take the lead on the problem of "extraterrestrial names". Komkov attached a very critical review of the status of the IAU lunar nomenclature methods: this had to be changed, under the UN authority and following the recommendations of its Working Group, yet to be elaborated. Would the IAU please collaborate and make suggestions to that effect? Also, there were legal problems for the applicability of the nomenclature decisions worldwide, which only the UN could solve.

This was the start of a long struggle between the IAU and the UN, which lasted over ten years, until 1982, when the UN dissolved its Working Group and finally recognized the authority of the IAU on assigning "extraterrestrial names". This important episode in the history of the IAU paradoxically was never brought up to the IAU top management, i.e., its Executive Committee, nor recorded in the IAU Transactions or any other official IAU document. This "hidden story" is reported here for the first time, based on the inspection of IAU archives in the Paris Secretariat, and of UN archives in its Dag Hammarskjöld Library in New York.

16.1 Act I: The UN Enters the Field of Naming Celestial Objects.

At the end of July 1971, Audouin Dollfus, a French astronomer of the Meudon Observatory, near Paris, received a strange letter. Sent on July 21, the letter came

T. Montmerle (✉)
Institut d'Astrophysique de Paris, Paris, France
e-mail: montmerle@iap.fr

© The Author(s), under exclusive license to Springer Nature Switzerland AG 2022

T. Montmerle, D. Fauque (eds.), *Astronomers as Diplomats*, Historical & Cultural Astronomy, https://doi.org/10.1007/978-3-030-98625-4_16

from the United Nations in New York, with a letterhead entitled: "*Group of Experts on Geographical names,*" and was signed "A.M. Komkov, Vice-President, UN Group of Experts on geographic names." It was addressed to Dollfus in his capacity of President of Commission 17 of the IAU: "*The Moon,*" with copy to Donald Menzel, Chairman of one of its three Working Groups (#1: "Lunar Nomenclature"). (Fig. 16.1).

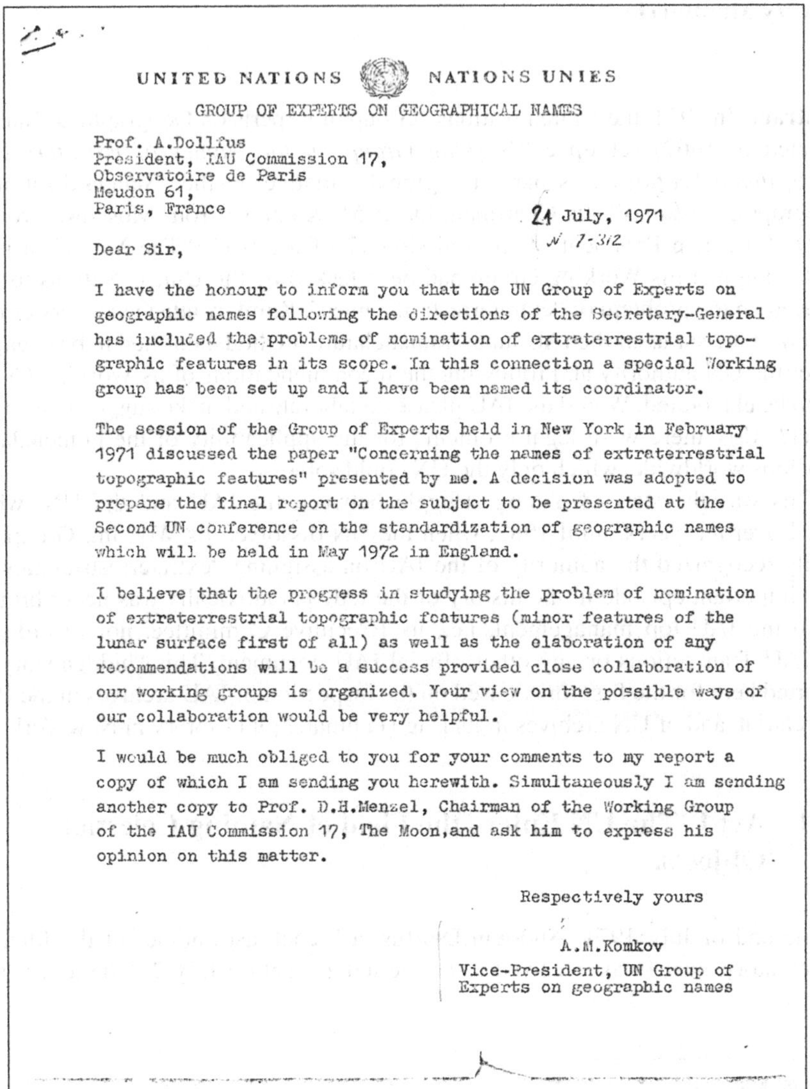

UNITED NATIONS NATIONS UNIES

GROUP OF EXPERTS ON GEOGRAPHICAL NAMES

Prof. A.Dollfus
President, IAU Commission 17,
Observatoire de Paris
Meudon 61,
Paris, France

24 July, 1971
N 7-312

Dear Sir,

I have the honour to inform you that the UN Group of Experts on geographic names following the directions of the Secretary-General has included the problems of nomination of extraterrestrial topographic features in its scope. In this connection a special Working group has been set up and I have been named its coordinator.

The session of the Group of Experts held in New York in February 1971 discussed the paper "Concerning the names of extraterrestrial topographic features" presented by me. A decision was adopted to prepare the final report on the subject to be presented at the Second UN conference on the standardization of geographic names which will be held in May 1972 in England.

I believe that the progress in studying the problem of nomination of extraterrestrial topographic features (minor features of the lunar surface first of all) as well as the elaboration of any recommendations will be a success provided close collaboration of our working groups is organized. Your view on the possible ways of our collaboration would be very helpful.

I would be much obliged to you for your comments to my report a copy of which I am sending you herewith. Simultaneously I am sending another copy to Prof. D.H.Menzel, Chairman of the Working Group of the IAU Commission 17, The Moon,and ask him to express his opinion on this matter.

Respectively yours

A.M.Komkov
Vice-President, UN Group of
Experts on geographic names

Fig. 16.1 The letter from the United Nations informing the IAU that it had included "the problem of nomination of extraterrestrial features" within its scope. (IAU archives)

Dollfus (1924–2010) was well-known: his main scientific interests rested in the study of the Sun, the Moon, and planets (primarily Mars); he had accomplished in particular several high-altitude balloon flights (up to 14,000 m in person) to diminish the effects of atmospheric absorption on certain UV spectroscopic observations. He was also familiar with IAU circles, having been President of Commission 16, "*Physical Study of Planets & Satellites*" (1958–1964), and was on this second term as President of Commission 17 (1967–1973). Although his style in English was perfect, his correspondence was frequently written in French.[1] Menzel (1901–1976), a well-known theoretician in solar studies, was the Director of the Harvard College Observatory, near Boston, and a good friend of the much younger Dollfus.[2]

In contrast, Komkov did not introduce himself,[3] except by beginning his letter with the following sentences:

> I have the honour to inform you that the UN Group of Experts on geographic names following the directions of the Secretary-General has included the problems of nomination of extraterrestrial topographic features in its scope. In this connection a special working group has been set up and I have been named its coordinator.

Attached to the letter was a draft report, entitled "*Concerning the names of extraterrestrial topographic features,*" which contained many criticisms of the IAU work and procedures in the field of nomenclature. Komkov had prepared it for a meeting of this "Group of Experts" in New York[4] in the preceding February, and he was asking Dollfus and Menzel for comments. After this exchange, Komkov explained that the report would be presented at the "*Second UN Conference on the standardization of geographic names,*" to be held the following year (May 1972) in London.[5] Komkov added, *"the elaboration of any recommendations will be a success provided close collaboration of our working groups is organized."*

In addition to a somewhat arrogant tone ("we lead, you follow"), the letter was thus frontally challenging the admittedly self-appointed, but well-established authority of the IAU in naming celestial objects since its creation in 1919. (Subsequent events would show that the adjective "extraterrestrial," used in Komkov's letter but not part of the usual astronomical vocabulary at the time, was to be understood as being restricted to the Moon and planets.)

[1] Due to the circumstances of its creation in 1919, the IAU has two official languages, English and French.

[2] Chapter 13 by Montmerle includes a few portraits of the protagonists of this story on the IAU side. I have not been able to find portraits on the UN side.

[3] His professional address used in correspondence indicated that he was "Head of Department of Geographical Names, Central Research Institute of Geodesy, Aerial Surveying and Cartography," in Moscow.

[4] Where the seat of the United Nations is located since 1952.

[5] UN Ref.: E/CONF.61/4: https://undocs.org/London_2_e_conf_61_4_en.pdf. The UNWG report, as well as the WGLN report answering it, and other related material, are available on-line: https://undocs.org/London_2_E_Conf61_4_Add1_e_226-237.pdf. However, some draft documents could be obtained only by contacting the UN Library.

It was the start of a long struggle, lasting over ten years, which paradoxically was never brought up to the IAU top management, i.e., its Executive Committee, and which is told here for the first time. This "hidden story" stems mainly from the inspection of the IAU and UN archives, respectively, in Paris (IAU Secretariat) in the course of my research about the Moon race,[6] and in New York (the United Nations' Dag Hammarskjöld Library).

16.2 Background: UN, IAU and "Extraterrestrial" Nomenclature Problems

The unexpected arrival of a UN Working Group on "Extraterrestrial Names" in the IAU horizon was found to have its origin around 1960, with the creation by the UN of a "Group of Experts on Geographical Names," under the auspices of the UN "*Department of Economic and Social Affairs*," a branch of its "Economic and Social Council".[7] The (obviously long-term) goal was to find a way to standardize geographical names across the globe (addressing issues like romanization, names from unwritten languages, etc.) and beyond, in particular new names in areas like the undersea floor and Antarctica. As a result, the various Committees and Working Groups would meet and present their Resolutions in the "*United Nations Conferences on the Standardization of Geographical Names*" ("UN Conferences" for short here) lasting three weeks, the first edition having been held in Geneva on Sep. 4–22, 1967.[8]

The periodicity was initially planned to be three years, but turned out to be five years instead. This would result in an interesting interleaved scheme between IAU General Assemblies (GA) and UN Conferences for passing Resolutions: IAU GAs took place in 1967 (Prague, Aug. 22–31), 1970 (Brighton, Aug. 18–27), 1973 (Sydney, Aug. 21–30), 1976 (Grenoble, Aug. 24-Sep. 2), 1979 (Montreal, Aug. 14–23), and 1982 (Patras, Greece, Aug. 17–26), while subsequent UN Conferences took place in 1972 (London, May 10–31), 1977 (Athens, Aug. 17-Sep. 7), and 1982 (Geneva again, Aug. 24-Sep. 14): during this period, the dates coincided only in 1967 and 1982 (see Table 16.1). Both Working Groups held their own meetings on a yearly basis (the UN Working Group always meeting in New York).

The Geneva Conference was presided by M.F. Burrill, who was a geographer (Executive Secretary of the Board on Geographical Names, US Dept. of the Interior). Among the national delegates was A.M. Komkov, for the USSR; ICSU (the International Council of Scientific Unions, in other words the "mother of

[6] Montmerle, this volume, Chap. 13. See also, e.g., Montmerle (2019a) for the origin of role of the IAU on celestial nomenclature.

[7] One of the six organs of the UN (which include the General Assembly and the Security Council). Note that UNESCO was never involved in this issue.

[8] UN Ref.: https://undocs.org/E/CONF.53/3.

Table 16.1 Calendar of Resolutions: IAU General Assemblies and UN "Conferences on the Standardization of Geographical Names"

Year	IAU General Assemblies	UN Conferences
1967	XIII (Prague)	1st (Geneva)
1968	-	-
1969	-	-
1970	XIV (Brighton)	-
1971	*First contact : UN letter*	-
1972	-	2nd (London) *First Resolution*
1973	XV (Sydney)	-
1974	-	-
1975	-	-
1976	XVI (Grenoble)	-
1977	-	3rd (Athens)
1978	-	-
1979	XVII (Montreal)	-
1980	-	-
1981	-	-
1982	XVIII (Patras)	4th (Geneva) *Final Resolution*

Shaded boxes indicate the years (1967–1970) prior to UN-IAU contacts. Yellow boxes indicate the start and end dates and milestones of the two organizations during their negotiations

Unions") was represented by two of its Unions, IGU and IUGG, and by SCAR.[9] Of the 20 Resolutions that were voted, three are relevant to our discussion: #1 "*recommends to the Economic and Social Council that a United Nations permanent committee on geographical names be created, consisting of experts, to provide for continuous co-ordination and liaison among nations to further the standardization of geographical names and to encourage the formation and work of regional groups;*" #2 "*recommends to the Economic and Social Council that the second United Nations Conference on the Standardization of Geographical Names be held not later than the last quarter of 1970*" (actually it took place in the second quarter of 1972, starting the five-year cycle); #8 (item B: "Maritime and Undersea features") "*recommends (. . .) (b) Establish means for the collection, approval and distribution by the United Nations of both a list of agreed terms and definitions for nomenclature of maritime and undersea features and an initial list of recommended geographical names for features requiring names.*"

The last item is the most important one, because it dealt with uninhabited regions (like the Moon and planets), and it would shape the spirit of the first contacts between the UN and the IAU. To this end, as explained in Komkov's report attached to his letter to Dollfus, a "*special working group on the names of extraterrestrial*

[9] IGU = International Geographical Union; IUGG = International Union of Geodesy and Geophysics; SCAR = Scientific Committee for Antarctic Research.

topographic features" (which I shall refer to here as "UNWG") was founded on Apr. 29, 1970. The UNWG then organized three meetings *"for making acquaintance with the state of the problem and for looking for possible ways of its solution."* Actually, this group had only three members: its President was M.F. Burrill (who was also the Chairman of the Group of Experts), and two Vice-Presidents, A.M. Komkov, and H.G. Lewis (UK). New members would be added in 1973, but Komkov appeared (with one important exception, see below) to be the only interface with the IAU and he never mentioned to the IAU the composition of his group: perhaps he did not want to disclose how small and specialized this group was. As far as I could find, they were all geographers, but of course, as discussed later, through the Group of Experts, they had likely access to other specialists (linguists, cartographers, etc.).

At any rate, it was clear that Komkov's group had only a second-hand approach to lunar (and martian) nomenclature, even though Komkov himself did a reasonably good job in his reports. As the Proceedings of the UN Conferences showed, this group was in fact a very small one, immersed in a much larger UN structure that dealt with all of the world's geographical areas. And this made discussions with the IAU Commission 17 (Dollfus) and its Working Group on Lunar Nomenclature (Menzel; WGLN for short here) difficult from the start. Although the term would never appear explicitly (for instance, in the correspondence between IAU members), in my opinion, the UNWG was somewhat considered as a group of "professional amateurs," which tended to be of a "polite arrogance," and this led, unsurprisingly, to a crisis a year later.

But the IAU was rich with its own experience. The revelation, for the first time in the history of mankind, of the far side of the Moon in the course of the Soviet *Luna-3* (1959) and *Zond-3* (1965) flybys, followed by the five American *Lunar Orbiter* missions (1966–1967), including two on a polar orbit, had revolutionized lunar science and posed new nomenclature problems. The first "map" of the far side of the Moon displayed 18 large features which were named by the Soviet Academy of Sciences: they were made public without consulting the IAU, in breach of the rules – but the IAU had little choice than accepting them at its 1961 General Assembly in Berkeley.[10] The same attitude towards naming repeated itself in 1965, after the *Zond-3* flyby, but met the opposition of Commission 17: an emerging crisis was avoided when the *Zond-3* data were merged with the first maps of the *Orbiters*, issued by NASA (see a 1967, 1:10,000,000 map of the far side of the Moon: Fig. 16.2).

The WGLN, under the vigorous leadership of D. Menzel, and the support of A. Dollfus, tasked itself with the naming of the largest craters of the far side (sizes larger than about 50–100 km). They used the same criteria that had been used in the official map and name catalogue of the near side of the Moon finalized by Mary Blagg and Karl Müller in 1935.[11] They contained nearly 700 entries corresponding to large craters and other large features, and they had been approved by the IAU at

[10] Transactions IAU vol.XI B (see the Reference list for bibliographic details on the Transactions.) More details in Chap. 13.

[11] Blagg and Müller (1935).

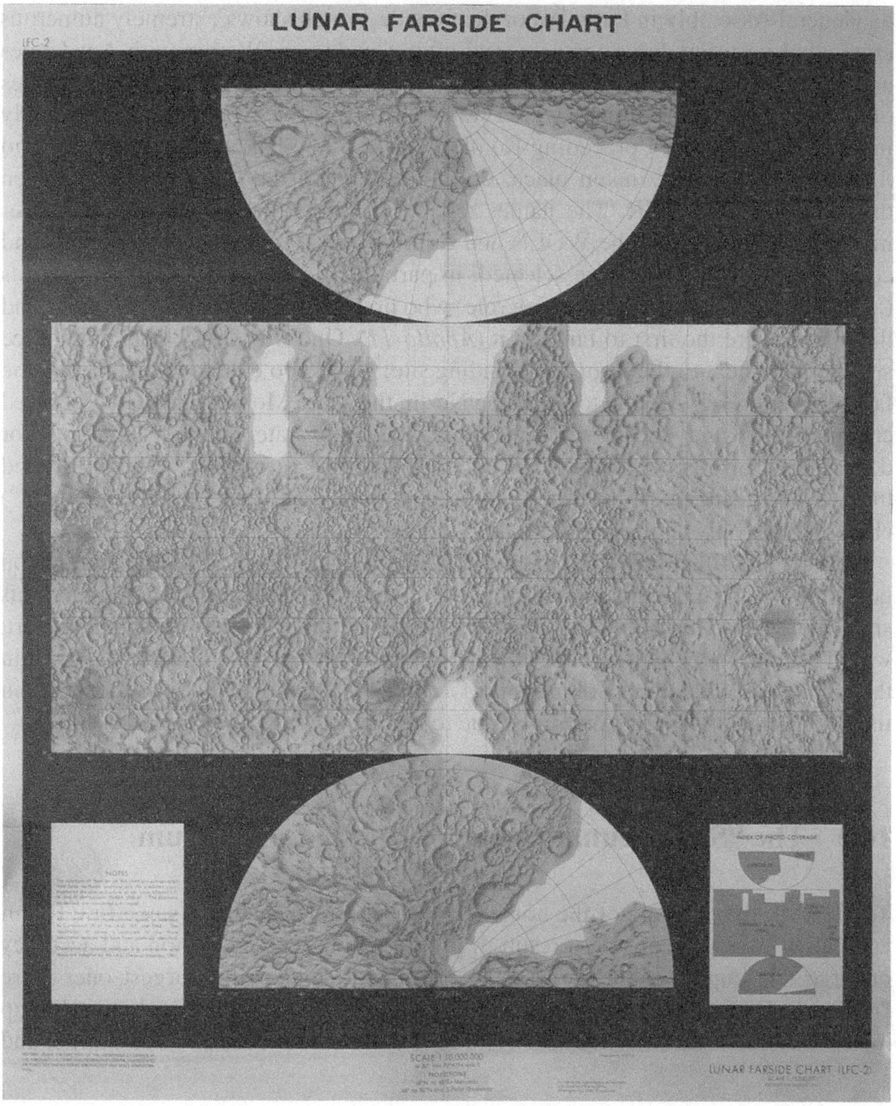

Fig. 16.2 NASA 1:10,000,000 chart of the far side of the Moon, published by NASA (1967), obtained by merging *Zond-3* data with *Lunar Orbiter* data (right-hand side quarter of the rectangle, including the large (900 km diameter) crater, filled with a "sea," called Mare Orientale). (IAU archives; original size: 59 cm × 73 cm)

its General Assembly in Paris.[12] For the far side, which shows extremely numerous craters but very few large features, and after a wide consultation with Academies and Learned Societies around the world, the WGLN drew a list of 513 new names of astronomers and scientists, which was approved at the 1970 General Assembly in Brighton,[13] i.e., just preceding Komkov's letter. The landings of *Apollo-11* and *Apollo-12* had already taken place, essentially ending the "Moon race" between the USA and the USSR. The names reflected also a "diplomatic" gesture (after much discussion within the WGLN and Commission 17!): to honor astronauts and cosmonauts, three areas were selected, in particular to honor the living astronauts (as unique exceptions to the naming rule) who first circled the Moon (*Apollo-8*), and those who were the first to land on it (*Apollo-11*). One was on the near side: three small craters next to the *Apollo-11* landing site;[14] and two extended features on the far side were reserved: one towards the North, the Mare Moscoviense, for deceased cosmonauts,[15] and one towards the South, a large crater named "Apollo," for deceased astronauts,[16] but also for the living, *Apollo-8* astronauts.[17] This illustrated that naming features on the Moon could also have a political (say, "diplomatic") character, not only "commemorative."

Altogether, after Brighton, about 1200 large craters and significant features on both sides of the Moon had been named. But what about the much more numerous smaller craters? That question, which Komkov had correctly identified in his report, would be extremely divisive, not only between the IAU and the UN, but also within the IAU itself, which, via Commission 17 and its WGLN, had pledged to answer in time for the next General Assembly, due to take place in Sydney in August 1973.

16.3 The "Small Lunar Crater" Naming Conundrum

To the IAU, the problem of the nomenclature of smaller features (say a few 10 km and below) was not new. As a result of progress in the magnification and accuracy of large telescopes (at the turn of the nineteenth century, the largest ones were "refractors," i.e., extrapolations of Galileo's instrument, basically a long tube, up to about 20 m, with one lens at each extremity), lunar features appeared sharper and

---

12 Transactions IAU vol. V.

13 Transactions IAU vol. XIV B; Menzel et al. (1971).

14 Named "Aldrin", "Collins" and "Armstrong" from West to East: the landing site itself was named "Statio Tranquillitatis." See Fig. 13.10 of Chap. 13.

15 At that time, only "Komarov." Others would be similarly honored later. Gagarin, the first man in space, who was killed in a jet fighter crash in 1968, had his own, large crater on the far side, not far from the lunar equator.

16 At that time, the three astronauts who lost their lives in the *Apollo-1* pad fire (1967): "Grissom," "White," "Chaffee." Later, the 14 astronauts of the Space Shuttles *Columbia* and *Discovery* would have craters within the same area named after them.

17 "Borman", "Anders", and "Lovell".

sharper, allowing to draw increasingly detailed maps. However, one should keep in mind that at the time astronomers would draw maps by hand (hence drawing talent and rendering of details mattered), while keeping an eye on the ocular (so eye quality mattered), and mechanically following the Moon's disk (so guiding accuracy of the telescope mattered too). This explained why, all along the nineteenth and early twentieth century there were sometimes significant differences from map to map, as well as differences in feature naming, a situation that, precisely, the IAU Commission 17, since its inception in 1919, sought to put into order.

The starting point was the most detailed, and widely used series of maps, with identifications written on them, produced by Beer & Mädler (1837).[18] Realizing that, aside from names (typically of astronomers and scientists) that had been attributed since the early maps of Giovanni Riccioli, a Jesuit priest, in the mid-seventeenth century, and adding others from various sources since then, Mädler invented a rather complex system to name smaller craters and other features (all dubbed "craters" here for short): to the name of the nearest large "parent" crater, add alphabetical letters of various types (capital or small, Roman or Greek, depending on secondary criteria: position measured or not, crater-like or not, etc.), to "satellite" craters, in order of increasing distance from the parent crater.

Too complex for practical use, what became to be known simply as "Mädler's system" was later simplified (even later than the Blagg & Müller catalogue and maps of 1935; see an example in Fig. 16.3) to adopt only capital letters (still in order of increasing distance) after the name of the parent crater. On paper, the great advantage of this system was its simplicity: in principle, it allowed to name with letters, even double letters, an infinite number of smaller and smaller features in the vicinity a given parent crater. In practice, however, this apparent simplicity had its limits: the "territory" centered on a given crater (ideally taken as circular) would inevitably collide with that of a neighboring crater, so how to attribute letters becomes problematic, hence the answer may be arbitrary; when smaller satellite craters are later discovered closer to the parent crater, letters get mixed up also. Adding the name of the parent crater (possibly making a random choice) before letters may remove the ambiguities, but then the maps may get too cluttered locally, at the expense of clarity of the details... These problems were unanimously recognized by astronomers, but the experts got used to it for decades, and as a matter of fact the "lettered crater names" version of Mädler's system was seldom used in practice.

All that changed with the progressive availability of (relatively) high-precision maps of the Moon (1:1,000,000 and 1:250,000 scale) produced by NASA,[19] starting

[18] Beer and Mädler (1837).

[19] According to Komkov's report, the Soviets also produced 1:5,000,000 maps in 1967 and 1969, and 1:1,000,000 maps in 1968, but they were published only in the Soviet Union. It is however possible that some of these data were incorporated into the NASA maps (see the example shown on Fig. 16.2), but in practice only the maps issued by NASA were used by the WGLN. In spite of its President being Russian, the UNWG apparently never used the Soviet maps, which was a major weakness in their understanding of the lunar mapping processes.

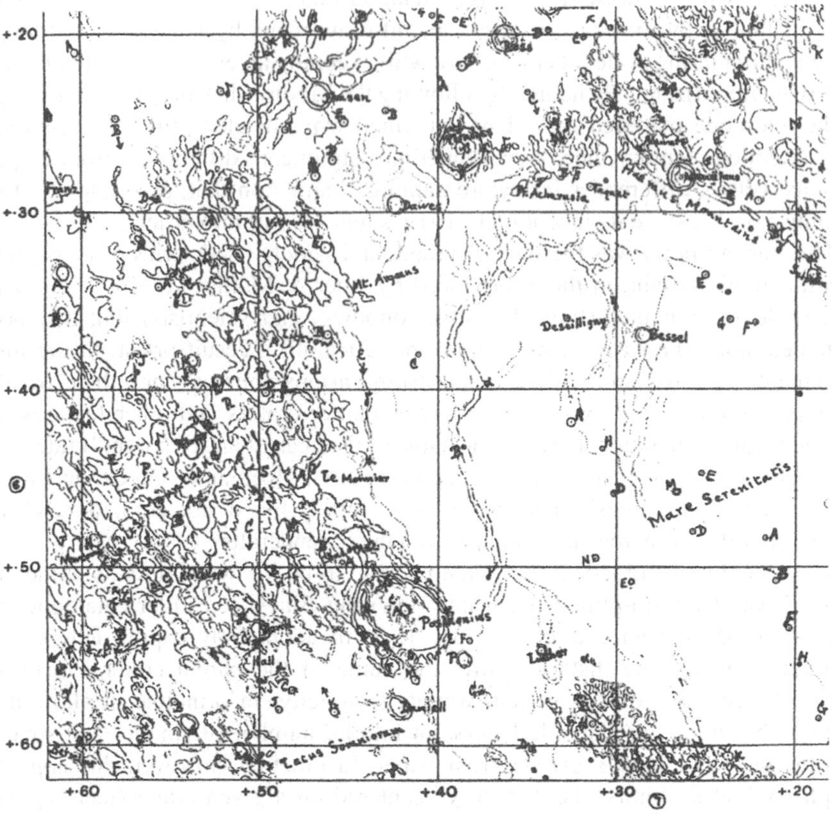

Fig. 16.3 A typical, hand-drawn map from Blagg & Müller (1935), showing crater names and "satellite craters" identified by letters, in the Mare Serenitatis region. Selected by Whitaker (1999). (North is down, i.e., inverted as seen in a telescope.). See also Fig. 16.4

in 1972 (Fig. 16.4), as a result of processing all available photographs (thousands of them) of the lunar surface taken at close range (from about 6000 km to sometimes as close as 40 km for the *Orbiters*). The spatial resolution on the lunar surface was typically less than 1 km!

Recognizing the need (but not necessarily the urgency: again, both the American and Soviet lunar programs were coming to an end) to find in priority "identifiers" for the maps (names, or something entirely different: some coding system, like coordinates, for instance?), basic questions had to be answered first:

(i) Should the nomenclature system be the same on both sides of the Moon, in particular given the large difference in crater populations?
(ii) If yes, which system should be applied for craters smaller than those already named: "Mädler's system," possibly revised, on both sides (i.e., in practice introducing it *ex nihilo* on the far side)?

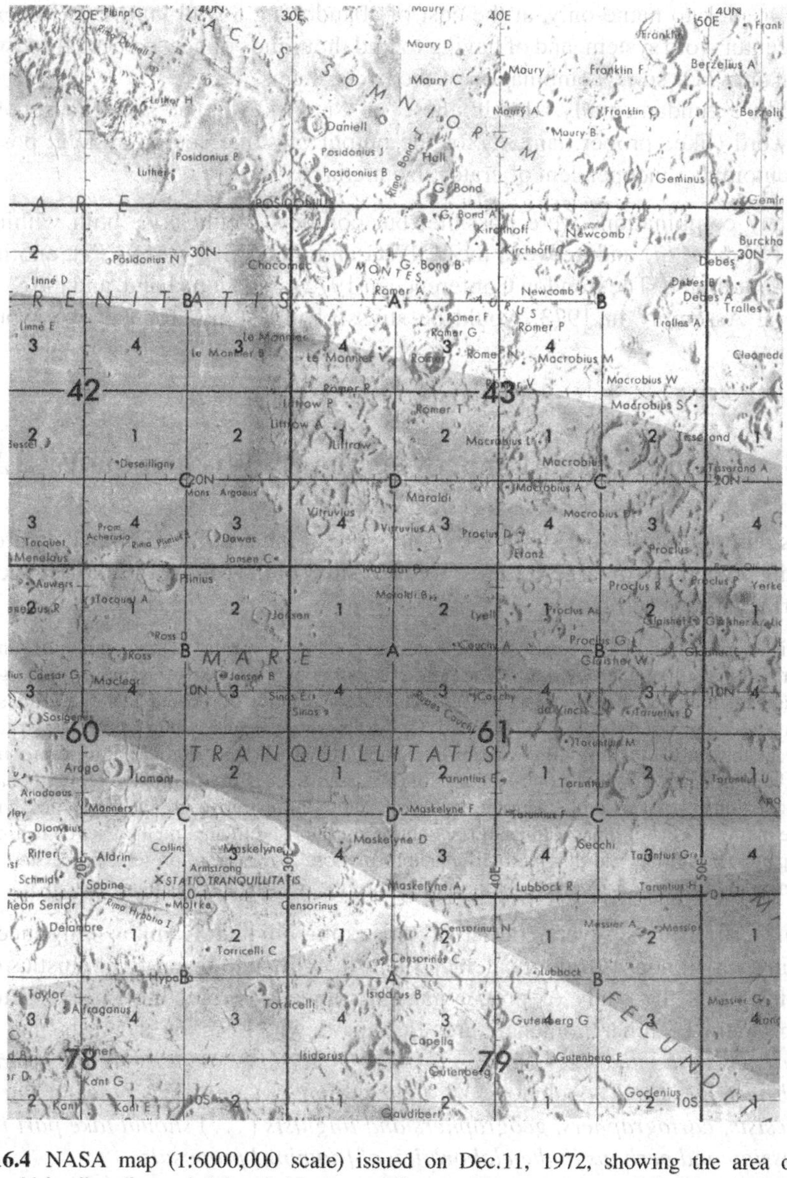

Fig. 16.4 NASA map (1:6000,000 scale) issued on Dec.11, 1972, showing the area of the *Apollo-11* landing (*lower left-hand side,* cross: "STATIO TRANQUILLITATIS"). The shaded areas correspond to the photographic coverage of *Apollo-15* (dark) and *Apollo-16* (light). The superimposed grid is the cartographic system introduced by NASA: large, numbered rectangles are the "Lunar Chart" ("LAC") sheet series (1,1,000,000), later called "*Regions;*" smaller, lettered ones are maps at 1:500,000 scale. The smallest, numbered ones correspond to a scale of 1:250,000 (later called "*Provinces*"). A feature or crater could then be approximately located by a combination of numbers and letters: e.g., "STATIO TRANQUILLITATIS" is inside *60C4*. "Satellite craters" are here named using a simplified Mädler system (where the name of the parent crater is explicitly mentioned in the label to avoid confusion). Note that the large crater Posidonius (located exactly at the upper corner between LAC 42 and 43; top of the figure) also appears in Fig. 16.3 (near coordinates: +50, +40), but here North is up. (IAU archives)

(iii) Or stick to name-only, at the cost of abandoning a well-known, if imperfect, century-old system, and of having to find thousands of new names? If so, which names? Or some combination of the two systems?
(iv) More fundamentally, should these be "names" in the strict sense of the word (like "proper names"), or the output of some coding system, possibly automatic, independent of crater size hierarchy?

Many combinations were possible, but conflicts would arise both within the WGLN (about *iii*) and between the UNWG and the WGLN (*iii* vs. *iv*), and in the same time frame. The London Conference in 1972, on the one hand, and the Sydney General Assembly in 1973, would be stressful milestones for the two Working Groups.

16.4 A Growing Tension between the UNWG and the WGLN

In his report, after having summarized the state of the art of IAU nomenclature and naming procedure (of large features) for both sides of the Moon, Komkov enunciates his criticisms (item 11):

> It is of great importance to stress that the "classical" system rules of nomination of lunar formations still applied by the IAU are not efficient enough in case of scanning and large-scale mapping of the Moon's surface. Therefore it cannot be denied that the criticism directed against it today is quite justified. C. Borkowski of Pennsylvania University writes: "Face to face with a lunar 'information explosion', classical lunar nomenclature is proving to be outmoded, slow, and controversial. Since it is now necessary to provide promptly data about lunar surface to all scientific, technical, and administrative groups concerned with the many aspects of lunar studies, it is essential to establish efficient methods of collecting, organizing, storing and distributing lunar information. However, at present, processing of lunar data is affected adversely by the structure of the existing lunar naming system".[20]

The criticism is severe, but it is interesting in that Komkov introduces an existing, albeit non-astronomical, idea (item *iv* above): to resort to linguistics (here "onomastics," the study of proper names)[21] as a possible 'third-party' actor in the problem of lunar nomenclature. Further down, Komkov suggests that *"the experience in nominating features in such geographic regions as the Antarctic can be useful,"* and concludes *"We believe that not only astronomers, but also geodesists, cartographers, geographers and linguists (. . .) should take part in the discussion and perhaps in the elaboration of [naming] proposals.*

Komkov ends his report by proposing that the UN Group of Experts and IAU Commission 17 (in practice, the UNWG and the WGLN) hold joint meetings,

[20] Paper by C. Borkowski on "*Lunese I*" presented at the tenth International Congress of Onomastic Sciences, Vienna, 1969. More below and excerpts in Appendix 1.

[21] https://icosweb.net/what-is-onomastics/.

and *"adopt appropriate recommendations at the second UN Conference on the standardization of geographical names in 1972."*

To Komkov and his report, Menzel replies on July 29, 1971: using a conciliatory tone, he *"expresses general agreement with the large majority of your recommendations."* While Menzel's agreement concerned rather minor issues on some crater names, he does mention his own attempt at devising *"a numerical system, which would designate the crater by a combination of letters and numbers, essentially based on latitude and longitude,"* adding *"But this system was objected to, largely because the positions of objects on the lunar surface are not known to that degree of accuracy. This has changed somewhat since, and perhaps this suggestion would be one for further discussion between our various groups."* He concludes by agreeing to the suggestion of joint meetings in view of the London Conference–which unfortunately never materialized.

Dollfus, on the other hand, did not reply directly to Komkov (letter of Nov. 30). Rather, he informed him of a recent initiative: the creation (based on a Dollfus's idea) of the *"Inter-Union Commission for Lunar Studies"* (IUCM),[22] which had held its first meeting two weeks before in Paris.[23] As for the WGLN, the IUCM suggested to add to it two experts designated by IGU and IUGS, but which, being under the ICSU umbrella, would not express any opinion about the UN and the UNWG. Dollfus added that he would be interested in knowing what would be the UN position (i.e., that of their Group of Experts) with regard to this evolution of ICSU's scientific unions about lunar studies.

So a third player had entered the game. In a reply (dated Jan. 10, 1972) to B. Levin, the Soviet member of the WGLN (and vice-president of Commission 17), who had exchanges with his compatriot Komkov about his letter and report, the IAU General Secretary, Cornelis (Kees) de Jager, from the University of Utrecht, gave his opinion on the situation so created: *"Your letter raises the extremely interesting but also complicated problem of the division of the fields of responsibility between the UN Special Working Group on the naming of extraterrestrial topographic features, the IAU, and the Inter-Union Commission on the Moon (IUCM)." "Briefly I think that the UN special WG should be satisfied in advising the UN Conference that the nomination of Lunar Features should be left to the IAU, in close cooperation with IUCM."* Yet de Jager expresses a concern regarding the Conference, only four months away: *"So far the IAU did not receive an official announcement of that Conference, and has not yet been invited to nominate representatives. Perhaps you, Dr. Levin, could insist with Professor Komkov that he invites IAU to send its representatives to the Conference."*

[22] Initially called the *"Inter-Union Commission on the Moon:"* the name was changed, but the acronym remained in use.

[23] It would be officially founded by ICSU in March 1972, after approval by all its participating Unions (IGU and IUGG, mentioned above; IUGS [Geological Sciences], URSI [radio science] and COSPAR [Committee for Space Research], as well as the IAU via the WGLN itself). (Ref.: Transactions IAU vol. XV B).

Time was starting to run out, but contact was kept. On Jan. 26, Komkov had met with Dollfus in Moscow, and sent him on Feb. 10 a letter summarizing his position. In addition to arguments already discussed on naming smaller features, he insisted on two points: (a) a legal one *("what representative international body should register, consider, coordinate recommended names with the organizations concerned and approve them for international use");* (b) concerning the IUCM, *"fully agrees"* on it to be *"the most competent and proper organization for working out the problem in its scientific aspect."* But the goal was to find a "universal" procedure that could be valid both for the Moon and the planets, so the IUCM could not have the overall competence. Komkov concluded *"we will be able to discuss this problem in detail and delimit our duties."*

Yet, it seemed that a different point of view was circulating: in a preparatory document bearing the letterhead of the UN "Center for Economic and Social Information," dated Feb.11, but with no author name, a section entitled "Moon and Mars Maps" started with

> Three countries, the United States, the United Kingdom, and the Soviet Union, have already been meeting to work out a system for naming extra-terrestrial features. (...) Soviet delegates at the three-nation meetings have noted that a significant number of names have already been applied to Soviet maps of the far side of the Moon. If the maps are to be issued within a reasonable period, *it is impossible to wait for approval from the International Astronomical Union.*

(The italics are mine.) In addition to the fact that the "countries" and "nations" mentioned were actually just the three members of the UNWG, it was striking that the IAU WGLN was not mentioned, and the IAU role was limited to "approving" the UN work, and, in this case, not even so! Looking at the rest of the document, which contains other sections on geography, sea and sea-bed, linguistic problems and the UN-sponsored 1:1,000,000 scale map of the world, it is easy to deduce that the document was written by the UN Group of Experts, not by Komkov, who appeared to have been overruled, perhaps by his President Burrill, in the views he had expressed to Dollfus.

This was at variance with another document: on Feb. 18, Levin wrote to de Jager a letter reporting his "negotiations" with Komkov, who assured him that the UNWG would recommend to the Conference: (a) *"to approve the IAU rules for names of large features;"* (b) *"to confirm that the naming should continue to be accomplished by the IAU;"* (c) *"to commission the IAU the preparation of rules for naming small topographic features."*

But unfortunately, on May 30, 1972, Resolution 21 was adopted by the London Conference,[24] and it made no such recommendations, in fact completely ignoring the IAU (Fig. 16.5): *"The Conference (...) Recommends that the United Nations Group of Experts on Geographical Names study the question of drafting an international convention on the standardization of extraterrestrial nomenclature, in co-operation with other competent international bodies."*

Something had gone seriously wrong.

[24] UN Ref.: E/CONF.61/4, pp. 13–14.

UNITED NATIONS NATIONS UNIES

INTEROFFICE MEMORANDUM

TO: Date: 195

THROUGH: FILE NO.

FROM:

SUBJECT:

EXTRACT FROM DRAFT REPORT OF THE CONFERENCE
Document E/CONF. 61/3, 30 May 1972

Resolutions
adopted by the 2nd United Nations Conference on the Standardization of Geographical Names (London, 10-31 May,1972)

RESOLUTION 21. Standardization of names of extra-terrestrial topographic features

The Conference,

Having discussed the present situation with regard to the naming of extraterrestrial topographic features,

Recognizing that greatly encreased lunar and planetary exploration, study and associated detailed large scale mapping require a new perspective on the naming of extraterrestrial features and a wider base for international agreement,

Recommends that the United Nations Group of Experts on Geographical Names study the question of drafting an international convention on the standardization of extraterrestrial nomenclature, in co-operation with other competent international bodies.

Fig. 16.5 Draft Resolution (#21) on extraterrestrial names adopted at the UN London Conference. (IAU archives)

16.5 The London Conference: Backstage

Let us go back in time. Both Komkov and Menzel (now having been invited and being the IAU/WGLN representative) were preparing their reports for the Conference.[25] Komkov essentially summarized his previous reports, and in his conclusions (item 10) wrote (Apr. 19, 1972):

[25] London Proceedings, op. cit.: https://unstats.un.org/unsd/geoinfo/ungegn/docs/2nd-uncsgn-docs/2uncsgn_econf61_L41.pdf.

> It is necessary to dwell on ways of organizing international collaboration in this matter. It seems expedient that the International Astronomical Union and its working groups on lunar and Martian nomenclature[26] would remain as the leading organization in studying scientific and technical problems." (...) "[The International Commission on Lunar Studies] is believed to be insufficient, as not only astronomers, but representatives of other sciences –geodesists, geophysicists, geomorphologists, cartographers, photogrammetrists, and others– take part in studying and mapping the Moon and other planets." (...) "As regards the legal aspects of the problem, the United Nations is the most competent and representative body for pursuing their solution.

Similarly, Menzel summarized the work accomplished by the WGLN before and after the Brighton General Assembly in 1970, and concluded with a firm IAU position (Apr. 21, 1972):

> We feel that the proliferation of working groups on extraterrestrial nomenclature would serve no useful purpose. Furthermore, I feel that the UN Special Working Group should agree to accept the recommendations approved by the International Astronomical Union in this particular area. To divide the responsibility among other commissions, other nations, and other unions would only produce chaos.

The Conference lasted three weeks, but Menzel stayed only one day, to give his presentation during a Special Session on Extraterrestrial Names, which took place on May 19. On May 26, presumably after his return to the USA, he wrote a report giving an account of the Session. Unfortunately, the document in the IAU archive, although signed "Menzel, Representative of the IAU," does not indicate to whom it was to be circulated (presumably to the WGLN, Commission 17 and Komkov), but it must have been an official, public document, since some time afterwards (on June 5), Menzel wrote some complementary "Notes" (implying "not to be circulated") to the IAU Executive (the President, B. Strömgren, the General Secretary, C. de Jager, and the future IAU President, L. Goldberg), Dollfus and the two other WGLN members (B. Levin and H. Masursky) and their consultants from Commission 17 (G. Kuiper and A. Anders), as well as the Chairman of the US National Committee for the IAU, D. Heeschen, and the National Academy of Sciences "Staff Officer for IAU Affairs," E. Dyer. Clearly, in parallel with the formal account of the Special Session, Menzel wanted to issue a warning message, not only to the IAU, but also, beyond the IAU, to high-level American academic officials.

[26] A Working Group of Commission 16 ("*Physical Study of Planets and Satellites*").

Menzel's report mainly discusses various proposals to the existing IAU naming system (for small features: not decided yet at that time), as presented by the UNWG (but which had not been discussed jointly with the WGLN before the Conference): (i) by Gurshtein & Shingareva (1970), from the USSR,[27] (ii) by Borkowski at the Vienna Conference on Onomastics (1969),[28] (iii) and a new one, by Waddell & Barsky, from the USA, presented at the London Conference under the title *"Contemporary Problems of Selenonymy"*.[29] These proposals essentially consisted in crater naming systems based on some coordinate "conversion code" from numbers to letters, with surprising results like (for a latitude-longitude combination): "Zeneze^{4}nonetoze^3" (Borkowski), or "BADQUIBHAT-NOBCAMPMOB" (Waddell & Barsky) (see Appendices A1 and A2 for proposal excerpts, comments and references). Even forgetting about the scientific problem of the accuracy of the coordinates, mentioned earlier, it was not surprising that, as Menzel remarked, astronomers would never support such cumbersome semantics, to replace a simple system based, one way or another, on commemorative names, even if they were to be very numerous . . .

But in the end, these discussions seemed trivial compared to the dissenting views expressed by Komkov and Menzel about the heart of the matter, which had not been seriously discussed before the Conference (except perhaps indirectly between Levin and Komskov in February): *"Finally,* [Komkov] *stated that whereas I* [Menzel] *had expressed the view that the United Nations should give advice to the IAU, he felt that the reverse position should be acceptable, that the IAU should give advice to the United Nations."* And Menzel to conclude: *"The cartographers make the maps, but the astronomers will use them. And hence the maps, to have a proper operational character, must have the final approval of the astronomers."* After which Komkov invited Menzel to participate in the work of the UNWG *"to represent the views of the IAU,"* which he accepted–but the invitation again never materialized.

In reality, the Session did not go smoothly at all. In his "Notes," Menzel writes in the first person and offers a less mundane view of the situation (the italics are mine):

[27] Unpublished; referred to as "Materials of the IAU, 1970" in Komkov's report to the Conference (op. cit.: Note 25) and elsewhere, but not found in the IAU archives. No publication on this topic could be found in ADS.

[28] See above, and Appendix 1.

[29] This paper was published in the Proceedings (op. cit., pp. 232–235), but, at least at the time of the Conference, neither in a scientific journal nor in other proceedings. No mention is made of the authors' institutions. See Appendix 2.

- *"Dr. Burrill, Chairman, interrupted my presentation, saying that my claims on behalf of the IAU were out of order.* Indeed their own Working Group would be delighted to have the advice of the IAU. He felt that the [UN] experts on geographical names were much more suited and experienced than astronomers in such problems. He also referred to a group named the International Commission on Outer Space Onomastics[30] and the Economic and Social Council of the United Nations."
- "Komkov gave a long address of which the chief element was his statement *that the astronomers have all too long had the field of lunar nomenclature to themselves. Now others have an interest and astronomers should recognize this fact.*"
- "Komkov and others have stated that they did not accept the arbitrary rule of the IAU."
- *"There was a considerable group in the United Nations that wished to take over the field of lunar nomenclature* and also to get into the into the problems presented by Mars, and other astronomical naming problems."
- "Russia tried to get NASA to agree to a bilateral agreement on lunar nomenclature, but NASA disagreed, deferring to the IAU."

And there was a final statement that *"if the IAU does not want to cooperate with the United Nations, there may be confrontation".*

But in the midst of this crisis situation, there was also a hope for future negotiations, which was very much in line with Menzel's own thinking: *"Throughout the discussion, it was clear that the conventional naming of craters after deceased astronomers, with few other scientists, principally physical scientists, was not satisfactory to the United Nations. They wanted to broaden the scope of recognition."* And the IAU had a decisive ally: according to Soviet participants, *"Russia tried to get NASA to agree to a bilateral agreement on lunar nomenclature, but NASA disagreed, deferring to the IAU."*

16.6 The IAU Fights Back

On June 15, 1972, Dollfus reacted to Menzel's London report and his "Notes," stressing again the potential role of IUCM, his brainchild, *"representing a definitely larger fan of disciplines than isolated IAU, and giving more weight."* He was disappointed with what he considered "a step backward" with respect to the discussions he had in January in Moscow with Levin and Komkov (which concurs with the idea that the UN Group of Experts, in particular his President Merrill, were considering that Komkov was "too soft" on the IAU). Then he suggested that IAU, emphasizing full support from IUCM, should continue the discussions.

Levin (June 26) noted that the problem of the border between IAU and UN responsibilities "is not solved" and asks again for the "point of view" of the IAU Executive Committee. Perhaps implicitly answering Levin's question on behalf of the EC (and certainly having discussed the issue with Strömgren), de Jager wrote to Menzel on Sep. 15. He had been recently elected President of COSPAR (a 1-

[30] Meaning in reality the International Council of Onomastic Sciences, which obviously Burrill was not familiar with. This is a reference to the paper by Borkowski, mentioned above.

year term), and in this capacity, he had attended a meeting of COPUOS (the United Nations "*Committee on the Peaceful Uses of Outer Space*"), held 4 days earlier in New York. COPUOS being a UN organization, he took this opportunity to mention de lunar nomenclature problem in his address, in the hope of passing a message to the UN (although not officially with a mandate from the IAU): in essence, the IAU authority is "*an axioma;*" the naming of lunar features should be organized by astronomers (not excluding outside contributions); an exception was the question of the "lunar visitors," who should be allowed to give names, just like explorers: "*what would be the legal advice of the UN in this case?*" From what I could find in the IAU archives, this address seems unfortunately to have had no effect, probably because de Jager's audience was composed mainly of space-oriented participants, who themselves certainly had no problem with de Jager's "axioma"....

However, whatever "axioma" the IAU authority may have been, another crisis unfolded shortly afterwards, during the winter of 1972–1973, in addition to the crisis with the UN, but this time *internal* to the IAU: the WGLN could not agree among themselves on a naming system for "small features." Referring to the itemized paragraph at the end of Sect. 16.3, while everyone quickly agreed that both sides of the Moon should have an identical naming systems (item *i*), and rejecting any automatic coding system (item *iv*), it concerned the item labeled (*ii*): For the near side, should the WG adopt the same system as for the far side, i.e., stick to finding names only, with no Mädler "letters," at the cost of abandoning the well-known, if imperfect, century-old system, the only one used so far in the literature, at the cost of having to find thousands of new names? If so, which kind of names? Or some combination of the two?

The detailed story of this painful crisis is told elsewhere.[31] Briefly told, it can be summarized as follows. There were in fact two schools: (*i*) the "LPL group," around G. Kuiper, the founder of the Lunar Planetary Laboratory,[32] and B. Levin, who did not want any change but accepted to work on a revision of the Mädler system, and at any rate refused to embark on finding names with undefined criteria, in addition refusing to include non-scientists (thus mechanically limiting the number of available decent names); and (*ii*) mainly Menzel and Dollfus, who, on the contrary, were ardent supporters of opening the name list to non-scientists (writers, artists, etc.), and critics of the Mädler system altogether, even revised, that they refused to introduce on the far side of the Moon. The weakness of their approach, however, was the problem of necessity: since admittedly not all craters could be named, which ones should be selected? This is where NASA came to the rescue: as mentioned above, its cartographers had devised a "combined map" plan, from 1:1,000,000 scale to 1:250,000 scale. What they needed was, first and foremost, at least one "identifier crater" per 1:250,000 scale map (see Fig. 16.4 above): if no crater was already named, then find a name for it; if the name was a Mädler-style, lettered "satellite

[31] Montmerle, this volume (Chap. 13). See also, e.g., Montmerle (2019a) for the origin of the role of the IAU on celestial nomenclature.

[32] Mainly the cartographers D.W.G. Arthur and E.A. Whitaker.

crater," so be it, it could be changed later. This meant finding a first batch of about 2000 names, a huge, but not impossible task. Then, thanks to NASA pressure, the crisis was over: the WGLN re-united (somewhat reluctantly), and eventually, thanks to Dollfus's organizational talent, came up with the Resolution that was adopted at the Sydney General Assembly. The new names would be added in the future.[33]

At any rate, returning to the UNWG, and of course unaware of the WGLN crisis, its next move came several months after the London Conference. On Jan. 29, 1973, Komkov sent out a "*Circular Letter N° 1*" to his Working Group, which had in the meantime become more international, with the addition of three new members (from Hungary, India, and Spain; specialty unknown) and a UN representative, presumably *ex-officio*. Komkov acknowledged correspondence with Menzel,[34] who informed him of the forthcoming publication [by NASA] of the 1:250,000 scale maps (Fig. 16.4), which *"will unavoidably require a significantly increased number of proper names of other kinds of designation of lunar features,"* adding: *"I think we shall be unanimous in the opinion that it is impossible to resolve the problem by efforts only of the astronomers disregarding the experience of the experts on geographical names and the cartographers, i.e. without a closest international co-operation of specialists of different fields of knowledge."*

Then, perhaps because of the arrival of new members in his WG, Komkov summarized once again the "problems" to be solved: a single naming system for the Moon and the planets? the same system for the near and far side of the Moon? if the "names" are proper names, what would be the criteria? which would be the approval procedures? what to do with "micro-features" seen from the ground (*Apollo, Lunokhod*)? And Komkov referred again to the London Resolution 21, which called for a "convention," i.e., some legal agreement between the UN and "competent international bodies" (undefined: scientific? administrative? other?), adding:

> The experts on the international law are of the opinion that since international scientific organizations, such as the IAU, IGU, and others, are non-governmental their decisions cannot be referred to as binding for official application in different countries.

The Circular was meant to set the UNWG in motion for the next (5th) Session of the UN Group of Experts on Geographical Names, due to be held on the following 5–16 March in New York. However, the report presented to the Group of Experts showed that they basically had nothing to propose, if only to reiterate their position with respect to the IAU, apparently focusing now on legal issues:

- "The naming of extraterrestrial features is a complex matter and one which properly falls within the orbit of the United Nations Group of Experts on Geographical Names. Nomenclature of extraterrestrial features must be acceptable to all nations (. . .) [which] places the matter firmly in the province of the United Nations Group of Experts on Geographical Names. The drawing-up of such systems of nomenclature was not appropriate to astronomers alone."

[33] Transactions IAU vol. XVI B.

[34] No record found in the archives.

- "There is the further question of the legal standing of names allocated to extraterrestrial features by non-governmental bodies. It was agreed that the status of the Group of Experts as a United Nations body and the standing of the experts as official representatives of their own linguistic/geographic divisions made the Group of Experts the body best constituted to deal with this question."
- Conclusion #6: "The Working Group, in accordance with resolution 21 of the London Conference, will continue its activity in drawing up a plan for international agreement on the standardization of the names of extraterrestrial features in co-operation with the international organizations (ICSU, IAU, etc.)."

The Group of Experts then made to the WG nine recommendations, which, unbelievably, completely ignored the ongoing IAU work and the astronomical reality, e.g.: *"#4. Systems for identifying small features will be investigated and tested; including methods based on the use of co-ordinates" "#5. The Working Group is charged with examining how best to achieve legal international status for names allocated to extraterrestrial features,"* etc. And Recommendation #7 states: *"In this connexion, the Chairman of the Group of Experts, Dr. M. Burrill, (...) will prepare a status report for the next General Assembly of the I.A.U."* Yet, at the time, the NASA maps were becoming available, and, for instance, the UN group could at the very least have used them to test various nomenclature systems...

In any case, on Aug.14, Burrill sent a letter to Strömgren and de Jager (spelled "de Jagger"!), using de Jager's Utrecht address.[35]

Beginning with a strange "Gentlemen" (usually one writes "Dear colleagues"), the letter listed, as an enclosure, the promised "status report" for the IAU, which was in fact the full report of the New York meeting (i.e., concerning all the activities of the Group of Experts),[36] with that of the UNWG embedded in it and essentially unchanged.

The second paragraph read, in way of introduction:

> The Group of Experts acknowledged that contacts with scientific organizations such as the International Astronomical Union and the International Committee on Onomastic Sciences might be improved. It was the general feeling that ultimate responsibility for the methods employed in name giving ought to lay with United Nation bodies because of their international standing. *All organizations that are occupied in name giving should be aware of this.*

(The italics are mine.) Then Burrill continues by invoking the issue of the "convention" mentioned in the Resolution 21, but using a totally abstruse formulation (excerpt):

> An [international] convention would probably cover both policy and procedure. Policy would identify both broad categories of names or other words deemed appropriate or inap-

[35] This was just a week before the start of the General Assembly, and it seems that the letter was received too late in Utrecht, then forwarded to Sydney: a handwritten note, likely by Strömgren, reads "seen in Sydney by Kees, the resolution is appended." This was the UN London Resolution 21, which was also re-typed in full in the letter.

[36] UN Ref.: https://unstats.un.org/unsd/geoinfo/UNGEGN/docs/6th-gegn-docs/6th-gegn_e.pdf (44 pages).

> propriate for the specific element in extraterrestrial names, and a subsequently expandable basic set of standard words to use as generic terms designating the kind of entity named. The policy would presumably be designed to apply in all extraterrestrial naming as far as practicable. (...).

It had become obvious beyond any doubt that the UN Group of Experts and the IAU did not speak the same language. The UNWG was going in circles month after month without proposing anything else than strange, non-astronomical idioms, without even trying to evaluate them as the experts they claimed to be, and being increasingly obsessed with so-called "legal" issues—as if implicitly acknowledging that no one, not even Komkov, was actually competent at all to discuss nomenclature as applied to astronomical objects.

This had to stop one way or another. Fortunately, overcoming its internal crisis mentioned above, Commission 17 had submitted a global "*Resolution on Lunar Nomenclature*" to the General Assembly, which adopted it. Then de Jager, on Aug.21, just a few days before he would step down at the end of his term as IAU General Secretary, wrote from Sydney to Komkov:

> (...) This [Resolution for the naming of lunar formations], being backed by the various interested international organizations of the Inter Union Commission [of the Moon] has now been accepted by the General Assembly of the International Astronomical Union. The naming as proposed will be used in the large lunar maps, at this time being printed by NASA, and will be accepted as a standard by all scientists of our Earth. (...) At the same time, the International Astronomical Union, in close cooperation with other organizations of the ICSU family, is now continuing its work with regard to the nomenclature of various other bodies of our planetary system.

With de Jager's letter (certainly written with the endorsement of Strömgren) as an answer to Merrill's own letter, and using easily understandable terms, the IAU deliberately put the UN in front of a "*fait accompli*"–at least as far as lunar nomenclature was concerned.

But much remained to be done: true, the Moon race was over, lunar missions were essentially terminated (and would not resume until two decades later), but now the interest was shifting towards missions to "rocky planets" of the solar system (Mars, Venus, Mercury), posing new organizational challenges to the IAU.

16.7 Act II: The Post-*Apollo*, "Planetary" Era

The Sydney General Assembly was not only the place where the IAU held a firm position against the UN way of considering "extraterrestrial" nomenclature (an event which went completely unnoticed, more in the last Section), it was also, following the Statutes, the moment when important organizational changes could be made.

First, the IAU leadership changed, as is the case after each triennium: the American Leo Goldberg succeeded the Dane Bengt Strömgren as IAU President, the Greek George Contopoulos succeeded the Dutch Kees de Jager. This turnover also concerned Commission 17: after two terms as President, the French Audouin Dollfus stepped down to be replaced by the British S. Keith Runcorn. But the main change concerned its Working Groups: a major reorganization took place between Commissions 17 ("The Moon") and 16 ("Physics of Planets and Satellites"), driven, precisely, by the change in priorities in space research.

The preceding triennium had seen the question of lunar nomenclature emerging not only as a scientific issue (if only as a tool to help science), but also as a cultural one, a topic of repeated controversies (in the aftermath of the *Apollo* missions) as to "what to name," "how to name," "who had the authority to name," etc., and after all: "what's in a name?" , to the extent that even the United Nations felt obliged to enter the game with the idea to rule the field, notwithstanding the century-old experience of astronomers. The Brighton-Sydney triennium had been very hectic in this respect, the WGLN having to deal, as we have seen, with two crises simultaneously: one, external, with the UNWG, the other, internal to the WGLN, about the naming processes and criteria, which staged unpleasant personal conflicts and left scars. And now was the dawn, after the dramatic "Moon era," of the post-*Apollo*, "planetary era." The time was thus ripe (in fact as the UNWG had proposed, but for different reasons) to acknowledge the fact that lunar and planetary nomenclatures should be treated together–albeit not necessarily in an identical manner (contrary to the UNWG approach).

Because of all these reasons, the WGLN was dissolved, and a new Working Group "*on Planetary System Nomenclature*" (a.k.a. WGPSN) was created, with two major differences: (*i*) *scientific:* the Moon and all solar system objects, present and future, were considered together; each object category (the Moon; planets as they were being explored, and their satellites; the outer solar system, etc.) had its own "Task Group" ("TG;" much in the way Commission 17 had itself comprised three Working Groups); (*ii*) *organizational:* recognizing the singularity (historical, cultural, social, etc.) of the problem of nomenclature, the WGPSN was directly put under the purview of the Executive Committee, with a permanent status (contrary to Commissions).[37] This meant in particular that, at least in the beginning, it had to regularly report on its activities directly to the EC, and major decisions (including nominations of officers, consultants, etc.) had to be endorsed or approved by the EC. In other words, promotion and independence (from a Commission) on one side, but "discipline" and a much-needed direct communication link with the IAU leadership on the other.

The WGSPN was formally approved the following year, during the 39th Executive Committee meeting (Aug. 27–29, 1974), held at Observatoire de Haute-Provence, in southern France, who appointed the Canadian Peter Millman as its first President (Fig. 16.6). As regards the "*Task Group on Lunar Nomenclature*" (TGLN), we find again the old guard of the WGLN (the Menzel-Dollfus duet, and well as H. Masursky), but also three new members (including Runcorn *ex-*

[37] The WGPSN still exists as an "Executive Committee Working Group," see https://www.iau.org/science/scientific_bodies/working_groups/98/.

Fig. 16.6 Peter Millman (1906–1990) at a meeting in Ottawa in 1964. (Courtesy Archives of the Royal Astronomical Society of Canada)

officio). Menzel was appointed Chairman, but resigned (see below) the following year; he was replaced by Millman as "acting President," which was an advantage in exchanges and negotiations with Komkov and the UNWG, because Millman was both WGSPN President (then directly in contact with the EC) and Chairman of the TGLN (then directly in contact with its members, notably Farouk El-Baz, who was Director of the National Air & Space Museum in Washington, acting as an interface with NASA).

After such important changes in structure and in membership turnover, the WGSPN experienced some difficulties at the beginning. A first meeting took place in Ottawa (Millman's home town) on June 27–28, 1974, to set up "basic principles for planetary system nomenclature," which were expressed in fairly general terms[38]: nomenclature should be "simple, clear and unambiguous," names chosen should be "single words, and expressed in the language of origin," take into consideration "the traditional aspects," etc., and "final approval of any selection is the responsibility of the IAU." A far cry from the UNWG recommendations... So general were the terms, in fact, that Dollfus (letter dated May 20, 1975) expressed criticisms to Millman:

> My personal worry about the efficiency of the WGSPN was in many respects shared by several other colleagues" (...) "the Ottawa meeting gave me the impression of discussing rather general principles of what nomenclature should be, when a large fraction of the

[38] Recalled in a later publication (Millman, 1976). See also below.

> members was already experienced..." (...) "concerning the Moon, though the preceding WG arrived at unanimously adopted decisions, resolutions endorsed

[i.e., the Sydney Resolution], *implementation work organized, and adoption by cartographic program* [NASA's "Region/Province" mapping scheme, presented as part of the Resolution], *the whole affair was unnecessarily shaken, the Chairman* [Menzel] *obliged to resign, and the overall structure and IAU efficiency questioned." (...) "I sincerely hope the WGSPN will now initiate the necessary action for reaching the coordinated scheme for Solar System objects. This, I hope, will be done on the basis of the work already carried out at IAU and agreed upon and as independently as possible of any group of pressure."*[39]

In parallel, on the UN side, the Group of Experts and the UNWG continued their activities, but the Sydney Resolution had a decisive effect. In preparation for the Sixth Session of the Group of Experts in New York (Mar. 5–26, 1975), Komkov wrote a report in which he simply acknowledged that: (*i*) Burrill's "official" letter to Strömgren and de Jager (sent just before the Sydney General Assembly, as mentioned above) had not been answered, and (*ii*) the IAU General Assembly had approved the Resolution on lunar nomenclature presented by Commission 17 on behalf of its Working Group.

Komkov did not add any comment: the IAU view on lunar nomenclature was thus de facto endorsed by the UNWG, and the adoption of this Resolution became a tipping point for the balance of power between the IAU and the UN.

On the other hand, Komkov's report rightly pointed out (as did Dollfus) a lack of coordination between the WGPSN's Task Groups ("over half on the Martian crater names were already adopted for the Moon") and other inconsistencies, and called for "permanent contacts" between the WGPSN and the UNWG.[40]

As a matter of fact, Millman had been invited by Komkov, and he could attend the Session where Extraterrestrial Names would be discussed (Mar. 14). In his "Summary Record" of the Session, Komkov noted in a conciliatory tone that

> There had not been entire agreement among the astronomers at the Ottawa meeting, and some of the solutions there proposed did not seem entirely satisfactory from a toponymic point of view. However, it is always easier to criticize than to be constructive. The need now [is] for the Group of Experts on Geographical Names Working Group, taking fully into account the views of the IAU which Mr. Millman would put forward, to make constructive proposals for the coming IAU [WGPSN] *meeting in Moscow, at which Mr. Millman and Mr. Komkov (USSR) would be present.*

Then came a simple sentence that completely changed the relations between the two Working Groups: *"The* [UNWG] *had concluded prior to the meeting* [i.e., the

[39] It is not quite clear what "group of pressure" Dollfus has in mind here. He may be alluding to the *Lunar and Planetary Laboratory* in Tucson (created by G. Kuiper, who had died in Dec. 1973), where a group around Arthur was still resisting the "anti-Mädler" approach for lunar nomenclature. See the discussion in Montmerle, this volume (Chap. 13), and Arthur (1976).

[40] One can note that no mention is made anymore of "onomastics" and other "linguistic" approaches to nomenclature: the principle of proper names, of various origins, was now adopted as a matter of principle, for the nomenclature of lunar and planetary features.

Session] *that the involvement with legal aspects which were many and complicated should be avoided.* (The italics are mine.)

On his side, Millman circulated on Apr. 25 his own report of the UN meeting, more centered on technical issues ("drafting an overall plan," romanization and transliteration, central archive and name bank, etc.), but also insisting in his preamble on the UNWG *"desire to stay clear, whenever possible, of the legalistic aspect of nomenclature problems."*

The next meeting of the WGPSN took place in Moscow on the following July. The minutes could not be found in the IAU archives, but on Oct.10, 1975, Millman reported to Contopoulos, the General Secretary, his satisfaction after the Executive Committee meeting (EC40) that had taken place on month before (Sep. 9–11) in Lagonissi (Greece): *"I was very pleased to learn from your letter of Sep. 18, 1975, that the IAU Executive Committee had accepted our resolutions from the Moscow Meeting of the IAU/WGPSN with only a few minor modifications. I will make changes in the Venus nomenclature resolution (...)"* (the resolutions concerned the Moon, Mercury, etc.). In addition, Millman noted the dissolution by ICSU of the IUCM[41] (which had been a temporary structure anyway, and which the IAU did not need anymore), and informed the Executive Committee that names would be solicited from the IAU National Committees and also from the UNWG, and that a *name bank* would be created *"to leave something for future generations of scientists."*

Other than that, the minutes of EC40 (item 8), perhaps in a way of recognition for the work of the WGPSN, note that *"In general, it was felt that the Executive Committee should only raise objections to the action of the Working Group if something seriously wrong is found"*–thus relieving the Executive Committee of inspecting perhaps cumbersome lists of planetary names... It was clear that a year after its first meeting, the WGPSN had successfully overcome its initial difficulties.

Once the main obstacles had been removed on the UN side (the "legal aspects" and exotic coding linguistics), the relations between the WGSPN and the UNWG went on more smoothly, the IAU having now clearly taken the leading role. In his *Icarus* article (accepted on Mar. 6, 1976),[42] Millman explained to the wide audience of the journal:

> The IAU/WGPSN has established lines of communication with the UN Working Group and with the UN Group of Experts. In this way we can assess the reaction of the nonscientist to nomenclature policy over a broad international spectrum, since the United Nations include almost three times the number of nations represented in the IAU.

As the Grenoble General Assembly was nearing (Aug. 24-Sep. 2, 1976), the main work of the WGPSN was to prepare its first global report to the Assembly. It turned out to be an impressive, exhaustive document of 49 pages, reproducing the reports of the three WG meetings, held successively in Ottawa (1974), Moscow (1975)

---

[41] Reported in the minutes of the EC40, item 19.

[42] *Op. cit. (Note 38)*

and Grenoble (on-site meeting, held on Aug. 30–31, 1976). All the decisions on nomenclature for the Moon[43] and planets (new names, identifications, rectifications, etc.) were presented in the form of resolutions, including tables, etc.[44] However, there was no mention of a role that the UNWG would have played.

The next major event took place in 1977: it was the Third UN Conference on Geographical Names, which took place in Athens (Aug. 17–Sep. 7). As usual, Komkov prepared a new report of the UNWG. After summarizing what had happened since the London Conference, five years before, and emphasizing the changes in the IAU structure with the new Working Group on Planetary System Nomenclature, Komkov notes:

> Thus, at present, favourable conditions for business and fruitful co-operation between the two Working Groups have been established. (...) Gradually the outline of the general system of nomenclature of extraterrestrial topographic features was elaborated. (...) we came to the conclusion that (...) the complicated legal aspects should not be touched.

Still, he writes:

> And now that the general system of planetary system nomenclature is outlined, it is expedient to define on broad international basis the procedure of introducing, considering and approving of proposals on names for the extraterrestrial topographic features as well as that of informing all the world scientific circles concerned.

In the end, the Athens Conference adopted Resolution 23, on "Extraterrestrial feature names," which stipulated:

- "The Conference".
  (...)
- "Realizing that those nations actively engaged in planetary exploration have developed certain procedures for producing names for maps, charts, and other publications,
- "Aware of the role played by the International Astronomical Union in the establishment of name lists and the application of names to features for international purposes,
- "Recommends that the United Nations Group of Experts on Geographical Names collaborate with the International Astronomical Union and other interested international bodies in programmes for the allocation of names to extraterrestrial features so that all requirements for names information may be met,
  (...)

Compared to Resolution 21 of the London Conference, it was clear that a very significant progress had indeed been achieved on the recognition of the IAU role by the UN.

This scenario of a combination of an IAU General Assembly and UN Conference repeated itself: the following IAU General Assembly took place in Montreal, Aug. 14–23, 1976. The WGSPN produced a 20-page report,[45] along lines comparable

[43] For the Moon, 164 new crater names were attributed in 1975 and 1976; two additional lists of 195 and 20 names were stored in a name bank. Similar lists were produced for planets and outer solar system objects.

[44] Transactions IAU Vol. XVI B, pp. 321–369.

[45] Transactions IAU vol. XVIB, pp. 285–304.

to the Grenoble report. Then, on the UNWG side, came the Fourth Conference, held in Geneva (fifteen years after the first), on Aug. 24 - Sep. 14, 1982, which in fact coincided almost exactly with the next IAU General Assembly, held in Patras (Greece) on Aug. 17–26.

Similarly, Komkov prepared his UNWG report for the Geneva Conference.[46] In anticipation of his conclusion, he first gave his assessment of the role of the UNWG, obviously directed at the Group of Experts:

> The attention of the International Astronomical Union (IAU) was drawn to the necessity of improving the system of naming extraterrestrial topographic features" (. . .) "All the above-said has contributed to the improvement of naming system of extraterrestrial topographic features as well as to the regulation of IAU activity in this field." (. . .)

"*The* [UN Group of Experts on Geographical Names] *has contributed greatly to the development of the system of extraterrestrial topographic features which is in use now." (. . .)*

> [The result] is indicated in the 'Preliminary Working Check-List of Topographic Names on Planetary Bodies' prepared by Peter Millman, the Chairman of the IAU/WGPSN, and published in February 1982 (Herzberg Institute of Astrophysics, Ottawa, Ontario, Canada). The Check-List includes 3362 topographic names of 18 planetary bodies (four planets and 14 satellites). (. . .) A 'Gazetteer of Planets and Satellites' will be compiled by IAU/WGSPN on the basis of the 'Preliminary Working Check-List'.

Then Komkov came to his conclusion:

> This work is in trusty hands of IAU/WGPSN which successfully fulfills its complicated and labour-consuming work on the basis of international co-operation." (. . .) "Thus, under the conditions existent now, the Group of Experts on Geographical Names and its Working Group on the names of extraterrestrial topographic features that made a contribution to the cause of regulation (normalization) of the naming of such topographic features, *may consider their tasks in this field to be fulfilled and in the future fully rely upon the activity of IAU/WGPSN.*

(The italics are mine.)

The resulting Resolution 13 that was passed by the Geneva Conference[47] is reproduced (in an adapted layout) in Fig. 16.7. The main sentences are, of course:

- *Noting that the naming of extraterrestrial features is done in a satisfactory way by the Working Group for Planetary System Nomenclature of the International Astronomical Union.*
- *Recommends that the Working Group on Extraterrestrial Features of the United Nations Group of Experts on Geographical Names should be dissolved.*

[46] UN Ref.: https://unstats.un.org/unsd/geoinfo/ungegn/docs/4th-uncsgn-docs/e_conf_74_L93.pdf.

[47] UN Ref.: https://digitallibrary.un.org/record/67708?ln=fr, p. 33.

FOURTH
UNITED NATIONS CONFERENCE
ON THE STANDARDIZATION
OF GEOGRAPHICAL NAMES

Geneva, 24 August–14 September 1982

13. Extraterrestrial feature names

The Conference,

Considering the statement on the present state of affairs made by the chairman of the Working Group on Extraterrestrial Features of the United Nations Group of Experts on Geographical Names (E/CONF.74/L.93),

Noting that the naming of extraterrestrial features is done in a satisfactory way by the Working Group for Planetary System Nomenclature of the International Astronomical Union,

1. Recommends that the Working Group on Extraterrestrial Features of the United Nations Group of Experts on Geographical Names should be dissolved;

2. Further recommends that the chairman of the Working Group on Extraterrestrial Features should continue to maintain liaison between the United Nations Group of Experts and the Working Group for Planetary System Nomenclature of the International Astronomical Union.

UNITED NATIONS

Fig. 16.7 The UN Resolution that recognized the IAU authority on "extraterrestrial names" (Geneva, 1982). (Layout adapted by the author: frame from the Proceedings cover, text of Resolution #13, p. 33)

16.8 Conclusion: The End of a Major Challenge Over the Authority of the IAU

So that was it. Started by the UN in the wake of the *Apollo* era, the challenge over the authority of the IAU in naming celestial objects (here those that could be observed and even visited in the course of space missions, hence belonging to the solar system) was over.

The UN had suddenly presented itself in 1971 as the only body entitled to successfully solve astronomical nomenclature problems, starting with the Moon, which astronomers had been studying and naming for centuries. In 1972, the second UN Conference on the Standardization of Geographical Names, which had decided unilaterally to include "extraterrestrial names" in its scope, did not even mention the IAU as an actor. After the IAU decided to challenge the UN and go on its own by appealing to the Sydney General Assembly in 1973 to approve the Resolution on Lunar Nomenclature proposed by its Commission 17 ("The Moon"), the tone changed. The UN eventually dropped the contentious issues that were not of concern to astronomers (legal aspects, "adapted" new languages, etc.) and progressively recognized the naming goals and procedures of the IAU. At its Third Conference in 1977, the UN was *"aware of the role played by the (IAU) in the establishment of name lists and the application of names to features for international purposes;"* finally, at its Fourth Conference five years later, the UN eventually closed the challenge, *"noting that the naming of extraterrestrial features is done in a satisfactory way by the Working Group for Planetary System Nomenclature of the International Astronomical Union."*

But what does this UN Resolution mean exactly? Since its creation in 1919, the IAU had appointed itself as the "guardian of the temple" of celestial names, starting with its Commission 17 "Lunar Nomenclature." In 1935, Blagg & Müller's catalogue and maps were considered as the "bible" of lunar nomenclature–at the time, the question of planetary features was still in its infancy, as exemplified by the errors of interpretation of Mars's "canals" by Schiaparelli and others. But the Cold War and the Moon race, and more generally space research, totally overwhelmed the picture. Naming lunar features became a political and social issue: the IAU authority was challenged from the start, with the naming by the Academy of Sciences of the USSR, without IAU prior approval, of the first features of the far side of the Moon discovered by *Luna-3* in 1959, then again after the *Zond-3* lunar flyby in 1965. It was challenged once more by the (American) public in 1968 after the *Apollo-8* orbiting lunar flight and its iconic "Earthrise" images; it was challenged by the astronauts themselves, either when flying low over the lunar surface, or on the ground, travelling up to 30 km "by car" in a complex terrain. Each time, even in crisis, the IAU restored its authority (at times with the decisive support of NASA, to be fair).

However, the UN challenge was different. Its Group of Experts on Geographical Names, as the title implies, had the naming authority, granted by the Union Nations, over the whole Earth. With the *Apollo* landings, they decided that, because the time had come that humans, having walked on the Moon, would soon walk on other celestial bodies, the Group was naturally entitled to extend its authority to other Earths. And unfortunately, they had preconceived, wrong ideas: owing to the simple fact that the "other Earths" are uninhabited, they have no historical names to start with, so why should the experience gained on the Earth (even in remote, but topologically similar, places like the undersea or the Antarctic) be imposed on extraterrestrial bodies?" However, the IAU Working Group was quick to realize that the UN Working Group had at least one point: the naming system of the IAU had a number of drawbacks, some known for a long time albeit workable (the Mädler lettering system), but more importantly, the name pool, restricted to astronomers and Moon-related scientists, was far too narrow. It was thus a breakthrough, but a surprisingly difficult decision within the IAU, to include non-scientists (writers, artists, etc.) in the naming possibilities, as well as names proposed by the explorers—the astronauts themselves.

In view of the fundamental importance of this challenge, and its international nature, it is really astonishing that *it has never been mentioned in the IAU official records.* The public ones are those of the General Assemblies (the *Transactions* of the IAU), published every three years; the "private" ones, initially confidential but now accessible online, are the Minutes of the Executive Committee meetings,[48] which take place every year. Unbelievably, *none* of these Minutes refer, be it in an "Any Other Business" agenda item, to the involvement of the United Nations in problems of nomenclature, not even after the creation of the WGPSN, which had by construction a direct access to the Executive Committee. The only exception I found was when the EC was asked to approve the nomination of A. Komkov as a "consultant" (EC43, 1977, item 7a), but there are no details about who he was or on his role. Also, we have seen that there had been ample correspondence between the IAU Working Group and General Secretaries, especially with C. de Jager in the 1971–1973 triennium. On this point, I had the privilege of exchanging emails with him in 2020 (he was 99 at the time!),[49] asking him why Strömgren never put the UN challenge on the agenda of EC Meetings: his reply to me (on Jan. 11, 2020) was:

> I have only vague memories to the UN approach but I knew my president, Strömgren, very well. He hated unnecessary correspondence and his reaction to the UN letter would have been "forget this letter; the moon is just the business of astronomers; let these political people deal with their own matters; these are already difficult enough.

Perhaps. But when the WGPSN was created, the next President was Leo Goldberg, whom, contrary to Strömgren, was a prolific letter writer and a very outspoken person. He, too, continued the official "omerta." Yet the issue was put forward on the public scene: as mentioned above, Millman had explained very clearly the "lines of communication" between the WGPSN and the UN Working Group in his 1976 *Icarus* article.

Whatever the reason, I think it was a mistake not to publicize, one way or the other, the UN Resolution of 1982. While this Resolution does not explicitly confer (or transfer?) the naming authority to the IAU, it does recognize the competence of the WGPSN, so much so that the UN dissolved its own Working Group to which it had conferred the same task. The reason for which it was a mistake is that *the IAU authority continues to be challenged,* even by astronomers. The case in point is the "*Uwingu*" affair, in which American colleagues involved in a fight with the IAU over the status of Pluto, categorized as a "dwarf planet" at the 2006 General Assembly in Prague, decided to open the naming of Mars craters to the public (for a fee), precisely claiming that the IAU did not have the "privilege" of naming celestial objects.[50]

---

[48] https://www.iau.org/administration/executive_bodies/executive_committee/ecminutes/.

[49] Unfortunately, C. de Jager passed away a year later, on May 27, 2021. His obituary is at https://www.iau.org/administration/membership/individual/8063/.

[50] See my own account of the story (Montmerle 2019b), which unfolded during my term as General Secretary (2012–2015).

It would have been a quick and short answer then to simply argue that "*the IAU authority to assign 'extraterrestrial names' had been officially recognized by a UN resolution in 1982.*" This resolution should now be explained and publicized so that no such unpleasant incident may happen again in the future.

Acknowledgments It is a pleasure to acknowledge, here also, the assistance of Mme Ginette Rude, Archivist at the IAU Secretariat in Paris until 2016, who directed my attention to the *Dollfus Files* and triggered my interest in this episode of the history of the IAU. I also warmly thank Ms. Hermine Diebolt and Ms. Susan Kurtas, from the United Nations' Dag Hammarskjöld Library in New York, for their kind help in sipping through the meanders of the UN's official documents.

I also owe posthumous thanks to Kees de Jager (1921–2021), science colleague and former IAU General Secretary, for interesting exchanges in 2020 about the role of "his" President Bengt Strömgren regarding the IAU-UN relations during their term (1973–1976).

10th International Congress of Onomastic Sciences. Proceedings. Disputations ad montium vocabula aliorumque nominum significationes pertinentes. Tom. II Wien, 1969.

X/3

SYNTAX AND SEMANTICS OF LUNESE I,

A MICROLANGUAGE FOR LABELING

TOPOGRAPHICAL FEATURES OF THE LUNAR SURFACE

CASIMIR BORKOWSKI

Department of Computer Science
Graduate School of Library and Information Sciences
The Knowledge Availability Systems Center

UNIVERSITY OF PITTSBURGH
Pittsburgh, Pennsylvania

399

X/3

are: 10°,N 12° (where "10°" stands for the longitude and "N 12°" for the latitude) and the digits of the coordinate reference "0", "1", "2", "3", "4", "5", etc. have been uniquely correlated with, respectively, the syllables "ze", "ne", "to", "te", "fo", "fi", etc., while letters "N" and "S", abbreviations for, respectively, "North" and "South" have been uniquely correlated with syllables "no" and "so", then the proper name for that feature shall be "Nezenoneto" -- where "Ne" stands for "1", "ze" for "0", etc.

Proper names derived from more precise coordinate information will be correspondingly longer. Thus, the coordinates: 010°,00',0;N 12°, 00',0 shall generate the Lunese I proper name: Zenezezezezenonetozezeze However, since a syllable string of this length would be unwieldy in normal use, procedures are available in Lunese I for abbreviating without loss of information overlong proper names. Thus, the abbreviated form of the above name shall be: "$Zeneze^{4}nonetoze^{3}$", where the exponents "4" and "3" indicate, respectively, a fourfold and a threefold repetition of the preceeding syllable.

The procedure adapted in Lunese I for expressing information about distinctive topographical characteristics of various portions of the lunar surface by means of a unique topographical term is conveniently simple. A description of the successive steps of this procedure follows.

Appendix 1 A new language ("*Lunese I*") proposed for lunar nomenclature by C. Borkowski (excerpts; IAU archives). The original reference is: *Disputationes ad montium vocabula aliorumque nominum significationes pertinentes.* Tenth International Congress of Onomastic Sciences. Vienna, 8–13 September 1969. *Abhandlungen – Proceedings – Actes.* Ed. H. H. Hornung. Vienna. Vol. 2 (pp. 397–400), 1969. 523 pp

CONTEMPORARY PROBLEMS OF SELENONYMY

Report presented by the United States of America*

In recent years an urgent need has arisen for the naming of lunar features. American and Soviet space probes have brought the moon into our "back yard" to be examined and analysed, and to have its surface formations identified by specific or specific and generic nomenclature.

At the present time, there are approximately 5,000 named features on the moon. This figure includes those letter features identified merely by the unimaginative expedient of adding letters to the name of the major physiographic unit in an area. Thus the smaller craters in the area of Sabine become "Sabine A", "Sabine B", and "Sabine C", and so on (see attached the section illustrated of the Ranger VIII Lunar Map).

A glance at the history of the naming of lunar features reveals the need for a uniform, internationally acceptable and self-sufficient system for deriving names for them.

As long ago as the seventeenth century, map-makers such as Jan Hevelius (1647) used the names of terrestrial continents, seas, and mountains to identify the larger plains, craters, ridges, mountains and other features on lunar maps. Thus, such names as "Alps", "Apennines" and "Carpathians" still survive on our maps of the moon today. Some map-makers, like van Langren, 1645, experimented with the naming of lunar features after deceased astronomers, scientists and other prominent personages. This proved to be more popular than the application of terrestrial names. As more powerful telescopes were built, astronomers were able to observe many additional configurations on the face of the moon and mappers soon depleted their reserve of deceased notables, turning to the use of names of living persons, often includng their own. Finally, they had to resort to the above-mentioned procedure of adding letters to the names.

A major step towards international acceptance of lunar feature names was taken in 1935 when the International Astronomical Union (IAU) approved the name

* The original text of this report, prepared by O. Waddell and C. Barsky, was contained in document E/CONF.61/L.107.

232

character, interest and policy. This often results in outcries of non-acceptance from other nations.

In order to resolve these problems, a new system is required to provide instant assignment of lunar names. This system must in no way interfere with the historico-classical selenonymy that is already in use, nor must it hinder any future decisions by the International Astronomical Union. It must, further, lend itself to unerring international communication, and at the same time be easily adaptable to automatic data processing, which has become an integral part of the space age.

The system we propose will satisfy these conditions and allow for the automatic naming of even very small lunar features (for example, craters less than 100 feet in radius). This system eliminates duplication, carries no national connotations, is limitless in scope, and should obviate the necessity of obtaining international agreement on each individual reference. The new system would

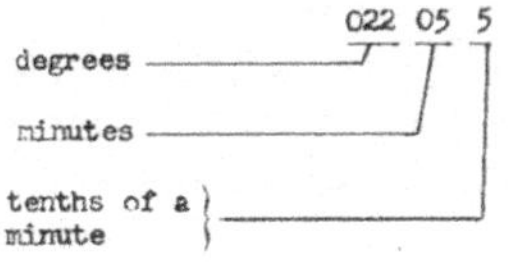

relieve those who are hard pressed in their search for new selenonymic identifications, and will eliminate the time, effort and expense that is now allotted to this activity.

The new system is based on the adoption of an internationally accepted system of lunar co-ordinates (the Selenographic Co-ordinate System, which appears on the section illustrated of the Ranger VIII Lunar Map), and the conversion of the co-ordinates in this system into phonetic units that will be used to construct lunar feature names.

In order to convert the numerical designations to phonetic equivalents, the following table has been constructed:

0	1	2	3	4	5	6	7	8	9	
B	C	D	F	G	H	J	K	L	M	first alphabet
N	P	Q	R	S	T	V	W	X	Z	second alphabet

The first alphabet will be used for converting numbers to phonetic equivalents, except in cases where two or more like numbers follow one another in the longitude or latitude readings, in which case the letter equivalents would alternate between the two alphabets for each succeeding like number. (For example, a reading of 333° 33.3′ would phoneticize as FRFRFR.)

The vowels A, E, I, O, U, and the semi-vowel Y are assigned no numerical values and are to be used to render the consonant-combinations pronounceable.

The full reading for Sabine B, therefore, is

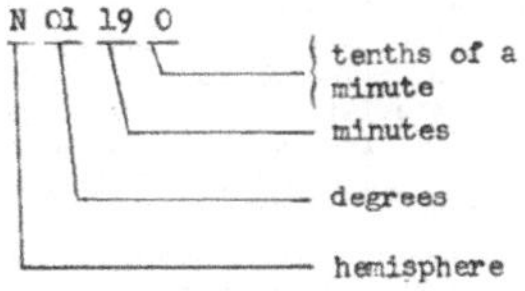

A conversion to letters would make

0	2	2	0	5	5		N	0	1	1	9	0
B	D	Q	B	H	T		N	B	C	P	M	B

Note that the designations N and S indicating the hemispheres retain their phonetic values.

The insertion of vowels into the BDQBHT NBCPMB combination evolved for Sabine B would result in some reasonably pronounceable sequence such as

233

BADQUIBHAT-NOBCAMPMOB

We recognize, of course, that at first sight such a derivation might seem lengthy and "unterrestrial", yet habit will overcome such preliminary objections as it

by the feature today, and DIMFEDHAB-SOCJIGHUB would have become CRÜGER.

The advantages of this system far outweigh its phonic drawbacks.

Appendix 2 Selenonymy, by O. Waddell & C. Barsky (1972). A coding language to transform lunar coordinates into pronounceable (?) names (excerpts). Published in the *Proceedings of the 2nd UN Conference on the Standardization of Geographical Names*, London, 1972 (UN Ref. E/CONF/0.61/4, pp. 232–234)

References

Arthur, D. W. G. (1976). A Dissenting Note. *Icarus, 27*, 571–573.

Beer, W., & Mädler, J. H. (1837). Allgemeine Selenographie [Full title: *Der Mond nach seinen kosmischen und individuellen Verhältnissen, oder allgemeine vergleichende Selenographie*; or: *The moon according to its cosmic and individual relationships, or general comparative selenography*]. Simon Schropp.

Blagg, M. A., & Müller, K. (1935), Named lunar formations (with W.H. Wesley, S.A. Saunder, and J.H.G. Franz, P. Lund, Humphries).

Menzel, D., Minnaert, M., Levin, B., Dollfus, A., & Bell, B. (1971). Report on Lunar Nomenclature by the Working Group of Commission 17 of the IAU. *Space Science Reviews, 12*, 136–186.

Millman, P. M. (1976). Topographic Nomenclature on Planetary Bodies. *Icarus, 29*, 155–157.

Montmerle, T. (2019a). In C. Sterken, J. Hearshaw, & D. Valls-Gabaud (Eds.), *Under One Sky: the IAU Centenary Symposium, IAU Symposium 349* (pp. 289–324). Cambridge University Press.

Montmerle, T. (2019b). In C. Sterken, J. Hearshaw, & D. Valls-Gabaud (Eds.), *Under One Sky: the IAU Centenary Symposium, IAU Symposium 349* (pp. 90–111). Cambridge University Press.

Whitaker, E. A. (1999). *Mapping and naming the moon* (p. 242). Cambridge University Press.

Thierry Montmerle has been General Secretary of the IAU (2012–2015). This made him familiar with IAU history and procedures, as well as with "backstage" information about the work of successive Executive Committees and Commissions. He participated in the IAU Centenary celebrations in 2019 by co-organizing the "Astronomers as Diplomats" conference in Paris, on which this book is founded. His scientific domain is space research in high-energy astrophysics, specializing in the role of energetic particles and photons (X-rays, gamma-rays) on the interstellar medium, and the resulting star and planet formation and evolution processes. After doctoral and post-doctoral studies in Montreal, Canada, he held a position at the Saclay Center for Nuclear Studies near Paris, France (operated by CEA, the French Atomic Energy Commission), where he spent most of his career.

Chapter 17
Diplomacy with the United Nations: Preparing for the *International Year of Astronomy*

Catherine Cesarsky and Claus Madsen

Abstract With a truly global reach, the *International Year of Astronomy* in 2009 arguably formed the largest and most comprehensive outreach activity ever undertaken in astronomy. We describe the arduous way from idea to reality with a focus on the political and diplomatic interactions that were necessary to win support for the idea outside the confines of astronomy. We describe the parties involved, the arguments that were put forward by the various stakeholders and the complex decision-making process that ultimately led to the approval by the United Nations, a formal requirement for an "International Year".

17.1 The Launch of an Idea

In 2002, at a lunch during a meeting of the IAU Executive Committee (May 6–8, St Petersburg; one of us, Catherine Cesarsky, was then one of the Vice Presidents) Franco Pacini,[1] IAU President at the time, suggested to a few of us to have 2009 declared by the United Nations an "International Year of Astronomy", 400 years after the first viewing of the sky through a telescope by Galileo—in analogy with, for instance, the "International Year of Physics" that was planned for 2005. We were all enthusiastic about it.[2] An IAU resolution, which we believe was written by him,

[1] 1939–2012 (see https://www.eso.org/public/announcements/ann12009/).
[2] The first discussion, at lunch, went unrecorded. It is here in record for the first time.

C. Cesarsky
AIM, CEA, CNRS, Université Paris-Saclay, Gif-sur-Yvette, France
e-mail: catherine.cesarsky@cea.fr
C. Madsen (✉)
Department of Physics and Astronomy, Aarhus University, Aarhus, Denmark
e-mail: claus.madsen@scienceuniverse.eu

© The Author(s), under exclusive license to Springer Nature Switzerland AG 2022

T. Montmerle, D. Fauque (eds.), *Astronomers as Diplomats*, Historical & Cultural Astronomy, https://doi.org/10.1007/978-3-030-98625-4_17

was approved by the Sydney General Assembly in July 2003.[3] It recommended that "the year 2009, the 400th anniversary of Galileo's accomplishments and the real birth of modern telescopic astronomy, be declared the 'Year of Astronomy', in which the potential of astronomy to enlighten and enrich humans will be brought to the largest possible audience all over the world". While building on this historical background, it would clearly become a tribute to astronomical research today, which is in a rapid development that—in its intellectual reach—in many ways deserves comparison with the Renaissance.

17.2 UNESCO Becomes Involved

As a next step, the IAU Executive Committee asked the government of Italy to request UNESCO to declare 2009 as the "IYA" as it soon became known (Fig. 17.1). In March 2004, the Ambassador of Italy, H.E. Francisco Caruso, sent a letter for this purpose to the Director-General of UNESCO, Kōichirō Matsuura. The first hurdle was to have the subject put on the agenda of the October 2005 UNESCO General Conference by the UNESCO Executive Board. This required to have as many countries as possible in support of this request. Franco Pacini himself was the conductor of the orchestra, and he rightly put to us that a mixture of big and small countries, covering several geographical areas, would be optimum. By then, Ron Ekers had become IAU President, and thus we had hopes for a strong support from Australia, but alas, despite Ron's prominence in his native country, his government was not favourable to the UN themed years. In fact, he had to struggle to avoid a too visible negative reaction from his country representatives. For Catherine Cesarsky, who had become President Elect in 2003, it was easier because her friend and well-known astronomer Jean Audouze was Vice President of the French UNESCO Committee, and he rapidly ensured French support. Norio Kaifu,[4] who was very influential in his country, swiftly obtained a very high level of support from the Japanese delegation. The UK was a particularly easy case, as Alec Boksenberg, a highly charismatic astronomer, was the Head of the National UNESCO Committee. In Portugal, with José Mariano Gago, a physicist, in place as Science Minister, it was easy for Franco Pacini himself to obtain a positive response. Other countries targeted were Brazil, through Beatriz Barbuy, who was extremely involved and effective; Canada through Jim Hesser, the Netherlands through Peter Bartel and George Miley, Austria through Josef Hron, and the USA, where Rick Fienberg, then editor of *Sky*

---

[3] Resolution B3 (Engvold 2007; see https://www.iau.org/static/resolutions/IAU2003_French.pdf and https://www.cambridge.org/core/services/aop-cambridge-core/content/view/0A05034C1D526CB10AA923740489EE93/S0251107X00005691a.pdf/chapter_iii_resolutions_of_the_general_assembly.pdf).

[4] 1943–2019, IAU President (2012–2015) (see https://www.iau.org/news/announcements/detail/ann19023/?lang).

Fig. 17.1 Catherine Cesarsky and Franco Pacini during the IAU Executive Committee meeting in Rome in April 2005 (Andersen et al., 2019). (Photo: Collection Hans Rickman.)

and Telescope, Bob Williams[5] and John Huchra were conducting an intense action together. In parallel, the IAU General Secretary, Oddbjørn Engvold, was tasked with writing to the IAU representatives of all the countries asking them to approach their UNESCO committee and try to reach their UNESCO Ambassador. In the end, the proposal to UNESCO was presented by Italy, co-sponsored by France, Japan, Brazil and the UK.

On 4 October 2005, Jean Audouze kindly relayed the minutes of the September UNESCO Executive Board meeting, in French. We thus learned that the UNESCO Director-General had received "with satisfaction" the proposal asking him to

[5] IAU President, 2009–2012.

support all efforts leading to the proclamation of 2009 as "International Year of Astronomy". However, the Director-General stressed that in the end, only the UN General Assembly can take this decision—and it would have to be taken at least one year before the designated year. This meant that the last possible date for us was the UN General Assembly commencing in the autumn of 2007. For this, a resolution first had to be approved by UNESCO, and then the countries that had been in favour at UNESCO had to support the proposal also at the UN General Assembly.

At the UNESCO General Conference in October 2005, the corresponding decision, "172 EX/Decision 57", was approved.[6] There had been many versions of the draft resolution, first by Franco Pacini himself, amended by Richard West, by Ron Ekers and others. The text approved by the UNESCO General Conference in October 2005 said:

> *The General Conference,*
>
> *Recognizing that the study of the universe has led to numerous scientific discoveries that have great influence not only on humankind's understanding of the universe but also on the technological, social and economic development of society,*
>
> *Aware that astronomy proves to have great implications in the study of science, philosophy, religion and culture,*
>
> *Noting the excellent opportunity to raise public awareness especially of the young generation on scientific issues and to demonstrate the links between science, education, culture and communication in the framework of the Space Education Programme and thematic initiative "Astronomy and World Heritage",*
>
> *Welcomes Executive Board decision 172 EX/57;*
>
> *Invites the Director-General to support all efforts leading to the proclamation of 2009 as the International Year of Astronomy;*
>
> *Recommends that the United Nations General Assembly at its 60th session adopts a resolution declaring 2009 as the United Nations International Year of Astronomy.*

It thus remained to obtain UN approval, and of course to get on with the work of organising the year activities. But after this, despite our exerting some pressure on our UNESCO representatives, there was a long pause . . . at least as seen from our side.

Only one issue, on UNESCO's initiative, could have started to move swiftly. On 3 October 2005, Francesco Bandarin, Director of the UNESCO World Heritage Centre, wrote to Ron Ekers, with Oddbjørn Engvold in copy,[7] to propose a partnership UNESCO-IAU devoted to "Astronomy and World Heritage". But alas, at the time, the IAU secretariat was dysfunctional because of a feud between two secretaries, and the letter was somehow ignored by IAU, but see later.

[6] https://unesdoc.unesco.org/ark:/48223/pf0000142311, and International Year of Astronomy 2009 Final Report (IAU 2010): https://www.astronomy2009.org/resources/documents/IYA2009_Final_Report/.

[7] Private Communication, C. Cesarsky Archive.

It was only at the end of June 2006 that the UNESCO Director-General wrote to the UN Secretary-General Kofi Annan, transmitting him officially the UNESCO resolution and asking him to submit it for approval at the UN General Assembly.

In July 2006, Walter Erdelen, Assistant Director-General for Natural Sciences, sent a letter to Ron Ekers, IAU President, at the IAU office. The IAU office was frantically preparing for the IAU General Assembly that was to take place in Prague on August 14–25, and somehow this letter was only "found" on 19 October; hence there had been no answer from IAU. It was, however, a very important letter. It informed IAU officially that the UNESCO General Conference had adopted the resolution on IYA, and confirmed that the Director-General had written to Kofi Annan. Importantly, it requested IAU to help gather UN support. Then the letter said:

> Moreover, as the UN specialized Agency with the mandate of promoting the natural sciences, UNESCO is prepared to take the lead upon proclamation by the UN General Assembly of the International Year of Astronomy 2009. In preparation for this eventual responsibility, the Natural Sciences Sector together with the Culture Sector and the Permanent Delegation of Italy in UNESCO, have formed a working group to make arrangements in support of the UN proclamation and in preparations for the celebration of the International Year of Astronomy. Recognizing the leadership role that IAU plays within the international community of Astronomy I would like to invite IAU to coordinate with the UNESCO working group on matters concerning the Year Of Astronomy 2009.

At the end of the Prague IAU General Assembly in 2006, Catherine Cesarsky became IAU President. Thanks to the UNESCO working group alluded to in this letter, and some efforts on IAU's part, we had at last the first meeting with UNESCO on the International Year of Astronomy, now named IYA2009, at the UNESCO headquarters in Paris, on August 29. UNESCO had been keen to invite representatives from ICSU,[8] IAF[9] and other institutions, so it was not just an IAU-UNESCO bi-lateral meeting.

The meeting was chaired by Robert Missotten, Secretary of the International Geoscience Programme of UNESCO and Chief of the Global Earth Observation Section. It was well attended. Participants included Jean Audouze, Alec Boksenberg, Beatriz Barbuy in addition to Catherine Cesarsky, who was also ESO Director-General. From the Italian side, unfortunately Franco Pacini could not come (it was the beginning of his ailment), but we had the then INAF[10] head, Piero Benvenuti. On the UNESCO side, we had the astronomer Ezio Bussoletti, who was the Italian delegate to UNESCO, and representatives of UNESCO entities interested in IYA2009, including Yolanda Berenguer from the Space Education programme and Anna Sidorenko-Dulom from the World Heritage centre. IAF was also represented.

At this meeting, we were given several documents and information on the action on the UNESCO side. We heard that the Italian Permanent Delegation to

---

8 International Council of Scientific Unions, of which the IAU is a member; now the International Science Council, since 2018. Based in Paris.

9 International Astronautical Federation, Paris.

10 Istituto Nazionale di Astrofisica, Rome.

UNESCO and the UNESCO working group, together, had drafted the resolution hoped to be approved by the UN General Assembly. This had been sent by the UNESCO Assistant Director-General for External Relations and Cooperations, Ahmed Sayyad, to all UNESCO Permanent Delegates and Observers, urging them to express support to IYA at the UN General Assembly. The draft resolution included the following paragraph:[11]

> 2. Designates the United Nations Educational, Scientific and Cultural Organization as the lead agency and the focal point for the Year to organize activities to be undertaken during the Year, in collaboration with the International Astronomical Union (IAU), Committee on Space Research (COSPAR) of ICSU, International Astronautical Federation (IAF), relevant entities of the United Nations system and other international and national institutions, organizations and associations throughout the world.

Not having read the letter by Erdelen, which mentioned the fact that UNESCO intended to take the lead in IYA2009, we were surprised and in effect put "*devant le fait accompli*": UNESCO would appear to lead IYA2009, according to the UN resolution which, we were told, could not be changed.

At this first IAU/UNESCO meeting, the subject of Astronomy and World Heritage was raised. A few days later, Catherine Cesarsky received a letter from Alex Gurshtein, President of the IAU Commission 41, "History of Astronomy", enclosing a copy of the missed letter from Francesco Bandarin of 3 October, while Anna Sidorenko-Dulom had sent this same letter to Beatriz Barbuy. This time an answer was sent to Francesco Bandarin, by the new IAU General Secretary Karel van der Hucht, expressing our high appreciation of this offer and our great interest in the matter. The joint activities on this subject became one of the Cornerstone Activities of IYA2009, chaired by Anna Sidorenko-Dulom, with the IAU work led by Clive Ruggles. This successful work still continues today, and gave rise, among other results, to two extensive "Thematic Studies" published in 2010[12] and 2017[13]: "Heritage Sites of Astronomy and Archaeoastronomy in the Context of the UNESCO World Heritage Convention", by Ruggles et al., and to the present-day IAU/UNESCO "Portal to the Heritage of Astronomy".[14]

17.3 Lobbying for the International Year of Astronomy

The most important decision at this first joint meeting was that, in addition to the letters sent by UNESCO to UN representatives, a visit of IAU representatives to Ambassadors of potential co-sponsoring nations of the draft resolution could

[11] Since this was a draft, there is no official record. Published here for the first time.

[12] https://www3.astronomicalheritage.net/index.php/thematic-study-1-contents.

[13] https://www3.astronomicalheritage.net/index.php/thematic-study-2-contents.

[14] See https://www3.astronomicalheritage.net/, and the article by C. Ruggles, this volume (Chap. 18).

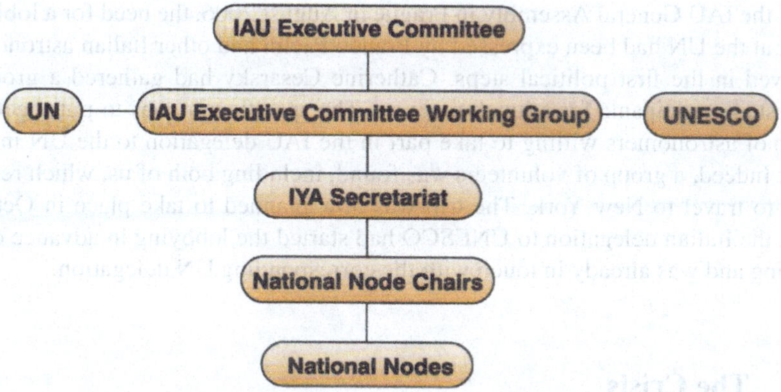

Fig. 17.2 This organigram presented the "internal" organisational set up. The national nodes acted as single-points-of-contact for each country, organising and coordinating the multitude of national and regional activities throughout the year

be useful, and that a delegation should be sent to present the case at the UN Headquarters. Said the minutes from the meeting:[15]

> According to the records of the 172nd Executive Board session, 19 UNESCO Member States had expressed support to the proposal and no objection from any Member State. However, many European Member States in general are reluctant to support the declaration of international years, therefore a pro-active awareness-raising campaign needs to be taken vis-à-vis the Missions in UN New York. Lobbying for the inclusion of the IYA proposal in the upcoming agenda of the UN General Assembly should be initiated immediately and throughout September.

In conclusion, it was agreed that the next (informal) meeting should take place after the official decision on the proclamation of the IYA had been made. All participants were reminded to start lobbying for the IYA immediately and to pursue it in New York, and we were asked to prepare our case. There was a great sense of urgency. It was felt that the trip to New York had to take place in September.

We swiftly prepared a nice brochure, led by Lars Lindberg Christensen, IAU Press Officer, to promote IYA2009, containing many of our ideas at the time. Since we were doing the work, we could also impose our points of view, and in this early brochure, we presented the organisation of the year with this diagram (Fig. 17.2):

For the rest of the IYA2009, IAU prepared beautiful brochures and a very interesting programme, and UNESCO and IAU agreed to appear at the same level in IYA documents and events.

[15] Minutes of the First Meeting on The International Year of Astronomy (2009), 29 August 2006, UNESCO HQ (Room XV), prepared by Robert Missotten, Secretary, International Geoscience Programme; Chief, Global Earth Observation Section, Division of Ecological and Earth Sciences, UNESCO.

At the IAU General Assembly in Prague in August 2006, the need for a lobbying effort at the UN had been expressed by Franco Pacini and other Italian astronomers involved in the first political steps. Catherine Cesarsky had gathered a group of interested participants for a lunch meeting. The specific aim was to put together a group of astronomers willing to take part in the IAU delegation to the UN in New York. Indeed, a group of volunteers was found, including both of us, which readied itself to travel to New York. The trip was now planned to take place in October. Also, the Italian delegation to UNESCO had started the lobbying in advance of the meeting and was already in touch with the corresponding UN delegation.

17.4 The Crisis

Then, on 16 October, a few days before the trip, we received an alarming e-mail from Ezio Bussoletti.[16] He explained the formal diplomatic procedures in great detail including how the national delegations act upon formal instructions from their capitals. He continued:

> As you see, the abovementioned steps need therefore a strong commitment of National Authorities.
>
> Unfortunately, despite the positions found and achieved here in UNESCO where I have presented the proposal, we have not found the same reaction in New York. Only Japan and US have formally informed us about their support while other countries have been silent. While this was expected from UK as they did not officially support the proposal but they unofficially informed us that they will not vote against, I am a bit perplexed by the silence of France. Evidently, despite the effort of our IAU French friends, the message from scientists did not reach the right political level.
>
> In addition to that we face, see for instance the mail from Ron [Ekers], the 'in principle' negative feeling of other countries as well as that of the EU Presidency who recalled us the position of Europe on the subject: here again unfortunately, the lobbying work has not been so far successful.
>
> All these considerations, the very short time left and the items in the agenda of Comm. II[17] which is presently focussed on items 'considered' more important, suggest not to try to launch the proposal this year as the risk of a rejection is too high. If this happens there will be no space for a further trial next year.
>
> On the contrary, as our Delegation indicates, a more articulated approach, starting early next year, will surely help to overcome all the problems encountered as it happened for both the Year of Physics and the Year of Planet Earth.

[16] Private Communication, C. Cesarsky Archive.

[17] The Second Committee of the UN General Assembly (The Economic and Financial Committee).

The mail then outlined a possible timetable for recovery, but ended on a sobering note:

> Finally let me recall the message we got from the UNESCO officials here who clarified that they cannot provide any support to IYA 2009 as the request must come and be supported by countries. The support will come from the Italian Delegation and, possibly, from some others.

After a few days of disturbing e-mails flying about, Catherine Cesarsky, in her new role as IAU President called for a crisis meeting of the Executive Committee. With participants from around the world, it was organised as a conference call by telephone. Claus Madsen, who at the time headed the ESO Public Affairs Department, participated from her office. After a brief discussion, it was agreed that Madsen be requested to come up with a strategy that could lead to a Resolution at the following UN General Assembly meeting (Autumn 2007), and in addition, coordinate the necessary work to achieve that.

17.5 The Recovery

After a "pit-stop" at Italy's Permanent Representation at UNESCO in Paris for a face-to-face meeting with Ezio Bussoletti to understand the Italian position somewhat better, Claus Madsen thus prepared an action plan—and set off to implement it. While it was clear that this would be an uphill struggle, on the positive side, he was given free hands to contact whomever considered necessary—and the signature of the IAU President on letters, when needed. He also solicited excellent support by his collaborator, Enikö Patkós, a young Hungarian lawyer, who before joining ESO had been a member of the Hungarian delegation to the UN Committee for the Peaceful Uses of Outer Space (Fig. 17.3). Thus, she was quite familiar with what to the uninitiated might appear as the somewhat arcane ways of the UN system.

In any event, we contacted, for example the Secretary-General for the Arab League, Amr Moussa, and Zalmay Khalilzad (a US citizen with Afghan roots), who interestingly was not only the Bush administration's representative to the UN but also a Sunni Muslim. Despite the somewhat precarious relationship of the US administration with parts of the Arab world, our thinking was that Khalilzad could therefore credibly act as an interlocutor with Arab diplomats, which was felt necessary given the importance of Arab culture and the role it has played in the history of astronomy.

Furthermore, not the least from George Miley's Universe Awareness outreach programme[18] (see below), we knew that moderate Arab governments were strongly interested in activities that could strengthen relations with the West, ameliorating tensions between the two world cultures.

[18] Of which Claus Madsen was co-chairman at the time.

Fig. 17.3 Enikö Patkós, at the time a lawyer in ESO's Public Affairs Department. (With kind permission)

In the subsequent weeks and months, Claus Madsen contacted and visited ambassadors and other senior officials in Paris, Rome, Prague, Brussels, Lisbon, Madrid, New York, Santiago de Chile and Cape Town. But the first station on the road—in January 2007—was Nordborg on the small island of Als in his native Denmark. At Danfoss Universe, a new Science Centre there, he met the Minister of Education, Bertel Haarder, and had the opportunity to discuss the Astronomical Year project with him. It was partly a chance encounter, since Madsen was a member of a jury assessing contributions to a science teaching contest and Haarder, as the responsible Minister, was supposed to hand over the prizes. However, Madsen already knew him personally from a visit to ESO in Santiago and knew that he was favourably versed towards astronomy, in general, and to ESO in particular. He immediately offered his help. Returning to his home in Garching late that evening, Madsen found a copy of an e-mail from the Minister to Denmark's ambassador to the UN, Ellen Margrethe Løj, in which he gave his backing to the idea. Haarder's support got us off to a good start and encouraged us to tackle the political headwind, which was quite strong in diplomatic circles and involved issues and motives that are not part of the normal internal discourse in astronomy.

Why did it cause such big problems to obtain a resolution on a subject that at best must rank quite low on the UN diplomatic agenda? The UN General Assembly is probably the most important meeting place and forum for world politics. This should be understood in the sense that here, (almost) all the world's nations and countries can voice their views. Unsurprisingly, it leads to exciting but sometimes also controversial debates. In Western countries, there has sometimes been a certain "fatigue" over endless—and sometimes despondent—discussions that they either considered too uncomfortable or a waste of time. The long period of Western hegemony has left many "unpaid" bills, understood by most, uncomfortable as they may be. But the world must also go on—that, too, is part of our dynamic Western culture. We will not pronounce judgment in that debate, but simply recognise the fact that there exists such tensions in international cooperation. As mentioned before, many Western countries had also grown tired of "International Years" because they saw them as having been abused and turned into instruments for political goals that they either could not or did not want to support. Therefore, on a proposal from Australia, Canada and Finland, the UN General Assembly in 2006 adopted a resolution that requested member states to exercise considerable restraint with regard to calling for new International Years. Both the EU and the US had endorsed this position and that a tightly worded resolution was introduced shortly before Italy had planned to propose to pronounce 2009 as the International Year of Astronomy.[19] Put bluntly: The IYA as a UN-approved initiative appeared to be dead in the water.

By the beginning of 2007, we were therefore in the situation that even if many countries easily felt sympathy for the idea that astronomy could be an interesting topic (also for an International Year), "for political reasons" they would not want to support it. Among the most sceptical countries were those in Northern Europe, including Denmark.

Our strategy therefore consisted in breaking the tight wall of rejection by convincing a "hard-line country" to change its position. It was possible with Denmark because Madsen knew the Minister of Education and knew that not only had he always paid much attention to quality in science education, but also he was open to new initiatives that could strengthen the interest in science among young people.

Furthermore, from his work linked to preparing the Czech accession to ESO (that was completed the same year, i.e. in 2007), Madsen also had good connections there. For example the brother of the key negotiating partner on the Czech side, Jan Palouš, was his country's UN ambassador. With recommendations from Palouš, Madsen met the Czech diplomat in New York and found him supportive of our cause. With Denmark and the Czech Republic "on our side", we had not just won over a hard-line country, but also one of the "new countries" in the EU (after the expansion of the EU membership in May 2004, the "old" EU Member States made much of taking into account the views and opinions of the "new" ones). But he also met with Finnish and French diplomats, who, however, referred to the UN resolution "against" international years and the EU decision to respect it. This was the feared "killer argument". The EU member states had committed to the so-called "Common Foreign and Security Policy" (CFSP), which of course also entails that, apart from exceptional circumstances, they will toe the same line at the United Nations. Then we took contact with "an old friend and colleague", José Mariano Gago, the particle physicist who had turned to politics and now served as Minister of Science in Portugal, whom Franco Pacini, as we saw, had already contacted three years before. As it happened, his country would hold the rotating six-month EU presidency during the second half of 2007—i.e. at the time when we planned to "go to the UN". He

[19] Resolution 61/185 "Proclamation of international years". To readers uninitiated to diplomatic language, the text might have appeared harmless, but for diplomats, the intention was hard to miss.

would have to convince his colleague in the Ministry of Foreign Affairs in Lisbon, of course. While this could not be taken for granted, we nevertheless felt that the snowball had begun to roll, also because Arab, Asian and Latin American diplomats appeared to be quite receptive to our cause.

Then we went to Brussels. In Brussels, Patkós and Madsen met with an experienced German diplomat, who on behalf of the EU Presidency of Germany during the first half of 2007 was dealing with the coordination of EU member state votes at the UN. For a while, we did not make much headway, but then he asked if we could provide an example of an activity that could take place during the IYA. We immediately jumped on the Universe Awareness Initiative (UNAWE),[20] the IAU programme designed to catch the minds and imaginations of underprivileged children across the world. At this moment, the mood of the diplomat changed. As it turned out, he was also involved in coordinating counter-terrorism measures. "That's good", he exclaimed. "We're desperately looking for peaceful and positive initiatives to prevent young people in precarious social conditions from falling into the fundamentalist, possibly terrorist trap. I can see how this can be of importance. I will discuss this with Berlin tonight". With this, we felt that Germany might come on board.

In parallel, we reached out to Brazil, Canada, Chile, Egypt, Indonesia, South Africa and Tunisia, while Norio Kaifu took care of Japan. We hoped that Canada would be able to convince the USA, and that South Africa could get more African countries behind the proposal. A somewhat curious situation occurred when Madsen, while staying in New York, paid a visit to the French mission, only to be met by a junior diplomat, who stuck to what had originally been the EU position of not endorsing the International Year proposals. Unlike Madsen, he had not been informed that Paris had changed its position. We now knew that we were finally ahead of the curve! The status of both the IAU and ESO of observership in the UN Committee for the Peaceful Uses of Outer Space also helped in facilitating discussions at the diplomatic level. At the same time, it was important to reassure the Italians, who continued to be wary.

In fact and quite understandably, the Italian Ministry of Foreign Affairs was very nervous about losing face by making a proposal that could not be adopted. Therefore they hesitated to produce a draft resolution text. But at the same time, we faced the problem that many of the countries we contacted would—despite encouraging sounds—only take a position upon having seen and evaluated the draft resolution text.

In the end, Madsen went to Rome to push for a draft resolution text to be circulated. At the meeting in Rome, the Italians began to gain confidence in the effort and shortly afterwards, in collaboration with ESO, the final draft resolution text was put together (see below). Once again, Patkós played an important role in the drafting exercise. What then followed, after the text had been released and circulated, was a "telephone action", calling numerous civil servants in various foreign ministries to

[20] Miley et al. (2005).

ensure that the text would not get stuck in a document pile of some desk officer, but would move through the chain of command quickly. For time was running out.

17.6 Going to New York

This time it also worked out with the delegation trip. On 31 October 2007, Beatriz Barbuy (then IAU Vice President), Catherine Cesarsky (IAU President), Kevin Govender (SALT),[21] Robert Williams (IAU President Elect) and Claus Madsen met at the Hayden Planetarium in New York to prepare the oral interventions at the UN, scheduled for the following day. The IAU delegation was subsequently augmented by Nella Le Moli and Leopoldo Benacchio (both INAF) as well as Luigi de Chiara, a seasoned diplomat from the Italian UN Mission, whose experience turned out to be crucial in this event (Fig. 17.4). The UN allowed us to hold a meeting over lunch, where all countries were invited. It was well attended, and we could talk about the wonders of astronomy and our intentions for the programme of IYA2009. Our presentations were very well received. In fact, some of the participants told us later that it had been the most interesting meeting they had participated in at the UN.

Later in the day, Italy submitted the official proposal to the General Assembly's Second Committee. The proposal was now supported by 30 countries. Notably there were small but remarkable diplomatic scoops. Among the co-sponsors were Algeria and Israel, as well as Armenia and Turkey (and Greece)—countries that rarely see eye to eye. Just one year before, the Italians had barely found more than a pair of allies and were *de facto* ready to give up on the project. Now things had changed. And so, the UN Second Committee delivered a consensus recommendation in favour of the proposal to the General Assembly. The resolution text put before the General Assembly as document A/RES/62/200[22] said:

> *The General Assembly,*
>
> *Recalling* its resolution 61/185 of 20 December 2006 on the proclamation of International Years,
>
> *Aware* that astronomy is one of the oldest basic sciences and that it has contributed and still contributes fundamentally to the evolution of other sciences and applications in a wide range of fields,
>
> *Recognizing* that astronomical observations have profound implications for the development of science, philosophy, culture and the general conception of the universe,
>
> *Noting* that, although there is a general interest in astronomy, it is often difficult for the general public to gain access to information and knowledge on the subject,

[21] SALT: South African Large Telescope, a 10m-class telescope located in the Karoo desert.

[22] https://library.un.org/sites/library.un.org/files/itp/a62-parti_0.pdf and International Year of Astronomy 2009 Final Report, IAU 2010 (ibid.).

Fig. 17.4 The IAU delegation that presented the IYA2009 initiative at the United Nations in New York on 1 November 2007. From left to right: Claus Madsen (ESO), Beatriz Barbuy (IAU Vice President), Robert Williams (IAU President Elect), Catherine Cesarsky (IAU President), Luigi di Chiara (Italian UN Mission), Kevin Govender (SALT), Nella Le Moli (INAF) and Leopoldo Benacchio (INAF). Photo IAU

Conscious that each society has developed legends, myths and traditions concerning the sky, the planets and the stars which form part of its cultural heritage,

Welcoming resolution 33 C/25 adopted by the General Conference of the United Nations Educational, Scientific and Cultural Organization on 19 October 2005 to express its support for the declaration of 2009 as the International Year of Astronomy, with a view to highlighting the importance of astronomical sciences and their contribution to knowledge and development,

Noting that the International Astronomical Union has been supporting the initiative since 2003 and that it will act to grant the project the widest impact,

Convinced that the Year could play a crucial role, inter alia, in raising public awareness of the importance of astronomy and basic sciences for sustainable development, promoting access to the universal knowledge of fundamental science through the excitement generated by the subject of astronomy, supporting formal and informal science education in schools as well as through science centres and museums and other relevant means, stimulating a long-term increase in student enrolment in the fields of science and technology, and supporting scientific literacy,

1. *Decides* to declare 2009 the International Year of Astronomy;
2. *Designates* the United Nations Educational, Scientific and Cultural Organization as the lead agency and focal point for the Year, and invites it to organize, in this capacity, activities to be realized during the Year, in collaboration with other relevant entities of the United Nations system, the International Astronomical Union, the European Southern

Observatory and astronomical societies and groups throughout the world, and, in this regard, notes that the activities of the Year will be funded from voluntary contributions, including from the private sector;

3. *Encourages* all Member States, the United Nations system and all other actors to take advantage of the Year to promote actions at all levels aimed at increasing awareness among the public of the importance of astronomical sciences and promoting widespread access to new knowledge and experiences of astronomical observation.

The last paragraph was clearly the door opener for the multitude of stakeholders to enlist support for their activities.

In the afternoon of 19 December (New York time), the proposal was finally adopted by the Plenary.[23] Claus Madsen sat at his computer in Garching that night. At 00:44 the long-awaited mail from New York ticked in: 2009 would be the world's International Year of Astronomy![24]

17.7 Lessons Learned

Of course, one might wonder why the case was not handled solely by the usual diplomatic actors and channels. However, in this particular case, our advantage was that we, on the one side, did not represent any specific country and therefore looked like "neutral" actors (diplomats are always suspicious of other countries' diplomats who pursue goals other than their own ones). On the other side, the IAU being a non-governmental organisation, contrary to the UN, we were not bound by national policies or bureaucratic constraints. We could act with agility, yet talk with a certain weight and with legitimacy as a highly respected international scientific organisation. In addition, ESO, an intergovernmental organisation (and therefore of the same legal standing as the UN), and equally respected scientific organisation acted as a principal partner during this phase. Thus, astronomy proved its value as a bridge between very different countries and (political) worldviews. In any event, the strategy worked.

To the lessons learned, however, we should add the realisation that when a benign science such as astronomy hits world politics, we enter into a very different environment and arguments that traditionally would carry the day within our scientific world, become seriously challenged. Understanding this wider world with its rivalling views, complex relations, and hierarchies then becomes important.[25] To mention just one example: The fact that support was relatively easily obtained at the

[23] https://www.iau.org/news/pressreleases/detail/iau0702/.

[24] In the end, 2009 was rich in "International Years": "Reconciliation", "Natural Fibres", "Human Rights Learning", and "Year of the Gorilla" (see https://www.un.org/en/sections/observances/international-years/).

[25] A discussion of these issues is provided in Madsen, C: "When the Dog Must Talk to the Cat—Communicating Science to Politicians or Science and Politics Thoughts About a Complex Relationship", proceedings, Conference on Future Professional Communication in Astronomy,

level of UNESCO did not at all automatically translate into similar support at the level of the UN General Assembly.

Before we end this tale, let us briefly return to the mutual actions of IAU and UNESCO in connection with IYA2009.

17.8 UNESCO-IAU: Continuation of the Relationship

After the UN approval was granted, we had a few more meetings with UNESCO. The most important point was the agreement by UNESCO to host the IYA2009 opening ceremony, which took place in the Paris Headquarters on January 15 and 16 and was an undisputed success. The UNESCO Director-General Kōichirō Matsuura gave an inspiring speech, saying: "This Year provides us with a fantastic opportunity to expand this knowledge, and enable all people to explore the wonders of the universe and appreciate the benefits of its study for society".[26]

At the IAU, in October 2006, we put in place the IYA2009 Working Group chaired by Catherine Cesarsky, which started delineating the activities and put in place a solid organisation. While the initiative on National Heritage remained under UNESCO chairmanship, all the other activities were run by IAU members or allies. The other groups summoned by UNESCO, such as COSPAR[27] and IAF, in the end intervened very little. ICSU even refused, to the end, to endorse IYA2009, despite the fact that Catherine Cesarsky was a member of their strategic committee. The reason for this refusal probably was that ICSU was heavily involved in the promotion of the International Polar Year (IPY, 2007/2008), which was a bona fide international scientific endeavour, and did not want to create any confusion between IPY and IYA. In the end, ICSU accepted to acknowledge the existence of IYA2009 on its website.

On the IYA2009 brochures and websites, after UN approval, IAU and UNESCO were shown as equal co-organisers of the year, under the UN (Fig. 17.5):

The accompanying text stated that "IAU is the initiator and the international leader of IYA2009". The text on UNESCO, however, was more generic.

In promoting the IYA2009, we had praised the initiative in the highest tones. Now it was up to the IAU and its partners to deliver. And they did with astounding success. A staggering 148 countries, with 73 organisational nodes and partners and 22 media partners, delivered a plethora of exciting, high-profile events reaching an

Académie Royale de Belgique, 2007, https://www.vub.be/STER/JAD/JAD13/jad13_5/jad13_5.pdf.

[26] https://cordis.europa.eu/article/id/30352-international-year-of-astronomy-2009-gets-underway.

[27] COSPAR : Committee for Space Research, based in Paris.

Fig. 17.5 IAU-UNESCO organisation of IYA2009

estimated 815 million people worldwide[28] (Fig. 17.6). To this should be added the legacy activities and the fact that the path to new international years promoting science and also involving astronomy, such as the International Year of Chemistry (2011), of Light and Light-based Technologies (2015), of the Periodic Table of Chemical Elements (2019), had been paved anew. Convincingly. By the IAU and the worldwide astronomical community, in close collaboration with amateur astronomers. Congratulations to All.

[28] https://www.iau.org/news/pressreleases/detail/iau1001/.

Fig. 17.6 The IYA Logo in some of the many language versions

References

Andersen, J., Baneke, D., & Madsen, C. (2019). *The International Astronomical Union — Uniting the Community for 100 years* (p. 376). Springer Nature Switzerland.

Engvold, O. (2007). Transactions of the International Astronomical Union, Vol. XXVB. In *Proceedings of the 25th General Assembly* (Sydney, Australia, July 12–26, 2003) (p. 33). Astronomical Society of the Pacific

Miley, G., Madsen, C., & Scorza de Appl, C. (2005). *Universe Awareness for Young Children*. ESO Messenger 121 (September issue).

Catherine Cesarsky has performed research activities in several central areas of modern astrophysics, interstellar medium, high-energy astrophysics, infrared astronomy, galaxy evolution.

Born in France, she holds a Physics degree from University of Buenos Aires, and a PhD in Astronomy from Harvard University (1971). She then worked at the California Institute of Technology. In 1974, she moved to France and joined the Commissariat à l'Energie Atomique (CEA). From 1985 to 1993, she was Head of Astrophysics at CEA, and from 1994 to 1999, she was Director of CEA basic research in physics and chemistry. From 1999 to 2007, she was the Director General of the European Southern Observatory. From 2009 to 2012, she was High Commissioner for Atomic Energy in France, advisor to the French government for science and energy issues. Since then, she is High-level Science Advisor at CEA. From 2006 to 2009, she was the IAU President, and supervised IYA2009.

Throughout her career, she held numerous functions related to her expertise. At present, she is Chairperson of the SKA Board.

Claus Madsen was born in Copenhagen in 1951 and holds an MSc in Science. After an initial stint in the ESO Sky Atlas Laboratory, dealing with astronomical photography, for more than three decades, Claus Madsen has worked professionally with science communication, international science policy and science diplomacy issues in various intergovernmental research organisations, mainly ESO, but also CERN and EFDA-JET. He represented ESO vis-à-vis the European Union for a five-year period and, for ten years, the same organisation in the function of Permanent Observer at the United Nations Committee for the Peaceful uses of Outer Space. He has also acted as an expert for the European Commission and national governments. Furthermore, he has coordinated EU-funded science projects and been a steering committee or board member in a number of science-related endeavours. Claus Madsen has published books and articles on science (astronomy), science communication and science policy and given talks and lectures on these topics around the world.

Claus Madsen is currently associated with the Department of Physics and Astronomy at Aarhus University. He is a member of the International Astronomical Union, the International PCST network, honorary member of the European Association for Astronomy Education and honorary member of the Austrian Society for Astronomy and Astrophysics (ÖGA^2).

Chapter 18
The IAU-UNESCO "Astronomy and World Heritage Initiative" and the Pic du Midi Observatory

Clive L. N. Ruggles

Abstract This chapter explores the involvement of the IAU in the promotion of astronomical sites as potential World Heritage. The most obvious such connection is in the context of the Astronomy and World Heritage Initiative (AWHI), a Thematic Initiative within UNESCO for which the IAU have been joint partners since 2008, and which has helped bring about a number of recent inscriptions onto the World Heritage List of sites relating to astronomy, both ancient and modern. But the IAU's involvement in efforts to promote astronomical World Heritage goes back before the creation of the AWHI, and the example of the Pic-du-Midi Observatory provides a link extending back to the earliest days of the IAU.

18.1 Before the AWHI: The Struve Geodetic Arc, World Heritage, and the IAU

The Struve Geodetic Arc was inscribed on UNESCO's World Heritage List in 2005 (UNESCO, 2005). This 2820 km-long triangulation network, stretching from Hammerfest on the northern coast of Norway down to the Black Sea, was constructed between 1816 and 1855 under the coordination of Friedrich Georg Wilhelm Struve, Professor of Mathematics and Astronomy at the University of Dorpat in the Governate of Livonia, now Tartu in Estonia, together with Carl Tenner, head of the Russian triangulation service. What had started as a project to produce a topographic map of Livonia developed into a coordinated effort by surveyors in several countries that was both scientifically and politically motivated: while its primary purpose was to obtain the first precise measurement of a long segment of a meridian, accurate mapping also provided benefits to state sponsors, such as determining territorial frontiers and improving trade links. The outstanding cultural significance of the Struve Arc lies in the use of triangulation to accurately determine

C. L. N. Ruggles (✉)
School of Archaeology and Ancient History, University of Leicester, Leicester, UK
e-mail: rug@le.ac.uk

© The Author(s), under exclusive license to Springer Nature Switzerland AG 2022
T. Montmerle, D. Fauque (eds.), *Astronomers as Diplomats*, Historical & Cultural Astronomy, https://doi.org/10.1007/978-3-030-98625-4_18

the size and shape of the Earth, an endeavour extending back to ancient Greek times (Cotte, 2010, p. 226). In the early nineteenth century it was well known that the Earth was not a perfect sphere but rather an oblate spheroid, flattened towards the poles; the challenge was to determine the Earth's radius at different latitudes and hence the precise extent of the flattening. The Struve Arc does not memorialise the pursuit of astronomy as such but rather the application of astronomy in the field of geodesy (ibid.: 224). Nonetheless, it certainly represented a milestone in the recognition of the heritage of science, being one of the first sites inscribed onto the World Heritage List solely because of its scientific value (Cameron, 2009, p. 48).

The 34 surviving nodes of Struve's network that have been recognised by UNESCO (out of an original 265) span ten modern countries: Norway, Sweden, Finland, Russia, Estonia, Latvia, Lithuania, Belarus, Ukraine, and Moldova. The Struve Geodetic Arc World Heritage Property is therefore both serial (multi-site) and transnational. For this reason, the development of a successful World Heritage nomination dossier also represented a considerable political achievement, since it required the cooperation of ten different State Parties (governments). The driving force, however, came from the scientific community, and in this the IAU played a key—albeit not the leading—role (Fig. 18.1).

The idea was first mooted at a conference in Tartu, Estonia, in 1993, and supporting resolutions and statements were sought from the International Federation of Surveyors (FIG) and the International Association for Geodesy (AIG) as well as the IAU. Thus, at the 22nd IAU General Assembly in The Hague in 1994, Commission 41 (History of Astronomy) proposed a resolution (B10) urging the Executive Committee of the IAU "to approach the governments of ... Norway, Sweden, Finland, Estonia, Latvia, Lithuania, Ukraine, Belarus, Poland [sic] and Moldova ... with a view to ... an approach to UNESCO to declare them to be World Heritage Sites" (IAU, 1996; owing to an error in the drafting, Poland was mistakenly substituted for Russia). The initiative was then led forward by the International Institution for the History of Surveying and Measurement, a subsidiary body of FIG. For a detailed account, see Smith (2005, pp. 4–7); see also Batten and Smith (2006).

It is no coincidence that during 2004, while the Struve Arc nomination was being considered by its advisory bodies, UNESCO convened a meeting of experts at their Regional Bureau for Science in Europe (ROSTE) in Venice to explore further how they should deal with heritage relating to astronomy. UNESCO's World Heritage Centre is continually striving to create a more "balanced, representative, and credible" World Heritage List and to identify types of heritage that are underrepresented. In order to encourage State Parties to submit heritage sites of a new type, they have set up a number of "Thematic Initiatives" that explore in detail how heritage sites of the type in question might best demonstrate the "outstanding universal value" (UNESCO, 2019, p. 20 [§§49–53]) needed to justify inclusion in the World Heritage List. Thus in March 2004, noting "increasing demands from State Parties for the evaluation and inscription on the World Heritage List of cultural sites that possess a connection to astronomy," they invited together experts in

Fig. 18.1 Map of the Struve Geodetic Arc. The 34 red points are the places registered in the UNESCO World Heritage Site. (Wikimedia Commons, Creative CommonsAttribution-Share Alike 3.0 Unported; https://commons.wikimedia.org/wiki/File:Struve_Geodetic_Arc-zoom-fr.svg)

the fields of archaeoastronomy, archaeology, history of astronomy, astronomy, and astrophysics to work alongside heritage staff to develop a framework for a new Thematic Initiative that was initially to be named "archaeoastronomical sites and observatories" but subsequently became the AWHI.

Tellingly, it was also noted in the invitation papers for the Venice meeting that "This is the first programme of UNESCO to make a direct link between science and culture". The World Heritage Centre had recognised that science in general was underrepresented on the List. In focusing on astronomy, they had focused on a field of very broad application, one obvious reason being that every human culture has a sky.

It is clear, then, that the IAU and several of its members were already much involved in World Heritage issues even before the AWHI was created. Another example of this was a collaboration between the University of Hamburg and ICOMOS Germany, which resulted in the publication of a report on the heritage of classical observatories from the Renaissance to the mid-twentieth century (Wolfschmidt, 2009). Indeed, one of the astronomers invited to the 2004 Venice meeting, Alec Boksenberg, went on to become Chair of the UK's National Commission for UNESCO.[1]

18.2 The UNESCO–IAU Astronomy and World Heritage Initiative

The UNESCO World Heritage Convention dates back to 1972 and each year UNESCO's World Heritage Committee meets (amongst other things) to inscribe new sites onto the World Heritage List. As of July 2021, UNESCO's World Heritage List contains 1154 properties (this term is now preferred to "sites" as it can also include landscapes), of which 897 are cultural, 218 are natural, and 39 are mixed (both cultural and natural). Only 43 are transnational.

It is not true to say that places with a connection to astronomy were entirely absent from the World Heritage List even prior to 2005. The point is that none of them was inscribed solely (or even in some cases at all) because of their astronomical connection. Thus, for example, the 40m-high meridian arc from Ulugh Beg's fourteenth-century observatory in Samarkand, Uzbekistan, was recognised as World Heritage not in its own right but as part of the broader "Samarkand—Crossroad of Cultures" World Heritage Site; likewise the Old Royal Observatory in Greenwich, UK, was listed as part of "Maritime Greenwich", and Pulkovo Observatory in Russia as part of the "Historic Centre of St Petersburg and related groups of monuments". Ancient examples include Newgrange Neolithic passage tomb in Ireland, famous for being lit up by the first light of the sun around winter solstice, which was included as part of "Brú na Bóinne—Archaeological Ensemble

[1] See Chap. 17 by Cesarsky and Madsen, this volume.

of the Bend of the Boyne", and Stonehenge in England, whose solstitial orientation had not been included amongst the reasons for listing "Stonehenge, Avebury and Associated Sites".

In 2005, UNESCO's World Heritage Committee formally requested the Director of the World Heritage Centre to "further explore the Thematic Initiative 'Astronomy and World Heritage'". The IAU, alongside UNESCO's advisory bodies and a number of appointed "national focal points" (representatives), was subsequently involved in developing an implementation strategy to improve the identification, conservation, and management of specific types of properties connected with astronomical observations and traditional astronomical knowledge. This was duly presented to and approved by the Committee in 2008 (UNESCO, 2008a, 2008b).

That year marked the run-up to the 2009 International Year of Astronomy, itself a joint endeavour between the IAU and UNESCO, and Astronomy and World Heritage was identified as one of twelve global "cornerstone projects" of activities centred on specific themes.[2] In the same year, the IAU's General Secretary Karel van der Hucht signed a "Memorandum of Understanding" between UNESCO and the IAU, a formal agreement whereby the two organisations undertook to work together from that point forwards in order to advance and implement the Initiative (IAU, 2008). The MoU was extended in 2013 by IAU General Secretary Thierry Montmerle (IAU, 2013), and then superseded in 2015 when the IAU entered into an "Official Partnership (Consultative Status)" with UNESCO, which provided a permanent basis under which the two bodies could continue jointly to run the initiative (Fig. 18.2).

The importance attached to astronomical heritage within the IAU increased speedily during those initial years. At the outset, the task of fulfilling the IAU's commitments under the MoU was assigned to a new Astronomy and World Heritage Working Group of Commission 41 (History of Astronomy) under the Chairmanship of Clive Ruggles, also at the time acting President of C41. Ruggles's professional standing as an archaeologist who had initially trained as an astrophysicist helped to forge strong links between the IAU and the archaeological and heritage communities.

The internal restructuring that took place during the 2012–2015 triennium further raised the profile of heritage within the IAU in two important ways. The creation of Division C (Education, Outreach, and Heritage) acknowledged heritage at the highest structural level within the IAU, while the creation of Commission C4 on World Heritage and Astronomy, replacing the Working Group, raised implementation activities to the Commission level (Montmerle, 2019). Ruggles became President of the new Commission C4 during 2015–2018.

Between 2008 and 2018, Ruggles worked alongside Anna Sidorenko from UNESCO's World Heritage Centre to jointly coordinate the initiative on behalf of the two organisations.

[2] See Cesarsky and Madsen, ibid.

Fig. 18.2 The IAU General Secretary, Karel van der Hucht, and the UNESCO Director General, Koïchiro Matsuura, after signing the Memorandum of Understanding between UNESCO and the IAU in 2008. (© IAU)

18.2.1 Achievements of the AWHI

The International Council on Monuments and Sites (ICOMOS) is UNESCO's advisory body for cultural heritage. It evaluates nominations for the inscription of cultural (as opposed to natural) properties on the World Heritage List. As part of its mission to promote the conservation, protection, and use of cultural monuments, buildings, and sites—and to encourage and help State Parties to submit viable and ultimately successful World Heritage nominations—ICOMOS periodically produces "Thematic studies", often in conjunction with partner organisations. These Thematic studies explore, and offer a context for, its evaluations regarding particular types of heritage (ICOMOS, 2019).

The first tangible product of the AWHI, a joint ICOMOS–IAU Thematic Study on the "Heritage Sites of Astronomy and Archaeoastronomy" (hereinafter "TS1"), was published in 2010 (Ruggles & Cotte, 2010). Its aims, in line with those of ICOMOS Thematic studies in general, were to identify the main characteristics of astronomical heritage sites from a World Heritage perspective, to examine a selection of representative case studies, to determine possible gaps in the World Heritage List, and to indicate (with reference to UNESCO's Operational Guidelines) the criteria under which such sites might be nominated for inscription on the List.

The case studies, again in line with established practice, could include sites already on the World Heritage List, on national Tentative Lists (publicly available lists of properties reported to UNESCO that a State Party intends to nominate in the future), or on neither, inclusion being no indication of the likely outcome if such a property were subsequently to be nominated for WHL inscription.

The scope of TS1 was very wide, ranging from Palaeolithic artefacts to the heritage of space exploration, and alongside its two editors (Ruggles for the IAU and Michel Cotte for ICOMOS) its 41 contributing authors included 15 astronomers together with 12 historians/historians of astronomy and 7 archaeologists/archaeoastronomers—many of whom were IAU members within Commission 41—alongside heritage specialists. Its 44 Case Studies included seven-stone antas, solar-aligned dolmens dating to the 4th millennium BC in Portugal and Spain; the thirteen towers of Chankillo, a monumental solar calendrical device from the third century BC in Peru; Beijing Ancient Observatory in China, dating from the fifteenth century; the Royal Observatory, Cape of Good Hope, South Africa, an iconic colonial observatory from the early nineteenth century; and the Einstein Tower in Potsdam, Germany, an early twentieth-century solar observatory that is also an architectural masterpiece.

The other main deliverable from the early phase of the AWHI was the UNESCO–IAU Portal to the Heritage of Astronomy (www.astronomicalheritage.net), which was launched in 2012 during the 28th IAU General Assembly in Beijing (IAU, 2012). This is a dynamic, publicly accessible repository containing detailed information about a range of astronomical heritage sites and artefacts. A "heritage finder" tool allows users to filter the information in various ways. Where a site is on the World Heritage List, this information is linked in directly from UNESCO's own World Heritage List website (whc.unesco.org/en/list). The Portal also contains a range of information aimed at State Parties interested in preparing a nomination dossier for a property connected to astronomy (Fig. 18.3).

The annual UNESCO World Heritage Committee meetings are accompanied by various "side events" that provide delegates with up-to-date information on relevant topics. The IAU took the initiative to organise one of these side events at the 39th WHC meeting in Bonn in 2015. Marking "eight years of fruitful collaboration between UNESCO and the IAU in promoting scientific heritage" it reviewed achievements to date, highlighted ongoing nomination projects, and introduced a draft version of a second Thematic Study. The event attracted government representatives from 19 countries, as well as officials from ICOMOS and the IUCN, UNESCO's advisory body for natural sites.

The second Thematic Study (hereinafter "TS2") was eventually published in 2017 (Ruggles, 2017). This focused on complex or problematic issues needing to be explored in more detail. Its eleven Case Studies were much longer than those in TS1, and structured in the form of segments of draft nomination dossiers, to give

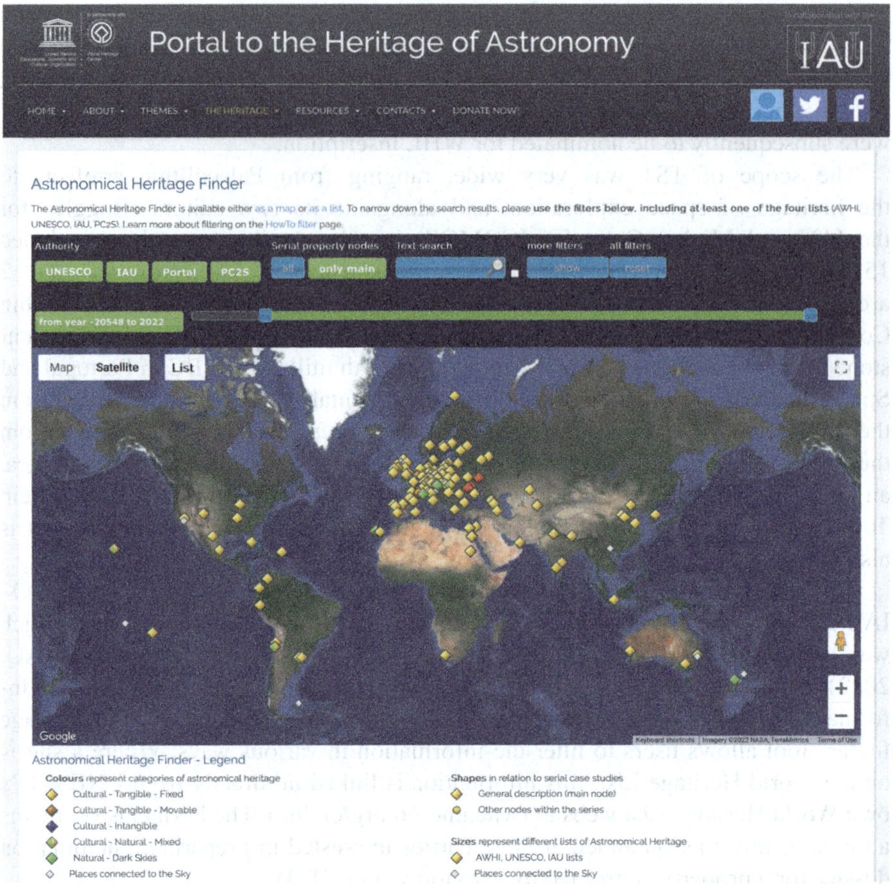

Fig. 18.3 Screenshot of the heritage finder on the UNESCO–IAU Portal to the Heritage of Astronomy (https://www.astronomicalheritage.net)

better guidance to any stakeholders considering whether that property might provide a viable World Heritage nomination in the future.

Ultimately, the success or otherwise of a Thematic Initiative will be measured by its effect on the World Heritage List. Did it result in new inscriptions relating to the type of heritage in question? The year 2010 saw the inscription of two astronomical World Heritage properties: the Jantar Mantar Observatory at Jaipur in India (UNESCO, 2010a) and Denfeng Observatory in China (UNESCO, 2010b). A year later, the solstitial alignments of Stonehenge and related monuments within the "Stonehenge, Avebury and Associated Sites" World Heritage Property in the UK, already on the List, were recognised retrospectively as attributes of its "outstanding universal value" (UNESCO, 2011). But it was the successful inscription in 2019

Fig. 18.4 Bentayga sanctuary, an archaeoastronomical site in central Gran Canaria forming part of the "Risco Caído and the Sacred Mountains of Gran Canaria" World Heritage Site. The photo shows light from the rising equinoctial sun falling upon a circular engraving in the sanctuary floor. (Photo: © Clive Ruggles)

of Jodrell Bank Observatory (UK) and Risco Caído and the Sacred Mountains of Gran Canaria (Spain)—a cultural landscape that combines symbolic, religious, and astronomical connotation and includes two important archaeoastronomical sites—that demonstrated that the AWHI had come of age (Ruggles, 2019). The thirteen towers of Chankillo in Peru were inscribed in 2021 and the Eise Eisinga Planetarium in the Netherlands is being considered for 2022. Other astronomical heritage sites appearing on national Tentative Lists promise further astronomical nominations in the near future (Fig. 18.4).

Another measure of the success of the AWHI is that it is now being taken forward into broader initiatives. An expert meeting held in the Canary Islands in 2018 brought together two UNESCO Thematic Initiatives and adopted a number of key recommendations relating to cultural heritage with both sacred/religious and astronomical values. In 2019, UNESCO launched a new Thematic Initiative on the Heritage of Astronomy, Science and Technology (TIHAST), focused more broadly upon the heritage of science in general but with the IAU still playing a key role.

18.3 Pic du Midi Observatory

Pic du Midi de Bigorre Observatory in France is featured as a major case study both in TS2 (Bourgeois, 2017) and on the Portal to the Heritage of Astronomy (UNESCO-IAU, 2017).

An issue of particular concern throughout the AWHI has been whether dark sky sites could be considered as World Heritage (Ruggles, 2013). The "Windows to the Universe" concept, introduced in TS1 (Marín et al., 2010) and elaborated in TS2, presented case studies of the world's most exceptional observatory sites together with some of its most pristine natural dark sky places. Giving the issue detailed consideration from a heritage perspective in TS2, Cotte (2017) concluded that, while dark sky areas cannot in themselves be considered as potential World Heritage Sites, the value of an unpolluted dark sky could certainly be recognised as contributing to broader cultural or natural qualities that might constitute "outstanding universal value".

The Pic du Midi case study featured alongside the "Windows to the Universe" observatories in TS2 as a stand-alone example offering, as Cotte described it (ibid.: 15), "a remarkable combination of cultural value (more than one century of continuous high mountain observation of the sky), identity landmark (one of the strongest and the most visible symbol[s] of the Languedoc region), natural value[s] amongst them quality of atmosphere and dark sky, and the ensemble offering an outstanding mountain landscape". Put another way, "It offers a veritable landscape and cultural icon, visible from far away, showing both the human presence at altitude and the scientific use of the mountains. The quality of its climate and atmospheric environment is responsible for both its daytime panoramas and its exceptional night sky" (Bourgeois, 2017, p. 180) (Fig. 18.5).

The case study describes in detail the development of the site, the exceptional science carried out there, and the tribulations of the directors struggling to ensure its

Fig. 18.5 Pic du Midi Observatory c. 1900. Taken from the summit (altitude 2876 m) towards the south-east. The "Baillaud Dome" would be erected on the rocky hillock to the right of the picture in 1906. Photograph by Eugène Trutat, part of the Fonds Eugène Trutat, preserved by the Muséum de Toulouse. (https://commons.wikimedia.org/wiki/File:MHNT_PHa_1521_E_018.jpg, Eugène Trutat, Public domain, via Wikimedia Commons)

survival and development. "140 years of life and science in the high mountains" are divided up as follows:

Period	Director	Key figures
From Genesis to the first stone (1865–1891)		Gen. Charles Nansouty; Xavier Voussenat
Systematic observations and the trials of the First World War (1892–1920)	Emile Marchand	Benjamin Baillaud
The "Dauzère years"—the first real threats for the Pic and scientific maturity (1920–1937)	Camille Dauzère	
The major projects (1937–1947)	Jules Baillaud	Bernard Lyot
The rise of astronomy and the international influence of the observatory (1947–1981)	Jean Rösch	
The dark years to come (1982–2000)	Michel Blanc	

As Bourgeois elaborates, the history and development of the Pic du Midi Observatory has primarily been guided by three factors:

- A cultural perception of mountains such as those in the Pyrenees as "a world created to defy temerity, elevate the soul, explore and discover" and which was prepared to take risks to rise to the challenge. This view, which took hold in the nineteenth century, underpinned the creation of the Ramond Society, a Pyrenean exploration company, and drove its members, led by General Charles Champion du Bois de Nansouty and the engineer Célestin-Xavier Vaussenat, to found the Observatory, against many odds. It also perhaps helps to explain some of the risks that were taken in order to maintain the observatory in later years.

- The exceptional nature of both the geographical location and the atmospheric quality for the purposes of astronomy and other sciences. The development from what, initially, was primarily a meteorological research and monitoring station into an internationally renowned research facility reached its apex during the prosperous Rösch years, starting with the arrival of British cosmic ray specialists in 1949 and their Nobel Prize-winning discovery of a new particle (the hyperon) in 1950, and culminating in the installation of the 2m Bernard Lyot Telescope in 1976.
- The continual threats to its existence. Added to the incessant challenges arising from the difficult mountain conditions and budgetary constraints were periodic crises, most notably that arising in the 1920s from its parlous state following neglect during the First World War, and—from the 1980s onwards—inexorable competition from other observatories being planned and built in even more exceptional locations (such as the Canary Islands, Hawai'i, and northern Chile). The "dark years" were eventually brought to an end by the extraordinary transformation of Pic du Midi into a major focus for tourism, a transformation that has ensured the continued survival of the observatory and its scientific work.

Although never a director at Pic du Midi itself, Benjamin Baillaud stands out as one of the key figures in its development. It was thanks to his initiative as director of Toulouse Observatory that the first permanent astronomical facility—the Baillaud Dome, 50 cm telescope and an accommodation building for astronomers—was installed on the Pic du Midi in 1908. Baillaud went on to become the first IAU President in 1919, as well as Director of the Paris Observatory (1908 to retirement in 1926).[3]

The Pic du Midi case study encapsulates the extraordinary heritage value of the Pic du Midi Observatory in the following statement: "The major active [observatory] sites (Hawai'i and the Canary Islands ... and the astronomical sites in the Chilean Andes ...) all have a relatively recent history ... While all these sites convey the feeling of exceptional human adventure, they do not convey the closeness to the historical roots of scientific exploration that is found at the Pic du Midi" (ibid.: 178). That said, whether or not the Pic du Midi Observatory itself is eventually recognised as World Heritage is not within the purview of the IAU. This can only happen if the French authorities should decide in the future to nominate it, and then only if UNESCO and its advisory bodies were satisfied that it is worthy of inscription.

[3] See Bougeret (2019), and Chap. 3 by D. Fauque and R. Fox, this volume.

18.4 Conclusion

In recent years, the term "cultural astronomy" (Ruggles & Saunders, 1993) has been used increasingly to embrace both archaeoastronomy and ethnoastronomy. But all astronomy (including modern astronomy and its historical roots) is ultimately cultural astronomy: it takes place in a cultural context (TS1: 262). In recognising this, astronomers (including those historians of astronomy and cultural astronomers who also form part of the IAU) have, over the past two decades, progressed from being facilitators and campaigners, seeking to spur others into taking action to protect important astronomical heritage; they have become an integral part of that action.

Moving forward, it appears that IAU members are set to adopt a leading role in the broader task of ensuring due recognition and protection for science heritage in general.

References

Batten, A. H., & Smith, J. R. (2006). Measuring the world: excursions in geodesy and astronomy. *Journal of Astronomical History and Heritage, 9*(1), 65–75.

Bougeret, J.-L. (2019). In C. Sterken, J. Hearnshaw, & D. Valls-Gabaud (Eds.), *Under One Sky: the IAU Centenary Symposium, IAU Symposium 349* (pp. 139–146). Cambridge University Press.

Bourgeois, N. (2017). Pic du Midi du Bigorre Observatory, France. In C. Ruggles (Ed.), *Heritage sites of astronomy and archaeoastronomy in the context of the UNESCO World Heritage Convention: A thematic study* (Vol. 2, pp. 141–186). Ocarina Books–ICOMOS.

Cameron, C. (2009). Recognizing science and technology at World Heritage Sites. *World Heritage, 54*, 44–51.

Cotte, M. (2010). Applied astronomy in modern times. In C. Ruggles & M. Cotte (Eds.), *Heritage sites of astronomy and archaeoastronomy in the context of the UNESCO World Heritage Convention: A thematic study* (pp. 224–228). ICOMOS–IAU.

Cotte, M. (2017). Windows to the universe and potential criteria in the World Heritage Convention context. In C. Ruggles (Ed.), *Heritage sites of astronomy and archaeoastronomy in the context of the UNESCO World Heritage Convention: A thematic study* (Vol. 2, pp. 7–16). Ocarina Books–ICOMOS.

IAU. (1996). *Resolution B10 approved at the IAU General Assembly* (The Hague, 1994): https://www.iau.org/static/resolutions/IAU1994_French.pdf, published in Transactions of the IAU, vol.XXIIB, Ed. I. Appenzeller, Kluwer Acad. Publ., pp. 46–47 (https://www.cambridge.org/core/services/aop-cambridge-core/content/view/29E2A369F381ECDE3A1D3E9CC971B71D/S0251107X0000852Xa.pdf/chapter_ii_twentysecond_general_assembly.pdf)

IAU. (2008). *UNESCO and the IAU sign key agreement on Astronomy and World Heritage Initiative*. Press release iau0808: https://www.iau.org/news/pressreleases/detail/iau0808

IAU. (2012). *Launch of UNESCO–IAU Astronomical Heritage Web Portal*. Press release iau1203: https://www.iau.org/news/pressreleases/detail/iau1203

IAU. (2013). IAU and UNESCO sign new agreement. Press release iau1203: https://www.iau.org/news/pressreleases/detail/iau1302/

ICOMOS. (2019). *Thematic studies for the World Heritage Convention.* https://www.icomos.org/en/about-the-centre/publicationsdoc/monographic-series-3/198-Thematic-studies-for-the-world-heritage-convention

Marín, C., Wainscoat, R., & Fayos-Solá, E. (2010). 'Windows to the Universe': Starlight, dark-sky areas and observatory sites. In C. Ruggles & M. Cotte (Eds.), *Heritage sites of astronomy and archaeoastronomy in the context of the UNESCO World Heritage Convention: A thematic study* (pp. 238–259). ICOMOS–IAU.

Montmerle, T. (2019). The IAU, from Commissions to Divisions… to Commissions. In C. Sterken, J. Hearnshaw, & D. Valls-Gabaud (Eds.), *Proc. IAU Symposium 349, Under One Sky: The IAU Centenary Symposium* (pp. 289–324). Cambridge University Press.

Ruggles, C. (2013). The recognition of 'Outstanding Universal Value' in relation to astronomy at dark-sky sites. In J. Hearnshaw, K. Pollard, & M. Head (Eds.), *Proceedings of the Third International Starlight Conference, Lake Tekapo, New Zealand* (pp. 5–13). Starlight Conference Organizing Committee.

Ruggles, C. (Ed.). (2017). *Heritage sites of astronomy and archaeoastronomy in the context of the UNESCO World Heritage Convention: A thematic study* (Vol. 2). Ocarina Books–ICOMOS. https://www.astronomicalheritage.net/index.php/thematic-study-2

Ruggles, C. (2019). Beyond stonehenge: Astronomical heritage. *Astronomy and Geophysics, 60*, 4.36–4.39.

Ruggles, C., & Cotte, M. (Eds.). (2010). *Heritage sites of astronomy and archaeoastronomy in the context of the UNESCO World Heritage Convention: A thematic study*. ICOMOS–IAU. https://www.astronomicalheritage.net/index.php/thematic-study-1

Ruggles, C., & Saunders, N. (1993). The study of cultural astronomy. In C. Ruggles & N. Saunders (Eds.), *Astronomies and cultures* (pp. 1–31). University Press of Colorado.

Smith, J. R. (2005). *The Struve Geodetic Arc*. Unpublished PDF available online at http://www.fig.net/organisation/perm/hsm/history_of/measurements/struve/struve_arc_smith_2005.pdf

UNESCO. (2005). *Struve Geodetic Arc*. whc.unesco.org/en/list/1187

UNESCO. (2008a). *Thematic Initiative "Astronomy and World Heritage": Integrated Implementation Strategy*. Document WHC–08/32.COM/INF.5C. Paris: UNESCO. whc.unesco.org/document/9959

UNESCO. (2008b). *World Heritage Committee Decision 32 COM 5*. whc.unesco.org/en/decisions/1756/

UNESCO. (2010a). *The Jantar Mantar, Jaipur.*whc.unesco.org/en/list/1338

UNESCO. (2010b). *Historic Monuments of Dengfeng in 'The Centre of Heaven and Earth'.*whc.unesco.org/en/list/1305

UNESCO. (2011). *Stonehenge, Avebury and Associated Sites.*whc.unesco.org/en/list/373

UNESCO. (2019). *Operational guidelines for the implementation of the world heritage convention*. UNESCO. whc.unesco.org/document/178167

UNESCO-IAU. (2017). *Pic du Midi de Bigorre Observatory, France.*https://www.astronomicalheritage.net/index.php/show-entity?identity=76&idsubentity=1

Wolfschmidt, G. (Ed.). (2009). *Astronomical Observatories: from Classical Astronomy to Modern Astrophysics*. ICOMOS (Monuments and Sites XVIII).

Clive L. N. Ruggles is Emeritus Professor of Archaeoastronomy in the School of Archaeology and Ancient History at the University of Leicester. He served as President of IAU Commission 41 (History of Astronomy) from 2009 to 2012 and Commission C4 (World Heritage and Astronomy) from 2015 to 2018. Over his career he has written numerous books, papers and articles on cultural astronomy and he was editor-in-chief of the 3-volume *Handbook of Archaeoastronomy and Ethnoastronomy* (Springer, 2014), a definitive source on theory, method and practice over the entire field. He was awarded the Agnes Mary Clerke Medal by the Royal Astronomical Society in 2017 for "a lifetime of distinguished work in the overlapping fields of archaeology, astronomy, and the history of science". Between 2008 and 2018 Clive was the IAU's co-ordinator for the UNESCO-IAU joint initiative on Astronomy and World Heritage (whc.unesco.org/en/astronomy) and he continues to direct the UNESCO-IAU Portal to the Heritage of Astronomy (www.astronomicalheritage.net).

Index

© The Author(s), under exclusive license to Springer Nature Switzerland AG 2022

T. Montmerle, D. Fauque (eds.), *Astronomers as Diplomats*, Historical & Cultural Astronomy, https://doi.org/10.1007/978-3-030-98625-4

Z

FSC
www.fsc.org
MIX
Papier | Fördert gute Waldnutzung
FSC® C083411